FOREST ENTOMOLOGY AND PATHOLOGY

Kruger Brentt
Publishers

FOREST ENTOMOLOGY AND PATHOLOGY

Gina Gunn | Harry Cedric
Kisty Bloom

Kruger Brentt
Publishers

2025

Kruger Brentt Publishers UK. LTD.
Company Number 9728962

Regd. Office: 68 St Margarets Road, Edgware, Middlesex HA8 9UU

ISBN: 978-1-78715-283-0

For information on all our publications visit our website at http://krugerbrentt.com/

PREFACE

"Forest Entomology and Pathology" delves into the intricate interactions between forest ecosystems and the complex organisms that inhabit them, focusing on insects and pathogens and their impact on forest health. This book serves as an essential resource for students, researchers, forest managers, and practitioners seeking to understand the dynamics of forest pests and diseases and their management strategies.

Forests play a vital role in supporting biodiversity, regulating climate, purifying water, and providing numerous ecosystem services essential for human well-being. However, forests face numerous threats, including insect pests and pathogens, which can cause significant damage to trees, disrupt ecosystem functions, and compromise forest resilience.

The primary objective of this book is to provide readers with a comprehensive overview of forest entomology and pathology, covering key topics such as insect biology and behavior, pathogen ecology and epidemiology, forest pest and disease identification, monitoring and detection methods, integrated pest management (IPM) strategies, and forest health assessment techniques. Each chapter offers insights, case studies, and practical guidance to help readers understand the complex interactions between forest organisms and their environment.

Contributions from leading experts in the field explore various aspects of forest entomology and pathology, including the biology and ecology of key insect pests and pathogens, the role of climate change and human activities in driving forest pest and disease outbreaks, and the development of sustainable management approaches to mitigate their impacts. Special emphasis is placed on the importance of interdisciplinary collaboration, stakeholder engagement, and adaptive management strategies in addressing forest health challenges.

As we confront the increasing threats posed by climate change, globalization, and land-use change, it is essential to adopt holistic and science-based approaches to forest pest and disease management. By integrating ecological knowledge, technological innovation, and community participation, we can enhance the resilience of forest ecosystems and ensure their long-term health and sustainability.

We extend our sincere gratitude to the contributors whose expertise and dedication have enriched this book. It is our hope that "Forest Entomology and Pathology" will serve as a valuable resource and catalyst for research, education, and action in the dynamic and vitally important field of forest health management.

Gina Gunn
Harry Cedric
Kisty Bloom

CONTENTS

01 RECENT DEVELOPMENTS IN THE SURVEILLANCE, EVALUATION, AND CONTROL OF FOREST PATHOGENS AND PESTS

This special issue was initiated in light of the United Nations' International Year of Plant Health in 2020. Its aim is to showcase the essential and inevitable steps involved in safeguarding forest health from pathogens and pests. These steps include identifying the "enemy," quantifying its impact, and devising effective strategies to combat it. These stages serve as milestones: gaining a deeper comprehension of the parasite's life-history techniques and effects enables us to advance towards the ultimate objective of disease and pest control.

One first and inevitable challenge, particularly when dealing with extraterrestrial parasites that are not well understood, is to precisely and clearly identify them [1]. Precise and prompt identification is also crucial for the subsequent use of efficient disease and pest control measures, as it enables us to gain a deeper understanding of the parasite's potential influence on woodland ecosystems. Multiple studies investigate this matter using both conventional and molecular methods. Mercado et al. [2] gave helpful instructions for accurately identifying Dendroctonus species that attack Pinus contorta Douglas ex Loudon in subalpine forests in the southern Rocky Mountains. The classification of these beetles, which are typically challenging to differentiate from similar species, has been enhanced by simplified identification methods that rely on morphological characteristics, attack patterns, and signals.

The progress in molecular-based diagnostics of infectious diseases is significantly improving our capacity to detect and analyze harmful substances, enabling more precise pathogen identification tests [3]. Ongoing substantial research endeavors are focused on identifying novel indicators for exceptionally precise diagnosis. A comparative genomics method was used to identify a novel biomarker, the RxLR effector gene PHYCI_587572, which is specific to the destructive and extremely polyphagous disease Phytophthora cinnamomi. In addition, a recombinase polymerase amplification-lateral flow dipstick assay that focuses on the PHYCI_587572 biomarker has demonstrated the ability to accurately detect P. cinnamomi isolates from various sources and geographic locations. This assay has shown high specificity, as it does not produce any false positive results when tested against 37 other oomycete and fungal species, including the closely related P. parvispora [4].

A recent advancement is the loop-mediated isothermal amplification (LAMP) technology, originally designed for diagnosing animal and human diseases [5]. However, it is now being used to address phytosanitary concerns and resolve taxonomic difficulties. A LAMP assay was created to definitively detect Dothistroma septosporum (G. Dorog.) M. Morelet, which is one of the causative agents of Dothistroma needle blight (DNB). DNB is a newly emerging disease that poses a threat to both wild and plantation pine forests. The assay demonstrated exceptional sensitivity, capable of detecting as low as 1 picogram of fungus DNA. Additionally, it exhibited remarkable specificity, effectively distinguishing this particular pathogen from the closely related Dothistroma pini Hulbary and Lecanosticta acicola (Thüm.) Syd. The number 6 is indicated by the numeral [6]. In order to determine the pathogen responsible for the initial epidemic of DNB in Italy, the LAMP method was utilized [7]. The initial occurrence of DNB at that location, almost fifty years ago (in the mid-1970s), was attributed to D. The identification of pini is solely based on micromorphological analysis [8]. Upon re-examination, the previously unverified report was subjected to pathogen testing using the LAMP method. This analysis identified the true cause of the early outbreak in the last century as D, rather than the previously suggested agent. Septosporum was likely introduced into the area by the cultivation of non-native Pinus radiata D. trees. Don.

After accurately identifying a harmful agent, the following stage is to expand the understanding of this species as much as possible [9]. Effective pest control requires a thorough understanding of the organism's fundamental characteristics and how its population changes over time. Both the citrus long-horned beetle, Anoplophora chinensis (Forster), and the gypsy moth, Lymantria dispar L., are invasive and destructive pests that can cause significant damage to various tree species in orchards, urban areas, and forests. Therefore, this study provides information on the temperature requirements of the citrus long-horned beetle and the important features that affect the gypsy moth's capacity to disperse. This knowledge is crucial for developing phenological models to guide management and eradication efforts [10,11]. Keena et al. [10] specifically assessed the adult survival, reproduction, and egg hatching of A. Experiments were conducted in a controlled laboratory environment to observe the development of chinensis at eight different constant temperatures. The results obtained from these experiments can be utilized to forecast the progression of various stages. Srivastava [11] conducted a comprehensive analysis of the existing literature regarding the differences in flying capability and flight distance among several subspecies of gypsy moths. The study also examined any changes in other relevant features. Efficient approaches for evaluating important characteristics of non-native pest populations, along with understanding of population dynamics in areas affected by invasions, provide essential information for effectively managing fresh introductions into unfamiliar environments. For instance, Robinett et al. [12] showed that white ash trees are still present in various forested areas in the central region of the Agrilus planipennis Fairmaire invasion site in Michigan, even after more than ten years since the invasion. This suggests that the populations of these beetles are still not reaching the maximum number that the sites can support.

In recent years, research efforts have been significantly enhanced by contemporary molecular and bioinformatic techniques. These efforts have been focused on examining the variety of fungi and pseudofungi in protected regions. Protected regions serve as repositories of biodiversity and are crucial for their preservation, thereby requiring specific attention. These undisturbed habitats are particularly valuable for investigating the ecology and dynamics of plant diseases in various vegetation types. A study conducted in a protected area in Eastern Sicily (Italy) investigated the variety and distribution of Phytophthora species in different types of habitats, including aquatic, riparian, and terrestrial. The survey identified a total of 11 Phytophthora species. Notably, the study found a connection between the presence of certain oomycete taxa and specific groups of vegetation types [13]. A separate study examined the effects and dissemination of the newly discovered charcoal canker pathogen Biscogniauxia mediterranea (De Not.) Kuntze in a preserved, low-lying forest (Castelfidardo Forest, central Italy) [14]. The charcoal canker pathogen had a severe influence on the sub-Mediterranean deciduous forest with mesophilic features. This forest is mostly dominated by oak species and has suffered greatly in recent decades. Additionally, it was discovered that the issue was connected to environmental pressures, specifically a lack of rainfall and prolonged drought. Other factors that made oaks more susceptible to decay included a lack of forest management, intense competition among fully grown trees, and the isolated location of this remaining forest.

Comprehending the mechanisms that control the dissemination and harmfulness of infections that result in significant ecological and economic harm remains a focal point of research in forest pathology. Differences in the physical features of the land, such as the kind of land (e.g., forests compared to previously used agricultural land), and variations in important ecological factors, have a significant impact on the vulnerability of trees and the way forest diseases spread [15]. Cheng et al. [16] discovered that in Japanese red pine (Pinus densiflora Sieb. et Zucc.) forests, the presence of Diplodia tip blight produced by Sphaeropsis sapinea (Fr.) Dyko and B. had a significant effect. The pathogen Sutton, which affects many conifer species from the Abies, Cedrus, Juniperus, Larix, Picea, Pinus, Pseudotsuga, and Thuja genera, shows variations based on stand types and vertical structure layers. The incidence of Diplodia tip blight disease was greater in the P. densiflora–conifer mixed forest compared to the P. densiflora–hardwood mixed forest or the pure P. densiflora forest. In addition, an increased variety of plants in the lower layer of vegetation led to heightened competition between juvenile Japanese red pines and shrub-herb plants, resulting in reduced resistance of the pine species to Diplodia tip blight [16].

The study conducted by Klavina et al. [17] presented proof that Norway spruce (Picea abies (L.) Karst.) plantations established in areas that were not previously forests become extremely vulnerable to the root rot pathogen Heterobasidion parviporum Niemelä and Korhonen after thinning. Additionally, the study found that in plantations growing in former pastures and meadows, the pathogen is able to spread extensively through large territorial clones, likely due to the lack of competition from other soil microorganisms. To prevent the spread of Heterobasidion infection to future trees through root contacts, it is crucial to use biological or chemical control agents to treat stumps during thinning operations. However, stumps serve as a main source of infection for Heterobasidion annosum s.l., as they allow the disease to propagate to healthy trees through the root connections by means of vegetative mycelium. The presence of Heterobasidion sp. in the air. Wood discs can be used to estimate spore deposits on the stumps of species that are vulnerable to them [18]. Brūna et al. [19] used wooden discs to examine the vulnerability of seven conifer species, including Larix sibirica Ledeb., P. abies, Picea sitchensis (Bong.), P. contorta, Pinus strobus L., and Pinus sylvestris L. and Pseudotsuga menziesii (Mirb.) were artificially inoculated with conidia of Heterobasidion. The terms “annosum sensu stricto (s.s.)” and “H.” are used. Parviporum can be studied either under controlled conditions or by exposure to natural airborne infection. The scientists noted that the infection rate of wood discs treated with conidial suspensions in a controlled environment was similar to the infection rate recorded when exposing the same species’ discs to natural airborne inoculum in spruce stands infested with H. Annosum or Heterobasidion. Small creatures. The significance of local spore sources in the dispersal range of the two forest pathogens was validated by the observation that the infection rate of wood discs declined as the distance from the inoculum sources (fruiting bodies) increased.

Effective control of forest pathogens and pests requires careful consideration of not only the pest itself, but also the other living organisms in the ecosystem and their interactions within multiple trophic levels. The significance of insects as carriers of disease-causing fungi is widely recognized [20,21]. However, there are still many elements of the interactions between insects and fungi that require further clarification. An abundance of Heterobasidion spp. Fruit bodies, such as mushrooms, grow on big P. abies logs that have been infected and left in stands following forest operations. These fruiting bodies with a high density of spores pose a significant hazard. Consequently, multiple studies advise the removal of huge, contaminated logs from the area. Nevertheless, these logs have a crucial role as a habitat for numerous fungi and invertebrates, including species that are scarce and at risk of extinction. Legzdina et al. [22] discovered a diverse group of invertebrates in Heterobasidion fruit bodies, rotting logs, and nearby soil. However, the scientists noted that more research is needed to understand the specific relationships between these organisms. In addition, Vissa et al. [23] investigated variations in the diversity of symbiotic mite populations linked to Dendroctonus ponderosae Hopkins, a financially detrimental bark beetle native to North America. They found that the diversity of mites that live in a mutually beneficial relationship with other organisms changed depending on the environment. They propose that this finding could be valuable for discovering important species interactions in forest ecosystems.

It is important to prioritize and enhance eco-sustainable control tactics in order to limit the negative effects of pest control measures on forest ecosystems. Implementing these solutions necessitates a higher level of skill and is more challenging to execute compared to conventional pesticide-based methods. Further research is required to gain a deeper understanding of how to enhance their effectiveness, while minimizing unwanted side effects. Bracalini et al. [24] conducted a study on the adverse consequences of mass trapping of Ips sexdentatus Böerner, specifically focusing on unintended captures of the target's natural predators. The researchers conducted experiments using slot traps that were equipped with mesh screens and escape windows in order to enhance their selectivity. These trap modifications were shown to be effective in addressing the issue, particularly for beetles that are larger than the intended target. This strategy should be implemented in conjunction with accurate mass-trapping scheduling to prevent periods when the target adult population is less active than the primary predator adult population. Given that biodiversity conservation is a fundamental aspect of sustainable forest management, it is crucial to assess the effects of all management practices on the entire biological community. Hartshorn [25] examined the effects of various forest management practices (such as clear cutting, burning, and chemical control of invasive species) on leaf litter arthropods.

These arthropods are highly abundant and diverse communities in forests and play a crucial role in providing ecosystem services, including decomposition, mineralization, pollination, and predation.

Microbial insecticides can be used in forest habitats where the use of pesticides is not possible. These biological agents are safe for both the individuals who use them and the animals that live in the woods, thanks to their high level of selectivity. Bacillus thuringiensis-based products are the most widely utilized environmentally friendly pesticides globally [26]. Bacillus thuringiensis var. The use of kurstaki (Btk) in combination with mating disruption has been found to be effective in controlling Thaumetopoea pityocampa Denis et Schiffermüller [27], an insect that poses a threat to forest, urban, and peri-urban pine trees. It is important to keep this insect under control as its larvae can also cause health issues for humans and pets. Additionally, Ruiu et al. [28] have suggested that Btk can be effectively combined with other biological products. The authors proposed using Btk in conjunction with LdMNPV, a species-specific multicapsid nucleopolyhedrovirus, to control the lepidopteran defoliator L. Disparities exist in multi-year integrated programs. Alternatively, certain fungal species have demonstrated significant hostility towards significant phytopathogenic fungi. Phlebiopsis gigantea is an effective biocontrol agent for the root rot of conifer trees caused by Heterobasidion annosum. Wit et al. [29] used partial DNA sequencing of elongation factor 1-alpha (EF1α) and in silico data to deduce the connections between representative isolates of P. gigantea and specific species of Basidiomycota. The acquired results suggest that the EF1α region is a reliable indicator for choosing the most competitive isolates of P. gigantea, which might potentially be used to reduce damage caused by root infections in forest trees.

This special issue is further enriched by two reviews that address two significant topics: the growing issue of forest ecosystem invasion by alien parasites and the occurrence of phytoplasma infections in forest trees. The former is a subject of current interest and widely popular, while the latter has received little attention, if not completely ignored. In the initial study conducted by Panzavolta et al. [30], the primary routes by which non-indigenous infections and pests enter forest ecosystems and uncontaminated areas were examined, along with the factors that influence their establishment and subsequent dissemination. This emphasizes the significance of promptly identifying, tracking, and eliminating detrimental species, while acknowledging the role of scientific investigation and the valuable input of volunteers in promptly identifying them (citizen science). Furthermore, it emphasizes the necessity of implementing government policies in order to properly address this global issue. Marcone et al. [31] conducted a review of the primary phytoplasma taxa that are responsible for causing illnesses in forest tree species in their second publication. These destructive bacteria, which do not have cell walls and are dependent on other organisms for survival, cause harm to several plant species around the world. However, they have received less research attention compared to phytoplasmas that infect fruit trees and other types of plant diseases. Thorough investigation of the vectors, plant host range, strain virulence, pathogenicity, and host tolerance/resistance is essential for effectively controlling phytoplasma infections.

The 21 papers included in this special issue provide an overview of the current research trends in forest health preservation. The diverse range of methods, spatial scales, methodological approaches, and issues addressed in contemporary forest pathology and entomology research demonstrate its adaptability and significance.

AT A GLANCE

Tree diseases and pests are essential elements of forest ecosystems. Through the process of killing and decomposing vulnerable trees, they effectively control the circulation of nutrients and energy, hence influencing the arrangement and makeup of forest ecosystems. Nevertheless, ecosystems can experience significant disturbances when the population density of these parasites surpasses their threshold of tolerance. Determining the source of disease and pest outbreaks, identifying their causal agents accurately and unambiguously, and comprehending their reproductive and dispersal patterns are all essential for implementing effective control measures. The research compiled in this special edition encompass a diverse array of subjects within the realm of forest pathology and entomology. Research encompasses various aspects, including the molecular identification of pathogens and pests, their surveillance and measurement in the field, the evaluation of their rate of growth, the analysis of their genetic diversity, the examination of the influence of plant diversity and ecosystem heterogeneity on the impact of pathogens and pests, and the development of strategies for disease and pest control. Case studies demonstrate the need of using creative research approaches to resolve taxonomic issues that were previously contentious. The diverse range of experimental methodologies and the breadth of scientific topics investigated demonstrate the current trends and relevance of contemporary forest health protection science.

REFERENCES

1. Moricca, S.; Bracalini, M.; Benigno, A.; Ghelardini, L.; Furtado, E.L.; Marino, C.L.; Panzavolta, T. Thousand cankers disease in *Juglans*: Optimizing sampling and identification procedures for the vector *Pityophthorus juglandis*, and the causal agent *Geosmithia morbida*. *MethodsX* 2020, *7*, 101174. [Google Scholar] [CrossRef] [PubMed]
2. Mercado, J.E. Improved Identification and New Records of *Dendroctonus* Bark Beetles Attacking *Pinus contorta* in the Subalpine Forest of the Southern Rocky Mountains. *Forests* 2020, *11*, 656. [Google Scholar] [CrossRef]

3. Rizzo, D.; Da Lio, D.; Bartolini, L.; Cappellini, G.; Bruscoli, T.; Bracalini, M.; Benigno, A.; Salemi, C.; Del Nista, D.; Aronadio, A.; et al. A duplex real-time PCR with probe for simultaneous detection of *Geosmithia morbida* and its vector *Pityophthorus juglandis*. *PLoS ONE* 2020, *15*, e0241109. [Google Scholar] [CrossRef] [PubMed]

4. Dai, T.; Wang, A.; Yang, X.; Yu, X.; Tian, W.; Xu, Y.; Hu, T. *PHYCI_587572*: An RxLR Effector Gene and New Biomarker in A Recombinase Polymerase Amplification Assay for Rapid Detection of *Phytophthora cinnamomi*. *Forests* 2020, *11*, 306. [Google Scholar] [CrossRef] [Green Version]

5. Mori, Y.; Kanda, H.; Notomi, T. Loop-mediated isothermal amplification (LAMP): Recent progress in research and development. *J. Infect. Chemother.* 2013, *19*, 404–411. [Google Scholar] [CrossRef]

6. Myrholm, C.L.; Tomm, B.D.; Heinzelmann, R.; Feau, N.; Hamelin, R.C.; McDougal, R.; Winkworth, R.C.; Ramsfield, T.D. Development of a Rapid Loop-Mediated Isothermal Amplification Assay for the Detection of *Dothistroma septosporum*. *Forests* 2021, *12*, 362. [Google Scholar] [CrossRef]

7. Aglietti, C.; Benigno, A.; Scali, E.; Capretti, P.; Ghelardini, L.; Moricca, S. Molecular-Based Reappraisal of a Historical Record of Dothistroma Needle Blight in the Centre of the Mediterranean Region. *Forests* 2021, *12*, 983. [Google Scholar] [CrossRef]

8. Magnani, G. Presence of *Dothistroma pini* on *Pinus radiata*. *Cellul. Carta* 1977, *28*, 37–42. [Google Scholar]

9. Moricca, S.; Bracalini, M.; Benigno, A.; Ghelardini, L.; Furtado, E.L.; Marino, C.L.; Panzavolta, T. Observations on the non-native thousand cankers disease of walnut in Europe's southernmost outbreak. *Glob. Ecol. Conserv.* 2020, *23*, e01159. [Google Scholar] [CrossRef]

10. Keena, M.A.; Moore, P.M.; Bradford, G. Effects of Temperature on *Anoplophora chinensis* (Coleoptera: Cerambycidae) Adult Survival, Reproduction, and Egg Hatch. *Forests* 2021, *12*, 432. [Google Scholar] [CrossRef]

11. Srivastava, V.; Keena, M.A.; Maennicke, G.E.; Hamelin, R.C.; Griess, V.C. Potential Differences and Methods of Determining Gypsy Moth Female Flight Capabilities: Implications for the Establishment and Spread in Novel Habitats. *Forests* 2021, *12*, 103. [Google Scholar] [CrossRef]

12. Robinett, M.A.; Poland, T.M.; McCullough, D.G. Captures of Emerald Ash Borer (*Agrilus planipennis*) Adults in Post-Invasion White Ash Sites with Varying Amounts of Live Phloem. *Forests* 2021, *12*, 262. [Google Scholar] [CrossRef]

13. Riolo, M.; Aloi, F.; La Spada, F.; Sciandrello, S.; Moricca, S.; Santilli, E.; Pane, A.; Cacciola, S.O. Diversity of *Phytophthora* Communities across Different Types of Mediterranean Vegetation in a Nature Reserve Area. *Forests* 2020, *11*, 853. [Google Scholar] [CrossRef]

14. Murolo, S.; Concas, J.; Salerno, A.; Maiorano, F.; Cingolani, L.; Carloni, F.; Moricca, S.; Romanazzi, G. Status of Charcoal Canker on Oak Trees at a Site of Community Importance: Case Study of the Relict Castelfidardo Forest (SIC Area IT520008, Castelfidardo, AN, Italy). *Forests* 2021, *12*, 1032. [Google Scholar] [CrossRef]

15. Moricca, S.; Benigno, A.; Oliveira Longa, C.M.; Cacciola, S.O.; Maresi, G. First documentation of life cycle completion of the alien rust pathogen *Melampsoridium hiratsukanum* in the Eastern Alps proves its successful establishment in this mountain range. *J. Fungi* 2021, *7*, 617. [Google Scholar] [CrossRef] [PubMed]

16. Cheng, Y.; Liang, J.; Xie, X.; Zhang, X. Effect of Plant Diversity on the Occurrence of Diplodia Tip Blight in Natural Secondary Japanese Red Pine Forests. *Forests* 2021, *12*, 1083. [Google Scholar] [CrossRef]

17. Klavina, D.; Bruna, L.; Zaluma, A.; Burnevica, N.; Polmanis, K.; Gaitnieks, T.; Piri, T. Infection and Spread of Root Rot Caused by *Heterobasidion parviporum* in *Picea abies* Stands after Thinning: Case Studies on Former Pasture and Meadow Lands. *Forests* 2021, *12*, 70. [Google Scholar] [CrossRef]

18. Möykkynen, T.; Von Weissenberg, K.; Pappinen, A. Estimation of dispersal gradients of S- and P-type basidiospores of *Heterobasidion annosum*. *Eur. J. For. Pathol.* 1997, *27*, 291–300. [Google Scholar] [CrossRef]

19. Brūna, L.; Lione, G.; Kenigsvalde, K.; Burņeviča, N.; Zaļuma, A.; Kļaviņa, D.; Gaitnieks, T.; Gonthier, P. Inferences on the Susceptibility of Wood of Different Tree Species to *Heterobasidion annosum* Sensu Lato Primary Infections and on the Range of Pathogen Spores Dispersal. *Forests* 2021, *12*, 854. [Google Scholar] [CrossRef]

20. Panzavolta, T.; Panichi, A.; Bracalini, M.; Croci, F.; Ginetti, B.; Ragazzi, A.; Tiberi, R.; Moricca, S. Dispersal and propagule pressure of *Botryosphaeriaceae* species in a declining oak stand is affected by insect vectors. *Forests* 2017, *8*, 228. [Google Scholar] [CrossRef] [Green Version]

21. Panzavolta, T.; Panichi, A.; Bracalini, M.; Croci, F.; Benigno, A.; Ragazzi, A.; Tiberi, R.; Moricca, S. Tree pathogens and their insect-mediated transport: Implications for oak tree die-off in a natural park area. *Glob. Ecol. Conserv.* 2018, *15*, e00437. [Google Scholar] [CrossRef]

22. Legzdina, L.; Spungis, V.; Burnevica, N.; Gaitnieks, T.; Menkis, A. Invertebrates in Fruitbodies of *Heterobasidion* spp., Infected *Picea abies* Logs and Adjacent Soil. *Forests* 2021, *12*, 1100. [Google Scholar] [CrossRef]

23. Vissa, S.; Mercado, J.E.; Malesky, D.; Uhey, D.A.; Mori, B.A.; Knee, W.; Evenden, M.L.; Hofstetter, R.W. Patterns of Diversity in the Symbiotic Mite Assemblage of the Mountain Pine Beetle, *Dendroctonus ponderosae* Hopkins. *Forests* 2020, *11*, 1102. [Google Scholar] [CrossRef]

24. Bracalini, M.; Croci, F.; Ciardi, E.; Mannucci, G.; Papucci, E.; Gestri, G.; Tiberi, R.; Panzavolta, T. *Ips sexdentatus* Mass-Trapping: Mitigation of Its Negative Effects on Saproxylic Beetles Larger Than the Target. *Forests* 2021, *12*, 175. [Google Scholar] [CrossRef]

25. Hartshorn, J. A Review of Forest Management Effects on Terrestrial Leaf Litter Inhabiting Arthropods. *Forests* 2021, *12*, 23. [Google Scholar] [CrossRef]

26. Fernández-Chapa, D.; Ramírez-Villalobos, J.; Galán-Wong, L. Toxic Potential of *Bacillus thuringiensis*: An Overview. In *Protecting Rice Grains in the Post-Genomic Era*; Jia, Y., Ed.; IntechOpen: London, UK, 2019; Available online: https://www.intechopen.com/chapters/67369 (accessed on 15 November 2021). [CrossRef] [Green Version]

27. Ferracini, C.; Saitta, V.; Pogolotti, C.; Rollet, I.; Vertui, F.; Dovigo, L. Monitoring and Management of the Pine Processionary Moth in the North-Western Italian Alps. *Forests* 2020, *11*, 1253. [Google Scholar] [CrossRef]

28. Ruiu, L.; Mannu, R.; Olivieri, M.; Lentini, A. Gypsy Moth Management with LdMNPV Baculovirus in Cork Oak Forest. *Forests* 2021, *12*, 495. [Google Scholar] [CrossRef]

29. Wit, M.; Sierota, Z.; Żółciak, A.; Mirzwa-Mróz, E.; Jabłońska, E.; Wakuliński, W. Phylogenetic Relationships between *Phlebiopsis gigantea* and Selected *Basidiomycota* Species Inferred from Partial DNA Sequence of Elongation Factor 1-Alpha Gene. *Forests* 2020, *11*, 592. [Google Scholar] [CrossRef]
30. Panzavolta, T.; Bracalini, M.; Benigno, A.; Moricca, S. Alien Invasive Pathogens and Pests Harming Trees, Forests, and Plantations: Pathways, Global Consequences and Management. *Forests* 2021, *12*, 1364. [Google Scholar] [CrossRef]
31. Marcone, C.; Valiunas, D.; Mondal, S.; Sundararaj, R. On Some Significant Phytoplasma Diseases of Forest Trees: An Update. *Forests* 2021, *12*, 408. [Google Scholar] [CrossRef]

02 ENHANCED IDENTIFICATION AND UPDATED RECORDS OF DENDROCTONUS BARK BEETLES INFESTING PINUS CONTORTA IN THE SUBALPINE FOREST OF THE SOUTHERN ROCKY MOUNTAINS

1. INTRODUCTION: AN OVERVIEW

Pinus contorta Douglas, also known as lodgepole pine, is found in the southern Rocky Mountains (SRM) of Colorado and southern Wyoming. It typically grows at elevations ranging from 2400 to 3200 m. In the subalpine zone, above 2800 m, it coexists with Picea engelmannii Parry ex Engelm, also known as Engelmann spruce. Within this moderate mountain forest, indigenous Dendroctonus bark beetles cause the demise of trees, resulting in modifications to the forest's structure and the composition of its species. The P. contorta tree is naturally targeted by two species of Dendroctonus bark beetles, namely D. ponderosae Hopk. (mountain pine beetle) and D. murrayanae Hopk. (lodgepole pine beetle). The extent of the influence on their host varies. During the most recent pandemic, D. ponderosae caused the death of almost 60% of the fully grown P. contorta trees in the SRMs, as documented in sources [2,3].

In contrast, D. murrayanae was observed to only impact tiny areas of trees, as stated in source [4]. The variation in the beetles' influence is a result of their distinct biological characteristics. Dendroctonus ponderosae infests all Pinus species in the Southern Rocky Mountains (SRM), but Dendroctonus murrayanae exclusively targets Pinus contorta. In addition, D. ponderosae is found across the SRMs, whereas D. murrayanae is restricted to areas above latitude 39.4°, indicating its preference for cooler, northern climes [5]. Furthermore, P. engelmannii faces a natural adversary called Dendroctonus rufipennis (Kirby), often known as the North American spruce beetle. Occasionally, D. ponderosae also attacks Picea spp., the species that P. engelmannii belongs to. The adult D. rufipennis and D. murrayanae have crimson elytra that stand out against their dark brown head and prothorax. These two species strongly resemble one other not just in color, but also in size, shape, and other exterior characteristics. Distinguishing between the two Dendroctonus species is challenging due to their comparable egg galleries, egg placement methods, and brood aggregation tendencies [6]. Consequently, the differentiation among the species of beetles is typically determined by the host that they infest [5,7]. Populations of Dendroctonus rufipennis have experienced a sudden increase in numbers over the past 20 years, resulting in the death of a significant amount of appropriate P. engelmannii trees in the SRM.

Within the SRM, the subalpine forest is found at elevations where the activity of D. ponderosae can be consistently reduced by cold temperatures [8]. As a result, there is limited evidence of D. ponderosae infestations on P. contorta trees at elevations higher than 3000 m [8,9,10]. However, in the previous epidemic, instances of D. ponderosae infestations were documented at altitudes over 3000 m, allowing for the investigation of their behavior in regions where they are not typically found. During the study conducted in the Roosevelt National Forest in northern SRM in Colorado, red-elytra Dendroctonus beetles that resembled D. murrayanae and D. rufipennis were lured to baited funnel traps (specifically 12-funnel Lindgren traps) that were used to monitor D. ponderosae activity. The traps were baited with Exobrevicomin-Myrcene-Transverbenol, which is a semiochemical produced by Synergy Semiochemicals. Identifying these beetles to species was challenging because of their similarity and their reaction to D. ponderosae lures, which included a component from one P. contorta tree. Incidents of comparable bark beetle infestations were also recorded in the southern regions of the SRM in Colorado, although the presence of D. murrayanae has not been confirmed. These bark beetles were consistently observed in mixed stands of P. contorta and P. engelmannii on every occasion. Therefore, determining whether it was D. murrayanae or D. rufipennis that caused the attacks on P. contorta in the subalpine woods of the region became an important topic.

The species Dendroctonus rufipennis has been observed preying on P. contorta in the SRMs during the periods of 1944-1949 and 1957, as shown in Figure 1. However, these findings have been disputed by taxonomic experts [5,7]. The occurrence of D. rufipennis infestations on P. contorta in the SRM is not well comprehended due to the challenge of distinguishing it from D. murrayanae infestations on the same host in subalpine forests. This study evaluated Dendroctonus beetles that were attacking P. contorta in the subalpine forests of Wyoming and Colorado in the SRM region. The aims of this study were to enhance the identification of mature Dendroctonus beetles and the detection of their attack signs and patterns, validate the identification of Dendroctonus species with red-elytra that attack P. contorta in subalpine forests, and uncover novel biological insights about these Dendroctonus species.

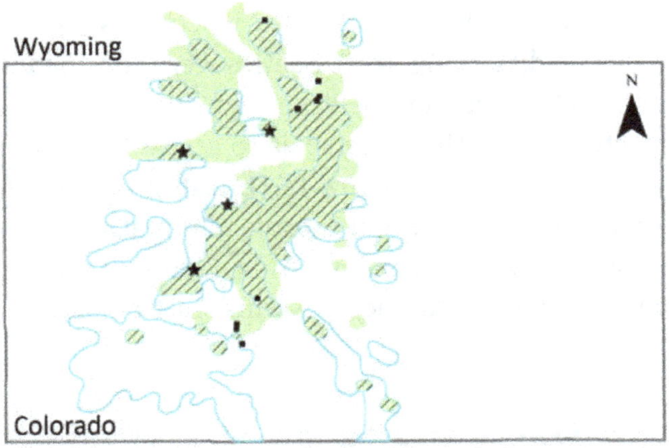

Figure 1. Southern Rocky Mountains sites were *D. rufipennis* was identified attacking *P. contorta*. Records from the 1940s and 1950s are indicated by stars, and squares indicate records made during this study. *P. engelmannii* (green) and *P. contorta* (blue) layers were modified from [13], trees intermix in the dashed areas.

2. MATERIALS AND METHODS

Sites in this study were in the SRMs, from southern Wyoming to southern Colorado (**Figure 1**). Yellowing trees in the spring or red ones in the fall with pitch tubes were used as evidence of trees infested by Dendroctonus beetles. In these areas, both live and dead adult beetles were collected. Live specimens were collected from live P. contorta as these attacked, while dead specimens were found trapped beneath the tree's bark and on the duff around the tree collar. Specimens were placed individually in 1.5 mL microcentrifuge tubes. In the laboratory, specimens were cleaned in an ultrasonic cleaner (100005, Sper Scientific) using warm water and mounted for examination. Previously determined diagnostic characters [**5,6,7,14,15,16,17**] were used to revise and determine the best diagnostic characters of collected specimens.

Voucher specimens from the Hopkins Collection, Rocky Mountain Station, Fort Collins, CO, and collected from *P. contorta* included: (n = 5) Rout N. F. (VII-48, Coll. Unk., No. 34220-L-6; (n = 5) Yeoman Campground, White River N. F. (IX-1947-Coll.

Wygant, No. 34220-L-7; (n = 4) Trappers Lake, White River N. F. (VII-1946, Coll. C.J. Hay, No 31409-E; (n = 2) Jumping Creek Campg., Helena, MT (Helena-Lewis and Clark N. F.) (VI-1989, Coll. J.B. Johnson, ID by Furniss), and collected from Pinus albicaulis (Whitebark pine) (Engelm.) in Wyoming: (n = 10) Bridger-Teton N. F. (labeled Dubois) (VII-1936, Coll. J. A. Beal, No. 17700-M-1).

A subsample of red-elytra Dendroctonus from Colorado and Wyoming were selected to examine male genitalia. In these, sex was determined by examining a secondary sexual character, i.e., the granules in the interstriae declivity, which are greatly reduced (nearly absent) in males in contrast to being abundant in females [**7**,**14**,**15**]. Before dissecting, male beetles were softened in warm water for five minutes, after which the abdomen was removed ventrally by pulling it using a pair of forceps posteriorly to the metacoxae to access to the aedeagus. Removed aedeagi were cleared in a solution of 10% KOH heated in a warm bath (55 °C) for 15 min to clear the sclerotized aedeagal capsule eliminating the need of removing the delicate endophallus [**18**] for its examination. Structures were examined with a Leica M16 stereomicroscope (Leica Microsystems, Castle Rock, CO, USA).

3. RESULTS

3.1. New Records of Dendroctonus spp. Attacking P. contorta in the Southern Rocky Mountains

Previous reports of D. rufipennis attacking P. contorta in the SRM where from Routt N. F., Grand Mesa N. F., and White River N. F. limiting records to North-Central Colorado west of the Continental Divide. Dendroctonus rufipennis was here detected attacking numerous P. contorta in subalpine forests from Medicine Bow N. F. in southern Wyoming and from Roosevelt N. F., Gunnison N. F., and Rio Grande N. F. in Colorado encompassing the entire SRM's latitudinal range (**Figure 1**) and the east side of the mountain range. Dendroctonus rufipennis is evidently affecting a larger number of trees than in 1940 and 1957, where it was reported to affect 250 and 140 trees, respectively. These findings make the occurrence of this event more widespread and potentially more impactful than previously reported. Dendroctonus adjunctus Blandf. has never been reported before attacking P. contorta. Attacks of this species were detected on a single site in the Rio Grande N. F. where it was found attacking three P. contorta together with D. rufipennis; therefore, treatment of this species is limited in this manuscript.

3.2. Diagnosing Dendroctonus Adult Beetles Attacking P. contorta in Subalpine Forest

The external color of the three Dendroctonus is a simple starting point to separate D. ponderosae from the two red-elytra Dendroctonus. Adult D. ponderosae has an even thorax and elytra color, whereas most adult D. murrayanae and D. rufipennis have a darker thorax contrasting with their reddish elytra (**Figure 2**). Although D. adjunctus is also even colored, this species is unique among the four species in having a vertical impression on its mid-upper frons. However, these color differences alone are not useful to diagnose a fraction of older D. rufipennis and D. murrayanae that have black elytra matching their head and thorax. Color should be used in conjunction with the length of dorsal setae, particularly of that in the elytral declivity where the two red-elytra beetles have longer setae than D. ponderosae (**Figure 2**), these two characters are evident to the naked eye on clean specimens. Low magnifications (< 7X) allow distinguishing the dull surface of the elytral declivity and the clearly impressed (flattened) second and third interstrial space that curve strongly towards the suture in D. ponderosae. This character distinguishes D. ponderosae from D. murrayanae and D. rufipennis and, as previously described [**14,15**], were referred to as striae II and III [**7,17**].

Figure 2. Three Dendroctonus bark beetles attacking P. contorta in the southern Rocky Mountains (SRM): (a) D. ponderosae distinguished from the other two by having equally colored thorax and elytra and shorter elytral declivity setae, (b) D. rufipennis examined in the SRM had duller exoskeletal surfaces than (c) D. murrayanae. Pictures by J. Mercado.

Whereas distinguishing D. ponderosae from the other two species was relatively easy, red-elytra species (i.e., D. murrayanae and D. rufipennis) are difficult to distinguish from each other. An early diagnostic character, the presence of punctures on the posterior half of the pre-episternal (synonyms: preëpisternal, proepisternal) area in D. murrayanae described as absent in D. rufipennis [**14**] (**Figure 3**) was examined. This morphological character proved to be confusing since (1) it was poorly defined, (2) its study requires the use of magnifications of 40X, and (3) punctures were present, but were more difficult to see in some D. rufipennis. Another character examined described by Wood [**7**], was the granules and punctures in the frons. He mentioned that "the relative number of punctures and granules offer the only reliable method of separating D. murrayanae from D. rufipennis."

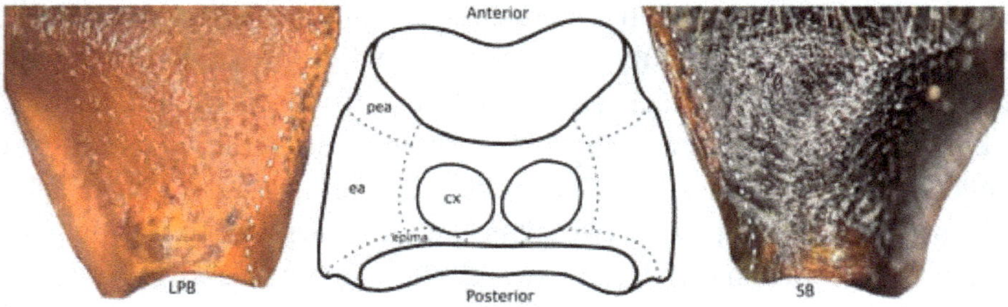

Figure 3. From anterior to posterior, the sides of a Dendroctonus thorax is divided into three areas [14]: (1), pre-episternal (pea), (2) episternal (ea), and (3) epimeral (epima). The character of the relative number of punctures at the posterior half of (pea) near white dashed lines, to distinguish D. murrayanae from D. rufipennis is difficult to use. Pictures by J. Mercado, thorax line drawing adapted from Hopkins [14].

The exoskeletal surfaces of Dendroctonus beetles, including the frons, are covered by semi-circular impressions called punctures, and by blunt cusps called granules, the number of these is never quantified, but puncture closeness and granule location and their spread are described between similar species. In both D. murrayanae and D. rufipennis, punctures are described as being very close; however, in the center of the frons (between the compound eyes), these are described as "distinct" in D. murrayanae and "largely obliterated" in D. rufipennis [5,7].

Wood [5,7] described that granules in the frons interpuncture space are more clearly separated or "isolated" in D. murrayanae than in D. rufipennis "coarsely granulate," but in this study granules were found isolated and dispersed in D. rufipennis as well. This study found that interpuncture space surfaces in D. rufipennis are usually rougher, making granules and punctures hard to discern, whereas these are usually smoother on D. murrayanae, making granules and punctures more apparent, especially in the middle of the frons. This character was found to be useful separating most D. rufipennis and D. murrayanae specimens, but some D. rufipennis had distinct mid-frons punctures; therefore, its use is recommended in addition to other characters. The smoother interspaces also helped to distinguish most D. murrayanae from D. rufipennis when examining other exoskeletal surfaces dorsally and laterally, such as elytra (**Figure 2**), from similar angles and at similar magnification. At low magnification (7.1X), examined D. murrayanae specimens in collection boxes appear glossier than those of D. rufipennis (**Figure 2**). It is recommended that the state of rough vs. smooth to describe the frons interpuncture space surface in the frons center is examined to help separate these species. The accuracy of this character should be studied further for specimens outside the SRM. A new diagnostic character found in this study was simple to use and works with the examined female D. rufipennis and D. murrayanae specimens. There is a sharp deflection of striae III away from the suture near striae IV and then towards the suture near striae VI in D. rufipennis. This deflection is present in D. murrayanae but it is smooth, making it less distinct (**Figure 4**). As described, the character was consistently present on all examined female specimens but not easily seen in some males, and its use in other regions merits further study. In males, or when the above characters are obscure or missing on a specimen in poor condition, the character to examine is the distinctly different endophallus [7] of D. rufipennis and D. murrayanae (**Figure 5**).

Figure 4. The striae (dashes) and interstriae in the declivity of D. rufipennis (left) and D. murrayanae (right). Striae III curves sharply (Sh) towards the suture opposite to striae VI in D. rufipennis, but smoothly (Sm) in D. murrayanae. Photos by J. Mercado.

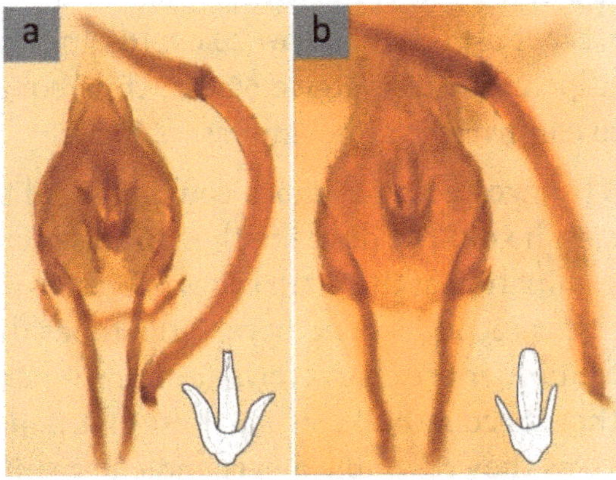

Figure 5. The male genital capsule of D. rufipennis (a) and D. murrayanae (b). An internal structure in the genital capsule, the endophallus, is seen through the median lobe (also inserts lower right corners). The endophallus is diagnostic for the two species when other means have been exhausted. Photos by J. Mercado. Endophallus line drawings adapted from Wood [7].

To improve diagnosing adult beetles of the similar red-elytra D. murrayanae and D. rufipennis attacking P. contorta in the SRM, the use of the simplified and improved couplet below is recommended.

Mid-frons punctures distinct; frons and elytra interpuncture spaces smooth; (females, difficult on males) declivity striae III turning sharply (angled) towards suture, opposite to striae VI; (males) genitalia distinctive. D. murrayanae Hopk.

Mid-frons punctures usually obscure; frons and elytra interpuncture spaces usually rough (rugose, granulate); (females, difficult on males) declivity striae III turning smoothly (arched) towards suture, opposite to striae VI; (males) genitalia distinctive. D. rufipennis (Kirby).

3.3. Diagnosing Dendroctonus Attacks Characteristics to Subalpine P. contorta

Separating D. rufipennis from D. murrayanae using only morphology is difficult; therefore, to complement adult insect determinations, available behavioral characteristics were reviewed, and a new helpful alternative is presented. These are also helpful in the absence of adult insects. The period and height from the duff of tree attack by the three Dendroctonus species are useful to identify the species. Reddish elytra Dendroctonus (i.e., D. murrayanae and D. rufipennis) attack earlier (June), whereas D. ponderosae attacks later (July and August). In the fall, D. rufipennis move to the lower 60 cm of the tree to overwinter [**5**], this is the area were D. murrayanae develops, making male genitalia examination needed for accurate species determination at that time. In June, attacks of D. rufipennis and D. murrayanae can be determined by the height at which they attack the tree. Attacks above 60 cm are made solely by D. rufipennis, while the attacks of D. murrayanae concentrate below that, but especially below 20 cm [**6**]. In July and August, both D. ponderosae and D. rufipennis attack above 60 cm, this requires examining other traits in late August to confirm the attacking insect.

Although Wood [7] suggested pitch tubes could be used to identify the attack of Dendroctonus species, he did not describe them in his monograph of the genus Dendroctonus. The examination of the pitch tubes of beetles attacking above 60 cm made in July and August is recommended here. Pitch tubes that sometimes form a slide on the lower outer border are characteristic of D. rufipennis (**Figure 6**). A few days after the attack, these become bulkier than those of D. ponderosae attacking the same tree, which differ as they resemble a perpendicular chimney (**Figure 6**). The characteristic sign was tested by predicting whether D. ponderosae or D. rufipennis was the attacking beetle. Out of 26 specimens collected with a chimney type pitch tube present, 58% were D. ponderosae (**Table 1**). In specimens collected with a slide/bulge-type pitch tube, all were D. rufipennis. The low accuracy predicting

D. ponderosae did not occur due to confusion with D. rufipennis but with D. adjunctus, performing undocumented attacks in P. contorta in southern Colorado. It is important to note that soft, sappy pitch tubes that somewhat resemble a slide can be made by D. ponderosae attacking P. contorta at lower elevation forests. This occurs when irruptive populations attack healthy trees with a strong quantitative oleoresin response, but a clear slide is not made by D. ponderosae. This should be used as the first step before verifying the insect and egg gallery and only in subalpine P. contorta/P. engelmannii forest type.

Figure 6. Distinguishing characteristics of Dendroctonus attacks to P. contorta in the southern Rocky Mountain subalpine forest include the shape of their pitch tubes and the length of their galleries. Clockwise from top left: (1) the chimney-like pitch tube of D. ponderosae is perpendicular to the tree with a near round outer border, (2) a mix of different pitch tube shapes with bulky (B) and slide-forming ones (S) by D. rufipennis and a chimney-shaped one (C) made by D. ponderosae, and (3) the turning notches in the short D. rufipennis egg gallery. Photos by J. Mercado.

Table 1. Quantified characters of Dendroctonus spp. bark beetles attacking subalpine P. contorta in the SRM during this study. Note: D. murrayanae Hopk. adults or their attacks were not found during this study.

	Attack Height		Pitch Tube Type		Egg Gallery Length		
Dendroctonus sp.	<20 cm	60–150 cm	chimney	slide/bulge	<13 cm	>23 cm	Total
D. adjunctus Blandf.	0 of 11	11 of 11	11 of 11	0 of 11	0 of 11	11 of 11	11
D. ponderosae Hopk.	0 of 16	16 of 16	15 of 16	1 of 16	0 of 16	16 of 16	16
D. rufipennis (Kirby)	0 of 75	75 of 75	0 of 75	75 of 75	78 of 75	0 of 75	75

Second, the identification accuracy of beetle attacks made above 60 cm increased by examining the length of egg galleries (n = 102). Dendroctonus rufipennis egg galleries averaged 13 cm, whereas those of D. ponderosae were longer than 23 cm, as previously described [**5**]. In the sampled egg galleries, 59.2% of these measuring over 16 cm in length corresponded to D. ponderosae (n = 16), and no D. rufipennis egg gallery measured over that. Moreover, 11 of the 27 long galleries studied were made by D. adjunctus, making this character useful for separating these two species from D. rufipennis in southern Colorado, although not from D. ponderosae. Dendroctonus rufipennis galleries also have short spurs on the sides along their short length that are used as turning sites for the adult beetle, which are absent in D. ponderosae egg galleries (**Figure 6**).

3.4. Documentation of D. rufipennis Reproduction in Live P. contorta and Verification of Museum Specimens from the 1940's Report of D. rufipennis Attacking That Host

The complete development of D. rufipennis adults in P. contorta was first suspected in 2014 from the collections of teneral adults from under the bark in Gunnison N. F. in southern Colorado, and in Medicine Bow N. F. in southern Wyoming where these were again collected from the same tree in July 2015 when a dead pupa was found. In mid-July of 2016, standing, live P. contorta in northern Colorado contained 12 D. rufipennis adults initiating galleries—three of which had laid eggs, which confirmed reproduction initiation attempts by the species. In about the same date, adults D. rufipennis were seen emerging a P. contorta in the Medicine Bow N. F. and simultaneously a P. contorta located 10 m from it was being attacked by D. rufipennis. Since there were no other trees with activity within a 30 m radius from these two trees, re-attack to P. contorta by D. rufipennis in that forest is suspected. As eggs, pupae, and teneral adults have been found under the bark of P. contorta reproduction in this host is considered confirmed from the SRM P. contorta/P. engelmannii subalpine forest.

Red-elytra Dendroctonus specimens collected from the 1940's D. rufipennis epidemic in Colorado (Wygant, White River N. F., 1947), were examined. These beetle specimens resembled D. murrayanae in their color pattern, but the examination of the male genitalia confirmed specimens represented D. rufipennis, confirming their prior attack to P. contorta in SRM.

4. DISCUSSION

While there have been prior reports of D. rufipennis attacking P. contorta in the SRM, doubts persisted regarding their accuracy. In his analysis of the genus Dendroctonus, Wood [7] analyzed a selection of red-elytra Dendroctonus specimens that were documented to have attacked P. contorta in the 1940s. After examining all potential instances of D. rufipennis attacks on P. contorta, it was concluded that the insect responsible was D. murrayanae [5,7]. Nevertheless, it remains uncertain if the specimens he examined were identical to those identified as D. rufipennis by Massey and Wygant [11] [19]. A limitation of the study conducted by Massey and Wygant [11] is that they solely rely on the diagnostic features outlined by Hopkins [14] to differentiate between D. rufipennis and D. murrayanae. Wood [7] found that these features were inadequate for distinguishing between these taxa without examining the male genitalia (endophallus). This investigation further proved the inadequacy, as all red-elytra beetles investigated from 2010 to 2016 were mistakenly identified as D. murrayanae using the key in [14].

While Massey and Wygant [11] may have been aware of a viable method to distinguish the two red-elytra Dendroctonus, they did not adequately communicate this in their study. Hence, only proficient taxonomists specializing in this specific group of beetles should utilize the characteristics and keys associated with them. Even then, it is crucial to exercise caution and possess a thorough comprehension of their constraints. Following Wood [7], it is advisable to rely on a thorough inspection of the male's endophallus as the most effective method for distinguishing between these two species based solely on adult morphology. This is particularly important when the recently described characteristic of having three elytral striae is unclear or absent. However, we discovered that only the endophallus and Wood's [5,7] essential couplet, which describes the clarity of the punctures on the center of the frons, provided useful diagnostic information. Uncertainties persist regarding the behavior of Dendroctonus bark beetles in subalpine P. contorta/P. engelmannii forest throughout their range. Firstly, what is the reason behind the aggressive behavior of D. rufipennis against this nonhost in the SRMs? In the absence of adequate primary hosts, Schmid and Frye [19] proposed that Dendroctonus species can attack nontypical hosts during outbreak conditions. In other forests where P. contorta and P. engelmannii are found together, D. ponderosae has taken advantage of the situation and attacked and developed in P. engelmannii [20] or the hybrid Picea engelmannii × glauca [21]. Instances of D. ponderosae pairs attacking P. engelmannii were also witnessed at one of the research sites in northern Colorado. Nevertheless, there were appropriate hosts located near both Dendroctonus species that were attacking nonhosts in Colorado. The potential for a nonhost attack in subalpine P. contorta/P. engelmannii forest type by D. rufipennis may arise from the following factors: (1) Simultaneous occurrences of large populations of both species can happen in locations where their hosts grow mixed together, causing confusion in the insects' sense of smell. (2) Both types of trees release a similar volatile compound, α-pinene, when they are under stress. (3) Even on trees previously infested by D. ponderosae, D. rufipennis is not repelled by any of the chemical signals emitted by that species. Furthermore, it is being questioned whether the action of D. rufipennis in subalpine P. contorta has been disregarded or misunderstood. Providing a definitive answer to this question is challenging. However, based on the author's expertise and inconsistencies found in earlier studies, there is evidence to imply that this is indeed a possibility [5,7,11]. Here, recommendations are offered to clarify this, and the author's continuing research is assessing the effects of this activity in the SRM. Does non-host assault by bark beetles have any biological purpose, in addition to other functions? Further work is warranted to explore the potential of reciprocal facilitation of several Dendroctonus species attacks on both species. According to suggestions, "primary bark beetles" who mostly attack living and visibly healthy trees, manage to survive by transforming into "secondary bark beetles". This means they attack trees that have already been affected by other bark beetles or are in a weakened state [22]. This study proposes a potential scenario where D. ponderosae, a type of bark beetle, attacks P. contorta trees at high latitudes or elevations in Canada. These trees have already been attacked by Ips mexicanus bark beetles in the past [23]. Therefore, it is justified to investigate the facilitation of these two species that attack together. Facilitation can result in the reproduction of a nonhost attacker, which simultaneously enhances the reproductive success of the native host attacker in its natural habitat. This phenomenon should be examined in the context of the SRM (Southern Rocky Mountains) with the species D. ponderosae and D. rufipennis. Laboratory investigations conducted in BC, Canada have demonstrated the successful reproduction of D. rufipennis. No developmental changes were observed in D. rufipennis reared in either P. contorta or white-P. engelmannii hybrid bolts of these species [24]. It appears that live P. contorta trees are better equipped to protect themselves against D. rufipennis attacks. This is indicated by the large number of dead D. rufipennis found near the base of attacked trees, in contrast to the small number of D. ponderosae detected over the years of observation. Nevertheless, the combined efforts of D. rufipennis attacks and the pheromones they emit, along with the defensive tree compounds released in response, may generate an olfactory signal that attracts D. ponderosae. This could enhance the vulnerability of the species initially targeted and potentially make D. rufipennis more susceptible later on by diminishing the trees' ability to defend against the primary insect pest. This could indicate an ecological adjustment that helps maintain unique populations of both species in this forest at high elevations. This study documents the presence of eggs and immature adults of D. rufipennis on P. contorta trees that have a few D. ponderosae attacks. This finding can be seen as a potential starting point for further investigation into this theory. Nevertheless, it is necessary to enhance the assessment of reproductive success in this particular host, as well as the environmental factors that enable its vulnerability to attack.

5. CONCISE SUMMARY

This paper presents valuable instructions for accurately identifying Dendroctonus species that attack P. contorta in subalpine forests in the SRM. This methodology utilizes behavioral and morphological characteristics to offer a straightforward method for identifying even the inadequately reported assault on P. contorta by D. rufipennis. All Dendroctonus bark beetles with red elytra that were observed attacking P. contorta in this investigation were identified as D. rufipennis. Although D. murrayanae may still be present in the SRM, our sampling from 50 cm to 2 m above the collar suggests that the species is rare in this area. The occurrences of atypical Dendroctonus attacking P. contorta in subalpine forests are either inadequately documented due to challenges in identifying these Dendroctonus, or increasing in frequency as a result of ongoing climate fluctuations. Streamlining the identification process of these species is an essential measure that equips us with the means to enhance our understanding of bark beetle disruptions in subalpine ecosystems. When adult male beetles are not present, the characteristics of their attack, along with a thorough inspection of their physical traits as outlined here, can assist in identifying the species responsible for the attack.

AT A GLANCE

Unusual and not well-understood attacks by Dendroctonus rufipennis (Kirby) on Pinus contorta Doug. ex Loudon were seen in the southern Rocky Mountains (SRM). The phenomena has been verified in all the areas that were researched. The reproduction of D. adjunctus Blandf. in the host is described for the first time, along with the initial attacks on the host. Enhanced detection and diagnostics of D. rufipennis will facilitate a more streamlined and effective identification of the species. It will enhance the ability of pest detection specialists and entomologists to detect pests, hence expanding our knowledge of the phenomenon within and beyond the established range. Context and Goals: Aside from D. ponderosae Hopk., other species of Dendroctonus also infested P. contorta trees that were growing alongside Picea engelmannii in the SRM' subalpine forest. Identifying these beetles proved to be challenging. The objective was to enhance the detection and differentiation of the species from other comparable Dendroctonus spp. that are attacking the same host, as well as to reveal biological insights regarding the phenomenon. Specimens of materials and Dendroctonus beetles that were attacking P. contorta were gathered from the entire SRM. The signs of their attack and their behavior were documented. The characteristics were modified based on existing literature, and additional traits were incorporated and evaluated. The identification of Dendroctonus bark beetles that attack P. contorta in the SRM was enhanced by utilizing updated and novel characteristics, such as indicators of attack, attack behavior, and physical traits of adult beetles. A more refined identification key couplet is provided to accurately differentiate D. murrayanae from D. rufipennis. In conclusion, simplified insect identifications that consider insect characteristics, attack patterns, and signs, as demonstrated here, enhance the detection of targeted insects and are accessible to users with varying levels of expertise. Accurate insect detections enhance our comprehension of their skills and the influence they exert on the woodland ecosystems we investigate, safeguard, and oversee worldwide.

REFERENCES

1. Peet, R.K. Forest vegetation of the Colorado front range. Vegetatio 1981, 45, 3–75. [**Google Scholar**] [**CrossRef**]
2. Klutsch, J.; Negron, J.F.; Costello, S.L.; Rhoades, C.C.; West, D.R.; Popp, J.; Caissie, R. Stand characteristics and downed woody debris accumulations associated with a mountain pine beetle (Dendroctonus ponderosae Hopkins) outbreak in Colorado. For. Ecol. Manag. 2009, 258, 641–649. [**Google Scholar**] [**CrossRef**]
3. Meddens, A.J.; Hicke, J.A. Spatial and temporal patterns of Landsat-based detection of tree mortality caused by a mountain pine beetle outbreak in Colorado, USA. For. Ecol. Manag. 2014, 322, 78–88. [**Google Scholar**] [**CrossRef**]
4. Forest Insect and Disease Conditions in the Rocky Mountain Region 2005. Available online: **https://www.fs.usda.gov/Internet/FSE_DOCUMENTS/stelprdb5279149.pdf** (accessed on 1 December 2016).
5. Wood, S.L. The Bark and Ambrosia Beetles of North and Central America (Coleoptera: Scolytidae), a Taxonomic Monograph, Great Basin Naturalist Memoirs, No. 6; Brigham Young University: Provo, UT, USA, 1982; pp. 150–203. [**Google Scholar**] [**CrossRef**]
6. Furniss, M.M.; Kegley, S.J. Biology of Dendroctonus murrayanae (Coleoptera: Curculionidae: Scolytinae) in Idaho and Montana and comparative taxonomic notes. Ann. Entomol. Soc. Am. 2008, 101, 1010–1016. [**Google Scholar**] [**CrossRef**]
7. Wood, S.L. A revision of the bark beetle genus Dendroctonus Erichson (Coleoptera: Scolytidae). Great Basin Nat. 1963, 23, 1–117. [**Google Scholar**] [**CrossRef**]
8. Amman, G.D.; Baker, B.H.; Stipe, L.E. Lodgepole Pine Losses to Mountain Pine Beetle Related to Elevation, Research Note INT-171; USDA/FS, Intermountain Forest & Range Exp. Sta.: Ogden, UT, USA, 1973; pp. 1–8.

9. DeLeon, D. The Biology and Control of the Black Hills Beetle (Dendroctonus ponderosae Hopk.), Summary of Studies in Colorado and Wyoming (1935–1938), Report to the Forest Insect Laboratory (Unpublished); USDA/FS: Fort Collins, CO, USA, 1939.
10. Mitton, J.B.; Ferrenberg, S.M. Mountain pine beetle develops an unprecedented summer generation in response to climate warming. Am. Nat. 2012, 179, E163–E171. [**Google Scholar**] [**CrossRef**] [**PubMed**] [**Green Version**]
11. Massey, C.L.; Wygant, N.D. Biology and Control of the Engelmann Spruce Beetle in Colorado, Circular No. 944; USDA/FS: Washington, DC, USA, 1954; pp. 1–35.
12. McCambridge, W.F.; Knight, F.B. Factors affecting spruce beetles during a small outbreak. Ecology 1972, 53, 830–839. [**Google Scholar**] [**CrossRef**]
13. USGS Digital Representations of Tree Species Range Maps from "Atlas of United States trees" by Elbert, L. Little, Jr. (and Other Publications). Available online: **https://gec.cr.usgs.gov/data/little/** (accessed on 15 December 2016).
14. Hopkins, A.D. Contributions Toward a Monograph of the Scolytid Beetles, Tech. Ser. 17, Part I: The genus Dendroctonus; USDA Bureau of Ent.: Washington, DC, USA, 1909; pp. 1–164.
15. Hopkins, A.D. Practical Information on the Scolytid Beetles of North American Forests, Bull. 83, Part I: Bark Beetles of the Genus Dendroctonus; USDA Bureau of Ent.: Washington, DC, USA, 1909; pp. 1–169.
16. Swaine, J.M. Canadian Bark Beetles, Part II. Preliminary Classification, with an Account of the Habits and Means of Control, Bulletin 14; Department of Agriculture: Ottawa, ON, Canada, 1918.
17. Bright, D.E., Jr. The Insects and Arachnids of Canada, Part 2: The Bark Beetles of Canada and Alaska, Coleoptera: Scolytidae; Canada Department of Agriculture: Ontario, ON, Canada, 1976; pp. 1–241.
18. Hopkins, A.D. Contributions toward a Monograph of the Scolytid Beetles, Part II: Preliminary Classification of the Superfamily Scolytoidea, Tech. Ser. 17; USDA Bureau of Ent.: Washington, DC, USA, 1915; pp. 1–232.
19. Schmid, J.; Frye, R. Spruce Beetle in the Rockies, Gen. Tech. Report RM-49; USDA/FS, Rocky Mountain Forest and Range Exp. Sta.: Fort Collins, CO, USA, 1977; pp. 1–38.
20. Amman, G.D.; Cole, W.E. Mountain Pine Beetle Dynamics in Lodgepole Pine Forests. Part II: Population Dynamics, Gen. Tech. Report INT-145; USDA/FS, Intermountain Forest and Range Exp. Sta.: Ogden, UT, USA, 1983; pp. 1–59.
21. Huber, D.P.; Aukema, B.H.; Hodgkinson, R.S.; Lindgren, B.S. Successful colonization, reproduction, and new generation emergence in live interior hybrid spruce Picea engelmannii × glauca by mountain pine beetle Dendroctonus ponderosae. Agric. For. Entomol. 2009, 11, 83–89. [**Google Scholar**] [**CrossRef**]
22. Koopmans, J.M. Spatial and temporal analyses of bark beetle population dynamics in southern British Columbia: Stand-level studies of the bole-infesting assemblage during eruptive transitions of mountain pine beetle, Dendroctonus ponderosae Hopkins. Master's Thesis, M.Sc., University of Northern British Columbia, British Columbia, Canada, 2008. [**Google Scholar**]

23. Smith, G.D.; Carroll, A.L.; Lindgren, B.S. Facilitation in bark beetles: Endemic mountain pine beetle gets a helping hand. Agric. Forest Entomol. 2011, 13, 37–43. [**Google Scholar**] [**CrossRef**]

24. Safranyik, L.; Linton, D.A. Brood production by three spp. of Dendroctonus (Coleoptera: Scolytidae) in bolts from host and non-host trees. J. Entomol. Soc. B. C. 1983, 80, 10–13. [**Google Scholar**]

03

A STUDY HAS IDENTIFIED A RXLR EFFECTOR GENE AS A NEW BIOMARKER IN A RECOMBINASE POLYMERASE AMPLIFICATION ASSAY FOR THE RAPID DETECTION OF PHYTOPHTHORA CINNAMOMI

1. INTRODUCTION: AN OVERVIEW

Phytophthora cinnamomi is a very damaging disease that affects around 5000 plant species, including numerous commercially significant crops [1,2,3]. One example of a widespread and highly damaging disease of avocados is root rot caused by P. cinnamomi. In California, USA, this disease is estimated to cost an annual loss of $40 million [4].

Phytophthora cinnamomi is responsible for inducing root and crown rot, canker, and dieback diseases in various woody ornamental crops, including azalea, boxwood, camellia, cedar, and rhododendron [5]. P. cinnamomi is a well-known invasive pathogen that affects both agriculture and horticulture, as well as natural plantings globally. P. cinnamomi, along with chestnut blight, played a significant role in the almost complete eradication of wild American chestnut trees in the eastern United States [6,7]. P. cinnamomi has irreversibly changed the makeup of plant communities and caused a significant decrease in biodiversity in the forests of south-west Western Australia. It has killed more than 3500 endemic plant species that are susceptible to it, and it also poses a threat to macrofungal and avian species [3,8,9,10]. P. cinnamomi has been detected in several provinces and municipalities in China, including Fujian [11,12,13], Hainan [14], Jiangsu [13], Shaanxi [15], Shanghai [16], and Zhejiang [13]. This poses a significant risk to the cultivation of avocado, blueberry, chestnut, kiwifruit, ornamental trees, as well as the overall well-being of natural forests.

Phytophthora species release many effector proteins that suppress host immunity and aid in plant infection. These include elicitins, tiny cysteine-rich toxins, transglutaminases, and RxLR effectors. RxLR effectors, which are identified by the presence of an Arg-Xaa-Leu-Arg (RxLR) motif, a signal peptide at the beginning, and frequently a Glu-Glu-Arg (EER) motif, play a role in transporting effector proteins from the plant apoplast to the cytoplasm [19,20,21]. Although the specific number and functions of P. cinnamomi RxLR effectors are yet unknown, a genomic investigation comparing P. infestans and P. cinnamomi has found at least 171 RxLR effector genes in the P. cinnamomi genome [3]. In addition, the P. cinnamomi genome contains only three out of the 10 RxLR effector homologues with known functional characteristics [3]. This suggests that there are distinct RxLR effector proteins and genes specific to P. cinnamomi.

Early identification is the initial and possibly most efficient measure in the hierarchy of strategies for controlling plant diseases. While chemical application and soil sanitation have the potential to decrease disease severity and inoculum level [5], continuous treatments may become expensive and ineffective once P. cinnamomi has already extensively spread in agricultural settings. Furthermore, addressing the issue in natural forests could be even more challenging due to the large scale of the problem. An alternative strategy for disease control involves promptly identifying the presence of the pathogen and subsequently eliminating it. It is cost-effective, environmentally friendly, and efficient when the pathogen is eliminated in the early stages. However, the effectiveness of the detection assay depends on the precision and swiftness of its results.

Molecular approaches have improved the detection assays of P. cinnamomi, although their accuracy has been questioned due to the emergence of novel infections. Prior molecular detection methods for P. cinnamomi involved conventional PCR [22,23,24,25,26], real-time PCR [27,28,29], and isothermal amplification assays [28,30]. Nevertheless, numerous assays have proven unsuccessful in distinguishing P. cinnamomi from P. parvispora [31], which is the species that is most closely related to P. cinnamomi within the genus Phytophthora [32,33,34].

Recombinase polymerase amplification (RPA) is a recently developed method used in the detection assays of Phytophthora species [28,35,36,37,38]. The RPA process utilizes three fundamental components: recombinase, DNA-binding proteins, and polymerase. Recombinases function by matching oligonucleotide primers with a homologous sequence in double-stranded DNA. Concurrently, DNA-binding proteins attach to the strands of DNA that have been moved aside, and polymerase initiates the production of a full double-stranded DNA molecule at the location where the primer has attached to the target DNA. The enzyme mixture undergoes exponential amplification of the target region within the template. RPA assays offer advantages over PCR and loop-mediated isothermal amplification (LAMP) assays in terms of shorter duration and simpler processes and equipment for amplifying DNA templates of Phytophthora species [36,37,38,39,40]. Furthermore, the use of lateral flow strips allows for the detection of amplified products in RPA, and the results can be easily interpreted without the need for specialized equipment. This indicates that combining RPA with lateral flow dipsticks (RPA-LFD) is a viable approach for creating a straightforward, fast, and precise method to detect plant pathogens in areas with limited resources. This study aimed to uncover a RxLR effector gene PHYCI_587572 by a comparative genomic method and create an RPA-LFD assay that targets this particular biomarker of P. cinnamomi.

2. MATERIALS AND METHODS

2.1. Isolate Selection and DNA Extraction

Phytophthora cinnamomi isolates were obtained from diseased roots of Pinus sp., Rhododendron simsii, and Camellia oleifera from various provinces in China. The roots were surface disinfected, dried, and placed on Phytophthora selective agar PARP-V8 agar [5]. Surface disinfestation of the diseased roots were accomplished incubation in 70% ethanol. When using 70% ethanol, infected roots were given single dips and then dipped in three vials containing autoclaved distilled water for 5–10 min per vial. Following surface disinfestation, roots were surface dried by placing them on filter paper. A total of 76 isolates were used in this study including 19 of P. cinnamomi, 40 samples representing 21 Phytophthora species, one of Globisporangium ultimum, and 16 other fungal species (**Table 1**). For DNA extraction, each isolate was cultured in 10% clarified V_8 juice (oomycetes) or potato dextrose broth (fungi) at 20–25 °C in the dark for 4–5 days. Genomic DNA (gDNA) was extracted from the harvested mycelia using a E.Z.N.A.® HP Fungal DNA Kit (Omega Bio-tek, Inc., Norcross, GA, USA) following the manufacturer's instructions. Concentrations of gDNA extractions were measured using a NanoDrop ND-1000 spectrophotometer (Thermo Fisher Scientific, Wilmington, DE, USA) then adjusted as needed. Extractions were stored at -20 °C until use.

Table 1. List of oomycete and fungal isolates used in this study and their detection results in the recombinase polymerase amplification-lateral flow dipstick (RPA-LFD) assay targeting the PHYCI_587572 biomarker.

(Sub) Clade	Species	Isolate	Host or Substrate	Location [a]	RPA-LFD [b]
7c	*Phytophthora cinnamomi*	Pci1	*Pinus* sp.	AH	+
		Pci2	*Rhododendron simsii*	JS	+
		Pci3	*Cedrus deodara*	JS	+
		Pci4	*Camellia oleifera*	JS	+
		Pci5	*Pinus* sp.	JS	+
		Pci6	*Rhododendron simsii*	AH	+
		Pci7	*Rhododendron simsii*	SD	+
		Pci8	*Cedrus deodara*	SD	+
		Pci9	*Cedrus deodara*	AH	+
		Pci10	*Pinus* sp.	SD	+
		JP-07-0035	*Pieris* sp.	Oregon, USA	+
		PCN-18-001	*Castanopsis* sp.	Taiwan	+
		PCN-18-078	*soil*	Taiwan	+
		JP-08-313	*Pieris* sp.	Oregon, USA	+
		JP-07-0376	N/A	Oregon, USA	+
		JP-09-325	N/A	Oregon, USA	+
		ATCC 15400	*Camellia japonica*	South Carolina, USA	+
		ATCC 15401	*Persea americana*	Puerto Rico	+
		CBS 144.22	*Cinnamomum burmannii*	Indonesia	+
7c	*P. parvispora*	CBS132771	*Arbutus unedo*	Italy	-
		CBS132772	*Arbutus unedo*	Italy	-
7a	*P. cambivora*	CBS 248.60	*Castanea sativa*	USA	-
	P. fragariae	CBS 209.46	*Fragaria × ananassa*	England, UK	-
	P. rubi	CBS 967.95	*Rubus idaeus*	Scotland, UK	-
7b	*P. melonis*	PMNJHG1	*Cucumis sativus*	JS	-
		PMNJHG2	*Cucumis sativus*	JS	-

(Sub) Clade	Species	Isolate	Host or Substrate	Location [a]	RPA-LFD [b]
		PMNJHG3	*Cucumis sativus*	JS	-
		PMNJDG1	*Benincasa hispida*	JS	-
		PMNJDG2	*Benincasa hispida*	JS	-
		PMNJDG3	*Benincasa hispida*	JS	-
		PMFJHL1	*Lagenaria siceraria*	FJ	-
		IMI 325917	*Cucumis* sp.	FJ	-
	P. sojae	P6497	*Glycine max*	Mississippi, USA	-
		Peng-R3	*Glycine max*	FJ	-
1	*P. cactorum*	Pcac1	*Malus pumila*	JS	-
		Pcac2	*Malus pumila*	JS	-
		Pcac3	*Rosa chinensis*	JS	-
	P. infestans	Pin1	*Solanum tuberosum*	FJ	-
		Pin2	*Solanum tuberosum*	YN	-
	P. nicotianae	Pni1	*Nicotiana tabacum*	YN	-
		Pni2	*Lycopersicum* sp.	JS	-
		Pni3	*Sophora sinensis*	JS	-
		Pni4	*Citrus* sp.	JS	-
2	*P. capsici*	Pcap1	*Capsicum annuum*	JS	-
3	*P. ilicis*	CBS 114348	*Ilex aquifolium*	The Netherlands	-
4	*P. palmivora*	Ppa1	Iridaceae	YN	-
4	*P. quercetorum*	15C7	Soil	South Carolina, USA	-
5	*P. castaneae*	CBS 587.85	Soil	Taiwan	-
6	*P. megasperma*	CBS 305.36	*Matthiola incana*	California, USA	-
	P. mississippiae	57J3	Irrigation water	Mississippi, USA	-
8	*P. drechsleri*	CBS 292.35	*Beta vulgaris* var. *altissima*	California, USA	-
		ATCC 56353	*Citrus sinensis*	Australia	-
	P. hibernalis	CBS 270.31	*Cirrus sinensis*	Setúbal, Portugal	-
	P. syringae	ATCC 34002	*Citrus* sp.	California, USA	-

(Sub) Clade	Species	Isolate	Host or Substrate	Location [a]	RPA-LFD [b]
	P. ramorum	ATCC MYA-2949	*Quercus agrifolia*	California, USA	-
10	*P. boehmeriae*	Pbo1	*Boehmeria nivea*	JS	-
		Pbo2	*Gossypium* sp.	JS	-
		Pbo3	*Boehmeria nivea*	JS	-
		Pbo4	*Gossypium* sp.	JS	-
Oomycete	*Globisporangium ultimum*	Gu1	Irrigation water	JS	-
Fungi	*Alternaria alternata*	LH1401	*Cucumis melo*	JS	-
	Aspergillus oryzae	Ao1	*Glycine max*	JS	-
	Cercospora kikuchii	Ck1	*Glycine max*	JS	-
	Colletotrichum gloeosporioides	Cg1	*Glycine max*	JS	-
	C. truncatum	Ct1	*Glycine max*	JS	-
	Diaporthe phaseolorum var. *caulivora*	DPC	*Glycine max*	JS	-
	Fusarium oxysporum	Fo1	*Pinus* sp.	JS	-
	Fusarium solani	Fs1	*Gossypium* sp.	JS	-
		Fs2	*Glycine max*	JS	-
	Fusarium equiseti	Fe1	*Pinus* sp.	JS	-
	Magnaporthe grisea	Guy11	*Oryza sativa*	French Guiana	-
	Nigrospora sphaerica	Ns1	*Glycine max*	JS	-
	Phakopsora pachyrhizi	Pa1	*Glycine max*	JS	-
	Phomopsis asparagi	Pas1	*Asparagus officinalis*	JS	-
	Rhizoctonia solani	Rs1	*Gossypium* sp.	JS	-
	Verticillium dahliae	Vda1	*Gossypium* sp.	JS	-

[a] Abbreviations of provinces in China: AH, Anhui province; JS, Jiangsu province; SD, Shandong; FJ, Fujian province; YN, Yunnan province; [b] Positive (+) or negative (-) reaction result in the RPA-LFD assay for detecting *P. cinnamomi*.

2.2. Genomic Sequences

The genomic sequences and gene models of P. capsici, P. cinnamomi, P. infestans, P. ramorum, and P. sojae were obtained via the MycoCosm site [41]. The subsequent comparative genomic analysis included P. asiatica, P. cactorum, P. cambivora, P. hibernalis, P. parvispora, P. syringae, and P. vignae.

2.3. Identification of A P. cinnamomi-Unique RxLR Effector Gene

To identify RxLR effectors, the genome sequence of P. cinnamomi was translated to open reading frames (ORFs). ORFs encoding < 50 amino acids were excluded. SignalP v3.0 was used to identify the N-terminal signal peptide [**42**] within remaining ORFs. Proteins with Hidden Markov Model (HMM) scores > 0.9 and without predicted transmembrane domain were predicted as secreted proteins. A regular expressions (REGEX) method was performed by script for identifying the (^\w{1,40}\w{1,96}R\wLR\w{1,40} [ED][ED][KR]) motif [**43,44**] within the predicted secreted proteins. The RxLR-EER domains of the candidates were aligned and then used to construct a HMM profile to search within the predicted secreted proteins using HMMER. Matches with a bit score of >0 were manually validated.

2.4. Construction of Binary Potato Virus X (PVX) Vectors

As no introns were predicted within PHYCI_587572, this gene was directly amplified using the gDNA of isolate Pci1 (**Table 1**). Each 25-μL PCR reaction include 1 μL of gDNA (100 ng per μL), 1 μL (10 μM) of each of primers PHYCI_587572-PVX-HA(infusion)-F and PHYCI_587572--PVX-HA(infusion)-R (**Table 2**), 1 μL of PrimeSTAR HS DNA Polymerase (Takara Bio, Kusatsu, Shiga, Japan), 4 μL of dNTP Mixure (2.5 mM), 10 μL of 5× PrimeSTAR Buffer, and 32 μL of nuclease-free water (nfH_2O). PCR was carried out using a Veriti 96-Well Thermal Cycler (Applied Biosystems, Singapore) following the program: an initial denaturation step at 95 °C for 3 min, 35 cycles of 95 °C for 15 s, 56 °C for 15 s, and 72 °C for 15 s, plus a final extension at 72 °C for 5 min. Each set of PCR reactions included a NTC. PCR products were examined in 1% agarose gel electrophoresis at 120 V for approximately 25 min. Agarose gel was stained by ethidium bromide and visualized on a transilluminator. The PCR assay was carried out three times. PCR products were digested with the SmaI restriction enzyme (NEB, R0141S) and cloned into the PVX vector pGR107 [**45**] to form pGR107:PHYCI_587572. The constructs were confirmed by sequencing at Genscript (Nanjing, China) using the primers LBA and LBB (**Table 2**). Recombinant binary plasmids were maintained and propagated in Escherichia coli strain JM109 grown in Luria-Bertani (LB) media amended with kanamycin (50 mg per mL) and tetracycline (12.5 mg per mL).

Table 2. Primers used in the construction of binary potato virus X (PVX) vectors, infiltration of Agrobacterium tumefaciens, and the recombinase polymerase amplification-lateral flow dipstick (RPA-LFD) assay targeting the PHYCI_587572 biomarker.

Assay	Name	Sequence (5′-3′)
PVX construction	PHYCI_587572-PVX-HA(infusion)-F	CTAGCATCGATTCCCGGGATGCTCTCGATGACCACAGCCTCC

Assay	Name	Sequence (5'-3')
	PHYCI_587572-PVX-HA(infusion)-R	CTCTAGAGGATCCCCGGGGAAATTCTCCTCGCGCGTG
Agrobacterium infiltration	LBa	CAATCACAGTGTTGGCTTGC
	LBb	GACCCTATGGGCTGTGTTG
RPA	PciRL587572F	GCGAGGCCCTCTCGATGACCACAGCCTCCAACCA
	PciRL587572R	[Biotin]TTGCTGCAGATATGTGCTGCTTGCCTGGACCATC
	PciRL587572P	[FAM]GAGGCAGTCGACGATGATGA[THF] TCCTCCGAAGATTCC[C3-spacer]

2.5. Agrobacterium Tumefaciens Infiltration

Constructs were introduced into A. tumefaciens strain GV3101 by electroporation [**46,47**]. Recombinants (pGR107:PHYCI_587572) were selected with tetracycline (12.5 mg per mL) and kanamycin (50 mg per mL). Individual colonies were verified in PCR using the primers LBa and LBb (**Table 2**).

Nicotiana benthamiana plants were grown in a greenhouse at a day/night temperature of 25/16 °C and a 16 h photoperiod for 4–6 weeks. Recombinant GV3101 cells were grown in LB media amended with kanamycin (50 mg per mL) for 48 h, harvested, washed with 10 mM $MgCl_2$ three times, resuspended in 10 mM $MgCl_2$ to a final OD_{600} of 0.4, then incubated at room temperature for 1 to 3 h prior to infiltration. For pressure infiltration, 100 mL of recombinant GV3101 cell suspension was infiltrated though a small nick into each of three infiltration sites per adaxial side of a N. benthamiana leaf using a syringe. After 0, 12, or 24 h, Agrobacterium tumefaciens cells carrying the Bax gene (pGR107:Bax) or inf1 gene (pGR107:inf1) were infiltrated into each of the three filtration sites. Agrobacterium tumefaciens cells carrying the enhanced green fluorescent protein (eGFP) gene in place of pGR107:Bax or pGR107:inf1 were infiltrated into three parallelly designated filtration sites of the same leaf after 0, 12, or 24 h as controls. Symptom development was monitored daily. Photographs were taken on 5 and 6 days after filtration of pGR107:Bax and pGR107:inf1, respectively. This experiment was conducted twice with three replicate leaves in each repeat.

2.6. RPA-LFD Assay

A set of RPA primers and probe targeting the P. cinnamomi-specific biomarker PHYCI_587572, including a forward primer PciRL587572F, a 5'-biotin-labeled reverse primer PciRL587572R, and a probe PciRL587572P labeled by fluorescein amidite (FAM), tetrahydrofuran (THF), and C3 spacer (**Table 2**), was designed according to the instruction for the nfo kit in the TwistAmp® DNA Amplification Kits Assay Design Manual. Primers and probe were synthesized by Sangon Biotech (Shanghai, China).

The RPA-LFD assay was performed following the procedures in the TwistAmp® nfo Kit Quick Guide (TwistDx Ltd., Cambridge, UK). Each RPA reaction in a total volume of 50 μL contained 1 μL of DNA template, 29.5 μL of rehydration buffer, 2.1 μL of each of primers PciRL587572F and PciRL587572R (10 μM), 0.6 μL of probe PciRL587572P (10 μM), 12.2 μL of nfH_2O, and 2.5 μL of magnesium acetate (280 mM). These reagents were added into reaction microtubes and mixed using a pipette. Freeze-dried pellet magnesium acetate was pre-added to the cap of the microtubes. After brief vortex and centrifugation, the reaction microtubes were incubated in a heating block set at 39 °C for 20 min. To visualize amplicons, 10 μL of RPA product was mixed with 90 μL of phosphate buffered saline with Tween 20 (PBST) running buffer. Then 5 μL of the RPA product per each reaction was added to the sample pad of a Milenia Genline HybriDetect 1 LFD (Milenia Biotec GmbH, Giessen, Germany). The LFD was then vertically immersed into 100 μL of PBST (Milenia Biotec GmbH, Giessen, Germany) in a sterile centrifuge tube for up to 5 min at room temperature (aver. 22 °C) until a clear control line was observed. The dipsticks yielded visible control lines, indicating valid tests. Test lines were visible on all dipsticks correlating to positive RPA reactions.A negative detection or NTC in contrast should not result in any signal or color at the position of the test line.

2.7. Evaluation of RPA Specificity and Sensitivity

The specificity of the RPA-LFD assay was evaluated by testing against all isolates listed in **Table 1**. One μL of gDNA (10 ng per μL) of each isolate was included in each 50 μL reactions. Each isolate was tested in triplicate.

To evaluate the sensitivity of the RPA-LFD assay, 10-fold serial dilution of gDNA of P. cinnamomi isolate Pci1 ranging from 100 ng to 100 fg (0.0001 ng) per μL were used as templates. Each concentration was evaluated in triplicate. Each set of reactions in the specificity and sensitivity evaluations included an NTC (nfH_2O in place of DNA template).

2.8. Detecting P. cinnamomi in Artificially Inoculated Pine Needles Using RPA-LFD

Pine needles were collected from a healthy deodar cedar tree (Cedrus deodara) at a landscape site of Nanjing Forestry University (Nanjing, China). They were artificially inoculated by P. cinnamomi. A 2 × 2 mm mycelial plug of P. cinnamomi isolate Pci1 was mixed with approximately 100 g of pine needles and incubated in each of triplicate sterile 250 mL flasks at 22 °C, 100% relative humidity, and a 12-h photoperiod for 5 days. Sterile agar plugs were mixed with pine needles to produce three non-inoculated control samples. Total DNAs were extracted from P. cinnamomi-colonized and non-inoculated samples using an NaOH lysis method. Briefly, 20 mg of plant tissues collected from the wound site of each Pine needle were placed into a 1.5-mL microtube containing 200 μL of NaOH (0.5 N). They were grinded for approximately 1 min until no large pieces of plant tissues were visible using a sterile tissue grinder pestle. Then, 5 μL of grinded tissues in NaOH were transferred to a new microtube containing 495 μL of Tris buffer (100 mM, pH 8.0). Two μL of the mixture were used as used as templates in the RPA-LFD assay. Purified gDNA (10 ng per μL) of isolate Pci1 and nfH_2O were used as a positive control and NTC, respectively, in two repeats of the experiment.

2.9. Detecting P. cinnamomi in Soil Samples Using the RPA-LFD Assay

A total of 13 soil samples were collected from five provinces in China (**Table 3**). A previous study found the presence of P. cinnamomi in all 13 samples [**30**]. Each soil sample was mixed well and divided into two equal subsamples. DNAs were directly extracted from the first set of 13 soil subsamples using a Plant DNA Mini Kit (Omega Biotek, Norcross, GA, USA). For the second set, a baiting method [**5**] using C. deodara leaves was deployed. After 5 days, DNA was extracted from leaf baits of each subsample using the NaOH lysis method. Both RPA-LFD and conventional PCR assays were performed using DNA extractions of both soil samples and leaf baits. Each DNA extraction was tested three times using each assay. Purified gDNA (10 ng) of isolate Pci1 was used in positive control reactions.

Table 3. Detection results of Phytophthora cinnamomi in 13 infested soil samples using the recombinase polymerase amplification-lateral flow dipstick (RPA-LFD) and PCR assays.

Assay	Name	Sequence (5′-3′)
PVX construction	PHYCI_587572-PVX-HA(infusion)-F	CTAGCATCGATTCCCGGGATGCTCTCGATGACCACAGCCTCC
	PHYCI_587572-PVX-HA(infusion)-R	CTCTAGAGGATCCCCGGGGAAATTCTCCTCGCGCGTG
Agrobacterium infiltration	LBa	CAATCACAGTGTTGGCTTGC
	LBb	GACCCTATGGGCTGTGTTG
RPA	PciRL587572F	GCGAGGCCCTCTCGATGACCACAGCCTCCAACCA
	PciRL587572R	[Biotin]TTGCTGCAGATATGTGCTGCTTGCCTGGACCATC
	PciRL587572P	[FAM]GAGGCAGTCGACGATGATGA[THF] TCCTCCGAAGATTCC[C3-spacer]
[a] Province abbreviations: AH, Anhui; JS, Jiangsu; SD, Shandong; FJ, Fujian; YN, Yunnan.		

3. RESULTS

3.1. Identification of A P. cinnamomi-Unique RxLR Effector Gene PHYCI_587572

A protein (Avh87), translated from the PHYCI_587572 gene located at scaffold_77:29462-29848, was found containing a signal peptide cleavage site at residue 20, an RLLR motif at 44–47, followed by an SEER motif at 63–66 (**Figure 1**). Sequences of the PHYCI_587572 gene and Avh87 were used to blast against the genome sequences and predicted ORFs of P. asiatica, P. cactorum, P. cambivora, P. capsici, P. hibernalis, P. infestans, P. parvispora, P. ramorum, P. sojae, P. syringae, and P. vignae, respectively. No homologs of PHYCI_587572 gene or Avh87 were found within these non-target species.

```
1         10         20         30         40         50         60         70
....|....| ....|....| ....|....| ....|....| ....|....| ....|....| ....|....|
MRLSQVLVVV TVSFLVTSEA LSMTTASNQA KIAKIPSADS PNQRLLRTHR EAVDDDDSSE DSSEERTISI

71        80         90         100        110        120       129
....|....| ....|....| ....|....| ....|....| ....|....| ....|....
SQMKAIVEEL GIDWKMVQAS STYLQQHAKY EQYQEKANAL MKAKTKSHGS PRITREENF
```

Figure 1. Amino acid sequences of avirulence homolog protein 87 (Avh87) derived from the PHYCI_587572 gene open reading frame of Phytophthora cinnamomi. Boxed residues 1–20: predicted signal peptide; 44–47: an RxLR (RLLR) motif; 63–66: a dEER (SEER) motif.

3.2. Suppression of Programmed Cell Death by Avh87

Avh87 encoded by the PHYCI_587572 gene consistently suppressed BAX- and INF1-mediated cell death among three replications between two repeats of the A. tumefaciens infiltration experiment. BAX (pGR107:Bax) could triggered apoptosis of N. benthamiana leaves when simultaneously infiltrated along with Avh87 (**Figure 2A**). However, infiltration leaves with pGR107:PHYCI_587572 12 or 24 h prior to that with pGR107:Bax inhibited the BAX-mediated cell death (**Figure** 2A). The inhibition of Avh87 appeared as more effective against INF1. As shown in **Figure** 2B, INF1-mediated cell death was suppressed when pGR107:Inf1 was infiltrated promptly following that with pGR107:PHYCI_587572. The eGFP protein in place of BAX or INF1 expressed as negative controls consistently induced minimum cell death (**Figure 2**) in both sets of experiments.

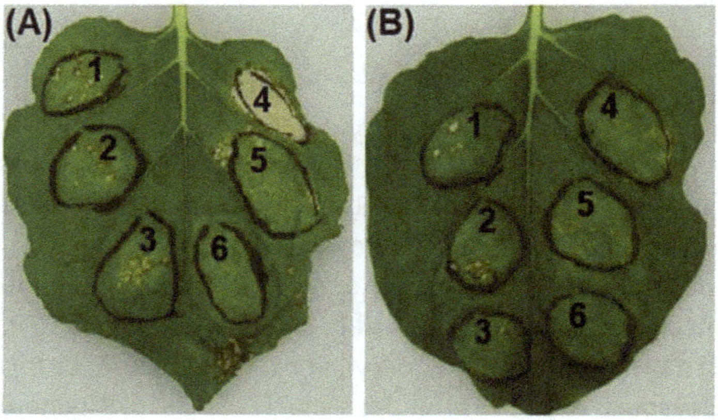

Figure 2. Suppression of BAX- and elicitin protein INF1-mediated cell death of Nicotiana benthamiana leaves infiltrated with Agrobacterium tumefaciens strains containing a PVX vector carry the PHYCI_587572 gene (Avh87) or a control gene (eGFP). (A): 1, Avh87→eGFP; 2, Avh87→12h→eGFP; 3, Avh87→24h→eGFP; 4, Avh87→BAX; 5, Avh87→12h→BAX; 6, Avh87→24h→BAX. (B): 1, Avh87→eGFP; 2, Avh87→12h→eGFP; 3, Avh87→24h→eGFP; 4, Avh87→INF1; 5, Avh87→12h→INF1; 6, Avh87→24h→INF1.

3.3. Specificity and Sensitivity of the RPA-LFD Assay

In the evaluation of specificity, identical results were obtained among three repeats of all isolates. All LFDs had a visible control line, indicating valid tests. Test lines were visible only on LFDs with genomic DNAs (gDNAs) of 19 P. cinnamomi isolates (**Table 1**). No test lines were observed on those of 22 other oomycete and 15 fungal species (**Figure 3**).

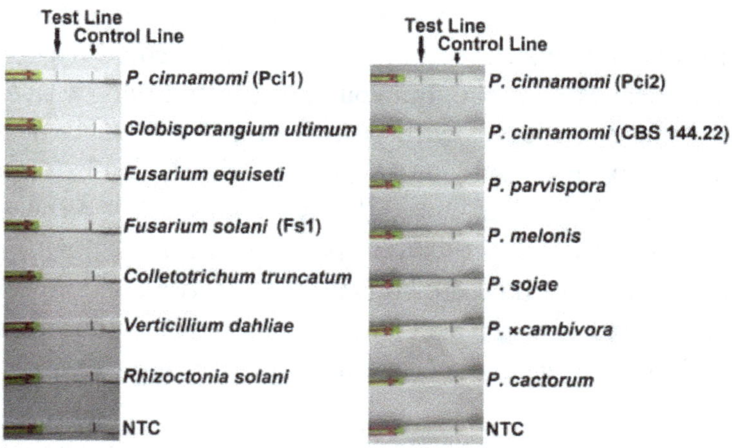

Figure 3. Detection results of the recombinase polymerase amplification-lateral flow dipstick assay targeting the PHYCI_587572 biomarker. Genomic DNAs (10 ng per 50-μL reaction) of Phytophthora cinnamomi isolates and other oomycete and fungal species. Nuclease-free water was used in place of DNA templates in no template controls (NTC).

In the sensitivity evaluation, all LFDs had a visible control line. Solid test lines were observed on LFDs of RPA reactions containing 0.1–100 ng of P. cinnamomi gDNA (**Figure 4**). They were also visible on LFDs of 0.01 to 0.001 ng of gDNA (**Figure 3**). No test lines were observed on those of 0.0001 ng (100 fg) of gDNA or no template control (NTC). These results were consistent across three repeats of the evaluation. In the comparison PCR assay, ≥1 ng of gDNA was required for consistent positive results using primers PciRL587572F and PciRL587572F. The result indicates that RPA-LFD assay was more sensitive than PCR by using LFD strips.

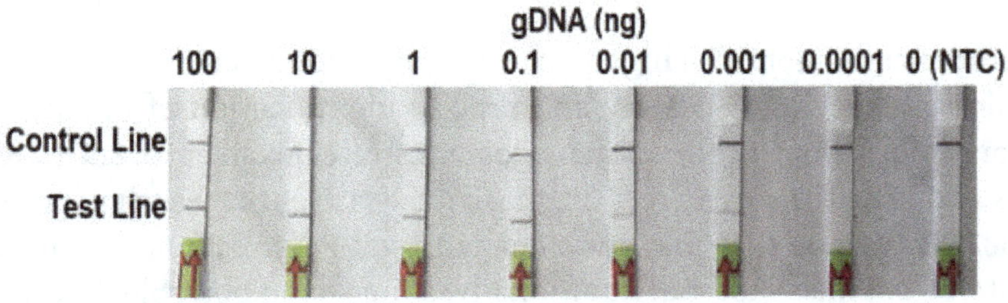

Figure 4. Evaluation results of sensitivity of the recombinase polymerase amplification-lateral flow dipstick assay targeting the PHYCI_587572 biomarker using a 10-fold serial dilutions of genomic DNA (gDNA) of Phytophthora cinnamomi isolate Pci1. Nuclease-free water was used in place of DNA templates in no template controls (NTC).

3.4. Detecting P. cinnamomi in Artificial Inoculated Pine Needles

All LFDs had a visible control line indicating valid tests. Phytophthora cinnamomi was detected in the total DNAs extracted from three P. cinnamomi-infested pine needle samples and the positive control, whereas not detectable in three non-inoculated samples or NTC (**Figure 5**). Identical results were recorded in both repeats of the experiment.

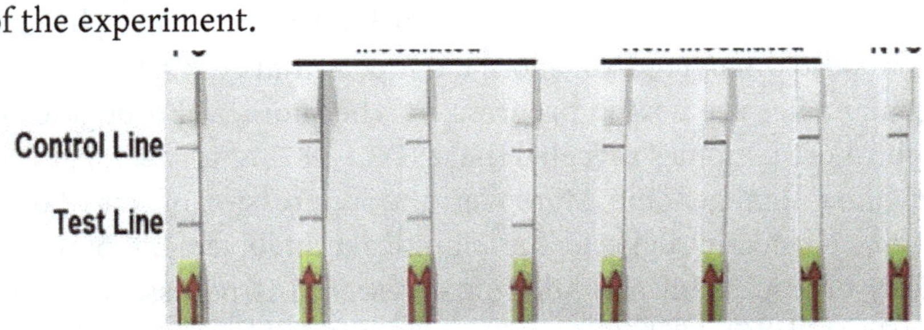

Figure 5. Detection of Phytophthora cinnamomi in artificially inoculated pine needles of Cedrus deodara using the recombinase polymerase amplification-lateral flow dipstick assay. Genomic DNA (10 ng per μL) of P. cinnamomi isolate Pci1 was used as the positive control (PC). Nuclease-free water was used in place of DNA templates in the no template control (NTC). Dipsticks of the first repeat are shown. Identical detection results were observed in the second repeat of the experiment.

3.5. Detection of P. cinnamomi in Infested Soil Samples

The RPA-LFD assay was applied to 13 P. cinnamomi-infested soil samples. Detection results were consistent among three repeats of the experiment. Using DNA samples directly extracted from soil samples, P. cinnamomi was detected in all 13 samples using the RPA-LFD assay, while it was detected in only six samples using PCR (**Table 3**). Using DNAs extracted from C. deodara leaf baits, both RPA-LFD and PCR assays detected P. cinnamomi in all 13 samples.

4. DISCUSSION

Phytophthora cinnamomi is a highly invasive plant pathogen that has a significant worldwide influence [3,6]. Timely and precise identification of P. cinnamomi is crucial for rapidly adopting illness management strategies and preventing excessive expenses resulting from incorrect diagnosis. The current research paper presents PHYCI_587572, a gene that may specifically target P. cinnamomi, identified by a comparative genomics method. Additionally, a quick and portable detection assay called RPA-LFD is introduced, which focuses on this new biomarker for identifying P. cinnamomi. The results of this study have significant consequences for the interactions between P. cinnamomi and its hosts, as well as for the control of this very destructive disease.

Although RxLR proteins have the potential to play a crucial role in the infection of plants by P. cinnamomi, the molecular function of only a limited number of RxLR effectors and genes has been identified so far [3]. The identification of genes responsible for the pathogenicity of oomycetes has been made easier due to the growing number of genome sequences available and the discovery of the RxLR-dEER motif. The PHYCI_587572 effector gene expressing Avh87 was discovered in the genomic sequence of P. cinnamomi during this research. Throughout this procedure, we also discovered 265 more RxLR potential genes. Efforts to ascertain their molecular roles are now in progress. Furthermore, a thorough examination of 12 Phytophthora genomes revealed that PHYCI_587572 is specifically associated with the P. cinnamomi genome. Moreover, the suppression of Avh87 on BAX- and INF1-induced cell death suggests the significant involvement of Avh87 in the pathogenicity of P. cinnamomi. Additional research is necessary to explore the biological role of PHYCI_587572 (Avh87) on the natural hosts of P. cinnamomi.

Effector genes play a crucial role in the pathogenicity of oomycetes, making them excellent candidates for the development of new biomarkers. The RxLR effector gene PHYCI_587572 discovered in this work was not found to have a homolog in other reference Phytophthora species. This indicates the distinct role it plays in the disease-causing ability of P. cinnamomi. However, due to the rapid evolution of many effector genes [21,48], they exhibit a greater degree of intraspecific polymorphism compared to universal housekeeping genes [49]. Therefore, using a new effector gene as a diagnostic biomarker may result in false-negative detections when dealing with mutated isolates of the target Phytophthora species. The study successfully used the RPA-LFD assay to detect all 19 P. cinnamomi isolates, including the ex-type isolate CBS 144.22, from various locations in China, the United States, Puerto Rico, and Indonesia. The assay specifically targeted the RxLR effector gene PHYCI_587572 and accurately identified the isolates from at least eight different host species in three provinces of China and two U.S. states (Table 1). This discovery suggests that the innovative RPA-LFD assay is highly inclusive when it comes to P. cinnamomi isolates. However, it is necessary to test more isolates in the future to ensure that it is inclusive.

Swift identification of pathogens is crucial for effectively controlling aggressive illnesses. RPA-based assays offer a significant benefit in terms of their quick reaction time and the need for less sophisticated equipment. The RPA-LFD assay in this study may be completed in 30 minutes, which is only half the time required by a LAMP assay [30] and a quarter of the time needed for PCR reactions. In addition, it does not necessitate the use of a thermal cycler or gel electrophoresis and imaging system, as its outcomes can be observed directly on lateral flow devices (LFDs). The RPA-LFD assay used fewer primers and a lower incubation temperature compared to the LAMP assay for P. cinnamomi [30]. The numerous benefits of the RPA-LFD assay make it a very suitable technique for detecting P. cinnamomi in both field settings and diagnostic facilities.

The creation of this innovative RPA-LFD assay has enhanced the sensitivity of isothermal detection assays for P. cinnamomi. According to the findings of this research, the RPA-LFD test was able to identify 0.001 nanograms of P. cinnamomi gDNA (equivalent to approximately 12 copies of nuclei) in a reaction system of 50 microliters. Our newly created LAMP assay is at least 200 times more sensitive than our previous assay, and 2000 times more sensitive than a PCR assay that uses primers F3 and B3 [30]. Nevertheless, the sensitivity of this assay seems to be lower than that of a qPCR-based assay previously described [29]. While the RPA-LFD test has the benefits of a quick turnaround time and less need for equipment compared to qPCR, additional research is necessary to enhance the sensitivity of isothermal assays for P. cinnamomi.

5. CONCISE SUMMARY

A novel RxLR effector PHYCI_587572 was found through bioinformatics analysis of the genomic sequence of P. cinnamomi. By comparing the genome sequences of 12 Phytophthora species, it was found that this effector gene is only present in P. cinnamomi. Therefore, it can be considered as an excellent biomarker for the development of diagnostic assays. This assay exhibited specificity towards P.

cinnamomi, but did not demonstrate detection capabilities for 37 other oomycete and fungal species, including P. parvispora, which is closely related to P. cinnamomi. Additionally, it was discovered that it had a sensitivity that was 200 times greater than a previously created LAMP assay. Furthermore, it identified the presence of P. cinnamomi in artificially infected leaves of Cedrus deodara. The findings of this study suggest that the RPA-LFD assay is a highly suitable technique for detecting P. cinnamomi in both field conditions and diagnostic laboratories. However, a more in-depth investigation is needed to fully comprehend the molecular and biological functions of PHYCI_587572 and its avirulence homolog protein Avh87.

6. AT A GLANCE

Phytophthora cinnamomi is a highly destructive pathogen that induces root and crown rot as well as dieback illnesses in about 5000 plant species. The precise and swift identification of P. cinnamomi is crucial in the ongoing efforts to prevent and control the disease. The study discovered a new effector gene, PHYCI_587572, which is exclusive to P. cinnamomi. This finding was made by a comparative genomic investigation of 12 Phytophthora species. The avirulence homolog protein 87 (Avh87) is distinguished by the presence of the Arg-Xaa-Leu-Arg (RxLR) motif. Avh87 inhibited the cell death caused by the pro-apoptotic protein BAX and the elicitin protein INF1 in Nicotiana benthamiana. Moreover, a detection technique was devised using a recombinase polymerase amplification-lateral flow dipstick to target this particular biomarker of P. cinnamomi. Although it effectively identified 19 P. cinnamomi isolates from throughout the world, this test failed to detect 37 other oomycete and fungal species, including P. parvispora, which is closely related to P. cinnamomi. Furthermore, it identified the presence of P. cinnamomi in artificially infected leaves of Cedrus deodara. In addition, the RPA-LFD assay demonstrated greater sensitivity compared to a traditional PCR assay, since it was able to detect as little as 2 picograms of genomic DNA in a 50-microliter reaction. The presence of P. cinnamomi was identified in 13 soil samples that were infested, resulting in a detection rate of 46.2% using PCR. The study's findings suggest that PHYCI_587572 is a distinct biomarker that can be used to detect P. cinnamomi. Further investigations are needed to determine the precise genetic background and biological role of PHYCI_587572 on the natural hosts of P. cinnamomi, despite its identification as an effector gene based on the RxLR motif of Avh87 and its avirulence activity on Nicotiana.

REFERENCES

1. Cahill, D.M.; Rookes, J.E.; Wilson, B.A.; Gibson, L.; McDougall, K.L. Phytophthora cinnamomi and Australia's biodiversity: Impacts, predictions and progress towards control. Aust. J. Bot. 2008, 56, 279–310. [**Google Scholar**] [**CrossRef**]

2. Jung, T.; Colquhoun, I.J.; Hardy, G.E.S. New insights into the survival strategy of the invasive soilborne pathogen Phytophthora cinnamomi in different natural ecosystems in Western Australia. For. Pathol. 2013, 43, 266–288. [**Google Scholar**] [**CrossRef**]

3. Hardham, A.R.; Blackman, L.M. Phytophthora cinnamomi. Mol. Plant Pathol. 2018, 19, 260–285. [**Google Scholar**] [**CrossRef**] [**PubMed**] [**Green Version**]

4. Coffey, M.D. Phytophthora root rot of avocado. In Plant Diseases of International Importance. Volume III. Diseases of Fruit Crops; Kumar, J., Chaub, H.S., Singh, U.S., Mukhopadhyay, A.N., Eds.; Prentice Hall: Englewood Cliffs, NJ, USA, 1992. [**Google Scholar**]

5. Erwin, D.C.; Ribeiro, O.K. Phytophthora Diseases Worldwide; APS Press: St. Paul, MN, USA, 1996. [**Google Scholar**]

6. Zentmyer, G.A. Phytophthora cinnamomi and the Diseases It Causes; American Phytopathological Society: St. Paul, MN, USA, 1980. [**Google Scholar**]

7. Freinkel, S. American Chestnut: The Life, Death, and Rebirth of a Perfect Tree; University of California Press: Berkeley, CA, USA, 2007. [**Google Scholar**]

8. Shearer, B.L.; Crane, C.E.; Cochrane, A. Quantification of the susceptibility of the native flora of the South-West Botanical Province, Western Australia, to Phytophthora cinnamomi. Aust. J. Bot. 2004, 52, 435–443. [**Google Scholar**] [**CrossRef**]

9. Anderson, P.; Brundrett, M.; Grierson, P.; Robinson, R. Impact of severe forest dieback caused by Phytophthora cinnamomi on macrofungal diversity in the northern jarrah forest of Western Australia. For. Ecol. Manag. 2010, 259, 1033–1040. [**Google Scholar**] [**CrossRef**]

10. Davis, R.A.; Valentine, L.E.; Craig, M.D.; Wilson, B.; Bancroft, W.J.; Mallie, M. Impact of Phytophthora-dieback on birds in Banksia woodlands in south west Western Australia. Biol. Conserv. 2014, 171, 136–144. [**Google Scholar**] [**CrossRef**] [**Green Version**]

11. Lan, C.Z.; Ruan, H.C.; Yao, J.A. First report of Phytophthora cinnamomi causing root and stem rot of blueberry (Vaccinium corymbosum) in China. Plant Dis. 2016, 100, 2537. [**Google Scholar**] [**CrossRef**]

12. Lan, C.Z.; Ruan, H.C.; Yao, J.A. First report of Phytophthora cinnamomi causing root rot of Castanea mollissima (Chinese chestnut) in China. Plant Dis. 2016, 100, 1248. [**Google Scholar**] [**CrossRef**]

13. Zheng, X.B.; Lu, J.Y. Studies on Phytophthora species in Fujian, Zhejiang, Jiangsu Provinces and Shanghai, China. Acta Mycol. Sin. 1989, 8, 161–168. [**Google Scholar**]

14. Zeng, H.C.; Ho, H.H.; Zheng, F.C. A survey of Phytophthora species on Hainan Island of South China. J. Phytopathol. 2009, 157, 33–39. [**Google Scholar**] [**CrossRef**]

15. Bi, X.Q.; Hieno, A.; Otsubo, K.; Kageyama, K.; Liu, G.; Li, M.Z. A multiplex PCR assay for three pathogenic Phytophthora species related to kiwifruit diseases in China. J. Gen. Plant Pathol. 2019, 85, 12–22. [**Google Scholar**] [**CrossRef**]

16. Zhou, X.G.; Zhu, Z.Y.; Lu, C.P.; Wang, S.J.; Ko, W.H. Phytophthora cinnamomi in Shanghai and it possible origin. Mycopathologia 1992, 120, 29–32. [**Google Scholar**] [**CrossRef**]

17. Wang, W.J.; Jiao, F.C. Effectors of Phytophthora pathogens are powerful weapons for manipulating host immunity. Planta 2019, 250, 413–425. [**Google Scholar**] [**CrossRef**] [**PubMed**] [**Green Version**]

18. Duclos, J.; Fauconnier, A.; Coelho, A.C.; Bollen, A.; Cravador, A.; Godfroid, E. Identification of an elicitin gene cluster in Phytophthora cinnamomi. DNA Seq. 1998, 9, 231–237. [**Google Scholar**] [**CrossRef**] [**PubMed**]

19. Dou, D.L.; Kale, S.D.; Wang, X.; Jiang, R.H.Y.; Bruce, N.A.; Arredondo, F.D.; Zhang, X.M.; Tyler, B.M. RXLR-mediated entry of Phytophthora sojae effector Avr1b into soybean cells does not require pathogen-encoded machinery. Plant Cell 2008, 20, 1930–1947. [**Google Scholar**] [**CrossRef**] [**Green Version**]

20. Wawra, S.; Belmonte, R.; Lobach, L.; Saraiva, M.; Willems, A.; van West, P. Secretion, delivery and function of oomycete effector proteins. Curr. Opin. Microbiol. 2012, 15, 685–691. [**Google Scholar**] [**CrossRef**] [**Green Version**]

21. Jiang, R.H.Y.; Tripathy, S.; Govers, F.; Tyler, B.M. RXLR effector reservoir in two Phytophthora species is dominated by a single rapidly evolving superfamily with more than 700 members. Proc. Natl. Acad. Sci. USA 2008, 105, 4874–4879. [**Google Scholar**] [**CrossRef**] [**Green Version**]

22. Kong, P.; Hong, C.X.; Richardson, P.A. Rapid detection of Phytophthora cinnamomi using PCR with primers derived from the Lpv putative storage protein genes. Plant Pathol. 2003, 52, 681–693. [**Google Scholar**] [**CrossRef**]

23. Schena, L.; Duncan, J.M.; Cooke, D.E.L. Development and application of a PCR-based 'molecular tool box' for the identification of Phytophthora species damaging forests and natural ecosystems. Plant Pathol. 2008, 57, 64–75. [**Google Scholar**] [**CrossRef**]

24. Trzewik, A.; Nowak, K.J.; Orlikowska, T. A simple method for extracting DNA from rhododendron plants infected with Phytophthora spp. for use in PCR. J. Plant Prot. Res. 2016, 56, 104–109. [**Google Scholar**] [**CrossRef**]

25. Williams, N.; Hardy, G.E.S.J.; O'Brien, P.A. Analysis of the distribution of Phytophthora cinnamomi in soil at a disease site in Western Australia using nested PCR. For. Pathol. 2009, 39, 95–109. [**Google Scholar**] [**CrossRef**]

26. Langrell, S.R.; Morel, O.; Robin, C. Touchdown nested multiplex PCR detection of Phytophthora cinnamomi and P. cambivora from French and English chestnut grove soils. Fungal Biol. 2011, 115, 672–682. [**Google Scholar**] [**CrossRef**] [**PubMed**]

27. Engelbrecht, J.; Duong, T.A.; van den Berg, N. Development of a nested quantitative real-time PCR for detecting Phytophthora cinnamomi in Persea americana rootstocks. Plant Dis. 2013, 97, 1012–1017. [**Google Scholar**] [**CrossRef**] [**PubMed**] [**Green Version**]

28. Miles, T.D.; Martin, F.N.; Coffey, M.D. Development of rapid isothermal amplification assays for detection of Phytophthora spp. in plant tissue. Phytopathology 2015, 105, 265–278. [**Google Scholar**] [**CrossRef**] [**Green Version**]

29. Kunadiya, M.B.; Dunstan, W.D.; White, D.; Hardy, G.E.S.J.; Grigg, A.H.; Burgess, T.I. A qPCR assay for the detection of Phytophthora cinnamomi including an mRNA protocol designed to establish propagule viability in environmental samples. Plant Dis. 2019, 103, 2443–2450. [**Google Scholar**] [**CrossRef**]

30. Dai, T.T.; Yang, X.; Hu, T.; Li, Z.Y.; Xu, Y.; Lu, C.C. A novel LAMP assay for the detection of Phytophthora cinnamomi utilizing a new target gene identified from genome sequences. Plant Dis. 2019, 103, 3101–3107. [**Google Scholar**] [**CrossRef**]

31. Kunadiya, M.B.; White, D.; Dunstan, W.A.; Hardy, G.E.S.J.; Andjic, V.; Burgess, T.I. Pathways to false-positive diagnoses using molecular genetic detection methods; Phytophthora

cinnamomi a case study. FEMS Microbiol. Lett. 2017, 364, fnx009. [**Google Scholar**] [**CrossRef**] [**Green Version**]

32. Yang, X.; Tyler, B.M.; Hong, C.X. An expanded phylogeny for the genus Phytophthora. IMA Fungus 2017, 8, 355–384. [**Google Scholar**] [**CrossRef**] [**Green Version**]
33. Scanu, B.; Hunter, G.C.; Linaldeddu, B.T.; Franceschini, A.; Maddau, L.; Jung, T. A taxonomic re-evaluation reveals that Phytophthora cinnamomi and P. cinnamomi var. parvispora are separate species. For. Pathol. 2014, 44, 1–20. [**Google Scholar**]
34. Martin, F.N.; Blair, J.E.; Coffey, M.D. A combined mitochondrial and nuclear multilocus phylogeny of the genus Phytophthora. Fungal Genet. Biol. 2014, 66, 19–32. [**Google Scholar**] [**CrossRef**]
35. Rojas, J.A.; Miles, T.D.; Coffey, M.D.; Martin, F.N.; Chilvers, M.I. Development and application of qPCR and RPA genus- and species-specific detection of Phytophthora sojae and P. sansomeana root rot pathogens of soybean. Plant Dis. 2017, 101, 1171–1181. [**Google Scholar**] [**CrossRef**] [**PubMed**] [**Green Version**]
36. Dai, T.T.; Yang, X.; Hu, T.; Xu, Y.; Zheng, X.B.; Jiao, B.; Shen, D.Y. Comparative evaluation of a novel recombinase polymerase amplification-lateral flow dipstick (RPA-LFD) assay, LAMP, conventional PCR, and leaf-disc baiting methods for detection of Phytophthora sojae. Front. Microbiol. 2019, 10, 1884. [**Google Scholar**] [**CrossRef**] [**PubMed**] [**Green Version**]
37. Yu, J.; Shen, D.; Dai, T.; Lu, X.; Xu, H.; Dou, D. Rapid and equipment-free detection of Phytophthora capsici using lateral flow strip-based recombinase polymerase amplification assay. Lett. Appl. Microbiol. 2019, 69, 64–70. [**Google Scholar**] [**CrossRef**] [**PubMed**]
38. Dai, T.T.; Hu, T.; Yang, X.; Shen, D.Y.; Jiao, B.B.; Tian, W.; Xu, Y. A recombinase polymerase amplification-lateral flow dipstick assay for rapid detection of the quarantine citrus pathogen in China, Phytophthora hibernalis. PeerJ 2019, 7, e8083. [**Google Scholar**] [**CrossRef**] [**Green Version**]
39. James, A.; Macdonald, J. Recombinase polymerase amplification: Emergence as a critical molecular technology for rapid, low-resource diagnostics. Expert Rev. Mol. Diagn. 2015, 15, 1475–1489. [**Google Scholar**] [**CrossRef**] [**Green Version**]
40. Daher, R.K.; Stewart, G.; Boissinot, M.; Bergeron, M.G. Recombinase polymerase amplification for diagnostic applications. Clin. Chem. 2016, 62, 947–958. [**Google Scholar**] [**CrossRef**]
41. Grigoriev, I.V.; Nikitin, R.; Haridas, S.; Kuo, A.; Ohm, R.; Otillar, R.; Riley, R.; Salamov, A.; Zhao, X.L.; Korzeniewski, F.; et al. MycoCosm portal: Gearing up for 1000 fungal genomes. Nucl. Acids Res. 2014, 42, D699–D704. [**Google Scholar**] [**CrossRef**]
42. Bendtsen, J.D.; Nielsen, H.; von Heijne, G.; Brunak, S. Improved prediction of signal peptides: SignalP 3.0. J. Mol. Biol. 2004, 340, 783–795. [**Google Scholar**] [**CrossRef**]
43. Haas, B.J.; Kamoun, S.; Zody, M.C.; Jiang, R.H.Y.; Handsaker, R.E.; Cano, L.M.; Grabherr, M.; Kodira, C.D.; Raffaele, S.; Torto-Alalibo, T.; et al. Genome sequence and analysis of the Irish potato famine pathogen Phytophthora infestans. Nature 2009, 461, 393–398. [**Google Scholar**] [**CrossRef**]
44. Whisson, S.C.; Boevink, P.C.; Moleleki, L.; Avrova, A.O.; Morales, J.G.; Gilroy, E.M.; Armstrong, M.R.; Grouffaud, S.; van West, P.; Chapman, S.; et al. A translocation signal for delivery of oomycete effector proteins into host plant cells. Nature 2007, 450, 115–118. [**Google Scholar**] [**CrossRef**]

45. Jones, L.; Hamilton, A.J.; Voinnet, O.; Thomas, C.L.; Maule, A.J.; Baulcombe, D.C. RNA-DNA interactions and DNA methylation in post-transcriptional gene silencing. Plant Cell 1999, 11, 2291–2301. [**Google Scholar**] [**PubMed**] [**Green Version**]

46. Hellens, R.P.; Edwards, E.A.; Leyland, N.R.; Bean, S.; Mullineaux, P.M. pGreen: A versatile and flexible binary Ti vector for Agrobacterium-mediated plant transformation. Plant Mol. Biol. 2000, 42, 819–832. [**Google Scholar**] [**CrossRef**] [**PubMed**]

47. Wise, A.A.; Liu, Z.; Binns, A.N. Three methods for the introduction of foreign DNA into Agrobacterium. In Agrobacterium Protocols; Wang, K., Ed.; Humana Press: Totowa, NJ, USA, 2006; pp. 43–54. [**Google Scholar**]

48. Finn, R.D.; Clements, J.; Eddy, S.R. HMMER web server: Interactive sequence similarity searching. Nucl. Acids Res. 2011, 39, W29–W37. [**Google Scholar**] [**CrossRef**] [**PubMed**] [**Green Version**]

49. Dong, S.M.; Qutob, D.; Tedman-Jones, J.; Kuflu, K.; Wang, Y.C.; Tyler, B.M.; Gijzen, M. The Phytophthora sojae avirulence locus Avr3c encodes a multi-copy RXLR effector with sequence polymorphisms among pathogen Strains. PLoS ONE 2009, 4, e5556. [**Google Scholar**] [**CrossRef**] [**Green Version**]

04

A RAPID LOOP-MEDIATED ISOTHERMAL AMPLIFICATION ASSAY HAS BEEN DEVELOPED FOR THE SPECIFIC DETECTION OF DOTHISTROMA SEPTOSPORUM

1. INTRODUCTION: AN OVERVIEW

Dothistroma needle blight, a disease caused by the pathogens Dothistroma septosporum and Dothistroma pini, causes substantial harm to pine trees in both natural and plantation environments worldwide [1]. Climate change has been identified as the cause of the growing losses in British Columbia caused by D. septosporum [2]. In the years 2012 and 2013, D. septosporum was discovered on Pinus contorta Dougl. var. latifolia Engelm., Pinus banksiana Lamb., and their hybrid in northern Alberta, Canada [3]. The presence of D. septosporum in Alberta and its potential to spread eastward through the boreal forest have prompted investigations into population genomics [4], host range [3], and, as detailed in this study, the development of rapid diagnostic methods for D. septosporum, the pathogen responsible for the disease in western Canada.

Ioos et al. [5], Langrell [6], and Schneider et al. [7] have provided descriptions of polymerase chain reaction (PCR)-based diagnostics for identifying D. septosporum DNA. These diagnostic tests are highly sensitive and capable of detecting the presence of the pathogen in diseased tissue. However, they can only be conducted in a well-equipped laboratory. Loop-Mediated Isothermal Amplification (LAMP) serves as a substitute for conventional PCR-based techniques. LAMP tests employ a set of four to six short DNA sequences called oligonucleotide primers, along with a

DNA polymerase enzyme, to facilitate the synthesis of new DNA strands by a process known as strand displacement. This synthesis occurs at a consistent temperature without the need for heat cycling. LAMP assays have been created for point-of-care diagnostics in medicine [9], agriculture [10], and significant forest pathogens [11]. Portable devices like the Genie® III developed by OptiGene (Horsham, West Sussex, UK) or the BioRanger instrument developed by Diagenetix Inc. (Honolulu, HI, USA) have been designed to enable LAMP assays to be performed in field settings.

In this study, we present the creation of a LAMP-based test that may be used to identify D. septosporum in laboratory cultures and efficiently and precisely detect D. septosporum in pine needles that are infected.

2. MATERIALS AND METHODS

2.1. Culturing of D. Septosporum

Pure cultures of D. septosporum were obtained from infected pine needles by excising a single acervulus and rolling it in a straight line across the surface of a 2% Malt Extract Agar (MEA) plate. Once spore germination was visible under a dissecting microscope (2–3 days at room temperature), a single germinated conidia was transferred to a fresh 2% MEA plate and incubated at room temperature. Alternatively, D. septosporum was isolated by surface sterilizing an infected needle in 10% bleach for 5 min to reduce surface contamination and then placing the infected needle on 2% water agar at room temperature to induce sporulation. An erupting conidial mass from an acervulus was then placed on fresh 2% water agar and a single hyphal tip was subcultured onto fresh 1% to 3% MEA [**3**]. Prior to DNA extraction, liquid cultures were prepared by inoculating 50 mL clarified V8 medium with mycelial plugs from actively growing cultures, then incubating these at 20 °C with occasional shaking for 6–8 weeks. V8 juice amended with 1.42 g $CaCO_3$ per 100 mL was clarified first by centrifugation for 15 min at 2500 rpm and then filtering the remaining supernatant 3 times through Whatman® qualitative filter paper, grade 4. To obtain the clarified V8 growth medium, clarified V8 juice was mixed with dH_2O in a 1:4 ratio.

Pure cultures of D. septosporum were obtained from infected Pinus attenuata Lemmon, Pinus attenuata × Pinus radiata D. Don, P. banksiana, P. contorta var. latifolia, P. contorta var. latifolia × P. banksiana, Pinus monticola Dougl. ex D. Don, and P. radiata, collected from Alaska, Alberta, British Columbia, California, Oregon, and New Zealand. Pure cultures of D. pini were obtained from infected Pinus mugo Turra, Pinus nigra Arnold, and Pinus ponderosa P. Laws. ex C. Laws. collected from Iowa, Montana, Nebraska, North Dakota, Ontario, and South Dakota. Pure cultures of L. acicola were obtained from infected P. banksiana, P. contorta var. latifolia, and P. contorta var. latifolia × P. banksiana collected from Alberta and British

Columbia. Cultures of Ceuthospora sp., Cyclaneusma minus (Butin) DiCosmo, Peredo and Minter "simile", C. minus "verum", D. pini, D. septosporum, Lophodermium conigenum (Brunaud) Hilitzer, Pestalotiopsis sp., Phytophthora pluvialis Reeser, Sutton and E. Hanson, and Strasseria geniculata (Berk. and Broome) Höhn. were also used.

2.2. Extraction of DNA from Cultures

In Canada, two DNA extraction methods were used to extract DNA from fungal cultures. The Qiagen DNeasy PowerPlant Pro kit (Qiagen, Valencia, CA, USA) was used to isolate DNA from mycelium harvested from liquid V8 cultures. Freeze dried mycelium was first ground with one 3 mm stainless steel bead in a SPEX® Mixer/Mill (SamplePrep, Metuchen, NJ, USA) for 30 s at setting 1/30. DNA was then extracted from the ground mycelium following the manufacturer's protocol with the following modifications: the optional phenolic separation solution was added at the fist extraction step and sample disruption by bead beating was replaced by a 10 min incubation at 65 °C. Alternatively, a simple Tris-EDTA extraction method was used to extract DNA from cultures growing on MEA. A single plug (5 mm diameter) was placed in microcentrifuge tubes with 400 μL of Tris-EDTA (10 mM Tris-1 mM EDTA, pH 8.0) solution, one 4 mm and three 2.8 mm stainless steel beads, and placed in a SPEX® 1600 MiniG® (SamplePrep, Metuchen, NJ, USA) bead beater for 3 min at 1500 rpm to disrupt the mycelium. After disruption the sample was centrifuged at 16,000× g for 3 min and the supernatant transferred to a new tube. DNA extraction method for all specimens is specified. In either case, DNA stock solutions were diluted 100-fold in sterile PCR-grade water to serve as a template for DNA amplification.

In New Zealand, DNA from cultures of Ceuthospora sp., C. minus "simile", C. minus "verum", D. pini, D. septosporum, L. conigenum, Pestalotiopsis sp., P. pluvialis, and S. geniculata was extracted using the FastDNA kit (MP Biomedicals, Solon, OH, USA) following the manufacturer's protocol. DNA of other Phytophthora spp. was provided by collaborators and the extraction method is unknown

2.3. ITS-Based Identification of Cultures

The nuclear ribosomal internal transcribed spacer (ITS) was PCR amplified using the total DNA extracted from putative isolates of D. septosporum and other fungi isolated from pine needles using the primers ITS-1F [**12**] and ITS-4 [**13**] following the protocol described in Feau et al. [**3**]. The PCR product was purified using the QIAquick PCR purification kit (Qiagen, Valencia, CA, USA), and sequenced by Macrogen (Seoul, Korea) or CHU de Québec—Université Laval Research Centre (Quebec, QC, Canada). To identify fungal cultures, ITS sequences were queried against the NCBI nr database using the BLASTn algorithm [**14**].

2.4. Extraction of DNA from Needles

DNA was extracted from pine needles using DNAzol® Direct (Molecular Research Centre, Inc., Cincinnati, OH, USA). Needles putatively infected by D. septosporum, as indicated by microscopic examination of conidia produced on needles that had been placed at room temperature in a sealed Petri plate with a moist paper towel to induce sporulation, as well as visually healthy needles, were selected for DNA extraction. Needles were either fresh and stored at 4 °C or dried and stored at -20 °C. When fresh, one to two infected needles, cut into 5 mm sections, were placed in a 1.5 or 2.0 mL centrifuge tube, with 100 µL of DNAzol® Direct. The needles were then ground with a SPEX® 1600 MiniG® (SamplePrep, Metuchen, NJ, USA) bead beater using one 4 mm and three 2.8 mm stainless steel balls at 1500 rpm for 3 min. When the needles were dry, one or two needles were ground using the bead beater prior to adding DNAzol® Direct. The DNAzol® and ground needle mixtures were incubated at room temperature for 15 min with occasional inversion. The tubes were then centrifuged at 16,000× g for 3 min to pellet debris, the DNA-containing supernatant was transferred to a fresh tube and a 100-fold dilution in sterile PCR-grade water was prepared to serve as template for DNA amplification. The same procedure was used to extract DNA from needles of P. contorta var. latifolia that were infected with Elytroderma deformans (Weir) Darker, Davisomycella ampla (Davis) Darker, Lophodermella concolor (Dearn.) Darker, Lophodermium seditiosum Minter, Staley and Millar, and Lecanosticta acicola (**Table S2**).

2.5. Development of LAMP Primers

The β-tubulin 2 gene was selected as a target for our LAMP assay as it has previously been shown to distinguish D. septosporum from close relatives [**5,7**]. A consensus of partial D. septosporum β-tubulin 2 gene sequences was generated from an alignment of GenBank accessions AY808205–AY808230 [**15**], assembled by Barnes et al. [**15**], that is broadly representative of D. septosporum diversity. The consensus sequence was then compared to the corresponding portions of the β-tubulin sequence data from L. acicola (AY808237, AY808238) [**15**] and D. pini (AY808231–AY808236) [**15**] to identify regions specific to D. septosporum and would therefore make suitable targets for the LAMP primers. LAMP primers for the D. septosporum-specific regions were developed using default parameters in the LAMP Designer 1.4 software package (PREMIER Biosoft, OptiGene, Horsham, West Sussex, UK). The top ranked primer set, as determined by the software, that distinguished D. septosporum and D. pini was selected (**Table 1**; **Figure 1**). The specificity of the F3/B3 primer pair and the component sequences of the FIP and BIP primers (F1c and F2, and B1c and B2, respectively (**Figure 1**)) were assessed using Primer-BLAST [**16**], with default parameters and the nr database for fungi.

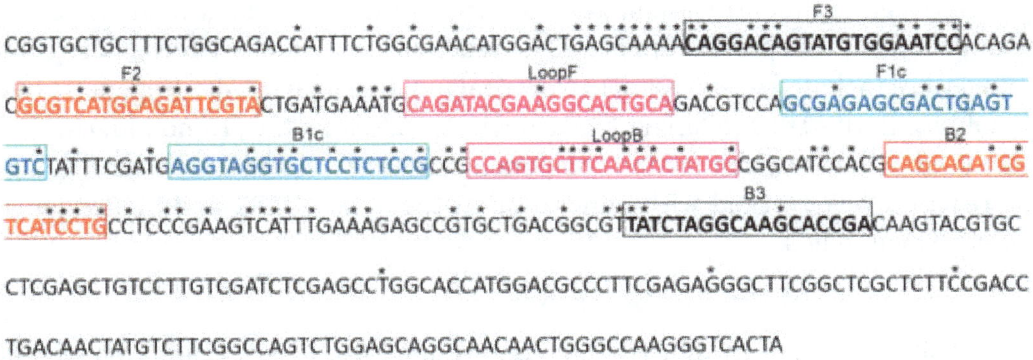

Figure 1. Position of LAMP primers based on consensus sequence of D. septosporum β-tubulin 2 gene sequences retrieved from GenBank (AY808205–AY808230). Primer FIP is a combination of F1c and F2, while primer BIP is a combination of B1c and B2. Differences from D. pini are noted by * symbol.

Table 1. LAMP primers specific to the β-tubulin 2 gene of Dothistroma septosporum.

Primer Name	Sequence 5′-3′
F3	CAGGACAGTATGTGGAATCC
B3	TCGGTGCTTGCCTAGATA
FIP	GACACTCAGTCGCTCTCGCGCGTCATGCAGATTCGTA
BIP	AGGTAGGTGCTCCTCTCCGCAGGATGACGATGTGCTG
LoopF	TGCAGTGCCTTCGTATCTG
LoopB	CCAGTGCTTCAACACTATGC

2.6. Preparation of the LAMP Reaction

Primers were first hydrated to a stock concentration of 500 μM using Tris-EDA diluted 10-fold with sterile PCR-grade water. A working stock of primer mix was then prepared that contained 0.8333 μM F3, 0.8333 μM B3, 1.666 μM LoopF, 1.666 μM LoopB, 3.333 μM FIP, and 3.333 μM BIP in PCR-grade water. Primers and template DNA were then combined in Optigene Isothermal master mix (ISO-001) (OptiGene, Horsham, West Sussex, UK) in the proportions of 0.6 master mix, 0.24 primer mix and 0.16 template DNA (i.e., for a 15 μL reaction, use 9 μL isothermal master mix, 3.6 μL of primer mix, and 2.4 μL of template DNA). The final primer concentrations for the LAMP reaction were 0.2 μM F3, 0.2 μM B3, 0.4 μM LoopF, 0.4 μM LoopB, 0.8 μM FIP, and 0.8 μM BIP. Template DNA from cultures or needles consisted of stock DNA diluted 100-fold in sterile PCR-grade water.

Primers, master mix and template DNA were combined with a final volume of 15 μL. The Genie® III instrument (OptiGene, Horsham, West Sussex, UK) was run at 65 °C for up to 60 min, and then annealing ramped from 98 to 80 °C at a rate of -0.05

°C/s for fragment identification. Both positive (pine DNA spiked with 0.1 to 0.01 ng of DNA from D. septosporum per 15 μL reaction) and a blank (sterile PCR-grade water) were included in each run. In New Zealand, the Smart-DART (Digenetix Inc., Honolulu, HI, USA) or BioRanger LAMP devices (Digenetix Inc., Honolulu, HI, USA) were used with the same reaction components as were utilized in Canada although the reaction was overlaid with mineral oil as the Smart-DART and BioRanger do not have heated lids.

2.7. Sensitivity of the LAMP Assay

The sensitivity of the LAMP assay was determined using serial dilutions of total D. septosporum DNA extracted from three different isolates, all of which were collected from the north Okanagan region of British Columbia from P. contorta var. latifolia. Total DNA concentration was first quantified using a Qubit™ 4 Fluorometer (Invitrogen, Waltham, MA, USA) and the Qubit™ dsDNA HS assay kit (Invitrogen, Waltham, MA, USA) and then standardized to 4.17 ng/μL. Serial dilutions were made from 4.17 ng/μL to 4.17×10^{-6} ng/μL and 2.4 μL of template DNA was added to each 15 μL LAMP reaction, providing a dilution series from 1 ng to 1.0×10^{-6} ng of template DNA in each reaction. The LAMP reaction was run for 60 min.

2.8. Specificity of the LAMP Assay

The specificity of the LAMP assay was assessed using DNA from cultured isolates of D. septosporum, D. pini, and Lecanosticta acicola, and using pine needles infected with L. acicola, Lophodermella concolor, Lophodermium seditiosum, and Elytroderma deformans. In New Zealand, the assay was tested against Phytophthora spp. and fungi associated with the foliage of Pinus radiata. Confirmation of the ability of the LAMP assay to detect D. septosporum within infected needles was conducted by using DNA extracted from a subsample of needles from which D. septosporum was previously isolated into pure culture and identified by ITS sequencing. In cases where the LAMP reaction was negative, the quality of the DNA was checked by PCR amplifying the ITS region to confirm that the DNA could be amplified.

2.9. Statistical Comparison of the DNA Extraction Methods

The annealing temperature and amplification time for D. septosporum-positive DNA samples extracted using the Qiagen DNeasy PowerPlant Pro kit and the simple Tris-EDTA methods were compared using the nonparametric Kruskal–Wallis test as the assumptions of equal variance and normal distribution were not met. The sample from New Zealand was excluded from the analysis as the annealing temperature and amplification time were not recorded.

3. RESULTS

Our D. septosporum LAMP assay has high specificity. DNA extracted from all isolates of D. septosporum included in the screening population produced positive amplification while no amplification occurred when the DNA of other fungi or Phytophthora spp. that were included in the screening population were used as template. (**Table 2**). Further, in silico analysis of the LAMP primers using Primer-BLAST suggested that all six primers were specific for D. septosporum. Detection times were significantly shorter for DNA samples extracted using the Qiagen method compared to the TE method ($H = 44.5345$; $p = 2.499 \times 10^{-11}$). In contrast, there was no significant difference between extraction methods in terms of annealing temperatures ($H = 1.7184$; $p = 0.1904$) (**Table 2**). Detection of D. septosporum from DNA extracted from infected needles took longer relative to DNA from cultures regardless of method; average detection time was 28.50 (SD ± 9.42) min although detection took as long as 54 min. Examples of positive and negative LAMP reactions are presented in **Figure 2**.

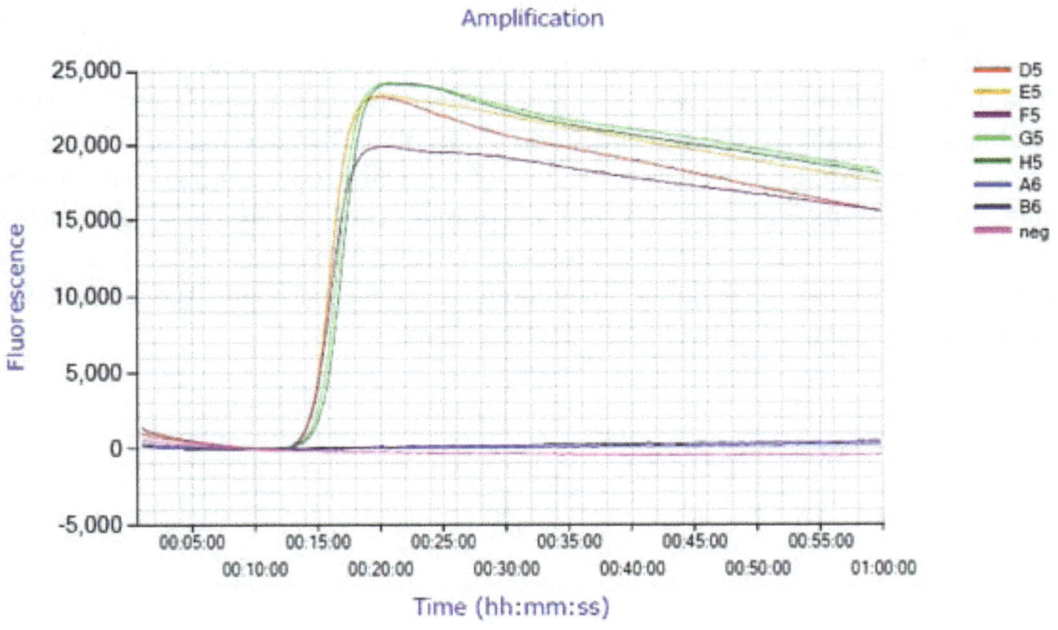

Figure 2. Output from LAMP assay to assess specificity of D. septosporum lamp assay. D5 = D. septosporum (DSEP-FLNRO-7M-Ne2-TAIGA-591); E5 = D. septosporum (DSEP-Whitecourt-Mountain-4077-2-Ne2); F5 = D. septosporum (DSEP-EM-Mt-14-Ne1-TAIGA-469); G5 = D. septosporum (DSEP-ABTIC-9-Ne1-TAIGA-590); H5 = D. septosporum (DSEP-CLG-2-Ne1-TAIGA-614); A6 = D. pini (DPIN-Brewer-Lake-Needle1); B6 = D. pini (DPIN-Brewer-Lake-Needle2); neg = sterile distilled water.

Table 2. Summary LAMP results based on amplification of template DNA extracted from pure cultures and needles. Cultures—Qiagen indicates results for DNA extracted using the Qiagen method and Cultures—TE indicates the results for DNA extracted using the TE method.

Origin and DNA Extraction Method	No. of Positive Amplifications/No. of True Positive Samples	No. of No Amplifications/No. of True Negative Samples	Average Annealing T °C (±SD)	Time (min) (±SD)
Cultures—Qiagen	40/40	55/55	88.57 (±0.13)	15.82 (±1.15)
Cultures—TE	25/25	3/3	88.62 (±0.06)	22.46 (±3.54)
Needles—DNAzol	27/27	5/5	88.55 (±0.14)	28.78 (±9.67)
All	92/92	63/63	88.58 (±0.12)	21.42 (±7.79)

In sensitivity testing, D. septosporum was consistently detected in less than 30 min when at least 2 pg of template DNA was included in the LAMP reaction. The detection limit of our LAMP assay was approximately 1 pg of template DNA (**Table 3**). Generally, as the concentration of template DNA decreased, the time to detection increased; for example, for 2 pg of template DNA detection took 18–26 min, whereas for 1 ng, detection took 14–16 min (**Table 3, Figure 3**). The LAMP assay detected the presence of D. septosporum within fresh and dried pine needles; however, if needles were damp chambered to induce pathogen sporulation for microscopic examination the LAMP reaction failed.

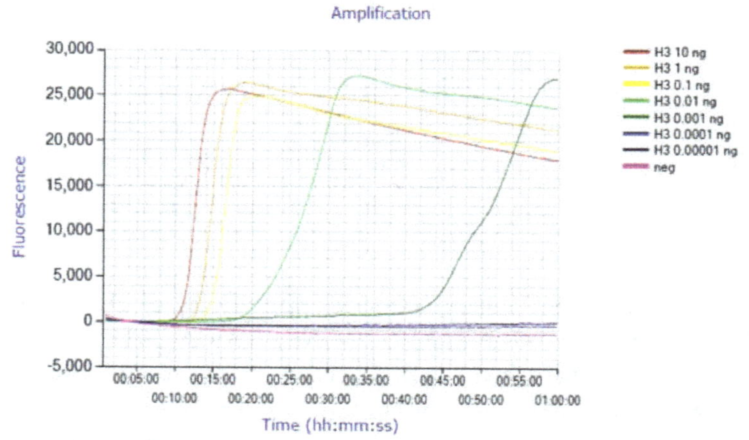

Figure 3. Determining the sensitivity of the LAMP reaction. Isolate H3 = D. septosporum (DSEP-Kingfisher-Ne1). Positive results for concentrations from 10 to 0.001 ng/μL. Negative results for 0.0001 ng/μL, 0.00001 ng/μL, and negative control (sterile distilled water).

Table 3. Sensitivity of LAMP assay determined by using serial dilutions of *D. septosporum* DNA.

Sample	ng Template DNA in 15 µL Reaction	Annealing Temp (°C)	Amplification Time (min)
Cherryville Site 2 Tree 2 Needle 2	10	88.67	13.30
	1	88.66	15.45
	0.1	88.73	18.00
	0.01	88.66	25.00
	0.009	88.59	18.00
	0.008	88.54	21.00
	0.007	88.64	21.45
	0.006	88.59	33.15
	0.005	88.57	20.30
	0.004	88.53	23.45
	0.003	88.53	23.45
	0.002	88.54	23.00
	0.001	-	-
	0.0001	-	-
	0.00001	-	-
Mabel Lake Tree 2 Needle 1	10	88.62	13.30
	1	88.62	15.45
	0.1	88.74	16.15
	0.01	88.71	19.30
	0.009	88.53	20.35
	0.008	88.54	25.29
	0.007	88.57	22.55
	0.006	88.68	23.56
	0.005	88.62	26.31
	0.004	88.64	20.19
	0.003	88.58	20.19
	0.002	88.55	25.13
	0.001	-	-
	0.0001	-	-
	0.00001	-	-
Kingfisher Needle 1	10	88.65	12.30
	1	88.67	14.30

Sample	ng Template DNA in 15 µL Reaction	Annealing Temp (°C)	Amplification Time (min)
	0.1	88.66	16.15
	0.01	88.68	28.00
	0.009	88.52	22.15
	0.008	88.57	21.30
	0.007	88.59	20.30
	0.006	88.62	24.30
	0.005	88.57	20.45
	0.004	88.57	21.30
	0.003	88.60	22.15
	0.002	88.61	18.30
	0.001	88.46	46.30
	0.0001	-	-
	0.00001	-	-
"-" indicates no amplification			

4. DISCUSSION

Dothistroma septosporum is a significant pathogen that causes Dothistroma needle blight in pine trees, affecting both natural and planted pine stands worldwide. This pathogen has comparable physical characteristics to the closely related species D. pini [15]. Additionally, DNA-based techniques have been created to distinguish between these two diseases [5,7]. Given that historical distribution records in Canada rely on morphology, it is imperative to employ DNA-based approaches in order to gain a comprehensive understanding of the regional spread of these infections. The discovery that D. septosporum was responsible for Dothistroma needle blight in Alberta inspired the creation of a LAMP diagnostic assay for this species [3,4].

The LAMP assay described in this study relies on the β-tubulin 2 gene, which has previously been demonstrated to differentiate D. septosporum from closely related species [5,7,15]. This study's findings provide more evidence that this specific genetic location is appropriate for diagnosing D. septosporum. The sample population of D. septosporum used in this study was geographically varied, including samples from different geographic subgroups identified by Capron et al. [4], as well as samples from Alaska, California, Oregon, and New Zealand. This indicates that the assay has a broad range of applicability.

Assessment of a dilution series indicated that the LAMP assay is highly sensitive and able to detect as little as 1 pg of pathogen DNA. The LAMP assay is not as sensitive as Langrell's [6] nested PCR assay that targets the ITS region. Schneider et al. [7] also found their ITS-based assay to be more sensitive than the qPCR assay of Ioos et al. [5] that targets β-tubulin 2. These authors suggested that this difference reflects target copy number; β-tubulin 2 is a single-copy gene whereas ITS has a high copy number [7]. The LAMP assay was as sensitive as the duplex-scorpion qPCR assay of Ioos et al. [5], but not as sensitive as their dual-labelled probe qPCR assay [5], despite both targeting β-tubulin 2. Advantages of the LAMP assay include the relatively easy to use equipment and speed of the assay; however, the assays of Ioos et al. [5] and Schneider et al. [7] can simultaneously detect D. pini and D. septosporum while the LAMP assay only detects D. septosporum.

The two methods used to purify DNA from cultures of D. septosporum resulted in significantly different time to detection by the LAMP assay; DNA extracted using the Qiagen method amplified more quickly than DNA extracted directly from mycelial plugs using the Tris–EDTA method. The DNA extracted using the Qiagen method is likely to have been associated with fewer impurities as therefore was a better template for amplification. The final concentration of DNA extracted by the Tris-EDTA method was also likely lower than DNA extracted using the Qiagen method. In contrast, there was no significant difference between the two extraction methods in terms of annealing temperature, indicating that extraction method did not influence specificity. Although the assay worked using DNA extracted with both methods, and is very sensitive, these results highlight the importance of using high-quality DNA, especially in situations where the presence of D. septosporum may represent a new distribution record.

Utilizing DNA retrieved from cultures resulted in a shorter time for pathogen diagnosis compared to utilizing DNA extracted from infected needles. The reason for this is probably the greater amount of D. septosporum DNA in extracts from cultures compared to the total DNA retrieved from infected needles. The assay effectively detected the presence of D. septosporum in fresh or dried needles held at different temperatures (room temperature, 4 °C, or -20 °C). However, when the needles were exposed to moisture and kept in a chamber at room temperature to encourage pathogen sporulation, the reliability of the assay was compromised. This result could be explained by an increase in the amounts of inhibitory substances or a decrease in the relative concentration of D. septosporum DNA caused by the growth of other fungus. However, the best outcomes were achieved when DNA was extracted from recently collected or appropriately preserved samples.

Dothistroma spp. can infect a total of 109 different species within the genera Pinus, Abies, Cedrus, Larix, Picea, and Pseudotsuga. Through the application of molecular techniques, it has been verified that D. septosporum is capable of infecting a total of 42 species of Pinus. Among these, 11 species have been demonstrated to also serve as hosts for D. pini. Among the 109 host taxa that have been discovered, susceptibility to D. septosporum varies from being mildly susceptible to very susceptible [1]. Furthermore, studies have demonstrated that D. septosporum has a hemi-biotrophic life cycle on P. radiata. It is capable of infecting and colonizing host tissue without causing any visible symptoms during the biotrophic phase, before transitioning to a necrotrophic phase [17]. If signs and symptoms of D. septosporum are observed in a species that is particularly susceptible to the disease, and in an area where the disease has been previously documented, it may not be required to confirm the presence of D. septosporum infection using a genetic assay. Nevertheless, in species that are somewhat vulnerable and where it is challenging to identify based on physical characteristics, or during a phase where the pathogen lives in a host without causing symptoms, or in regions where the disease has not been previously observed, the LAMP assay can quickly confirm the presence of infection. Integrating a LAMP diagnostic test into monitoring efforts could enhance disease control by allowing for early detection and intervention, even before the pathogen has become widespread. One benefit of the LAMP technique is that the amplification temperature remains consistent, which makes the technology well-suited for point-of-care applications. Genie® III, a portable device, enables the execution of DNA-based diagnostics in non-laboratory environments. Our test is compatible with a point-of-care diagnosis and produces a positive reaction within a timeframe of 30 to 60 minutes. Currently, we are enhancing techniques for preparing samples in the field to facilitate on-site testing.

5. AT A GLANCE

A Loop-Mediated Isothermal Amplification (LAMP) assay was created to identify the pine pathogen Dothistroma septosporum (G. Dorog.) M. Morelet. The LAMP assay's specificity was evaluated by testing it on several types of fungi found in pine needles, such as Dothistroma pini Hulbary and Lecanosticta acicola (Thüm.) Syd. The test only amplified the DNA of D. septosporum. The assay demonstrated high sensitivity, capable of detecting as little as 1 picogram of total D. septosporum DNA. This assay allows for the quick testing of DNA isolated from diseased host needles for the presence of D. septosporum. It can be performed using relatively simple equipment, without the requirement for a fully equipped molecular biology laboratory.

REFERENCES

1. Drenkhan, R.; Tomešová-Haataja, V.; Fraser, S.; Bradshaw, R.E.; Vahalík, P.; Mullett, M.S.; Martín-García, J.; Bulman, L.S.; Wingfield, M.J.; Kirisits, T.; et al. Global geographic distribution and host range of Dothistroma species: A comprehensive review. For. Path. 2016, 46, 408–442. [**Google Scholar**] [**CrossRef**]
2. Woods, A.; Coates, K.D.; Hamann, A. Is an unprecedented Dothistroma needle blight epidemic related to climate change? Bioscience 2005, 55, 761–769. [**Google Scholar**]

3. Feau, N.; Ramsfield, T.D.; Myrholm, C.L.; Tomm, B.; Cerezke, H.F.; Benowicz, A.; Samis, E.; Romano, A.; Dale, A.L.; Capron, A.; et al. DNA-barcoding identification of Dothistroma septosporum on Pinus contorta var. latifolia, P. banksiana and their hybrid in northern Alberta, Canada. Can. J. Plant Pathol. 2020, 1–8. **[Google Scholar]** **[CrossRef]**
4. Capron, A.; Feau, N.; Heinzelmann, R.; Barnes, I.; Benowicz, A.; Bradshaw, R.E.; Dale, A.; Lewis, K.J.; Owen, T.J.; Reich, R.; et al. Signatures of post-glacial genetic isolation and human-driven migration in the Dothistroma needle blight pathogen in western Canada. Phytopathology 2020, 111, 116–127. **[Google Scholar]** **[CrossRef]** **[PubMed]**
5. Ioos, R.; Fabre, B.; Saurat, C.; Fourrier, C.; Frey, P.; Marçais, B. Development, comparison, and validation of real-time and conventional PCR tools for the detection of the fungal pathogens causing brown spot and red band needle blights of pine. Phytopathology 2010, 100, 105–114. **[Google Scholar]** **[CrossRef]** **[PubMed]** **[Green Version]**
6. Langrell, S.R.H. Nested polymerase chain reaction-based detection of Dothistroma septosporum, red band needle blight of pine, a tool in support of phytosanitary regimes. Mol. Ecol. Resour. 2011, 11, 749–752. **[Google Scholar]** **[CrossRef]** **[PubMed]**
7. Schneider, S.; Jung, E.; Queloz, V.; Meyer, J.B.; Rigling, D. Detection of pine needle diseases caused by Dothistroma septosporum, Dothistroma pini and Lecanosticta acicola using different methodologies. For. Path. 2019, 49, e12495. **[Google Scholar]** **[CrossRef]**
8. Notomi, T.; Okayama, H.; Masubuchi, H.; Yonekawa, T.; Watanabe, K.; Amino, N.; Hase, T. Loop-mediated isothermal amplification of DNA. Nucleic Acids Res. 2000, 28, e63. **[Google Scholar]** **[CrossRef]** **[PubMed]** **[Green Version]**
9. Moehling, T.J.; Choi, G.; Dugan, L.C.; Salit, M.; Meagher, R.J. LAMP diagnostics at the point-of-care: Emerging trends and perspectives for the developer community. Expert Rev. Mol. Diagn. 2021, 21, 43–61. **[Google Scholar]** **[CrossRef]** **[PubMed]**
10. Hariharan, G.; Prasannath, K. Recent advances in molecular diagnostics of fungal plant pathogens: A mini review. Front Cell Infect. Microbiol. 2020, 10, 600234. **[Google Scholar]** **[CrossRef]** **[PubMed]**
11. Aglietti, C.; Luchi, N.; Pepori, A.L.; Bartolini, P.; Pecori, F.; Raio, A.; Capretti, P.; Santini, A. Real-time loop-mediated isothermal amplification: An early-warning tool for quarantine plant pathogen detection. AMB Expr. 2019, 9, 50. **[Google Scholar]** **[CrossRef]** **[PubMed]**
12. Gardes, M.; Bruns, T.D. ITS primers with enhanced specificity for basidiomycetes—application to the identification of mycorrhizae and rusts. Mol. Ecol. 1993, 2, 113–118. **[Google Scholar]** **[CrossRef]** **[PubMed]**
13. White, T.J.; Bruns, T.D.; Lee, S.B.; Taylor, J.W. Amplification and direct sequencing of fungal ribosomal RNA genes for phylogenetics. In PCR Protocols: A Guide to Methods and Applications; Innis, M., Gelfand, D., Shinsky, J., White, T., Eds.; Academic Press: San Diego, CA, USA, 1990; pp. 315–322. **[Google Scholar]**
14. Altschul, S.F.; Gish, W.; Miller, W.; Myers, E.W.; Lipman, D.J. Basic local alignment search tool. J. Mol. Biol. 1990, 215, 403–410. **[Google Scholar]** **[CrossRef]**
15. Barnes, I.; Crous, P.W.; Wingfield, B.D.; Wingfield, M.J. Multigene phylogenies reveal that red band needle blight is caused by two distinct species of Dothistroma, D. septosporum and D. pini. Stud. Mycol. 2004, 50, 551–565. **[Google Scholar]**

16. Ye, J.; Coulouris, G.; Zaretxkaya, I.; Cutcutache, I.; Rozen, S.; Madden, T. Primer-BLAST: A tool to design target-specific primers for polymerase chain reaction. BMC Bioinform. 2012, 13, 134. **[Google Scholar]** **[CrossRef]** **[Green Version]**

17. Kabir, M.S.; Ganley, R.J.; Bradshaw, R.E. The hemibiotrophic lifestyle of the fungal pine pathogen Dothistroma septosporum. For. Path. 2015, 45, 190–202. **[Google Scholar]** **[CrossRef]**

05 REEVALUATING THE HISTORICAL RECORD OF DOTHISTROMA NEEDLE BLIGHT IN THE MEDITERRANEAN REGION USING MOLECULAR TECHNIQUES

1. INTRODUCTION: AN OVERVIEW

Dothistroma needle blight (DNB) is a globally prevalent and significant disease that affects the foliage of pine trees [1,2]. The disease is caused by two closely related ascomycete fungus, namely Dothistroma septosporum (Dorog.) Morelet (also known as Scirrhia pini Funk & Parker, with the teleomorph being Mycosphaerella pini Rostr.), and Dothistroma pini Hullbary (with an unknown teleomorph) [3]. These pathogens cause identical symptoms, which start as water-soaked lesions and subsequently turn into little brown-to-reddish spots. Under suitable conditions, black conidiomata develop within these areas. The spots eventually enlarge to cover the entire diameter of the needle, forming horizontal bands. This is why the condition is also known as "red band needle blight". Infected needles undergo necrosis and eventually detach [3,4].

Moisture, rainfall, and temperature are the main elements that contribute to the occurrence and intensity of DNB (Diplodia pinea) in both plantations and nurseries [5]. The primary elements that contribute to the disease at the population level are the lack of air movement and high levels of humidity. The disease has a greater effect on the lower branches of pine trees, particularly those that are shady and located closer to the ground. Pines growing in dense stands exhibit greater susceptibility compared to pines growing in more open stands. Young pines in open stands, where the canopy conditions foster the growth of dense understory vegetation, are especially vulnerable. Furthermore, pines growing within stands are consistently more susceptible than those growing on the periphery [6]. Infection on a single tree propagates from the lower branches to the highest regions of the canopy. When the climate is conducive (with high relative humidity and temperatures around 16-18 °C), the illness can be quite severe and lead to a decrease in photosynthesis, widespread loss of leaves, slowed growth, and death of the plant. The extensive prevalence of the disease across vast regions of the northern and southern hemispheres, in diverse climatic and ecological settings, demonstrates the remarkable ability of DNB pathogens to adapt. Climate simulations have indicated that the agents of DNB are capable of persisting in several climates, including sub-arctic, temperate, continental, subtropical, and dry tropical regions [7].

Prior to the 1970s, the illness was predominantly associated with causing significant harm to Monterey pine (Pinus radiata D. Don), a rapidly growing species commonly utilized in plantations in the southern hemisphere, particularly in Tanzania and Kenya. The organism responsible for the disease was identified as D. pini Hulbary [4]. The persistent rise in demand for woody material has fueled the expansion of P. radiata plantations on a global scale, resulting in a swift and occasionally catastrophic outbreak of DNB. This disease emerged in Chile, New Zealand, as well as central and southern Africa. Currently, the disease is found in the northern hemisphere, specifically in regions such as British Columbia, California, and Oregon. However, it has not shown any significant signs of becoming an epidemic [4]. Additionally, DNB was observed in Central France and Southern Europe, specifically in Spain, ex-Yugoslavia, and Romania. In these regions, Dothistroma was documented in P. radiata and black pine (Pinus nigra J.F.Arnold) plantations [8,9,10].

In Italy, after the war, there was a rise in the establishment of plantations for fast-growing tree species to provide wood. Specifically, there was an increase in P. radiata plantings in the central-southern regions. Magnani [11], with the endorsement of mycologists Ana Magan in Spain and Morelet in France, documented the occurrence of D. pini Hulbary in the southern region of the peninsula. The report originated once more from a man-made plantation of the non-indigenous P. radiata.

The initial documentation of DNB in Italy was located in the heart of the Mediterranean region, specifically in the southern part of the Apennine mountains, known as Calabria in southern Italy. The outbreak site, San Pietro di Caridà, is located along the mountain range, in an area that experiences a microclimate distinguished by elevated humidity and substantial yearly precipitation [12]. At this location, the warm and moist air masses originating from the Ionian sea and the Tyrrhenian sea, which are separated by the narrow strip of land known as southern Calabria, converge on the ridge, resulting in turbulent conditions and frequent rainfall.

In this study, we conducted a survey in a historical outbreak site in southern Italy to determine if the foliar disease known as Dothistroma needle blight (DNB) is still present in the area. Our objectives were to definitively identify the causal agent of DNB and assess the current host species of the pathogen. This is important because the taxonomy of the genus Dothistroma has been revised since the disease was first reported, and DNB is now considered an emerging global issue for pine trees. Accurately identifying the causal agent was important both in terms of taxonomy and epidemiology. Initially, the sickness was attributed to D. pini only based on micromorphological study [11]. Subsequent study, however, revealed that the only reliable way to accurately identify the agent(s) of DNB is through the use of molecular approaches [13,14]. Specific assays based on loop-mediated isothermal amplification (LAMP) have enabled faster and more sensitive differential identification of the pine pathogens D. pini, D. septosporum, and Lecanosticta acicola [15]. Given the extensive occurrence of D. septosporum in Calabria on Pinus nigra subsp. laricio in the Sila National Park, which is about 150 km north [16,17], the final goal of the study was to investigate the possibility of the pathogen spreading from the introduced P. radiata to other pine species in this area. The location is located within the natural habitat of Corsican pine (Pinus nigra subsp. laricio (Poir.) Maire), which is found in a few populations scattered between Corsica and Southern Italy [18]. These populations serve as repositories for the genetic diversity of the species [19,20]. P. nigra subsp. laricio has been observed to have a high vulnerability to DNB in the Sila National Park [17] and in several other areas beyond its natural distribution range, such as continental France, the United Kingdom, and New Zealand, where the species is cultivated in man-made plantations [21,22,23,24,25,26].

2. MATERIALS AND METHODS

2.1. Study Area and Plant Material

Sampling was performed in the fall of 2019 in the Prateria district, in the Municipality of San Pietro di Caridà, Province of Reggio Calabria (Calabria, southern Italy), in which the samples observed by Magnani [**11**] had also been collected in 1977. The Prateria district, in the Calabrian Apennine, extends between 800 and 1100 m above sea level, in a mountainous area of transition between the southern side of the Serre mountains and the northern side of the Aspromonte massif. Monitoring in search of the old DNB outbreak site focused on the south-facing slopes that drain into the Metramo lake (38°28′ N, 16°13′ E, about 900 m above sea level), an artificial water reserve built at the beginning of this century for irrigation purposes (**Figure 1**). The vegetation, compared to the descriptions from the 1970s, when there were in the area extensive plantations of P. radiata and P. nigra subsp. laricio cultivated since the middle of the twentieth century for the production of wood, is now very different. Broad-leaved trees prevail, mainly beech (Fagus sylvatica L.), which grows in extensive pure high forests and to a lesser extent in coppice forests, covering all the mountains around. There are, however, scattered in the beech forest, nuclei of P. nigra subsp. laricio and still some remnants of the old plantations of P. radiata that have been cut down over time.

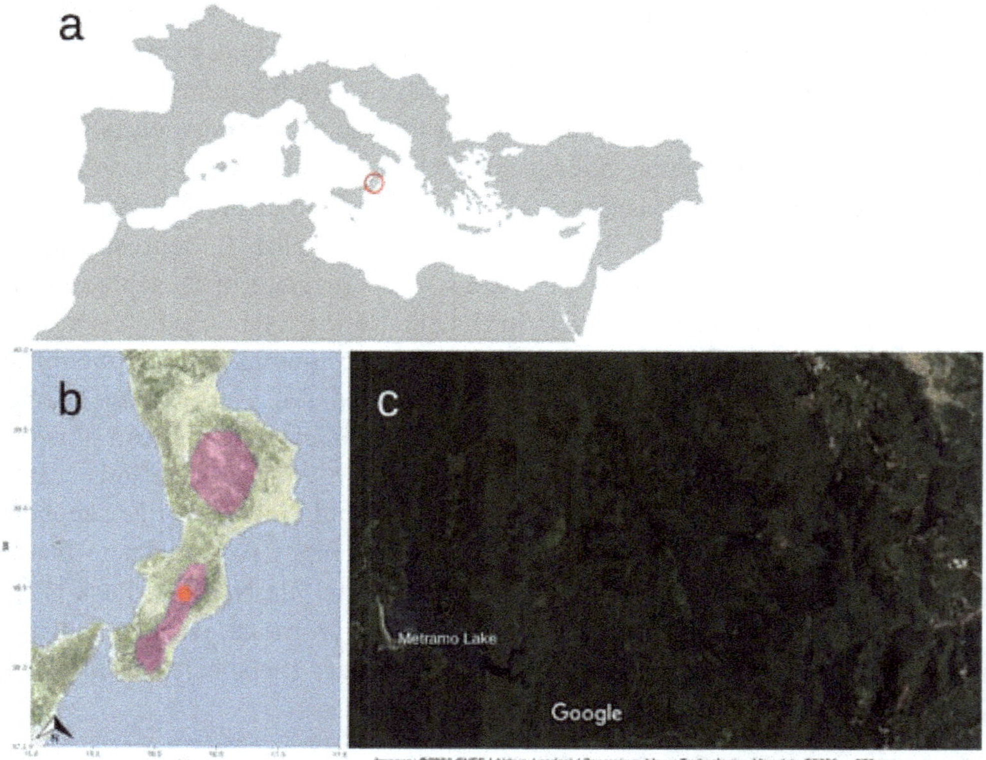

Figure 1. Map position of the study area of Prateria (red symbols) in the center of the Mediterranean region (a,b) on the south-facing slope of the Serre mountains towards the Metramo Lake (San Pietro di Caridà, Reggio Calabria, Italy) (c). The approximate local distribution range of Pinus nigra subsp. laricio (adapted from the European Forest Genetic Resources Programme, EUFORGEN http://www.euforgen.org/species/pinus-nigra/ (accessed on 3 July 2021)) is shown in magenta coloring (b). Maps were drawn using the sf [27] and ggplot2 [28] packages in R [29].

Needle samples were collected from adult plants of P. radiata and P. nigra subsp. laricio that showed typical DNB symptoms such as defoliation, reduced growth, and characteristic red bands on the needles, with or without the presence of fruiting bodies attributable to Dothistroma on the basis of simple visual inspection with the naked eye. Samples were taken from six plant groups (4 groups of P. nigra subsp. laricio and 2 groups of P. radiata) at least 250 m apart in an area of approximately 4 square kilometers (**Figure 1**c). Needle samples, bagged in sealed envelopes (two separate envelopes for each pine group and each containing several dozen needles from different pine plants of the same species), were transported to the laboratory and kept refrigerated until further handling.

2.2. Isolation of Fungi

In the laboratory, symptomatic needles that clearly showed deep red bands, orange-red-brown needle apices, and green bases were examined under the stereomicroscope (Optica Lab, Geass, Torino, Italy). In order for fungal fruiting bodies to mature, needles were incubated at room temperature for 5 days in Petri dishes containing moist blotting paper. Conidia were mounted in lactophenol cotton blue and examined under a light microscope (Axiophot, Carl Zeiss) at up to ×40 magnification.

In a laminar flow cabinet, the needles were surface-sterilized by gently wiping with 70% ethanol, and conidiomata were excised from tissues with a fine-pointed scalpel under the stereomicroscope (×10 magnification) and crushed into a drop of sterile distilled water. The dispersion (10 µL) was taken with a pipette and streaked onto a 90 mm Petri dish containing 1.5% Malt Extract Agar (MEA) plus streptomycin (0.050 gL^{-1}). Three streaks (30 µL) were made on each dish. Dishes were incubated in the dark at 20 °C according to species requirements [**30**].

2.3. DNA Extraction from Mycelia and Pine Needles

The mycelium of axenic cultures was grown on 1.5% MEA plus cellophane into 90 mm Petri dishes and maintained in the dark at 20 °C. After 7–15 days, ca. 70 mg (fresh weight) of mycelium was scraped from the cellophane surface and put in a 2 mL Eppendorf tube. Similarly, fragments (2–5 mm long) of symptomatic P. radiata and P. nigra subsp. laricio pine needles (ca. 70 mg fresh weight obtained by pooling together pieces of different symptomatic needles from the envelope corresponding to a sample) were transferred into 2 mL microcentrifuge tubes (12 samples for DNA extraction in total). Needle samples were frozen at -20 °C overnight and ground for 1 min at 30 Hz with two 3 mm sterile steel beads in a MM400 mixer mill (Retsch GmbH, Haan, Germany).

DNA was extracted from fungal mycelia following the CTAB (cetyl trimethylammonium bromide) protocol described in [**31**], while DNA was extracted from pine needles using the Nucleospin Plant II kit (Macherey-Nagel, Düren, Germany) following the manufacturer's instructions. Total DNA concentrations were estimated using a NanoDrop® ND-1000 spectrophotometer (Thermo Fisher

Scientific, Waltham, MA, USA). Eluted DNA samples were kept at -20 °C until further processing.

2.4. LAMP on DNA Extracts from Needles

Each DNA extract from pine needles was tested by LAMP reaction using the species-specific assays previously optimized for D. septosporum, D. pini and L. acicola [**15**]. Each run was performed separately on a portable Genie III instrument (OptiGene Limited, Horsham, UK) according to conditions described in [**15**]. Each sample was tested in duplicate, including in each test, positive controls (containing external target DNA extracts from D. septosporum, D. pini or L. acicola cultures) and two no-template controls (NTC, in which 5 µL of pure water were used instead of DNA), to ensure effective LAMP reaction conditions and verify the absence of contamination.

2.5. Conventional PCR and Sequencing on Mycelial DNA Extracts

DNA from mycelia of four isolates was amplified at the beta-2-tubulin (tub2) gene region using the Btub2Fd/Btub4Rd primer pairs [**32**]. The expected PCR product size was ca. 431 bp. Each PCR reaction was performed in a final volume of 25 µL using a MB series thermocycler (Cheimika, Italy). Each sample was amplified in duplicate, including in each run no template controls (NTC), in which DNA was substituted by dd-water and positive controls composed by D. septosporum DNA.

PCR reaction cycle was as follows: a first step of 3 min at 95 °C, 40 cycles composed of (i) 30 s at 95 °C, (ii) 30 s at 54 °C, (iii) 30 s at 72 °C, a final step of 7 min at 72 °C. Each reaction mixture contained 19.25 µL of dd-water, 2.5 µL of 1× DreamTaq Green Buffer (Thermo Fisher Scientific, Waltham, USA), 0.5 µL PCR primer mixture (each at a final concentration of 50 µM), 0.5 µL of 10 mM dNTPs, 0.25 µL of 5 U* μL^{-1} DreamTaq polymerase (Thermo Fisher Scientific, Waltham, USA) and 2 µL of template DNA. For each sample, 5 µL of post-amplification products were visualized by electrophoresis runs in 1% agarose gel (Sigma-Aldrich, St. Louis, MO, USA) using 1× Tris-acetate-EDTA (TAE) as a buffer and ethidium bromide (0.5 $\mu g\ mL^{-1}$) for staining. DNA fragments were purified from the PCR products by using NucleoSpin® Gel and PCR Clean-up (Macherey-Nagel, Düren, Germany), and were sent for Sanger sequencing to StarSEQ® GmbH (Mainz, Germany). Chromatograms were checked and nucleotide sequences trimmed using Unipro EUGENE [**33**]. Sequences were aligned using T-coffee [**34**] and alignments visualized with Mview [**35**] through the EMBL-EBI Job Dispatcher Web Services [**36**]. Identity searches of consensus sequences were performed with the Blastn tool in the NCBI GenBank database (**http://www.ncbi.nlm.nih.gov**, accessed on 7 may 2021).

Phylogenetic analyses were performed using a qiime2 version 2021.4 pipeline [**37**]. De novo multiple sequence alignment was performed with MAFFT version 7

[38] with default settings. Sequences for the outgroup were downloaded from the National Center for Biotechnology Information (NCBI: **https://www.ncbi.nlm.nih.gov**, accessed on 3 July 2021). Three strains of Lecanosticta were included in the analyses (L. variabilis, L. longispora and L. acicola). In order to reduce alignment errors that may introduce noise and confound phylogenetic inference, alignments were masked (i.e., filtered) before phylogenetic analysis, eliminating alignment columns that were phylogenetically uninformative or misleading [39].

Max gap frequency value was set equal to 1, retaining all columns regardless of gap character frequency. The minimum conservation value was set equal to 0.4, retaining a column only if it contains at least one character that is present in at least 40% of the sequences. Phylogenetic analyses were conducted using RAxLM (randomized accelerated maximum likelihood) version 8 [40] with the addition of rapid bootstrapping support values under the gamma model [41]. The tree with maximum likelihood score was selected, setting a seed for the parsimony starting tree equal to 1993, and a seed for rapid bootstrapping equal to 2756. Alignment had 164 distinct alignment patterns, and the proportion of gaps and completely undetermined characters in this alignment corresponded to 20.00%. RAxML rapid bootstrapping and a subsequent maximum likelihood (ML) search were performed using 1 distinct models/data partitions with joint branch length optimization, executing 1000 rapid bootstrap inferences and thereafter a thorough ML search.

3. RESULTS

3.1. Symptoms, Signs and Conidial Micromorphology

Typical DNB symptoms were present on all needle samples taken in six different groups of pines, two of P. radiata and four of P. nigra subsp. laricio, on the south-facing slopes towards Lake Metramo at elevations ranging from about 900 to 1100 m a.s.l. Trees of both species were defoliated, with transparent crowns and typical lion-tailed branches. These sampling points represent, in the first case, remains of old plantations of the introduced P. radiata, and in the second case, naturally regenerated trees of the native P. nigra subsp. laricio.

Under the stereoscope, necrotic brown spots and black stroma were visible beneath the needle epidermis and typical black conidiomata in the red bands and discolored areas appeared after a few days of incubation in high humidity conditions. Conidia had morphological characteristics consistent with attribution of the species to D. septosporum or D. pini [13], i.e., they were thin-walled, hyaline, smooth, fusiform to short-clavate, straight or more frequently curved, 2–3(5) septate, and measured 12–40 × 2–3 μm.

3.2. LAMP Assays

DNA extracts from pine needles with typical DNB symptoms, with or without fruiting bodies, all tested positive for D. septosporum and negative for both D. pini and L. acicola by applying LAMP optimized assays. Each sample that tested positive for D. septosporum on the portable instrument was amplified in about 20 min with amplification times (tamp), quality of amplification curves and fluorescence (F) (tamp range: 22–23 min, F range: 15,000–30,000) comparable to those obtained on the real time PCR instrument used during the optimization of the assay **[15]**. No amplification curves were obtained from negative samples. All positive controls were amplified (D. septosporum tamp 20–22 min, F 10,000–30,000; D pini tamp 15–22 min, F 20,000–25,000; L. acicola tamp 21–23 min, F 15,000–25,000), confirming the effectiveness of each run. An example of the reaction kinetics is given in **Figure 2.**

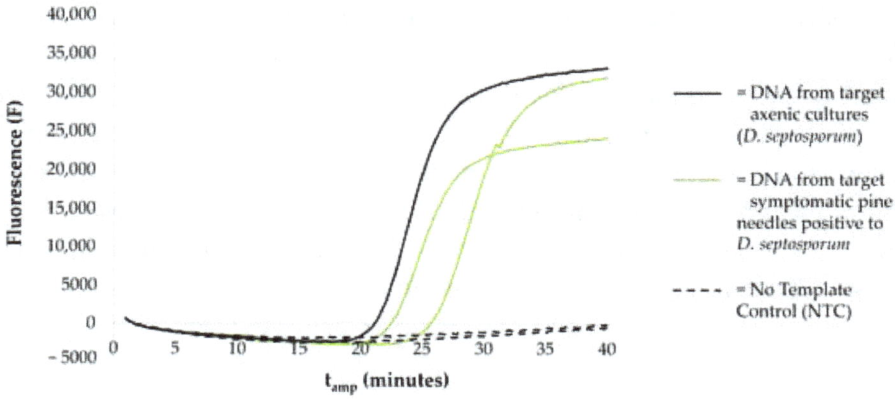

Figure 2. Selection of kinetics showing amplification results of the LAMP assay for the detection of D. septosporum [15] on a portable instrument (Genie III OptiGene, Limited, Horsham, UK).

3.3. Isolation of Fungi and Identification by Sequencing

Numerous dark brown or blackish colonies resembling those of Dothistroma species were obtained from conidia from all the groups of sampled pines. Colonies grew slowly, as is typical of Dothistroma, and stained the substrate red for dothistromin production after 1–2 weeks of growth. Mycelial plugs from 15 axenic colonies were cut from the agar and placed into microeppendorf tubes (1.5 mL), covered with sterile distilled water and stored at 4 °C.

Partial sequences of the tub2 gene region were obtained from four isolates from Pinus nigra subsp. laricio. Amplicon size was ca. 431 bp as expected. After quality check, trimming and alignment, sequences comprised 297–372 base pairs due to poor quality sequences requiring more trimming in some cases. Blast alignment in the GeneBank database showed that all four fungal isolates could be classified as D. septosporum (99.35%–100% sequence identity, best hit D. septosporum strain

CMW44656). Sequences were deposited in GenBank under accession numbers (MZ594908-MZ594911).

Dothistroma spp. included in phylogenetic analyses formed a cluster independent from Lecanosticta spp., which were selected as an outgroup (**Figure 3**). Sequences from the study site (RCDS001, RCDS003, RCDS007, RCDS009) were placed inside the Dothistroma septosporum clade, supporting the conclusion that these isolates belong to this species.

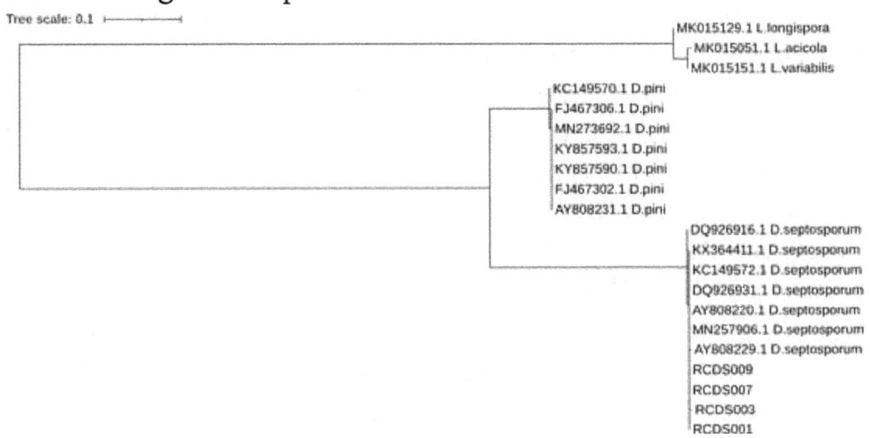

Figure 3. Phylogenetic tree with maximum likelihood score (branch support computed out of 1000 rapid bootstrapped trees) based on partial tub2 gene sequences of isolates RCDSS001, RCDS003, RCDS007, RCDS009 from Pinus nigra subsp. laricio in Prateria (San Pietro di Caridà, Reggio Calabria, Italy) and of Dothistroma species isolates retrieved from GenBank.

4. DISCUSSION

Following the initial discovery of DNB in the mountains of southern Calabria [11], the cultivation of P. radiata has significantly declined in southern Italy [42]. Presently, after forty-four years, only small clusters of pine trees, particularly the native P. nigra subsp. laricio, can be found sporadically within extensive beech forests in the same area as the initial report. With the advancements in molecular techniques, we can now differentiate between species that look the same and identify hidden pathogens. Using these techniques, we were able to verify the accuracy of Magnani's record [11] by conducting our investigation at the same location. Consequently, we confirmed that his observations, which were never validated before, were indeed accurate. Indeed, the significant alteration of the Dothistroma genus, including the renaming of its species [13], occurred well after the release of Magnani's paper. This study confirms that, even after fifty years of establishing P. radiata plantations, Dothistroma still exists in the area. However, trees showing obvious signs of the disease are spread out or found in small, isolated clusters that are physically separated from each other by non-host species.

We have confirmed that the pathogen is still present in the area. Our study showed that the recently released LAMP diagnostic tests, which can distinguish between infections caused by DNB agents and L. acicola in pine needles, may be effectively used on the portable instrument (Genie III Optigene, UK). This outcome promotes the potential utilization of these substances in practical applications, once a refined method for extracting them from raw materials is developed. The use of LAMP diagnostics and sequencing in the tub2 gene region has shown that the pathogen discovered today in the research area belongs to the species known as D. septosporum. Considering the unique climate conditions in the area caused by the geography, its location between two seas, and the presence of the artificial lake Metramo, which contribute to higher atmospheric humidity compared to the surrounding areas and promote the survival and spread of the pathogen, it is reasonable to assume that the population detected today is likely descended from the one found in 1977. However, the available data does not provide definitive confirmation of this assertion.

The significance of *P. radiata* in the area's silviculture has significantly diminished over the past 50 years due to changes in afforestation policy and forest management practices. However, the threat presented by D. septosporum to native pines remains a severe concern. Indeed, within the scope of this study, the pathogen was discovered on the indigenous species P. nigra subsp. laricio, just like in the Sila Massif [15], which is situated around 150 km north of the current discovery in Calabria. The remaining natural populations of this indigenous pine species are currently conserved in Calabria and Sicily, as well as a few on the French island of Corsica. The existence of the disease thus poses a risk to the genetic assets of the species.

Reports of *D. septosporum* outside of Italy in the central southern Mediterranean area (specifically Spain, Greece, and Turkey) are infrequent and limited to pine plantations. However, the fungus has been found in natural forests in Montenegro and southwestern Turkey. The scientific discourse regarding the geographical source of D. septosporum is extensive and contentious [13,49,50,51,52], partially due to inadequate sampling of the fungus in certain areas within its present distribution. A recent study utilized microsatellite markers to genotype a significant number of samples from various locations worldwide, revealing that Eastern Europe (specifically the Baltics and Western Russia) and Western Asia are likely the original regions where the fungus D. septosporum originated. The study also suggests that Turkish populations may have evolved from ancestral populations through a separation that occurred prior to the colonization of Central Europe. The analysis conducted by Mullet and colleagues [53] did not include a sufficient number of fungal isolates from the epidemic in the Sila Massif to reconstruct the evolutionary history of that population. Additionally, the presence of D. septosporum in the locations of the current investigation had not yet been confirmed. The current status of whether D. septosporum is native to these places remains uncertain.

The significant vulnerability of P. nigra subsp. laricio when planted outside its natural habitat and exposed to D. septosporum indicates a lack of mutual evolutionary adaptation with the pathogen. Likewise, the significant harm found on this pine tree in its original habitat in the Sila Massif [17] also suggests a recent introduction. However, other potential reasons, such as a climate change-induced environment that promotes disease, cannot be ruled out [54]. Nevertheless, it is justifiable to presume that the populations of D. septosporum recently discovered on this particular pine species in its original habitat in Southern Italy could represent recent invasions. As far as we know, there is only one documented case of DNB in Corsica. This case involves a single instance of D. septosporum found in P. nigra subsp. laricio, which is mentioned in reference [53]. However, this information has not been added to the EPPO database yet. You can find more information about the distribution of DNB in the EPPO database at this link: https://gd.eppo.int/taxon/SCIRPI/distribution (accessed on 19 July 2021). EPPO has stated that the pathogen was not present in Corsica until at least 2011, when a monitoring program was carried out. No further updates have been recorded since then. Hence, the discovery in Corsica may potentially be the result of a recent importation. There are no existing records documenting the presence of the disease in Sicily, even though natural populations of P. nigra subsp. laricio can be found there. Conducting a comparative genomic investigation of the Calabrian populations of the fungus could provide valuable insights into the historical distribution of the infection in Southern Europe.

5. CONCISE SUMMARY

DNA-based methods are becoming more commonly employed to reinforce and expand traditional taxonomic connections, as well as to unravel complex taxonomic histories and define species boundaries. The fungus first classified as D. pini based only on micromorphological characteristics in the first documented Italian epidemic of the disease has been molecularly identified as D. septosporum, providing a clear example. The implementation of the LAMP approach, a fast and highly precise diagnostic tool, will enable us to properly and specifically identify the causative agents of DNB in the Mediterranean region. Additional genetic study could offer a more comprehensive understanding of the present spread of the organisms responsible for red band needle blight in the area, as well as knowledge about the disease's origin.

6. AT A GLANCE

This study involved the use of species-specific Loop mediated isothermal AMPlification (LAMP) diagnostic assays, followed by sequencing of fungal isolates at the beta-2-tubulin (tub2) gene region. The aim was to verify a previously reported but unconfirmed case of Dothistroma needle blight (DNB) in Monterey pine (Pinus radiata D. Don) trees in the southernmost region of Italy.

The account originates from the mid-1970s and was written before the molecular-based taxonomic revision of the genus Dothistroma, which established the currently accepted species. In autumn 2019, we collected symptomatic needles from Monterey pine and Corsican pine (Pinus nigra subsp. laricio (Poir.) Palib. ex Maire) in the same location where the initial discovery was made. Diagnostic tests confirmed the existence of Dothistroma septosporum (Dorogin) M. Morelet on both types of pine trees. Through our research, we achieved the following objectives: (i) confirmed the presence of the disease; (ii) determined the exact taxonomic identity of the causal agent present at the site; (iii) verified the effectiveness of the species-specific LAMP diagnostic protocol we recently developed for Dothistroma, which can be used with a portable field instrument; and (iv) demonstrated that the pathogen also affects the native P. nigra subsp. laricio, a species that is highly susceptible to the disease and is found in the mountains of Calabria, where the genetic resources of this species are conserved. An examination of the genetic makeup of the uncommon populations of D. septosporum in the central Mediterranean region and in the original habitat of P. nigra subsp. laricio could provide insights into the historical movement of the disease in southern Europe and enhance our understanding of the threat it poses to the preservation of indigenous pine species.

REFERENCES

1. Drenkhan, R.; Tomešová-Haataja, V.; Fraser, S.; Bradshaw, R.E.; Vahalik, P.; Mullett, M.S.; Martín-García, J.; Bulman, L.S.; Wingfield, M.J.; Kirisits, T.; et al. Global geographic distribution and host range of Dothistroma species: A comprehensive review. For. Pathol. 2016, 46, 408–442. [**Google Scholar**] [**CrossRef**]

2. Mullett, M.S.; Adamson, K.; Bragança, H.; Bulgakov, T.S.; Georgieva, M.; Henriques, J.; Jürisoo, L.; Laas, M.; Drenkhan, R. New country and regional records of the pine needle blight pathogens Lecanosticta acicola, Dothistroma septosporum and Dothistroma pini. For. Pathol. 2018, 48, e12440. [**Google Scholar**] [**CrossRef**]

3. Barnes, I.; van der Nest, A.; Mullett, M.S.; Crous, P.W.; Drenkhan, R.; Musolin, D.L.; Wingfield, M.J. Neotypification of Dothistroma septosporum and epitypification of D. pini, causal agents of Dothistroma needle blight of pine. For. Pathol. 2016, 46, 388–407. [**Google Scholar**] [**CrossRef**]

4. Gibson, I.A.S. Dothistroma blight of Pinus radiata. Annu. Rev Phytopathol. 1972, 10, 51–72. [**Google Scholar**] [**CrossRef**]

5. Woods, A.; Coates, K.D.; Hamann, A. Is an unprecedented Dothistroma needle blight epidemic related to climate change? BioScience 2005, 55, 761–769. [**Google Scholar**] [**CrossRef**] [**Green Version**]

6. Barnes, I.; Kirisits, T.; Akulov, A.; Chhetri, D.B.; Wingfield, B.D.; Bulgakov, T.S.; Wingfield, M.J. New host and country records of the Dothistroma needle blight pathogens from Europe and Asia. For. Pathol 2008, 38, 178–195. [**Google Scholar**] [**CrossRef**]

7. Watt, M.S.; Kriticos, D.J.; Alcaraz, S.; Brown, A.V.; Leriche, A. The hosts and potential geographic range of Dothistroma needle blight. For. Ecol Manag. 2009, 257, 1505–1519. [**Google Scholar**] [**CrossRef**]

8. Morelet, M. Une maladie des pins, nouvelle pour la France, due à Scirrhia pini Funk et Parker et à son stade conidien Dothistroma pini Hulbary. Bulletin Mens. Soc. Linn. Lyon 1967, 36, 361–367. [**Google Scholar**] [**CrossRef**]

9. Arthaud, J. Scirrhia pini (Ascomycetes, Dothideaceae) present in the southwest [of France]. Bull. Soc. Linn. Bordx. 1972, 2, 123–124. [**Google Scholar**]

10. Fernandez de Ana Magan, F.J. Red Band Disease of Pinus Radiata; Comunicaciones I.N.I.A. Serie Proteccion Vegetal; Ministerio De Agricultura, Instituto Nacional de Investigaciones Agrarias: Madrid, Spain, 1975; Volume 3, p. 16. Available online: **http://libros.inia.es/libros/product_info.php?products_id=1113** (accessed on 22 July 2021).

11. Magnani, G. Presence of Dothistroma pini on Pinus radiata. Cellul. Carta 1977, 28, 37–42. [**Google Scholar**]

12. Fratianni, S.; Acquaotta, F. The climate of Italy. In Landscapes and Landforms of Italy; Soldati, M., Marchetti, M., Eds.; World Geomorphological Landscapes; Springer International Publishing: Basel, Switzerland, 2017; pp. 29–38. [**Google Scholar**] [**CrossRef**]

13. Barnes, I.; Crous, P.W.; Wingfield, B.D.; Wingfield, M.J. Multigene phylogenies reveal that red band needle blight of Pinus is caused by two distinct species of Dothistroma, D. septosporum and D. pini. Stud. Mycol. 2004, 50, 551–565. [**Google Scholar**]

14. Ioos, R.; Fabre, B.; Saurat, C.; Fourrier, C.; Frey, P.; Marcais, B. Development, comparison, and validation of real-time and conventional PCR tools for the detection of the fungal pathogens causing brown spot and red band needle blights of pine. Phytopathology 2010, 100, 105–114. [**Google Scholar**] [**CrossRef**] [**Green Version**]

15. Aglietti, C.; Meinecke, C.D.; Ghelardini, L.; Barnes, I.; van der Nest, A.; Villari, C. Rapid detection of pine pathogens Lecanosticta acicola, Dothistroma pini and D. septosporum on needles by probe-based LAMP assays. Forests 2021, 12, 479. [**Google Scholar**] [**CrossRef**]

16. Marchi, G.; Ghelardini, L. Outbreak of Dothistroma septosporum on Corsican Pine in Southern Italy. J. Plant Pathol. 2017, 99, S53. Available online: **https://iris.unito.it/retrieve/handle/2318/1651431/368356/Masino%20A.%20et%20al.%20-%20JPP%2099%20%28supp.%20XXIII%20convegno%29.pdf** (accessed on 7 July 2021).

17. Ghelardini, L.; Aglietti, C.; Loria, F.; Cerboneschi, M.; Gionni, A.; Goti, E.; Maresi, G.; Moricca, S.; Marchi, G. Dothistroma needle blight in protected pine forests in Italy. Manag. Biol. Invasions 2020, 11, 689–702. [**Google Scholar**] [**CrossRef**]

18. Farjon, A. Pinus nigra subsp. laricio. IUCN Red List Threat. Species 2013, e.T20453493A20453502. [**Google Scholar**] [**CrossRef**]

19. Bonavita, S.; Vendramin, G.G.; Bernardini, V.; Avolio, S.; Regina, T.M.R. The first SSR-based assessment of genetic variation and structure among Pinus laricio Poiret populations within their native area. Plant Biosyst. 2016, 150, 1271–1281. [**Google Scholar**] [**CrossRef**]
20. Plutino, M.; Pollastrini, M.; Avolio, S.; Bernardini, V. Tree inventory data of Pinus nigra J.F.Arnold subsp. laricio (Poir.) Maire in southern Italy. Ann. Silv. Res. 2018, 42, 43–45. [**Google Scholar**] [**CrossRef**]
21. Villebonne, D.; Maugard, F. Rapid development of Dothistroma needle blight (Scirrhia pini) on Corsican pine (Pinus nigra subsp. laricio) in France. In La Sante des Forets, Annual Report 1998; Les Cahiers du Département de la santé des forêts (DSF); DERF: Paris, France, 1999; Volume 1. [**Google Scholar**]
22. Bulman, L.S.; Gadgil, P.D.; Kershaw, D.J.; Ray, J.W. Assessment and control of Dothistroma needle-blight. Res. Bull. 2004, 229, 1176–2632. [**Google Scholar**]
23. Sturrock, R.N.; Frankel, S.J.; Brown, A.V.; Hennon, P.E.; Kliejunas, J.T.; Lewis, K.J.; Worrall, J.J.; Woods, A.J. Climate change and forest diseases: Climate change and forest diseases. Plant Pathol. 2011, 60, 133–149. [**Google Scholar**] [**CrossRef**]
24. Fraser, S.; Brown, A.V.; Woodward, S. Intraspecific variation in susceptibility to Dothistroma needle blight within native Scottish Pinus sylvestris. Plant Pathol. 2015, 64, 864–870. [**Google Scholar**] [**CrossRef**]
25. Fraser, S.; Woodward, S.; Brown, A.V. Inter- and intraspecific variation in susceptibility to Dothistroma needle blight in Britain. How susceptible are Pinus sylvestris and Pinus contorta? For. Pathol. 2016, 46, 534–546. [**Google Scholar**] [**CrossRef**]
26. Mullett, M.; Peace, A.; Brown, A. Persistence of Dothistroma septosporum on abscised pine needles and its implications for disease management. Plant Dis. 2016, 100, 1271–1277. [**Google Scholar**] [**CrossRef**] [**PubMed**] [**Green Version**]
27. Pebesma, E. Simple Features for R: Standardized Support for Spatial Vector Data. R J. 2018, 10, 439–446. [**Google Scholar**] [**CrossRef**] [**Green Version**]
28. Wickham, H. ggplot2: Elegant Graphics for Data Analysis, 2nd ed.; Springer: New York, NY, USA, 2016; p. 260. [**Google Scholar**] [**CrossRef**]
29. R Core Team. R: A Language and Environment for Statistical Computing; R Foundation for Statistical Computing: Vienna, Austria, 2020; Available online: **https://www.R-project.org/** (accessed on 8 July 2021).
30. Mullett, C.M.; Barnes, I. Dothistroma Isolation and Molecular Identification Methods. In Detection and Diagnostics of Dothistroma; COST ACTION FP1102 Determining Invasiveness and Risk of Dothistroma, Training School May 2012, Brno, Czech Republic. Available online: **https://www.forestresearch.gov.uk/documents/305/DIAROD_052012_Isolation_and_indentification_97fNCCI.pdf** (accessed on 7 July 2021).
31. Carter-House, D.; Stajich, J.E.; Unruh, S.; Kurbessoian, T. Fungal CTAB DNA Extraction. Protocols.io 2020. [**Google Scholar**] [**CrossRef**]
32. Woudenberg, J.H.C.; Aveskamp, M.M.; de Gruyter, J.; Spiers, A.G.; Crous, P.W. Multiple Didymella teleomorphs are linked to the Phoma clematidina morphotype. Persoonia 2009, 22, 56–62. [**Google Scholar**] [**CrossRef**] [**Green Version**]
33. Okonechnikov, K.; Golosova, O.; Fursov, M.; the UGENE Team. Unipro UGENE: A unified bioinformatics toolkit. Bioinformatics 2012, 28, 1166–1167. [**Google Scholar**] [**CrossRef**] [**Green Version**]

34. Notredame, C.; Higgins, D.G.; Heringa, J. T-coffee: A novel method for fast and accurate multiple sequence alignment. J. Mol. Biol. 2000, 302, 205–217. [**Google Scholar**] [**CrossRef**] [**Green Version**]

35. Brown, N.P.; Leroy, C.; Sander, C. MView: A web-compatible database search or multiple alignment viewer. Bioinformatics 1998, 14, 380–381. [**Google Scholar**] [**CrossRef**]

36. Madeira, F.; Park, Y.M.; Lee, J.; Buso, N.; Gur, T.; Madhusoodanan, N.; Basutkar, P.; Tivey, A.R.N.; Potter, S.C.; Finn, R.D.; et al. The EMBL-EBI search and sequence analysis tools APIs in 2019. Nucleic Acids Res. 2019, 47, W636–W641. [**Google Scholar**] [**CrossRef**] [**Green Version**]

37. Bolyen, E.; Rideout, J.R.; Dillon, M.R.; Bokulich, N.A.; Abnet, C.C.; Al-Ghalith, G.A.; Alexander, H.; Alm, E.J.; Arumugam, M.; Asnicar, F.; et al. Reproducible, interactive, scalable and extensible microbiome data science using QIIME 2. Nat. Biotechnol. 2019, 37, 852–857. [**Google Scholar**] [**CrossRef**] [**PubMed**]

38. Katoh, K.; Standley, D.M. MAFFT Multiple Sequence Alignment Software Version 7: Improvements in performance and usability. Mol. Biol. Evol. 2013, 30, 772–780. [**Google Scholar**] [**CrossRef**] [**PubMed**] [**Green Version**]

39. Lane, D. 16S/23S rRNA Sequencing. In Nucleic Acid Techniques in Bacterial Systematics; Stackebrandt, E., Goodfellow, M., Eds.; John Wiley and Sons: New York, NY, USA, 1991; pp. 115–175. [**Google Scholar**]

40. Stamatakis, A. RAxML version 8: A tool for phylogenetic analysis and post-analysis of large phylogenies. Bioinformatics 2014, 30, 1312–1313. [**Google Scholar**] [**CrossRef**] [**PubMed**]

41. Stamatakis, A.; Hoover, P.; Rougemont, J. A Rapid Bootstrap Algorithm for the RAxML Web Servers. Syst. Biol. 2008, 57, 758–771. [**Google Scholar**] [**CrossRef**] [**PubMed**]

42. Ciancio, O.; Garfì, V.; Iovino, F.; Menguzzato, G. Il pino insigne nell'arboricoltura da legno: Analisi di interventi realizzati nell'Italia meridionale. Ital. For. Mont. 2006, 61, 217–231. Available online: **http://ojs.aisf.it/index.php/ifm/article/view/36** (accessed on 7 July 2021). [**CrossRef**] [**Green Version**]

43. Tsopelas, P.; Barnes, I.; Soulioti, N.; Wingfield, M.J. Dothistroma septosporum identified in Greece on Pinus brutia and Pinus nigra plantations. Plant Dis. 2013, 97, 1247. [**Google Scholar**] [**CrossRef**]

44. Doğmuş-Lehtijärvi, H.T.; Lehtijärvi, A.; Oskay, F.; Aday Kaya, A.G.; Örtel, E.; Datumani, A. Dothistroma needle blight in Turkey. In Book of Abstract, IUFRO WP 7.02.02 Foliage Shoot and Stems Diseases: Biosecurity in Natural Forests and Plantations, Genomics and Biotechnology for Biosecurity and Forestry; Cerno Hora and Brno,: Czech Republic, 20–25 May 2013; Mendel University: Brno, Czech Republic, 2013; pp. 68–69. [**Google Scholar**]

45. Tunalı, Z.; Doğmuş-Lehtİjärvİ, H.T.; Oskay, F. Detection of Fungal Needle Disease Agents of Turkish Pine (Pinus brutia Ten.) forests in Burdur Province Using Molecular Techniques. Süleyman Demirel Üniversitesi Fen Bilimleri Enstitüsü Dergisi; 2018; Volume 22, pp. 628–636, (In Turkish with English Abstract); Available online: **https://www.cabdirect.org/cabdirect/abstract/20203105032** (accessed on 7 July 2021).

46. Oskay, F.; Tunalı, Z.; Lehtijärvi, A.T.; Doğmuş-Lehtijärvi, H.T.; Woodward, S.; Mullett, M. Distribution and genetic diversity of Dothistroma septosporum in Pinus brutia forests of south-western Turkey. Plant Pathol. 2020, 69, 1551–1564. [**Google Scholar**] [**CrossRef**]

47. Lazarević, J.; Davidenko, K.; Millberg, H. Incidence of Dothistroma septosporum in different pine forests in Montenegro. Mycol. Montenegrina 2014, 17, 119–131. [**Google Scholar**]
48. Lazarević, J.; Davidenko, K.; Millberg, H. Dothistroma needle blight on high altitude pine forests in Montenegr. Balt. For. 2017, 23, 294–302. [**Google Scholar**]
49. Adamson, K.; Mullett, M.S.; Solheim, H.; Barnes, I.; Müller, M.M.; Hantula, J.; Vuorinen, M.; Kačergius, A.; Markovskaja, S.; Musolin, D.L.; et al. Looking for relationships between the populations of Dothistroma septosporum in northern Europe and Asia. Fungal Genet. Biol. 2018, 110, 15–25. [**Google Scholar**] [**CrossRef**]
50. Dale, A.L.; Lewis, K.J.; Murray, B.W. Sexual reproduction and gene flow in the pine pathogen Dothistroma septosporum in British Columbia. Phytopathology 2011, 101, 68–76. [**Google Scholar**] [**CrossRef**] [**PubMed**] [**Green Version**]
51. Mullett, M.S.; Brown, A.V.; Fraser, S.; Baden, R.; Tubby, K.V. Insights into the pathways of spread and potential origins of Dothistroma septosporum in Britain. Fungal Ecol. 2017, 26, 85–98. [**Google Scholar**] [**CrossRef**]
52. Capron, A.; Feau, N.; Heinzelmann, R.; Barnes, I.; Benowicz, A.; Bradshaw, R.E.; Dale, A.; Lewis, K.J.; Owen, T.J.; Reich, R.; et al. Signatures of post-glacial genetic isolation and human-driven migration in the Dothistroma needle blight pathogen in Western Canada. Phytopathology 2021, 111, 116–127. [**Google Scholar**] [**CrossRef**] [**PubMed**]
53. Mullett, M.S.; Drenkhan, R.; Adamson, K.; Boroń, P.; Lenart-Boroń, A.; Barnes, I.; Tomšovský, M.; Jánošíková, Z.; Adamčíková, K.; Ondrušková, E.; et al. Worldwide genetic structure elucidates the Eurasian origin and invasion pathways of Dothistroma septosporum, causal agent of Dothistroma needle blight. J. Fungi 2021, 7, 111. [**Google Scholar**] [**CrossRef**] [**PubMed**]
54. Woods, A.J.; Martín-García, J.; Bulman, L.; Vasconcelos, M.W.; Boberg, J.; La Porta, N.; Peredo, H.; Vergara, G.; Ahumada, R.; Brown, A.; et al. Dothistroma needle blight, weather and possible climatic triggers for the disease's recent emergence. For. Pathol. 2016, 46, 443–452. [**Google Scholar**] [**CrossRef**] [**Green Version**]

06 THE IMPACT OF TEMPERATURE ON THE SURVIVAL, REPRODUCTION, AND EGG HATCH OF ANOPLOPHORA CHINENSIS (COLEOPTERA: CERAMBYCIDAE) ADULTS

1. INTRODUCTION: AN OVERVIEW

Temperature is a crucial abiotic element that impacts various aspects of insect life-history processes, including survival, development, and reproduction. By comprehending the way insects react to temperature, it becomes feasible to assess their potential geographic distribution and create phenology models that can forecast the timing of crucial stages necessary for implementing survey and management strategies. Developing predictive methods for species such as Anoplophora chinensis (Forster) (Coleoptera: Cerambycidae) is crucial, especially because they become invasive when carried over international commerce channels, either in live plants or solid wood packaging materials. A. chinensis is primarily distributed in the same geographic area as its closely related species, Anoplophora glabripennis (Motschulsky). However, A. chinensis has been observed in warmer regions of southern Asia and is absent in certain northern regions of China and far east Russia [1]. Anoplophora chinensis has a large distribution in China, Japan, and Korea, and is sometimes found in Vietnam, Malaysia, Indonesia, and the Philippines [2,3].

This organism is a polyphagous xylophage that targets 73 distinct genera of deciduous trees, surpassing A. glabripennis in its range. Additionally, there have been reports of it attacking certain coniferous trees of the Cryptomena and Pinus genera [4]. The beetle is the primary wood-infesting pest of citrus trees in Japan and China. It also causes significant damage to Casuarina equisetifolia L. shelterbelts in China [5,6,7]. Within its invaded habitat, A. chinensis exhibits a preference for trees belonging to the genera Acer, Betula, and Corylus. However, it has successfully undergone its entire life cycle on 24 other species of trees, and partially finished development on an additional five genera [3,8]. Female insects deposit their eggs in locations they have gnawed into the bark of the lower trunk or visible roots of the tree. The young insects, in their initial developmental stages, consume food beneath the bark, while the more mature insects consume food in the xylem tissue [3]. The larvae inflict harm to the tree's vascular system and generate structural vulnerabilities that may result in the tree's demise. Adults are present during the late spring, summer, and early fall, contingent upon the prevailing climatic conditions in the area.

According to a recent update, Anoplophora malasiaca (Thomson) is now regarded a synonym of A. chinensis. This conclusion was reached based on the similar reproductive system characteristics shared by both species [2]. Nevertheless, there is ongoing debate regarding this synonymy due to the ability to distinguish between the two based on the presence of two white spots on the pronotum and their mitochondrial COI region haplotypes [9,10]. The initial incursion of A. chinensis was discovered in the Lombardy region of Italy in 2000. However, throughout the past two decades, further invasions have been detected in Italy, France, the Netherlands, Croatia, and Turkey [11]. The infestations in the Netherlands and Croatia have been linked to the presence of live Acer palmatum Thunberg plants imported from China, whereas the infection in France originated from beetles that emerged from bonsai trees imported from Asia [11,12]. In 2001, a discovery was made outside a greenhouse in Tukwila, Washington, United States, which was likewise linked to imported bonsais [13]. By analyzing the mitochondrial DNA haplotypes, it was discovered that the infestations in the Lazio and Lombardy regions of Italy were comparable to Japanese populations, but the infestations in Tuscany were more similar to Chinese populations [14]. Hence, the invasive populations might derive from regions where the two original species are present and could have distinct biological reactions to temperature and other characteristics. Given that all instances of invasive infestations or potential ones (in Washington State, US) have led to the implementation of eradication measures, it is crucial to determine whether there are any distinctions among these original species. To simplify matters, A. malasiaca will be denoted as "A. chinensis from Japan".

Information regarding the impact of temperature on the life cycle of A. chinensis adults and eggs is available. At ambient temperatures, adults from a Japanese population of A. chinensis survived for an average of 77.6 ± 20.3 days. During this time, they fed on citrus twigs for sustenance and used bolts for egg-laying. The maturation period lasted for an average of 9.8 ± 1.4 days. Female adults deposited an average of 193.8 ± 65.2 eggs, with the highest number of eggs being laid 30 days after emerging from the pupal stage. Approximately 90% of these eggs successfully hatched [7]. A separate study conducted on Japanese adults revealed variations in maturation feeding (the eating necessary to attain sexual maturity), lifespan, and reproductive capacity at a temperature of 24 °C when various hosts were employed [15]. The fecundity of a separate group of Japanese females was seen to vary between 60 and 140 eggs when they were provided with Acer, Citrus, and Platanus bolts to lay on in outdoor cages [10]. The average number of eggs laid by Chinese A. chinensis females when fed on Melia azedarach L. was 176, and their lifespan was 92 days (103 days for males). However, when they were fed Casuarina sp., they only lay 56 eggs and survived for 62 days (63 days for males) [1]. According to estimates, it takes a total of 184 degree-days for 50% of Japanese A. chinensis eggs to hatch when the anticipated minimum temperature (Tmin) of 6.7 °C is used [7].

This study aimed to ascertain the reaction of A. chinensis adults and eggs to various consistent temperatures. Subsequently, the mathematical correlations between temperature and the survival of adult A. chinensis, their reproduction, and the hatching of their eggs were established. This was done to aid in the creation of phenology models and to compare these correlations with those previously documented for A. glabripennis. Lastly, the consequences for management and eradication programs are examined.

2. MATERIALS AND METHODS

2.1. Populations and Temperatures

The populations used in this study were established from larvae that were shipped on 26 February 2018, from the USDA Agricultural Research Service (ARS), European Biological Control Laboratory in Montferrier-sur-Lez, France, under permit to the USDA Forest Service quarantine facility in Ansonia, CT. Fifty larvae from the fifteenth ARS laboratory generation of an invasive A. chinensis population from Lombardy, Italy (41.58° N and 87.42° W) were used to start the USDA Forest Service (FS) colony. The Italian colony was initiated by ARS in 2002 and augmented each year with freshly collected adults. The Italian colony was determined to have originated from Japan based on molecular tests. Individuals used in this study were from the sixteenth or seventeenth Italian laboratory generations.

Fifty larvae from the fifth ARS generation of a native A. chinensis population from Yancun, Guangdong Province, China colony (40.45° N and 73.45° W) were used to start the FS colony. The Chinese population was started by ARS in 2012 with 180 larvae from China. Individuals used in this study were from the sixth or seventh laboratory generations. Voucher specimens of both populations were deposited at the Entomology Division, Yale Peabody Museum of Natural History, New Haven, CT.

Eight constant temperatures were used for these studies: 5, 10, 15, 20, 25, 30, 35, and 40 °C. Temperature fluctuations in the environmental chambers rarely exceeded 1 degree above or below the set value, and a photoperiod of 16:8 (L:D) h was used. Humidity was passively maintained by placing open reservoirs of water ($32 \times 28 \times 14$ cm^3) in the bottom of the chambers or by humidifiers that were part of the chamber (15, 20 and 25 °C). The humidity in the chambers averaged 80 ± 10%, 85 ± 15%, 70 ± 10%, 80 ± 5%, 60 ± 5%, 45 ± 5%, 68 ± 4%, and 55 ± 10% RH at 5, 10, 15, 20, 25, 30, 35, and 40 °C, respectively. Actual humidity within the rearing containers was likely higher than in the chamber, due to the presence of fresh host material.

2.2. Mating Adults and Obtaining Eggs

Adults used in these studies came from larvae reared on an artificial diet, either the A. glabripennis diet designated "AG2" [**16**] or a drier A. chinensis diet that was a modification of a host-free diet developed for A. glabripennis [**17,18**]. Pre-pupae were placed in 50 mL plastic centrifuge tubes (with two pinholes in the lid and a piece of damp paper towel) and held in the dark at 25 °C until pupae formed, and the adults emerged. Callow adults were held for four more days under the same conditions as the pre-pupae/pupae to allow the cuticle to fully sclerotize before they were sexed, weighed, and assigned to a temperature regime. Chinese adults (15 females and 16–18 males) were initially assigned evenly to the 15, 20, 25, and 30 °C temperature regimes, and then ten of each sex were assigned evenly to the 10, 35, and 40 °C temperature regimes.

Virgin females in the 15–35 °C temperature regimes were held individually in 3.8-L glass jars with Acer rubrum L. (red maple) twigs (3–7 mm diameter with leaves removed) as a food source and A. rubrum bolts (3–7 cm diameter and 22 cm long) with both ends waxed as an oviposition substrate. A virgin male was added to the female's jar 0–11 days later (i.e., as soon as a female became available at the same temperature). Males that were waiting for a mate were held at their assigned temperatures in 950 mL glass jars with fresh twigs as a food source. All adults held at 10 and 40 °C were held individually in 950 mL glass jars with fresh twigs as a food source. After twelve weeks at 10 °C (presumably enough time for the females to maturation feed), the three males that were still alive were placed with females in the larger glass jars to see if the females would chew pits or lay eggs. Twigs and bolts were cut twice a month and were stored at 4 °C and ≥80% RH until used in the weekly food and bolt changes. Folded paper towels were placed in the bottoms of the jars to collect frass and excess moisture. Two holes (1–2 mm diameter) in the plastic jar lids allowed airflow.

The oviposition bolts were replaced twice a week (30–35 °C), weekly (15–25 °C), or monthly (10 °C) until a female died. Oviposition bolts held at 15–35 °C were checked daily for oviposition sites. The date of each new oviposition pit chewed in the bark was recorded, and the pit marked to distinguish the day it appeared. When the oviposition bolts were removed from the jars, all pits chewed in the bark, including both pits that obviously contained eggs and those that did not, were counted, and the bolt diameter was measured.

The bark was stripped off many of the daily marked bolts the day they were removed from the jar to obtain the eggs for the egg hatch part of the study. After the bark was stripped off, the eggs were carefully removed from the bolt by lifting a thin strip of wood with the egg on it to prevent breaking the eggshell. Eggs attached to the small pieces of wood were placed individually in wells of a 24-well tissue culture plate that was held in a water box at the temperature at which the egg was laid or to which the egg was moved. Ten eggs maximum, or half of the eggs at minimum, that were laid by females at 15 °C were kept at that temperature. For females held at other temperatures, a minimum of 10 eggs was held at the oviposition temperature, and the others were moved as needed.

To minimize temperature shifts for eggs that had already begun developing, one-day-old eggs were only moved to temperatures that were ±5 °C of the laying temperature, but eggs laid the same day were moved to other temperatures as needed. Eggs laid more than one day before the bolt was removed from the mating jar were held at the oviposition temperature. Eggs laid by Chinese females held at 25 or 30 °C were moved to any temperature, while eggs laid at 15 °C were only moved to 20–30 °C and eggs laid at 20 °C were only moved to 10–35 °C. This was because there were fewer eggs laid at the cooler temperatures. Two to ten eggs from a Chinese female were held at each temperature 15–30 °C, and 1–3 eggs were held at each temperature 5–10 °C and 35–40 °C.

Once enough eggs were obtained from each mating pair for the egg hatch part of the study, the rest of the bolts from each female were held at 25 or 30 °C for 7–10 days to allow the eggs to hatch before stripping the bark off the bolt. This provided a better measure of the percentage of the eggs that were fertile, since removing the eggs could damage or kill them. Eggs were checked daily for 4, 4, 4, 6, 9, 12, 30, and 30 weeks at 40, 35, 30, 25, 20, 15, 10, and 5 °C, respectively. These time periods were chosen to ensure that all eggs that could hatch at a given temperature did. Eggs that had not hatched after the given time period at 5–20 °C were moved to 25 °C and observed for an additional four weeks.

Adults from the invasive Italian population were more limited in numbers, so only 8–10 females and 11–14 males were assigned on a rotating basis to the 15, 20, 25, and 30 °C temperature regimes. All the same feeding and holding methods were used as detailed above for the Chinese population adults and eggs. Since Italian females also laid fewer eggs, their eggs were only held at 15–30 °C. Eggs laid at these temperatures were moved using the same scheme as detailed for the Chinese eggs.

2.3. Adult Survival and Reproduction

Mating pairs and single adults were checked daily for mortality, with the only exception that single adults at 10 °C were checked three times a week. When the female in the pair died, the male was moved to a 950 mL glass jar and maintained until death. When the male in the pair died, the female was maintained in the mating jar until death. For each temperature at which females laid eggs, the mean number of eggs laid per week per live female during that week was calculated. When the female died, she was dissected to determine if she had any fully formed eggs in her ovarioles that she had not oviposited. The following dependent variables were analyzed in PROC GLIMMIX [**19**]: Female and male weight at four days, female and male longevity, days to the first egg, fecundity, maximum weekly oviposition, a week at which maximum oviposition occurred, mean weekly oviposition pits per female, mature unlaid eggs at female death, and percentage hatch on bolts held at 25–30 °C. A gamma distribution with a log link was used for all the variables except the percentage hatch for which a beta distribution with a logit link was used and the unlaid eggs in the female at death for which a Poisson distribution with a log link was used. The two populations were analyzed separately for each variable. The model used temperature, population (Italy or China), and the interaction between the two as fixed effects. Differences among means were determined using the Tukey–Kramer test with $\alpha = 0.05$ [**19**].

The relationship between longevity and temperature was separately fit for each sex and population combination to a linear model ($y = bx + a$), and the upper threshold (Tmax) was calculated to be the point at which the line crossed the x-axis. The relationship between fecundity and temperature for the Chinese population was fit using the following:

$$y = \begin{cases} at(t - Tmax)(Tmin - t)^{\frac{1}{b}} \text{ for } Tmax \leq t \leq Tmin \\ \text{otherwise } 0 \end{cases} \tag{1}$$

where y = fecundity, t = temperature °C are the known values, and the others are estimated. The a is an empirical constant, b modulates the slopes, and Tmax and Tmin are the upper and lower temperature thresholds, respectively [**20**]. Nonlinear convergence for the model was based on the Marquardt method (PROC NLIN [**19**]). The optimum temperature for Equation (1) was calculated by equating the first derivative to zero and solving for t.

The relationship between temperature (t) and a number of days from female emergence to first oviposition was fit using the following:

$$y = at^b \qquad (2)$$

where a and b are empirical constants, and nonlinear convergence was based on the Marquardt method (PROC NLIN [**19**]). The percentage of females that oviposited and produced progeny was also calculated for each temperature and population combination.

2.4. Egg Hatch

Two dependent variables were analyzed using PROC GLIMMIX [**19**]: Days to hatch and percent hatch on bolts held at 25–30 °C. The model treated temperature, population, and temperature by population as fixed effects, whereas maternal family was treated as a random effect. Differences among means were determined using the Tukey–Kramer test with $\alpha = 0.05$ [**19**]. A gamma distribution with a log link was used for days to hatch, and a beta distribution with a logit link was used for the percentage hatch on bolts. The percentage hatch of eggs laid at or moved to various temperatures after being removed from under the bark was also calculated.

To describe the relationship between temperature and rate of hatch for eggs from the combined populations, Equation (1) was altered by changing y = 1/days to hatch, and the curve was fit using the Marquardt convergence method (PROC NLIN [**19**]). The cumulative distributions of time to egg hatch at 15, 20, 25, and 30 °C and degree days to egg hatch for all eggs were calculated using the estimated Tmin from Equation (1). The cumulative proportion of eggs hatching (P) over accumulated degree-days (DD) was described using a Gompertz function:

$$P = \exp\left(-\exp\left(-bDD + a\right)\right) \qquad (3)$$

in which a and b are the lag and the rate of increase, respectively ([**21**], PROC NLIN and Marquardt convergence method, [**19**]).

3. RESULTS

3.1. Adult Survival and Reproduction

At the initiation of the temperature treatments, adult females from China (186.5 ± 28.9 mg) were significantly heavier than females from Italy (145.3 ± 25.0 mg) ($F = 39.96$, $df = 1, 93$, $p < 0.0001$). There was also a significant difference ($F = 40.21$, $df = 1, 139$, $p < 0.0001$) in the average mass of males from the Chinese population (127.5 ± 22.8 mg) and the Italian population (105.1 ± 21.0 mg). The mass of the beetles held at the different temperatures for either sex did not differ (females: $F = 0.46$, $df = 6, 93$, $p = 0.8351$; males: $F = 1.17$, $df = 6, 139$, $p = 0.3274$).

Both male and female longevity increased with decreasing temperature and Italian individuals of both sexes tended to survive a shorter time than Chinese individuals (**Table 1**). Temperature and population both had significant effects on both female (temperature: $F = 126.6$, $df = 6, 113$, $p < 0.0001$; population: $F = 39.3$, $df = 1, 113$, $p < 0.0001$) and male longevity (temperature: $F = 127.8$, $df = 6, 139$, $p < 0.0001$; population: $F = 60.2$, $df = 1, 139$, $p < 0.0001$), but the interaction between the two did not (**Table 1**). Female and male longevity fit a linear model, but variation increased at the lower temperatures so that the confidence of the fit (adjusted R2) was lower, especially for the Italian population (**Figure 1**, **Table 2**). At 10 °C the Chinese population appeared to have two biotypes, one that was dominated by males that survived about 100 days and another that was dominated by females that survived about 375 days. The estimated Tmin for each sex and population combination for longevity and the parameters for the fitted models are given in **Table 2**.

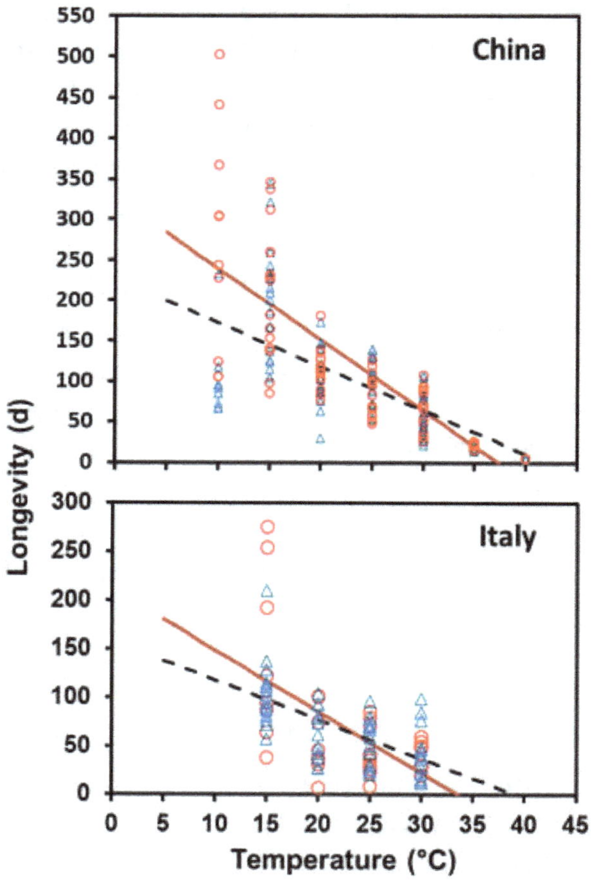

Figure 1. Relationship between temperature and longevity for male and female A. chinensis from China and Italy. Observed male data are open triangles, observed female data are open circles, predicted male dashed line, and predicted female solid line.

Table 1. Effects of temperature on Anoplophora chinensis adult survival and reproduction ((mean ± SE (n) or percentage (n)). Means were compared between each temperature and population combination for each parameter.

Parameter	Population	Temperature (°C) 10	15	20	25	30	35	40	Statistics Population * Temperature
Female Longevity (d)	China	272.7 ± 35.6a (10)	205.9 ± 21.9ab (15)	116.1 ± 12.4c (15)	83.0 ± 8.8cd (15)	66.0 ± 7.0de (15)	21.6 ± 2.8f (10)	5.1 ± 0.7g (10)	F = 0.3681, df = 3, 113 *p* = 0.4928
	Italy	*NA*	141.5 ± 20.6bc (8)	55.0 ± 8.0de (8)	44.6 ± 5.8e (10)	40.6 ± 5.9de (8)	*NA*	*NA*	
Male Longevity (d)	China	101.1 ± 12.2bc (10)	201.1 ± 19.2a (16)	109 ± 10.1b (17)	102.6 ± 9.2b (18)	62.8 ± 5.7cd (18)	21.3 ± 2.6e (10)	5.1 ± 0.6f (10)	F = 0.91, df = 3, 139 *p* = 0.4358
	Italy	*NA*	97.7 ± 10.4bc (13)	68.0 ± 7.8bcd (11)	57.2 ± 6.1d (13)	41.4 ± 4.2d (14)	*NA*	*NA*	
Fecundity	China	0.0 (10)	40.1 ± 8.6abcd (15)	76 ± 16.3cd (15)	92.2 ± 19.7abc (15)	118.0 ± 30.9a (15)	0.0 (10)	0.0 (10)	F = 0.49, df = 3, 69 *p* = 0.6895
	Italy	*NA*	15.8 ± 5.4d (8)	29.8 ± 11cd (8)	24.0 ± 8.9bcd (10)	26.0 ± 8.8ab (8)	*NA*	*NA*	
Max Oviposition/ Wk	China	*NA*	6.8 ± 1c (15)	13.2 ± 1.9ab (15)	18.7 ± 2.6ab (15)	26.3 ± 4.6a (15)	*NA*	*NA*	F = 2.01, df = 3, 69 *p* = 0.1210
	Italy	*NA*	5.3 ± 1.2c (8)	8.2 ± 2bc (8)	7.8 ± 1.9bc (10)	8.7 ± 1.9bc (8)	*NA*	*NA*	
Week Max Oviposition Occurred	China	*NA*	10.1 ± 0.8a (15)	5.8 ± 0.4bc (15)	4.8 ± 0.4c (15)	4.2 ± 0.4cd (15)	*NA*	*NA*	F = 2.04, df = 3, 69 *p* = 0.1164
	Italy	*NA*	8 ± 0.9ab (8)	2.8 ± 0.4d (8)	2.8 ± 0.4d (10)	2.7 ± 0.3d (8)	*NA*	*NA*	
Time First Egg Laid (d)	China	*NA*	54.3 ± 3.3a (15)	23.9 ± 1.4b (15)	14.7 ± 0.9cd (15)	14.2 ± 1.1cd (15)	*NA*	*NA*	F = 5.26, df = 3, 69 *p* = 0.0025
	Italy	*NA*	39.2 ± 3.7a (8)	20.0 ± 2.1bc (8)	20.0 ± 2.1bc (10)	12.8 ± 1.2d (8)	*NA*	*NA*	
Mature Eggs in Female at Death	China	0.00 ± 0.00ab (10)	0.93 ± 0.25a (15)	1.73 ± 0.33ab (15)	1.67 ± 0.33ab (15)	2.67 ± 0.41ab (15)	0.10 ± 0.10b (10)	0.00 ± 0.00ab (10)	F = 3.81, df = 3, 113 *p* = 0.0121
	Italy	*NA*	2.25 ± 0.57a (8)	2.38 ± 0.55a (8)	2.40 ± 0.50a (10)	1.25 ± 0.38a (8)	*NA*	*NA*	

Parameter	Population	Temperature (°C) 10	15	20	25	30	35	40	Statistics Population * Temperature
Weekly Oviposition Pits Chewed/ Female	China	*NA*	13.9 ± 1.5bc (15)	27.7 ± 3a (15)	38.8 ± 4.2a (15)	46.4 ± 6.2a (15)	24.1 ± 4.5 ab (10)	*NA*	F = 2.42, df = 3, 80 *p* = 0.0718
	Italy	*NA*	8.4 ± 1.3c (8)	12.4 ± 2bc (8)	15.3 ± 2.1bc (10)	36.6 ± 6.3a (8)	*NA*	*NA*	
% Hatch on Bolts Held at 25–30 °C	China	*NA*	24.4 ± 7.1b (15)	59.6 ± 8.1ab (15)	77 ± 6.8a (15)	40.4 ± 11.8ab (15)	*NA*	*NA*	F = 2.28, df = 3, 46 *p* = 0.0921
	Italy	*NA*	12 ± 7b (7)	88 ± 11.9ab (5)	36 ± 14.8ab (6)	50 ± 23ab (7)	*NA*	*NA*	
% Females Laying Eggs	China	0 (10)	100 (15)	100 (15)	100 (15)	100 (15)	0 (10)	0 (10)	F = 0.49, df = 3, 69 *p* = 0.6905
	Italy	*NA*	87.5 (8)	62.5 (8)	60 (10)	87.5 (8)	*NA*	*NA*	
% Females with Progeny	China	0 (10)	60 (15)	100 (15)	100 (15)	100 (15)	0 (10)	0 (10)	F = 0.49, df = 3, 69 *p* = 0.6907
	Italy	*NA*	25 (8)	50 (8)	50 (10)	37.5 (8)	*NA*	*NA*	

With rows, means followed by the same letter are not significantly different based on the Tukey–Kramer test with α = 0.05. *NA* = parameter not applicable either because no females were held, or no eggs were laid at these temperatures.

Once Chinese females started laying eggs, they generally continued laying eggs until about five weeks before they died, while Italian females stopped laying at about two weeks before death at 30 °C. The time to first oviposition increased with decreasing temperature with the maximum being 12 weeks at 15 °C (**Figure 2**). The fecundity of A. chinensis from China peaked at 30 °C and declined with both increasing and decreasing temperatures, while the fecundity of females from Italy did not vary much over the 15–30 °C temperature range (**Table 1, Figure 2**). This effect of temperature on fecundity was significant ($F = 2.93$, $df = 3, 69$, $p = 0.0394$) and so was the effect of population ($F = 31.47$, $df = 1, 69$, $p < 0.0001$), but the interaction between the two was not (**Table 1**). Both the percentage of Italian females that laid and produced progeny were also lower than that for Chinese females, further reducing their total reproduction over the 15–30 °C temperature range (**Table 1**).

Equation (1) described the relationship between the mean fecundity of Chinese females and temperature reasonably well based on the adjusted R2 value (**Table 2, Figure 3**A). No curve was fit for the Italian population because fecundity varied little over the range of temperatures where oviposition occurred.

Table 2. Parameter values for linear and non-linear models were used to describe the relationships between temperature and selected life-history values for *Anoplophora chinensis*.

Response Variable	n	T_{Max}	T_{Min}	a	b	F	df	p	Adjusted R^2	Temp. Range (°C)	Optimum Temp. (°C)
China Female Longevity (d) [a]	90	41.18	---	326.9 ± 18.7	-8.77 ± 0.71	150.21	1, 89	<0.0001	0.626	10–40	---
Italy Female Longevity (d) [a]	34	33.57	---	211.8 ± 38.1	-6.31 ± 1.63	14.89	1, 32	0.0005	0.296	10–40	---
China Male Longevity (d) [a]	98	41.89	---	226.1 ± 14.8	-5.40 ± 0.57	90.34	1, 97	<0.0001	0.477	15–30	---
Italy Male Longevity (d) [a]	53	39.15	---	158.8 ± 16.8	-4.06 ± 0.72	31.65	1, 52	<0.0001	0.366	15–30	---
China Mean Fecundity [b]	6	35.00 ± 0.00	9.21 ± 2.40	0.07661 ± 0.02420	1.87 ± 0.64	58.52	2, 4	0.0169	0.979	10–35	29.4
China Mean % Egg Hatch on bolts [b]	5	30.77 ± 0.42	10.22 ± 0.55	0.00079 ± 0.00020	1.79 ± 0.38	416	1, 4	0.0368	0.992	10–30	25.5
Mean Time to First Oviposition [c]	8	---	---	10805.4 ± 10369.1	-2.02 ± 0.34	105.8	2, 6	<0.0001	0.880	15–30	---
Hatch Rate (1/d) [b]	1031	38.19 ± 1.29	12.7 ± 0.51	0.000036 ± 0.000015	1.18 ± 0.19	7201	4 , 1027	<0.0001	0.529	15–35	29.2

Parameter values are means followed by standard errors. In the table and the models, T_{Max} = upper temperature threshold and T_{Min} = lower temperature threshold. The temperature range used did not always include all the data points, since some were above or below the thresholds based on preliminary plots. [a] model $y = ax + b$. [b] model $y = \begin{cases} at(t - T_{max})(T_{min} - t)^{\frac{1}{b}} \text{ for } T_{max} \leq t \leq T_{min} \\ \text{otherwise } 0 \end{cases}$

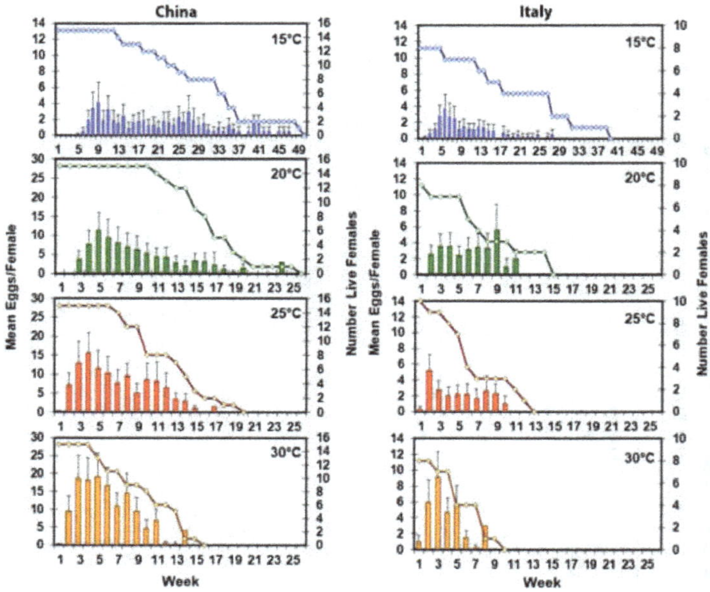

Figure 2. Mean (±SE) weekly oviposition and female survival for female A. chinensis from China (Left) and Italy (Right) held at four constant temperatures.

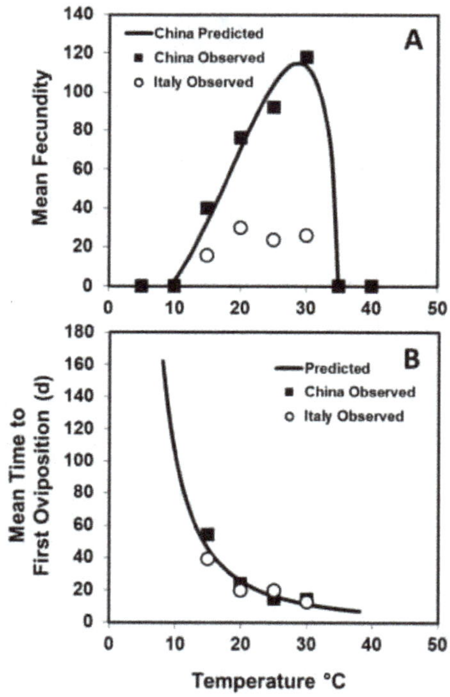

Figure 3. Relationships between temperature and mean fecundity (A) and time to oviposition (B) for A. chinensis from China (solid squares) and Italy (open circles). The solid line in graph A is for only the Chinese population, and in graph B it is for the two populations combined.

There were significant effects of temperature ($F = 8.33$, $df = 3, 69$, $p < 0.0001$) and population ($F = 23.6$, $df = 1, 69$, $p < 0.0001$) on mean maximum weekly oviposition, peaking at 26 eggs per week at 30 °C and declining below that temperature for Chinese females and remaining constant at about eight eggs per week for Italian females (**Table 1**). The maximum weekly oviposition recorded for any female was 12, 22, 26, and 35 eggs for Chinese females and 12, 15, 17, and 19 eggs for Italian females at 15, 20, 25, and 30 °C, respectively. The week that the maximum weekly oviposition occurred was significantly affected by temperature ($F = 40.32$, $df = 3, 69$, $p < 0.0001$), but not population ($F = 43.77$, $df = 1, 69$, $p < 0.0001$); oviposition peaked sooner at higher temperatures than at lower ones (**Table 1**).

The number of days to first oviposition was significantly affected by the interaction between temperature and population ($F = 5.26$, $df = 3, 69$, $p = 0.0025$), generally declining with increasing temperature (**Table 1**). Equation (2) described the relationship between temperature and number of days to first oviposition well with an adjusted R^2 of 0.88 (**Figure 3**B). The average number of oviposition pits chewed weekly per female was significantly affected by both temperature ($F = 23.03$, $df = 4, 80$, $p < 0.0001$) and population ($F = 40.09$, $df = 1, 80$, $p < 0.0001$), peaking at 25 °C and declining both above (Chinese population) and below those temperatures (**Table 1**).

The interaction between temperature and the population the female came from significantly ($F = 3.81$, $df = 3, 113$, $p = 0.0121$) affected the number of mature unlaid eggs found in the female at death. At temperatures between 15 and 30 °C there was no difference in the number of unlaid eggs for Italian females (**Table 1**). Chinese females had no unlaid eggs at 10 and 40 °C, and only one female had a single matured egg in her ovarioles at death at 35 °C. Chinese females at 15 °C had fewer unlaid eggs than those at 30 °C. The maximum number of unlaid mature eggs found in females from both populations was eight eggs.

3.2. Egg Hatch

Temperature had a significant effect ($F = 392.34$, $df = 4, 897$, $p < 0.0001$) on number of days to egg hatch, but population did not ($F = 0.13$, $df = 1, 897$, $p = 0.7202$). The number of days to egg hatch for the populations combined was lowest at 30 °C and increased both above and below that temperature (**Table 3**). The relationship between hatch rate (1/d) and temperature was fit to Equation (1) with good confidence ($R^2 = 0.529$, **Table 2**, **Figure 4**B). The Tmin and Tmax for hatch rate appear to be good estimates, since no hatch occurred at 5, 10, or 40 °C, although only eggs from the Chinese population were used at these temperatures (**Table 3**).

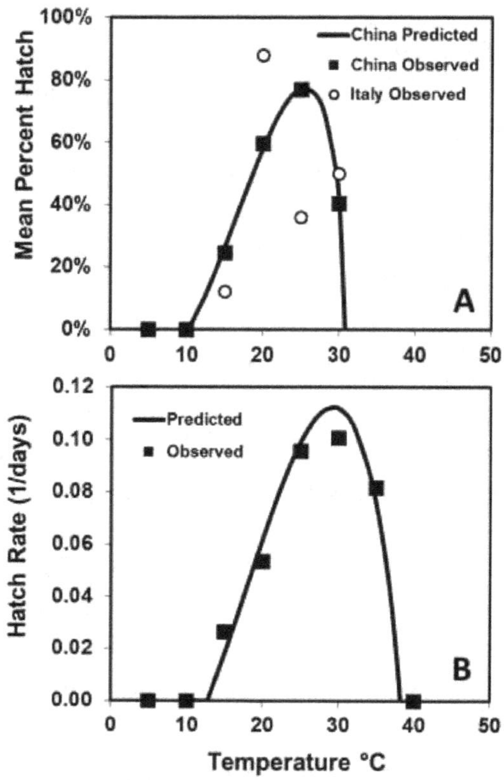

Figure 4. Relationships between temperature and mean percentage egg hatch (A) and mean hatch rate (1/d) (B) for A. chinensis. In graph A, observed points are solid squares (Chinese) and open circles (Italian), and the predicted line is only for the Chinese population. Results are pooled across populations in graph B (solid squares are mean observed points).

Table 3. Effects of temperature on Anoplophora chinensis egg hatch (mean ± SE (n)).

Parameter	Population	Temperature 5	10	15	20	25	30	35	40
Time to Egg Hatch (d)	Both	NA	NA	38.0 ± 1.2a (34)	18.7 ± 0.4b (125)	10.5 ± 0.1cd (770)	9.9 ± 0.2d (96)	12.3 ± 0.9c (6)	
% Hatch of Eggs Removed from Under the Bark	China	0.0 (51)	0.0 (90)	11.7 (291)	30.9 (369)	60.5 (1138)	27.3 (333)	5.5 (109)	0.0 (51)
	Italy	NA	NA	0.0 (57)	12.1 (91)	52.9 (155)	7.4 (68)	NA	NA

Time to egg hatch values followed by the same letter are not significantly different ($F = 392.34$, df = 4, 897, $p < 0.0001$) based on the Tukey–Kramer test with $\alpha = 0.05$ [19]. The data for the Chinese and Italian populations were pooled for time to hatch. NA = parameter not applicable, since no eggs were held or hatched at these temperatures. Some eggs laid the day the bolt was removed, or the previous day, were moved to other temperatures and held for hatch. Other eggs were held at the oviposition temperature.

The cumulative proportions of the number of days to hatch at 15 to 35 °C (populations pooled) is shown in **Figure 5**A. The first hatch occurred at 31, 9, 5, 5, and 11 days and the last at 50, 30, 19, 16, and 15 days for eggs held at 15, 20, 25, 30, and 35 °C, respectively. The cumulative proportion versus accumulated degree-days was fit to Equation (3) with high confidence (R^2 = 0.994; **Figure 5**B). Degree-day accumulations of 125.5 (95% CI 125.3–125.6) and 165.6 (95% CI 164.4–167.0), are predicted for 50% and 90% of the eggs to hatch, respectively (**Figure 5**B, 0.5 and 0.9 values).

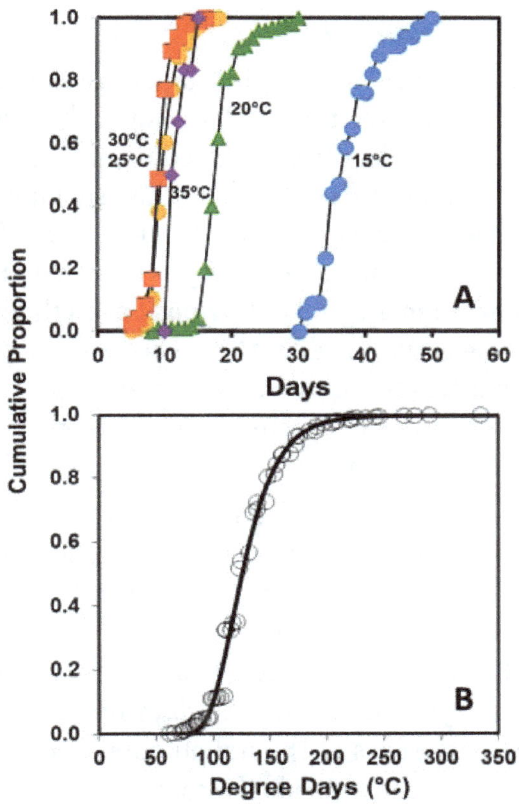

Figure 5. Observed cumulative frequency distributions for A. chinensis egg hatch times (d) at five constant temperatures (°C) (A). Cumulative proportion of A. chinensis eggs hatching over accumulated degree-days (B). The solid line in the bottom graph represents the fit to the equation: Predicted proportion hatching = exp (–exp (0.0469 (±0.00139) X accumulated degree-days + 5.5186 (±0.1683))); adjusted R^2 = 0.994.

The percentage hatch of eggs laid at or moved to temperatures for each population is given in **Table 3**. Because the percentage hatch for eggs removed from under the bark seemed low, the percentage hatch for eggs laid at different temperatures then held on bolts at 25–30 °C was used for developing the relationship between percentage hatch (as a measure of fertility) and temperature.

Percentage hatch on bolts was significantly affected by temperature ($F = 5.85$, $df = 3, 63$, $p = 0.0014$), but not population or the intersection between the two (**Table 1**). Percentage hatch was highest at 25 °C for the Chinese population and 20 °C for the Italian population and declined as temperature increased or decreased (**Table 1**, **Figure 4**A). The relationship between temperature and mean percentage hatch of Chinese eggs on bolts was fit with high confidence ($R^2 = 0.992$) by Equation (1), but no curve was fit for the Italian egg data, because the maximum likelihood estimator did not converge (**Figure 4**A, **Table 2**).

4. DISCUSSION

The observed adult activity over the range of temperatures assessed was comparable to what was seen for A. glabripennis [**22,23**]. Adults held at 15–25 °C exhibited the full range of normal behaviors (resting, feeding, mating, and ovipositing), just more slowly at cooler temperatures. At 10 °C Chinese adults moved very slowly, fed little, and did not chew oviposition sites. At 30–35 °C adult behavior increased, and they would attempt to walk or fly away when their twigs and bolts were changed. At 35 °C, Chinese females did not oviposit even though they chewed some oviposition pits. Chinese adults held at 40 °C were more lethargic, fed little, and did not live long. The attempts to move away from the higher temperatures fit with the observations that A. glabripennis seeks shaded areas to rest when the temperature is >29 °C in China [**24**].

Adult longevity declined with the increase in activity at higher temperatures. This is likely due to increased metabolism at higher temperatures which has been found to reduce the life span in insects [**25**]. The longevity at 20–25 °C on A. rubrum for both sexes from the Chinese population is comparable to that recorded on Melia azedarach (91.8 days female and 103.0 days male) and higher than recorded on Casuarina sp. (61.6 days female and 63.0 days male) at room temperature in China [**1**]. Conversely, both sexes of the Italian population on A. rubrum at 25 °C lived for a shorter time than did individuals from a Japanese population at 24 °C on either Citrus unshiu (Yu.Tanaka ex Swingle) or Salix schwerinii E. Wolf [**15**]. This highlights the effects of host plants on longevity and suggests that caution should be used when applying predictions solely based on a single host in the laboratory. Female A. chinensis from both populations lived longer than A. glabripennis females at temperatures ≤15 °C, but comparable lengths of time at higher temperatures [**23**]. However, the male longevity of individuals from the Chinese population was comparable to that of A. glabripennis individuals over the 10–35 °C range [**23**]. The longer survival of female A. chinensis than A. glabripennis at cooler temperatures is surprising, since the native range of A. glabripennis extends further north than that of A. chinensis, but fits with the fact that A. glabripennis has a higher metabolic rate than A. chinensis at cold temperatures [**17**].

The timing of first oviposition (13–17 days to first egg at 24 °C depending on the host) and the shape of the weekly oviposition curve (bell-shaped with a long right tail) are comparable to the findings of this study for A. chinensis [**1**,**7**]. The mean fecundity (92 for Chinese and 24 for Italian at 25 °C), however, was lower than that of either Chinese females on M. azedarach (176 eggs) or field-reared Japanese females on citrus (193.8 eggs) [**1**,**7**]. This is most likely due to the host used in the studies, since the host has been shown to affect fecundity in Anoplophora spp. [**15**,**26**]. The increasing number of oviposition sites chewed per week, maximum oviposition per week, and fecundity for females from the Chinese population with increasing temperatures 15–30 °C was to be expected because of increased efficiencies of assimilation and conversion of ingested food at higher temperatures [**27**]. What was not expected was a lack of temperature response for reproductive output for the Italian population, and the causes for this are unknown. Compared to A. glabripennis, the A. chinensis Chinese females laid more eggs across the entire temperature range of 15–30 °C, and fecundity peaked at a higher temperature [**23**]. The higher fecundity could have been because A. chinensis were, on average, heavier than A. glabripennis females, and fecundity has been correlated with size in both species [**7**,**28**]. The ability of A. chinensis to handle higher temperatures better than A. glabripennis is also suggested by its native range extending further south, even though much of their ranges overlap [**2**]. At the more extreme temperatures, the pattern of reproductive arrest seen for A. chinensis is comparable to that found for Hylotrupes bajulus L. [**29**] and A. glabripennis [**28**]. At 10 °C and ≥35 °C, females laid no eggs, and there were no fully developed eggs in their ovarioles at death. This was either due to a lack of egg maturation or eggs being reabsorbed within these females to use as an energy source.

Eggs of most insect species will hatch over a 15 °C range of constant temperatures, and under varying temperatures like occur in nature, egg hatch may occur 3–4 °C below that range [**30**]. As temperatures approach (within 2–3 °C), the Tmax and Tmin, percentage hatch declines dramatically [**30**]. Anoplophora chinensis eggs from the Chinese population appear to be an exception, as eggs hatched over a broader range of constant temperatures (15–35 °C), although the predicted Tmin and Tmax were only 3 °C below the temperatures at which hatch percentage declined sharply. However, the cold tolerance of A. chinensis eggs seems to be less than that of A. glabripennis, since none of the A. chinensis eggs that were held at 5–10 °C for 200 days and then were moved to 25 °C were able to hatch, while eggs of A. glabripennis did hatch under similar conditions. This would suggest that A. chinensis eggs would not be able to overwinter, unless the more sheltered tree bole and roots location they use provides more protection from colder temperatures than is afforded A. glabripennis eggs laid under the bark up higher on the tree. However, care should be taken in making this assumption, since removing the eggs from the bark can damage A. chinensis eggs, and we did not confirm this by holding eggs under the bark at the low temperatures.

The percentage of Italian eggs laid at 20 °C and hatched at 25 °C was similar to that for Japanese A. chinensis eggs laid and hatched at 20 °C in Citrus sp. Bolts (90%) [7]. The lower hatch percentages seen for eggs removed from bolts was probably due to damage sustained when removing them, since there were reductions for all temperatures, and the eggshells of A. chinensis broke more easily than those of A. glabripennis. The percentage hatch of the Chinese population eggs kept on bolts was comparable to A. glabripennis egg hatch when removed from bolts at 15–30 °C. However, unlike A. glabripennis, some A. chinensis eggs that were removed from bolts were able to hatch at 35 °C, possibly another indication that A. chinensis may be better adapted for warm climates than A. glabripennis.

The predicted number of degree-days to 50% hatch of A. chinensis eggs was 166 DD which is comparable to previous estimates for Japanese A. chinensis (184 DD) [7], but considerably less than the 239 DD required by A. glabripennis [**23**]. The Tmin of 12.7 °C is higher than the 6.7 °C estimated for Japanese A. chinensis and slightly higher than the 9.7 °C estimated for A. glabripennis [**7,23**]. In addition, the Tmax is quite a bit higher for A. chinensis than A. glabripennis. The lack of hatch for A. chinensis eggs held at 10 °C substantiates the Tmin estimated in this study and suggests that either the previous estimate was low or that some Japanese populations of A. chinensis may have a lower threshold.

Diurnal and seasonal temperature variations in the lower bole and exposed roots of trees where A. chinensis eggs are laid could be substantially different from ambient or from tree trunks and branches where A. glabripennis eggs are laid. Previous studies have shown that phloem temperatures in the upper bole and branches can be much higher than ambient (averaging 2–10 °C higher and maximum 21 °C higher), especially when the tree section is exposed to direct sunlight [**31,32**]. However, other factors can also alter within-tree temperatures, such as reflected and thermal radiation, wind chills, and soil temperature around the roots. In fact, the deeper in the trunk or under soil, the more insulated the wood is, and the slower the temperatures may respond to ambient changes [**32**]. When using the estimated DD and Tmin for predicting when A. chinensis eggs will hatch, ambient air temperatures will likely not provide accurate estimates. Further work to determine exactly how air temperature differs and temperature changes diurnally and seasonally in the root zone of trees is needed to accurately predict egg hatch timing and potential effects on percentage hatch for A. chinensis.

The impact of temperatures on the adults can be seen if the temperatures from the most northern and southern invasive A. chinensis populations in Europe are compared. Monthly mean temperatures in Rome, Italy are 17, 21, 23, 24, 21, and 17

°C during May, June, July, August, September, and October, respectively, while the corresponding temperatures in Rotterdam, Netherlands were 12, 15, 17, 17, 14, and 11 °C, respectively (**https://weatherbase.com**, accessed 2 April 2021). For Rome, Italy, during the same time period, the maximum temperatures averaged 22, 26, 29, 29, 26, and 22 °C, the maximum recorded temperature was 37 °C for July-September, and there are, on average, four days a year that the temperature exceeds 32 °C. Thus, Rome experiences temperatures at which the beetles can survive and reproduce well from June to August, and summer temperatures rarely reach >32 °C, so it should not reduce adult survival or suspend oviposition. Average temperatures in Rotterdam, however, barely reach levels at which beetles will oviposit June–August, and even high averages only reach ≥20 °C in July and August—so although the beetles may survive a long time, they will lay few eggs. This may partially explain the total of 25 exit holes seen in the Westland, Netherlands, infestation and the relative ease of eradicating that infestation, while the infestation in Rome is larger and more difficult to eradicate [**11**]. Average summer temperatures in major citrus-growing areas in California like Visalia would be near-optimal, but average highs and maximum recorded temperatures could suspend oviposition and reduce adult longevity part of the time. Citrus-growing areas in Florida (e.g., Bartow) would be near-optimum for beetles, and even the maximum recorded temperatures would rarely reach the Tmax for adult survival or result in the suspension of oviposition. These data on temperature effects on reproduction and survival of adults and eggs of A. chinensis, if combined with temperature response data for larvae, could be used to develop maps of the potential geographical range for this species and stage-based phenological models to predict the timing of life stages. Such information is critical for informing management and eradication programs, both in the native and introduced ranges.

5. CONCISE SUMMARY

The present study found that temperature had a substantial influence on all the life-history characteristics of the Chinese population and most of the Italian population characteristics, except for fecundity and mean maximum oviposition per week. The impact of temperature on the lifespan of both genders varied throughout the populations, with Chinese folks exhibiting greater longevity compared to Italian individuals. In addition, adults belonging to the Chinese community had more weight compared to individuals from the Italian population. There were variations in the impact of temperature on lifespan between males and females. Females from the Italian population had a lower rate of oviposition and progeny production compared to females from the Chinese population. There were no discernible disparities observed in the time it took for the first oviposition or the rate at which eggs hatched across the range of temperatures that were examined.

The mathematical correlations between temperature and the life-history factors yielded estimations of the minimum and/or maximum temperature thresholds as well as the ideal temperatures for several of the assessed qualities. These links will enhance the creation of phenology models and enhance estimates of prospective geographic ranges to assist managers in implementing control and eradication efforts.

6. AT A GLANCE

Anoplophora chinensis, sometimes known as the Forster beetle, is a highly invasive species that poses a threat to several tree species found in orchards, urban areas, and forests. The study assesses the adult survival, reproduction, and egg hatch of A. chinensis from Italy and China under controlled laboratory circumstances. The evaluation is conducted at eight consistent temperatures ranging from 5 to 40 °C. The projected maximum temperature (Tmax) for achieving lifespan was 42 °C for females and 33 °C for males from China, whereas for males from Italy it was 42 °C and for females it was 39 °C. The estimated maximum temperature (Tmax), minimum temperature (Tmin), and temperature at which fecundity is highest (optimum temperature) were 35, 9, and 29 °C, respectively. The females deposited eggs within a temperature range of 15–30 °C, and the eggs successfully hatched within a temperature range of 15–35 °C. The time it took for the first oviposition to occur rose significantly from 13 days at a temperature of 30 °C to over 300 days at a temperature close to 10 °C. The projected minimum temperature for egg hatch was 13 °C, the maximum temperature at 38 °C, and the ideal temperature was 29 °C. The highest estimated percentage of hatch occurred at a temperature of 26 °C, with a maximum temperature (Tmax) of 31 °C and a minimum temperature (Tmin) of 10 °C. The findings suggest that beetles can survive and reproduce in summer temperatures across many latitudes. However, when temperatures reach or exceed 35 °C, the ability to lay eggs stops and the survival rate of adult beetles decreases. Furthermore, female individuals have the ability to persist until the autumn season, albeit with a reduced reproductive output characterized by a diminished number of eggs that may not successfully undergo the hatching process. The temperature responses of A. chinensis can be utilized to construct phenological models that can accurately anticipate the timing of stages for management or eradication purposes.

REFERENCES

1. Huang, J.; He, X.; Ding, B.; Xu, D.; Gao, M. Study on development of pupa and adult supplement nutrient of Anoplophora chinensis. Sci. Silvae Sin. 2003, 39, 148–152. [**Google Scholar**]
2. Lingafelter, S.W.; Hoebeke, E.R. Revision of Anoplophora (Coleoptera: Cerambycidae); Entomological Society of Washington: Washignton, DC, USA, 2002; p. 236. [**Google Scholar**]
3. Haack, R.A.; Herard, F.; Sun, J.H.; Turgeon, J.J. Managing invasive populations of Asian longhorned beetle and citrus longhorned beetle: A worldwide perspective. Annu. Rev.

Entomol. 2010, 55, 521–546. **[Google Scholar]** **[CrossRef]** **[Green Version]**

4. Sjöman, H.; Östberg, J.; Nilsson, J. Review of host trees for the wood-boring pests Anoplophora glabripennis and Anoplophora chinensis: An urban forest perspective. Arboric. Urban For. 2014, 40, 143–164. **[Google Scholar]**
5. Ge, X.Z.; Zong, S.X.; He, S.Y.; Liu, Y.T.; Kong, X.Q. Areas of China predicted to have a suitable climate for Anoplophora chinensis under a climate-warming scenario. Entomol. Exp. Appl. 2014, 153, 256–265. **[Google Scholar]** **[CrossRef]**
6. Cai, S.; He, X.; Gurr, G.M.; Zeng, L.; Huang, J.; Ji, B. Artificial diet and temperature regimes for successful rearing of the citrus longhorned beetle, Anoplophora chinensis (Coleoptera: Cerambycidae). Linye Kexue Sci. Silvae Sin. 2016, 52, 141–149. **[Google Scholar]**
7. Adachi, I. Reproductive-biology of the white-spotted longicorn beetle, Anoplophora malasiaca Thomson (Coleoptera, Cerambycidae), in citrus trees. Appl. Entomol. Zool. 1988, 23, 256–264. **[Google Scholar]** **[CrossRef]** **[Green Version]**
8. Hérard, F.; Ciampitti, M.; Maspero, M.; Krehan, H.; Benker, U.; Boegel, C.; Schrage, R.; Bouhot-Delduc, L.; Bialooki, P. Anoplophora species in Europe: Infestations and management processes. EPPO Bull. 2006, 36, 470–474. **[Google Scholar]** **[CrossRef]**
9. Ohbayashi, N.; Ogawa, J.; Su, Z.-H. Phylogenetic analysis of the lamiine genus Anoplophora and its relatives (Coleoptera, Cerambycidae) based on the mitochondrial COI gene. Spec. Bull. Jpn. Soc. Coleopterol. Tokyo 2009, 7, 309–324. **[Google Scholar]**
10. Iwaizumi, R.; Arimoto, M.; Kurauchi, T. A study on the occurance and fecundity of white spotted longicorn, Anoplophora malasiaca (Coleoptera: Cerambycidae). Res. Bull. Plant Prot. Jpn. 2014, 50, 9–15. **[Google Scholar]**
11. Herard, F.; Maspero, M. History of discoveries and management of the citrus longhorned beetle, Anoplophora chinensis, in Europe. J. Pest Sci. 2019, 92, 117–130. **[Google Scholar]** **[CrossRef]**
12. Eschen, R.; Gregoire, J.C.; Hengeveld, G.M.; de Hoop, B.M.; Rigaux, L.; Potting, R.P.J. Trade patterns of the tree nursery industry in Europe and changes following findings of citrus longhorn beetle, Anoplophora chinensis Forster. Neobiota 2015, 26, 1–20. **[Google Scholar]** **[CrossRef]** **[Green Version]**
13. Chambers, B. Citrus Longhorned Beetle Program, King County, Washington: Environmental Assessment. April 2002. Available online: **https://www.aphis.usda.gov/plant_health/ea/downloads/clb.pdf** (accessed on 2 April 2021).
14. Strangi, A.; Binazzi, F.; Peverieri, G.S.; Roversi, P.F. The Anoplophora chinensis (Forster) (Coleoptera Cerambycidae lamiinae) outbreaks in Italy: A possible geographical origin. Redia 2017, 100, 175–179. **[Google Scholar]**
15. Fujiwara-Tsujii, N.; Yasui, H.; Tanaka, S. Comparison of fecundity and longevity of Anoplophora malasiaca (Coleoptera: Cerambycidae) adults fed on three different host-plants. Entomol. Sci. 2016, 19, 201–206. **[Google Scholar]** **[CrossRef]**
16. Keena, M.A. Pourable artificial diet for rearing Anoplophora glabripennis (Coleoptera: Cerambycidae) and methods to optimize larval survival and synchronize development. Ann. Entomol. Soc. Am. 2005, 98, 536–547. **[Google Scholar]** **[CrossRef]** **[Green Version]**
17. Javal, M.; Roques, A.; Roux, G.; Laparie, M. Respiration-based monitoring of metabolic rate following cold-exposure in two invasive Anoplophora species depending on acclimation

regime. Comp. Biochem. Physiol. A Mol. Integr. Physiol. 2018, 216, 20–27. [**Google Scholar**] [**CrossRef**] [**PubMed**]

18. Dubois, T.; Hajek, A.E.; Smith, S. Methods for rearing the Asian longhorned beetle (Coleoptera: Cerambycidae) on artificial diet. Ann. Entomol. Soc. Am. 2002, 95, 223–230. [**Google Scholar**] [**CrossRef**]
19. SAS_Institute. Sas/Stat User's Guide, Version 9.4; SAS Institute: Cary, NC, USA, 2015. [**Google Scholar**]
20. Briere, J.F.; Pracros, P.; Le Roux, A.Y.; Pierre, J.S. A novel rate model of temperature-dependent development for arthropods. Environ. Entomol. 1999, 28, 22–29. [**Google Scholar**] [**CrossRef**]
21. Brown, R.F.; Mayer, D.G. Representing cumulative germination.2. The use of the weibull function and other empirically derived curves. Ann. Bot. 1988, 61, 127–138. [**Google Scholar**] [**CrossRef**]
22. Zhou, J.; Zhang, K.; Lu, Y. Study on adult activity and behavioral mechanism of Anoplophora nobilis Ganglbauer. Sci. Silvae Sin. 1984, 20, 372–379. [**Google Scholar**]
23. Keena, M.A. Effects of temperature on Anoplophora glabripennis (Coleoptera: Cerambycidae) adult survival, reproduction, and egg hatch. Environ. Entomol. 2006, 35, 912–921. [**Google Scholar**] [**CrossRef**] [**Green Version**]
24. Xiao, G. Forest Insects of China; Forest Research Institute, Chinese Academy of Forestry: Beijing, China, 1980. [**Google Scholar**]
25. Slansky, F.; Scriber, J.M. Food consumption and utilization. In Comprehensive Insect Physiology, Biochemistry, and Pharmacology; Kerkut, A., Gilbert, L.I., Eds.; Pergamon: Oxford, UK, 1985; Volume 4, pp. 87–163. [**Google Scholar**]
26. Smith, M.T.; Bancroft, J.; Tropp, J. Age-specific fecundity of Anoplophora glabripennis (Coleoptera: Cerambycidae) on three tree species infested in the United States. Environ. Entomol. 2002, 31, 76–83. [**Google Scholar**] [**CrossRef**] [**Green Version**]
27. Haack, R.A.; Slansky, F. Nutritional ecology of wood-feeding coleoptera, lepidoptera, and hymenoptera. In Nutritional Ecology of Insects, Mites, and Spiders; Slansky, F., Rodriguez, J.G., Eds.; John Wiley and Sons, Inc.: New York, NY, USA, 1987; pp. 449–486. [**Google Scholar**]
28. Keena, M.A. Anoplophora glabripennis (Coleoptera: Cerambycidae) fecundity and longevity under laboratory conditions: Comparison of populations from New York and Illinois on Acer saccharum. Environ. Entomol. 2002, 31, 490–498. [**Google Scholar**] [**CrossRef**] [**Green Version**]
29. Dürr, H.J.R. The Morphology and Bionomics of the European Houseborer, Hylotrupes bajulus (Coleoptera: Cerambycidae); The Government Printer Pretoria, Republic of South Africa: Pretoria, South Africa, 1957; Volume 4.
30. Howe, R.W. Temperature effects on embryonic development in insects. Annu. Rev. Entomol. 1967, 12, 15–42. [**Google Scholar**] [**CrossRef**] [**PubMed**]
31. Bolstad, P.V.; Swift, L.; Collins, F.; Regniere, J. Measured and predicted air temperatures at basin to regional scales in the southern Appalachian mountains. Agric. For. Meteorol. 1998, 91, 161–176. [**Google Scholar**] [**CrossRef**]
32. Derby, R.W.; Gates, D.M. Temperature of tree trunks-calculated and observed. Am. J. Bot. 1966, 53, 580–587. [**Google Scholar**] [**CrossRef**]

07 ASSESSING GYPSY MOTH FEMALE FLIGHT ABILITIES: EXAMINING POTENTIAL VARIATIONS AND DETERMINATION TECHNIQUES

1. INTRODUCTION: AN OVERVIEW

The gypsy moth, scientifically known as Lymantria dispar (Lepidoptera: Lymantriidae), is a forest bug that may inflict significant damage to ecosystems and individual trees by repeatedly stripping them of their leaves [1,2]. This species is also responsible for enormous economic losses due to the resulting damage [2]. According to Liebhold et al. [3], the caterpillars of gypsy moth populations in North America have the ability to consume more than 300 shrubs and tree types, with around 100 being their favorite choices. Moreover, Keena and Richards [4] discovered that gypsy moths from Europe and Asia were capable of colonizing North American tree species without any previous contact with the host plants. This poses a significant risk to temperate forests worldwide.

There are three different types of gypsy moth that have been identified: the European gypsy moth (Lymantria dispar dispar), the Asian gypsy moth (Lymantria dispar asiatica), and the Japanese gypsy moth (Lymantria dispar japonica) [5,6]. The European gypsy moth (EGM) is indigenous to Eurasia, ranging from Portugal in the west to the Ural Mountains in the east [7]. The introduction of this subspecies into North America occurred in 1869, and it has since become well-established in numerous forests across the Eastern United States and Canada [8,9]. The Asian gypsy moth (AGM) is naturally found in a region that stretches from the Ural Mountains in the east, through Russia, China, Korea, and into certain parts of Japan. On the other hand, the Japanese gypsy moth (JGM) is only found on the main islands of Japan and the Kuril Islands of Russia [5].

The European gypsy moth (EGM) was initially brought into the northeastern state of Massachusetts, in the United States, in 1869 by a French mathematician with the purpose of conducting breeding studies with a native silkworm. At this moment, gypsy moth larvae managed to flee and successfully settled in the nearby region. Two decades after its initial establishment, a significant epidemic took place, capturing national attention. Despite extensive eradication efforts conducted by local and federal organizations, the gypsy moth persisted in spreading throughout the forests of the Eastern United States and Canada during the course of the following century [10]. In the United States, the rate at which EGM spread was on average 9.45 km/year from 1900 to 1915, 2.82 km/year from 1916 to 1965, and 20.78 km/year from 1966 to 1990 [11]. According to McManus and Csoka [10], the higher rate of expansion in the latter half of the 19th century may be attributed to three factors: (1) a decline in the utilization of DDT and other chemical pesticides; (2) the presence of adjacent forests primarily consisting of oak species (Quercus spp.), which are the preferred hosts of the EGM; and (3) significant outbreaks in the 1970s and early 1980s that intensified the insect's spread. Following the discovery of the significance of stratified dispersal in the propagation of EGM [11], a barrier-zone management initiative known as "Slow the Spread" was implemented, resulting in a decrease in the yearly spread rate to less than 4 km [12]. In addition, outbreaks in established areas are currently more effectively controlled through a density-dependent mechanism by two diseases that specifically target the hosts. These pathogens are the fungus Entomophaga maimaiga Humber, Shimazu, and Soper (belonging to the Zygomycetes order: Entomophthorales) and a nucleopolyhedrosis virus known as LdNPV [13,14].

Furthermore, the most efficient means of managing gypsy moth infestations in Europe is now through the airborne dispersal of Bacillus thuringiensis Berliner serovar kurstaki (Btk) [15]. Gypsy moth populations in established locations typically have outbreaks at intervals of around five to ten years [16], which typically persist for a duration of one to three years [17,18]. The longest durations of outbreaks were documented as 12.5 and 13.0 years in Poland and the Czech Republic, respectively [19]. The incidence of the epidemic declines in a gradient from the northwest to the southeast regions of Europe. In Central Europe, the time is roughly 12 years, whereas in the Mediterranean and Balkan regions, it ranges from seven to eight years [20]. A study conducted by Hlasny et al. [21] discovered a significant correlation between epidemic times and geographical gradients. The investigation also indicated considerable spatial and temporal variations in the outbreak patterns in the areas surrounding the Carpathian Mountains.

The invasion of EGM has ecological repercussions. Typically, the EGM shows a tendency to favor specific tree species during outbreaks. These tree species experience higher rates of defoliation compared to species that are exclusively devoured by certain larval stages or species that are rarely, if ever, ingested by any larval stage. After losing their leaves, the weakened trees become more vulnerable to other factors, such as boring insects and parasitic fungi, which eventually lead to the death of the trees [10]. As a result, EGM has led to a reduction in the prevalence of their favored host species, particularly oaks, in forests of Eastern North America. This might potentially lead to a lasting change in the types of species present in these forests [23].

The EGM invasion of North America is also linked to significant economic expenses. The costs often arise from the subsequent origins: The economic impacts of tree mortality or reduced growth include the loss of timber value, decreased residential property value, household expenses for tree care, local government costs for tree removal and replacement or treatment, and federal government expenditures for research, regulation, management, and outreach. Aukema et al. [24] calculated that the EGM (Emerald Ash Borer) caused $254 million in annual damages, with a significant share ($120 million) attributed to the decline in residential property values.

The Annual General Meeting (AGM) was not brought into North America until a far later time. The initial documented introduction took place in 1991, in close proximity to the Vancouver port, located in British Columbia, Canada [25]. Subsequently, separate Annual General Meetings (AGMs) were found in the United States, specifically in Washington and Oregon, as well as in other regions of British Columbia. Ships originating from Eastern Russia are believed to have carried AGM larvae to ports located around the Pacific Northwest Coast. The infestations were swiftly eliminated with the use of treatment and trapping methods. However, after the first introduction, AGM has been identified and eliminated on at least 20 different instances across the United States [25]. In 2020, a campaign was implemented in Western Washington State to eliminate recently discovered instances of EGM with AGM genetic features, as well as the Hokkaido gypsy moth (Lymantria umbrosa Butler) [26].

If Asian Giant Hornets (AGMs) were to settle in North America, there is a strong likelihood that the resulting economic and ecological harm would surpass that of the European Gypsy Moth (EGM) for various reasons (Figure 1). While both male AGM and EGM can fly, only AGM females possess the ability to fly, whilst EGM females do not [7]. Research has additionally discovered that Asian and Japanese female gypsy moths exhibit a preference for light sources, resulting in the frequent deposition of eggs on ships and automobiles [27,28]. Furthermore, AGMs exhibit a broad geographical distribution that includes diverse forest environments. Introducing AGMs from different origins could result in progeny with a broader range of hosts compared to the original populations [4]. Gypsy moth populations exhibit variations in both the larval and egg responses to temperature. The necessary degree of coldness for an egg mass to hatch differs among populations of all subspecies. Therefore, the introduction of genotypes that have a lower requirement for chilling into areas that are already infested could lead to an increase in variance in the hatching of eggs within the population. This could also result in the development of a strain that does not undergo diapause in locations that are not yet infested but have moderate temperatures [29]. Certain AGM larvae exhibit a prolonged development period and reach larger sizes, perhaps resulting in more extensive defoliation even at lower population densities compared to EGM larvae [30]. Nevertheless, when the duration of growth and development extends, there is a greater likelihood for population dynamics to be influenced by natural adversaries and various other circumstances.

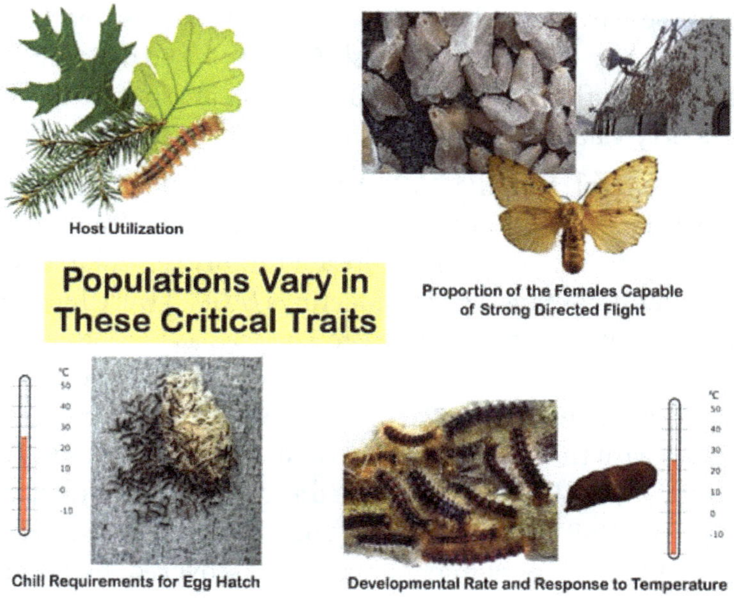

Figure 1. Gypsy moth populations vary in several key traits that can alter the dispersal capabilities or their potential to cause damage in novel habitats.

The flight ability and flight distance of female gypsy moths are crucial criteria in estimating their spread. Although existing spread and distribution modelling approaches have relied on generic flight distance estimations [31], incorporating more particular information might enhance the accuracy of future projections regarding possible dispersion and spread. Nevertheless, there is no clear demarcation indicating the flying capacity of individual subspecies of gypsy moths. Although separating gypsy moths into three subspecies can be beneficial for descriptive purposes, it is important to note that features can be transferred between subspecies through hybridization [32]. Therefore, it is crucial to analyze the ability to fly within specific populations rather than restricting the investigation to a subspecies level. The main goal of this communication is to record the existing knowledge about the flight ability and prospective flight distances of female gypsy moths from various populations. In addition, we consider this information on flight capability in relation to other crucial elements in a program aimed at avoiding the colonization and dispersal of female organisms capable of flight in new environments. This involves examining the present capacity to differentiate populations with females capable of flying and analyzing the impact of female flight on the population dynamics (establishment and expansion) in new environments.

2. GYPSY MOTH POPULATION DYNAMICS (ESTABLISHMENT AND SPREAD) AND POTENTIAL IMPACTS OF FEMALE FLIGHT

The invasion of gypsy moths is extensively researched and well-documented in North America [33]. Novel introductions and establishments are frequently tracked using comprehensive surveys. This data is then utilized to gain insights into population dynamics and other characteristics that influence establishment, such as dispersal capacity, environmental tolerance, and host preferences [9].

The population dynamics of gypsy moths are intricate since there are several natural enemies that control population density or decrease outbreak populations [34]. The populations of gypsy moths are mostly controlled by two distinct entomopathogens, namely Entomophaga maimaiga and the nucleopolyhedrosis virus LdMNPV, soon following outbreaks [35]. Outbreaks have been documented in regions where the moth's chosen host species make up at least 20% of the total species present [36]. Nevertheless, there are several additional crucial elements that play a role in shaping the population dynamics of the species. Land usage is one of the contributing elements. A study conducted by Lentini et al. [37] revealed that outbreaks tend to occur more frequently in forests and woods that have been overexploited, with a canopy cover rate of less than 25%, compared to natural ones. In addition, the population dynamics of gypsy moths may be influenced by management methods [15]. In a study conducted by Mannu et al. [15], it was discovered that applying Btk (Bacillus thuringiensis var. kurstaki) early during the population growth of gypsy moths resulted in a shift in the peak abundance of the moths. This shift caused the outbreak peak to be delayed by two to three years. On the other hand, applying Btk during the culmination phases of the moths was found to be more successful in suppressing their populations.

Recent research has revealed that the establishment and distribution patterns of gypsy moths vary significantly at the forefront of their range [33,38]. Gray et al. [34] examined the correlation between gypsy moths, forest composition, stand architecture, and natural enemies. The researchers discovered that the arrangement of the habitat did not have an impact on the populations of natural enemies. Additionally, they observed that the period before an eruption was distinct from the period before establishment, mostly because of a high rate of successful mating. Furthermore, they observed a significant level of mating success and adult female survival during the pre-eruptive period in all of the locations and habitat structures. This suggests that when population densities are very low before foundation, there may be difficulty in finding mates, leading to the Allee effect. The Allee effect is a phenomenon where individual fitness is positively related to population density. This effect may operate as a limiting factor in the establishment of freshly created gypsy moth populations. In this situation, a member of a species that is affected by the Allee effect will undergo a decline in its overall reproductive success when the population density is low. When it comes to invasive species, if they are introduced at a low density, there is a chance that Allee effects will affect how the invasion progresses. This could result in longer delays, slower spread, and a lesser likelihood of establishing a population [40].

Mate-finding failures, which contribute to the Allee effect, play a significant role in the dispersal and establishment of gypsy moths in populations of flightless females in North America [41,42]. Recent research has revealed that both Allee effects and stratified diffusion contribute to the spreading of gypsy moths into new locations. Consequently, strategies derived from these results have been implemented at the forefront to impede the expansion of their population. Gypsy moth colonies that are geographically separated and have females who cannot fly are susceptible to Allee effects, which can lead to their extinction due to their low population size. Mating success appears to be a significant density-dependent element that influences sexually active sporadic populations [43]. Sharov et al. [44] discovered a considerably greater probability of mating in populations with higher densities. The fecundity of mature gypsy moths declines significantly after two days, allowing only a narrow window of opportunity for mating before death [45]. Therefore, the impact of female mobility on low-density populations may vary.

The presence of flight-capable females may reduce the significance of the Allee effect. Females capable of flight are attracted to lights, such as the moon or artificial light sources, or bright-colored surfaces, in order to deposit their eggs. Unmated females had a lower frequency of flight compared to mated females, as they prioritize calling for a mate first [46] (Figure 2). Nevertheless, unmated virgins will eventually relocate in order to attract a partner. Given that both genders are drawn to illumination, female individuals that have not yet mated may enhance their likelihood of locating a partner by flying towards adjacent sources of light, where males will also gather [27]. After mating, female insects take flight and gather around light sources in order to deposit their eggs, resulting in the formation of localized groups of individuals capable of flying. These concentrations may exist either within an area that is already infested with a mixture of people from different flights, or in areas that are located ahead of the normally afflicted region, if they are traveling into new habitats. These nearby accumulations could merge and accelerate the rate of growth, similar to how small populations ahead of the advancing front in flightless female populations enhance the speed of invasion into new territories. However, in the case of females capable of flight, the number of these minor populations located outside the main region of infestation may be greater, and they could be situated farther away from the forefront of the infestation. There is already evidence of this phenomenon occurring. It was shown that female flight capability was maintained at higher levels, even up to 70%, compared to random mating within European populations where both flighted and non-flighted types coexisted. When individuals from populations with females capable of flying mate with individuals from populations where females cannot fly, the first-generation females can only glide downwards. Additionally, only approximately 15% of the female offspring resulting from matings between these hybrids are capable of powerful and purposeful flight [32].

Figure 2. Schematic showing the sequence of mate finding, mating, female flight, and oviposition for populations with flight-capable or flightless females. Note that flight-capable virgin females will fly after calling for a few hours, when no mating occurs, but they will resume calling in the new location. Strong directed female flight to lights or light-colored objects occurs after mating, and females will congregate, in order to oviposit.

In addition to the Allee effect, climate suitability is also an important factor in determining the establishment probability of gypsy moths when suitable hosts are present [31]. In the past, various authors have used phenology (Gypsy Moth Life

Stage model) and climate matching models (MaxEnt) to assess the establishment probability of gypsy moths [**31,47**]. These studies show that the potential southern margin of gypsy moth spread in the USA will be limited by insufficient chilling that terminates diapause, while exposure to extreme cold temperatures will result in egg mortality in the northern regions of North America [**48**]. Moreover, cold winter temperatures were found to be associated with reduced summer trap captures of EGMs [**49**].

Interestingly, May et al. [**50**] found no evidence of metabolic adaptation to colder environments based on their comparisons between northern and southern populations of North America gypsy moths. Moreover, Tobin et al. [**51**] found that the average development rates of North American gypsy moths were faster in areas experiencing higher temperatures despite larvae being exposed to repeated supraoptimal temperatures. Here, Tobin et al. [**51**] provide evidence of ongoing gypsy moth southern range margin retraction and suggest higher optimal temperatures during larval and pupal development period to be associated with gypsy moth range margin retraction. The authors further suggest that higher temperatures could exceed the positive effects on gypsy moth population growth. Moreover, recent works by Faske et al. [**52**] and Banahene et al. [**53**] provide evidence of local physiological adaptation under warmer climates and occurrence of strong temperature-dependent selective pressures in the southern populations of gypsy moth, respectively.

Just like their EGM counterparts, the larvae and pupae of gypsy moth from Asian origins with flight-capable females may struggle to survive and develop in regions that experience extended periods of temperatures ≥ 30 °C [**30**]. The developmental rates and responses to temperature of many AGM populations are very similar to those of EGM populations, with the exception of the most slowly developing ones from the L. dispar japonica subspecies [**30**]. In addition, the variation between AGM populations in both host utilization and egg chill requirements is already present between EGM populations [**4,29**], so establishment potential will depend in part on the traits of the source AGM population and the climate and hosts present in the area where it is introduced. AGM populations should be able to establish in most cases where EGM can establish, based on modeling that has been done [**54**].

Environmental cues may also impact flight behavior in gypsy moths. Temperature and light intensity may play important roles in determining flight propensity [**55**]. Iwaizumi and Arakawa [**56**] collected egg masses from four different localities in Japan: Chiba, Kobe, Hachinohe, and Tomakomai. From the reared larvae, female flight duration under yellow and black light sources was examined. The authors observed yellow fluorescent light that cut out ultraviolet radiation had a suppressive effect on female flight. The relatively high light intensity of 10 lux most likely suppressed the nocturnal flight behavior of the tested females. Iwaizumi and Arakawa [56] found that females from three of the populations flew an average of 200 m, while females from the Tomakomai population had a mean flight distance of roughly 600 m.

Chen et al. [57] measured the flight propensity of female AGM from seven different populations under three combinations of light intensities. Three of the populations originated in China, two from Russia, and one each from Japan and South Korea. Seventy-two percent of all females tested had the ability of directed flight. The authors also found that the range of light intensities used within the study (0.05, 0.10, and 0.40 lux) did not have a significant effect on flight propensity. This agrees with research by Charlton et al. [55] that found female gypsy moth flight to begin when light intensity falls below 2 lux.

3. VARIATION IN FEMALE FLIGHT CAPABILITY AND FLIGHT DISTANCES

3.1. Female Flight Capability

Keena et al. [46] investigated the propensity and capability of flight in female gypsy moths from populations established in Russia and North America, as well as the flight capability of hybrids bred from these two populations. While the majority of female moths from the Russian populations were able to fly, females gathered from the population established in the United States were not flight capable. To further analyze how hybridization could affect flight capability, the authors bred four new populations and assessed flight abilities. Parents used during breeding, as well as the percentage of offspring capable of sustained flight (FL) and those capable of gliding flight (GF), are shown below in **Table 1.**

Table 1. Origin and flight-capacity percentages of North American, Russian, and hybrid gypsy moths [18].

Origin		Sustained Flight	Gliding Flight
Female	Male		
Russia	Russia	88%	11%
Russia	North America	0%	51%
North America	Russia	2%	65%
North America	North America	0%	0%

Gypsy moths with both parents from North America were not capable of sustained or gliding flight. However, over 50% of the crosses of North American and Russian populations were able to glide. This suggests that if individuals from flight-capable populations become established in North America and mate with the currently established gypsy moth populations, it is likely that the first-generation offspring would have substantially improved gliding abilities. The possibility of sustained flight amongst hybrids is less established. While only 2% of the Russian-North American crosses were able to fly within this study, a separate investigation found that 8% of captive hybrid females were able to attain sustained flight if they were fed foliage, while no hybrids were flight-capable when reared on an artificial diet [**58**]. This suggests that female flight capacity of gypsy moths of all strains may be underestimated in studies if an artificial diet is used.

Female gypsy moth flight capacity has been further explored in a study that cross-bred gypsy moths that have established in North Carolina, USA, that are incapable of flight with individuals from Mineralni, Russia, that have a high capacity for ascending flight [**32**]. The authors compared the parents, reciprocal F1 hybrids, double reciprocal F2 hybrids, and all potential backcrosses. Heritability of female flight was found to be at least 0.60, while relative wing size had a heritability of 0.70. However, overall, flight capacity is determined by a combination of behavior, flight musculature, and wing size, which is controlled by multiple genes. The authors go on to suggest the following:

"In a freely hybridizing population, the amount of flight capability maintained would depend on several factors: initial ratio of flight capable to flightless females, costs versus fitness of flight in the particular environment, propensity of different hybrids to mate, etc. Should females with full flight capability be introduced into North America in an area where the flightless females are already established, the populations would hybridize, and the ability of L. dispar to spread could be increased".

Conversely, if a population of gypsy moths with flight-capable females were to become established in an area of North America where gypsy moth populations were not present, flight-capable females could remain in the population for many generations [**32**]. Introduction of AGM strains into North America in locations with established EGMs would most likely create hybrid populations. The probability of hybridization will be highest when both introduced and native populations have similar developmental responses to temperature and numbers of instars, since these traits can have an impact on the seasonal dynamics of the gypsy moth. These hybrid populations would have reduced flight capability, as compared to their Asian parentals [**46**], and may also incorporate other undesirable traits from the invading population, depending on the genetics of the trait. For example, the phenotype that requires less chill before eggs are able to hatch is a dominant trait that, once introduced, will increase in the population [**29**]. Subsequent introductions of other AGM strains could further increase the EGM population's flight capacities and introduce other traits that could alter the population's biology and behavior further.

3.2. Female Flight Distance

The nocturnal flight activities of female gypsy moths from Yokohama City and Chiba City, Japan, were assessed by Iwaizumi et al. [**56**]. Female moth behavior was monitored in a cage under different photoperiodic conditions. Mean flight distances of 226 and 269 m were observed for virgin females and mated females, respectively. However, the maximum recorded value of 746 m for virgin females is closer in agreement with Liebhold et al. [**59**], who estimated that female gypsy moths in Kanazawa, Japan, had a maximum flight distance per night of around 1 km.

Yang et al. [**60**] evaluated the effect of mating status and age on the flight activity of female gypsy moths from seven populations within China. Using a computer-linked flight mill, the authors found flight activity peaked when females were one day old and decreased thereafter. However, no observable difference was seen between mated and unmated females. Total flight difference varied among the populations. One-day-old females from Guizhou, China, had a mean flight distance of 7.5 km, with one individual even reaching 10.67 km, while one-day-old females from Inner Mongolia flew an average of 3.95 km. However, the authors caution that these distances may not represent entirely realistic conditions, as the tethering used within the flight mill experiment most likely increases flight distances. It is also important to consider the life cycle of gypsy moths in estimating their spread. The time females spend in the moth stage is, in part, dependent on how soon they mate, as individuals die after depositing their eggs. Gypsy moth adults live no more than three days, and generally only one day [**61**].

To accurately predict the likely rate of spread of an AGM outbreak in North America, information on potential flight distances is needed. **Table 2** summarizes all the literature on flight distances for studies conducted with gypsy moth populations from China, Japan, and Eastern Russia. By using data collected at ports in the Republic of Korea, Japan, and Russia [**62,63**], minimal potential flight distances can be estimated. Average distances between general egg collection areas and forest edge were 1.7 km in the Republic of Korea, 3.1 km at Japanese ports, and 0.5 km in Russia. Flight distances for all ports are shown in **Table 3.** These numbers could provide a rough estimate for flight distances for these populations, but because the actual distances between the forests and the ports differ between countries, care should be taken in using this data.

Table 2. Flight distances of gypsy moth populations in China, Japan, and Eastern Russia.

Authors	Year	Population	Distance	Notes
Rozkhov and Vasilyeva	1982	Asia (unspecified)	100 km	
Baranchikov	1986	Russia Far East	3–5 km	
Savotikov et al.	1995	Asia (unspecified)	20–40 km	
Liebhold et al.	2008	Japan (Kanazawa)	≤1 km	max distance, one night
Iwaizumi et al.	2010	Japan (Yokahoma City, Chiba City)	746 m	max distance, one night, virgin female
			511 m	max distance, one night, mated female
			226 m	mean distance, one night, virgin female
			269 m	mean distance, one night, mated female
Iwaizumi and Arakawa	2010	Japan (Tomakomai)	659 ± 335 m	mean distance, one night
		Japan (Hachinohe)	188 ± 202 m	mean distance, one night
		Japan (Kobe)	356 ± 351 m	mean distance, one night
		Japan (Chiba)	255 ± 116 m	mean distance, one night
Yang et al.	2017	China (Jining, Inner Mongolia)	3.95 ± 0.29 km	mean distance, one night; one-day-old females; measured with flight mill
		China (Sandeli, Liaoning)	6.63 ± 1.40 km	mean distance, one night; one-day-old females; measured with flight mill
		China (Yanzikou, Beijing)	5.56 ± 1.16 km	mean distance, one night; one-day-old females; measured with flight mill
		China (Longhua, Hebei)	4.03 ± 0.99 km	mean distance, one night; one-day-old females; measured with flight mill
		China (Lianyungang, Jiangsu)	5.79 ± 1.44 km	mean distance, one night; one-day-old females; measured with flight mill
		China (Liuan, Anhui)	6.54 ± 1.12 km	mean distance, one night; one-day-old females; measured with flight mill
		China (Xifeng, Guizhou)	7.50 ± 2.28 km	mean distance, one night; one-day-old females; measured with flight mill

Table 3. Potential flight distance of gypsy moth populations from the Republic of Korea, Japan, and Russia.

Country	City Port	Distance from Port to Forest Edge (km)	Notes
Republic of Korea	Donghae	2	
Republic of Korea	Okgye	1	
Republic of Korea	Incheon	2	
Republic of Korea	Pyongtaek	2–5	Depends on size of forest needed

Country	City Port	Distance from Port to Forest Edge (km)	Notes
Republic of Korea	Busan	2	
Republic of Korea	Pohang	2	
Republic of Korea	Ulsan	3	
Republic of Korea	Gunsan	1–3	Depends on size of forest needed
Republic of Korea	Mokpo	1–2.5	Depends on size of forest needed
Republic of Korea	Gwangyang	1.5	
Republic of Korea	Yeongilman	1.5	
Republic of Korea	Onsan	2–4.5	Depends on size of forest needed
Republic of Korea	Daesan	1	
Japan	Kokura	3–5	Depends on size of forest needed
Japan	Ube	4	
Japan	Oita	3–8	Depends on size of forest needed
Japan	Hirosihima	3–4	Depends on size of forest needed
Japan	Matsunaga	2	
Japan	Tsuruga	1.5	
Japan	Kanazawa	1–7	Depends on size of forest needed
Japan	Chiba	5–8	Depends on size of forest needed
Japan	Fushiki	7	
Japan	Toyama-shinko	6–14	Depends on size of forest needed
Japan	Sakata	0	Port directly next to forest
Japan	Hachinohe	5	
Japan	Nagahama	4.5	
Japan	Aomori	4	
Japan	Hakodate	0.5–4	Depends on port location used
Japan	Tomakomai	3	
Japan	Otaru	1	
Russia	Vladivostok	3.5–9	9 k to closest trap; 3.5 k to closest forest edge

Country	City Port	Distance from Port to Forest Edge (km)	Notes
Russia	Nakhodka	0–3	3 k to closest trap; 0 k to closest forest edge
Russia	Vostochny	0.5	
Russia	Olga	0	Port directly next to forest
Russia	Slavyanka	1	
Russia	Zarubino	0	Port directly next to forest
Russia	Posyet	0	Port directly next to forest
Russia	Plastun	0	Port directly next to forest
Russia	Vanino	0.2	
Russia	Kozmino	0	Port directly next to forest
Russia	Korsakov	0.5	

All of these factors combined suggest that it may be appropriate to use a range of techniques when estimating flight capacity and flight distances of AGMs. Further research is required to classify all populations, using genetic markers and flight distance to improve predicted rate of spread accuracy associated with introduced gypsy moths in novel environments.

4. IDENTIFICATION OF GYPSY MOTH SUBSPECIES AND SPECIFIC TRAITS WITHIN POPULATIONS

4.1. Genetic and Genomic Analyses of Subspecies and Populations

Understanding the underlying mechanisms of flight capacity is an important goal. Various approaches have been used, from physiological inquiries to genomics. The ability to identify subspecies is of vital importance in stopping the spread of flight-capable AGMs. However, easily observed morphological differences between subspecies are minimal [**64**]. Genetic methods of identification have been developed which help to differentiate between subspecies and populations [**65,66,67**] and therefore aid in determining flight capability and potential flight distance.

Gypsy moth genetic variation and distribution of female flight has been described by Keena et al. [7]. Gypsy moths were obtained from 46 locations around the world, representing the current established range of the species. Female flight capability, pre-flight behaviors, female muscle strength, and female wing size were evaluated for all 46 strains. For 31 of these strains, the authors determined the mitochondrial DNA (mtDNA) haplotype, the nuclear DNA genotype at one locus, and the microsatellite genotype at four loci. Female gypsy moths that were capable of directed flight came from Northeastern Europe, Siberia, and Asia. Wing musculature and size were found to be useful in predicting female flight capacity within populations. An analysis of all DNA haplotypes and genotypes was able to place 94% of individuals into their general geographic origin: Asian, Siberian, European, or North American. However, a clear genetic distinction was not found between European strains that differed in flight-capable females.

By analyzing microsatellite loci and mitochondrial DNA sequences, Wu et al. [**68**] detected four genetic clusters of Lymantria dispar, which correspond to the general place of origin for the named gypsy moth subspecies—Western Europe, the Asian continent, and the islands of Japan. The fourth cluster was found for gypsy moth populations established in North America and was most likely caused by a population genetic bottleneck after initial establishment on the continent. Wu et al. [**68**] also found evidence of continuing hybridization between subspecies, with high levels of genetic variation in East Asian gypsy moth populations. Analysis of genome-wide single nucleotide polymorphisms (SNPs) is a powerful approach that also enables clear distinction of subspecies and geographic variants, while revealing introgression near the geographic boundaries between subspecies [**69**]. These polymorphisms could be translated into genomic bio-surveillance tools [**70**].

Receptors in the olfactory systems of insects have been shown to be important in flight orientation [**71**]. Thus, it could be possible to differentiate between populations containing flight-capable and flightless gypsy moth females by examining olfactory genes. McCormick et al. [**72**] studied the differences between olfaction-related genes in gypsy moth females from one flightless and two flight-capable populations. The flightless population, considered to comprise EGMs, came from New Jersey, USA, while the flight-capable populations had origins in Northern Japan and Far Eastern Russia and were considered to comprise JGMs and AGMs, respectively. By using next-generation sequencing, they identified 115 chemosensory genes and conducted a principal component analysis to determine gene-expression patterns. The olfactory-related genes of JGMs appeared more divergent, whereas the AGM and EGM populations shared many commonly expressed genes. This suggests that these differences are not related to flight capability [**72**].

Lymantria genomes have been sequenced [**73**,**74**], and analyses revealed that the genomes of Lymantria are among the largest in the lepidoptera [**73**]. Genetic comparisons of AGM and EGM genomes found differences in metabolic pathways which suggest strong signatures of energy-related pathways in both subspecies, dominated by metabolic functions related to thermogenesis [**73**]. Divergent proteins associated with muscle contraction could possibly help to explain differences in wing musculature and, therefore, flight [**74**]. Further functional studies are needed to confirm these predictions.

4.2. Physical Differences among Subspecies and Populations

Physical characteristics of insects play a large role in determining flight capacity. In order for insects to fly, total flight muscle mass must make up a minimum of 12% to 16% of total body mass [**75**]. Several studies have sought to examine the physical characteristics of female gypsy moths between flight-capable and flightless populations.

Keena et al. [**32**] evaluated the muscle strength and wing size of female gypsy moths from North American and Russian populations, as well as various crosses between the two strains. Muscle strength was tested by inverting females onto their backs and evaluating the ease with which they were able to right themselves by beating their wings. Over 90% of the females from the Russian population were able to right themselves within one or two quick wingbeats, while the majority of North American females were unable to do so. More than 50% of the hybrid females were able to right themselves, although many individuals were significantly challenged with the movement. Female wing measurements varied considerably among the populations and crosses. The wings of Russian females were greater in size than those of hybrids, which in turn were greater than those of the individuals from North America.

To further explore the relationship between wing characteristics and female flight capability within gypsy moth populations, Shi et al. [**76**] conducted a morphometric analysis of 821 gypsy moths from eight different strains. Individuals of both sexes were sourced from one site in the United States, two sites in Europe, two sites in Russia, two sites in China, and one site in Japan. Body mass, length, width of fore- and hindwings, wing area, wing aspects, and wing loads were measured for all individuals. Average wing loading was significantly lower, while average wing area was larger, for females from flight-capable populations. The authors of the study created a model, using female wing load and forewing length, that correctly predicted female flight capability of the populations 97% of the time. This information could be especially useful if used in conjunction with reliable molecular identification methods in determining the flight capability of invasive gypsy moth populations.

4.3. Identification of Flight-Capable Populations

No single tried and true technique exists to identify flight-capable AGM populations and individuals. However, the two methods described above show the most promise. The model developed by Shi et al. [**76**] could be used to predict flight capability of a population (native or introduced) by measuring female forewing length and wing loading. A disadvantage of using this method is that it is necessary to capture an adult female gypsy moth. While male gypsy moths can be captured with the use of traps baited with a synthetic sex pheromone called "disparlure" [**77**], female gypsy moths remain elusive. Black-light traps can capture both male and female AGMs but are generally not considered specific enough for detection purposes [**25**]. Another method of capturing female gypsy moths involves the collection of egg masses, followed by the rearing of juveniles into adulthood [**46**]. However, the additional time this would take could hamper detection and eradication efforts.

An alternative method of identifying flight-capable gypsy moth populations was employed by Picq et al. [**69**]. They used single nucleotide polymorphism (SNP) markers to identify the geographic origins of gypsy moth samples. With this method, genetic material could be taken from both female and male moths, to determine geographic origin. This information could then be compared to known flight-capability information for gypsy moth populations. The disadvantage of this is that it does not directly predict flight capacity. A combination of this approach with genetic markers associated with flight by genome-wide association studies, or by functional analyses, could provide solutions in the future [**78**].

5. CONCISE SUMMARY

Gypsy moths from various populations have been and are still being introduced into new settings, where they have the ability to establish themselves and spread. The establishment and dispersal potential of gypsy moth populations are influenced by a set of features, including the ability of females to fly, their exploitation of hosts, and their requirements for egg cooling. It is evident that these characteristics vary throughout populations, including within each subspecies of gypsy moth. Therefore, introducing gypsy moths from different parts of the world into places where the species already exists could lead to undesirable alterations in gypsy moth biology. Furthermore, the inclusion of female organisms capable of flight, even if they are from Europe, could potentially improve their ability to spread and may necessitate adjustments to the existing management procedures. Hence, to assess the potential risks of introducing gypsy moths into a new ecosystem, it is necessary to gather data not only on the population's origin but also on its biological characteristics. A method that can quickly determine the flight capacity of freshly imported female gypsy moths is essential for any system designed to assess the dangers associated with their introduction. However, we currently lack such a tool. Additional expedient methods for evaluating the usage of larval hosts, specifically conifers, and the chilling requirements of eggs would be valuable, but are currently unavailable. Despite extensive knowledge of gypsy moths in various regions throughout the world, our understanding is still insufficient to effectively prepare for or address their entrance into unfamiliar environments.

6. AT A GLANCE

The Asian gypsy moth is being introduced into new settings and is frequently intercepted in North America. These subspecies are expected to be more dangerous to the woodlands and urban habitats of North America compared to the established gypsy moths (Lymantria dispar dispar L.) since they have a higher ability for female flight. Asian gypsy moth populations exhibit significant variations in various important characteristics, such as the flight skills of female moths. This message first discusses the probable effects of female flying, together with other important characteristics, on the ecology and distribution of this insect. This study offers a comprehensive review of the existing literature on the differences in flight capability and flight distance among gypsy moth populations. It also examines variations in other important traits and explores potential methods of identification, with a particular focus on the Asian subspecies Lymantria dispar japonica Motschulsky and Lymantria dispar asiatica Vinkovskij. Presently, there exist useful technologies for determining the broad source of introduced gypsy moth populations. However, these tools do not furnish sufficient information for the efficient management of introductions. The essential characteristics of gypsy moths vary throughout populations, even within each subspecies. Therefore, introducing gypsy moths from other parts of the world into places where gypsy moths already exist could lead to undesirable alterations in the biology of gypsy moths. Additionally, the inclusion of females capable of flying seems to have the potential to improve a population's ability to spread out and may necessitate adjustments to the management procedures employed for females that cannot fly. Hence, there is a need for efficient techniques to evaluate crucial characteristics in introduced populations in order to effectively prepare for or address fresh introductions into unfamiliar environments.

REFERENCES

1. Gninenko, Y.I.; Orlinskii, A.D. Outbreaks of Lymantria dispar in Russian forests during the 1990s. EPPO Bull. 2003, 33, 325–329. [**Google Scholar**] [**CrossRef**]
2. Bigsby, K.M.; Ambrose, M.J.; Tobin, P.C.; Sills, E.O. The cost of gypsy moth sex in the city. Urban For. Urban Green. 2014, 13, 459–468. [**Google Scholar**] [**CrossRef**]
3. Liebhold, A.; Gottschalk, K.; Muzika, R.M.; Montgomery, M.E.; Young, R.; O'Day, K.; Kelley, B. Suitability of North American Tree Species to the Gypsy Moth: A Summary of Field and Laboratory Tests; USDA For. Serv; Northeastern Experiment Station: Radnor, PA, USA, 1995; Volume GTR-NE-211, pp. 1–34. [**Google Scholar**]
4. Keena, M.A.; Richards, J.Y. Comparison of survival and development of gypsy moth Lymantria dispar L. (Lepidoptera: Erebidae) populations from different geographic areas on North American conifers. Insects 2020, 11, 260. [**Google Scholar**] [**CrossRef**] [**PubMed**]
5. Pogue, M.G.; Schaefer, P.W. A Review of Selected Species of Lymantria (Hübner 1819) Including Three New Species (Lepidoptera: Noctuidae: Lymantriinae) from Subtropical

and Temperate Regions of Asia; North AmericaForest Health Technology Enterprise Team: Morgantown, WV, USA, 2007; Volume FHTET-2006-07, pp. 201–217. [**Google Scholar**]

6. Djoumad, A.; Nisole, A.; Zahiri, R.; Freschi, L.; Picq, S.; Gundersen-Rindal, D.E.; Sparks, M.E.; Dewar, K.; Stewart, D.; Maaroufi, H.; et al. Comparative analysis of mitochondrial genomes of geographic variants of the gypsy moth, Lymantria dispar, reveals a previously undescribed genotypic entity. Sci. Rep. UK 2017, 7. [**Google Scholar**] [**CrossRef**] [**PubMed**] [**Green Version**]
7. Keena, M.A.; Cote, M.J.; Grinberg, P.S.; Wallner, W.E. World distribution of female flight and genetic variation in Lymantria dispar (Lepidoptera: Lymantriidae). Environ. Entomol. 2008, 37, 636–649. [**Google Scholar**] [**CrossRef**] [**Green Version**]
8. Gray, D.R. Hitchhikers on trade routes: A phenology model estimates the probabilities of gypsy moth introduction and establishment. Ecol. Appl. 2010, 20, 2300–2309. [**Google Scholar**] [**CrossRef**]
9. Tobin, P.C. Space-time patterns during the establishment of a nonindigenous species. Popul. Ecol. 2007, 49, 257–263. [**Google Scholar**] [**CrossRef**]
10. McManus, M.; Csoka, G. History and impact of gypsy moth in North America and comparison of recent outbreaks in Europe. Acta Silv. Lign. Hung. 2007, 3, 47–64. [**Google Scholar**]
11. Liebhold, A.M.; Halverson, J.A.; Elmes, G.A. Gypsy moth invasion in North America: A quantitative analysis. J. Biogeogr. 1992, 19, 513–520. [**Google Scholar**] [**CrossRef**]
12. Roberts, E.A.; Tobin, P.C.; Wu, J. Decision Support System for the Gypsy Moth Slow-the-Spread Program. Available online: **http://yt.ento.vt.edu/old-da/** (accessed on 17 January 2021).
13. Hajek, A.E. Pathology and epizootiology of Entomophaga maimaiga infections in forest Lepidoptera. Microbiol. Mol. Biol. Rev. 1999, 63, 814–835. [**Google Scholar**] [**CrossRef**] [**Green Version**]
14. Dwyer, G.; Dushoff, J.; Yee, S.H. The combined effects of pathogens and predators on insect outbreaks. Nature 2004, 430, 341–345. [**Google Scholar**] [**CrossRef**] [**PubMed**]
15. Mannu, R.; Cocco, A.; Luciano, P.; Lentini, A. Influence of Bacillus thuringiensis application timing on population dynamics of gypsy moth in Mediterranean cork oak forests. Pest Manag. Sci. 2020, 76, 1103–1111. [**Google Scholar**] [**CrossRef**] [**PubMed**]
16. Haynes, K.J.; Liebhold, A.M.; Johnson, D.M. Spatial analysis of harmonic oscillation of gypsy moth outbreak intensity. Oecologia 2009, 159, 249–256. [**Google Scholar**] [**CrossRef**] [**PubMed**]
17. Johnson, D.M.; Liebhold, A.M.; Tobin, P.C.; Bjørnstad, O.N. Allee effects and pulsed invasion by the gypsy moth. Nature 2006, 444, 361–363. [**Google Scholar**] [**CrossRef**]
18. Liebhold, A.; Elkinton, J.; Williams, D.; Muzika, R.M. What causes outbreaks of the gypsy moth in North America? Popul. Ecol. 2000, 42, 257–266. [**Google Scholar**] [**CrossRef**]
19. Johnson, D.M.; Liebhold, A.M.; Bjornstad, O.N.; McManus, M.L. Circumpolar variation in periodicity and synchrony among gypsy moth populations. J. Anim. Ecol. 2005, 74, 882–892. [**Google Scholar**] [**CrossRef**]
20. Weiser, J. Patterns over place and time. In Epizootiology of Insect Diseases; Fuxa, J.R., Tanda, Y., Eds.; Wiley: New York, NK, USA, 1987; pp. 215–244. [**Google Scholar**]
21. Hlásny, T.; Trombik, J.; Holuša, J.; Lukášová, K.; Grendár, M.; Turčáni, M.; Zúbrik, M.; Tabaković-Tošić, M.; Hirka, A.; Buksha, I.; et al. Multi-decade patterns of gypsy moth

fluctuations in the Carpathian mountains and options for outbreak forecasting. J. Pest Sci. 2016, 89, 413–425. [**Google Scholar**] [**CrossRef**]

22. Davidson, C.B.; Gottschalk, K.W.; Johnson, J.E. Tree mortality following defoliation by the European gypsy moth (Lymantria dispar L.) in the United States: A review. For. Sci. 1999, 45, 74–84. [**Google Scholar**]
23. Morin, R.S.; Liebhold, A.M. Invasive forest defoliator contributes to the impending downward trend of oak dominance in eastern North America. Forestry 2016, 89, 284–289. [**Google Scholar**] [**CrossRef**] [**Green Version**]
24. Aukema, J.E.; Leung, B.; Kovacs, K.; Chivers, C.; Britton, K.O.; Englin, J.; Frankel, S.J.; Haight, R.G.; Holmes, T.P.; Liebhold, A.M.; et al. Economic impacts of non-native forest insects in the continental United States. PLoS ONE 2011, 6. [**Google Scholar**] [**CrossRef**]
25. Animal Palnt Health Insepction Service [APHIS]—United States Department of Agriculture. Asian Gypsy Moth Survey and Response Guidelines. Available online:**https://www.aphis.usda.gov/plant_health/plant_pest_info/gypsy_moth/downloads/AGMSurveyResponseGuidelines.pdf** (accessed on 17 January 2021).
26. Washington State Department of Agriculture-Plant Protection Division Pest Program. Cooperative Gypsy Moth Eradication Program Snohomish County. Washington State Environmental Assessment; Washington State Department of Agriculture: Olympia, WA, USA, 2020.
27. Wallner, W.E.; Humble, L.M.; Levin, R.E.; Baranchikov, Y.N.; Carde, R.T. Response of adult lymantriid moths to illumination devices in the Russian Far-East. J. Econ. Entomol. 1995, 88, 337–342. [**Google Scholar**] [**CrossRef**]
28. Schaefer, P.W.; Strothkamp, K.G. Mass flights of Lymantria dispar japonica and Lymantria mathura (Erebidae: Lymantriinae) to commercial lighting, with notes on female viability and fecundity. J. Lepid. Soc. 2014, 68, 124–129. [**Google Scholar**]
29. Keena, M.A. Inheritance and world variation in thermal requirements for egg hatch in Lymantria dispar (Lepidoptera: Erebidae). Environ. Entomol. 2016, 45, 1–10. [**Google Scholar**] [**CrossRef**] [**PubMed**] [**Green Version**]
30. Limbu, S.; Keena, M.; Chen, F.; Cook, G.; Nadel, H.; Hoover, K. Effects of temperature on development of Lymantria dispar asiatica and Lymantria dispar japonica (Lepidoptera: Erebidae). Environ. Entomol. 2017, 46, 1012–1023. [**Google Scholar**] [**CrossRef**] [**PubMed**]
31. Srivastava, V.; Griess, V.C.; Keena, M.A. Assessing the potential distribution of Asian gypsy moth in Canada: A comparison of two methodological approaches. Sci. Rep. UK 2020, 10, 22. [**Google Scholar**] [**CrossRef**] [**PubMed**] [**Green Version**]
32. Keena, M.A.; Grinberg, P.S.; Wallner, W.E. Inheritance of female flight in Lymantria dispar (Lepidoptera: Lymantriidae). Environ. Entomol. 2007, 36, 484–494. [**Google Scholar**] [**CrossRef**] [**Green Version**]
33. Tobin, P.C.; Blackburn, L.M. Long-distance dispersal of the gypsy moth (Lepidoptera: Lymantriidae) facilitated its initial invasion of Wisconsin. Environ. Entomol. 2008, 37, 87–93. [**Google Scholar**] [**CrossRef**] [**Green Version**]
34. Gray, R.H.; Lorimer, C.G.; Tobin, P.C.; Raffa, K.F. Preoutbreak dynamics of a recently established invasive herbivore: Roles of natural enemies and habitat structure in stage-specific performance of gypsy moth (Lepidoptera: Lymantriidae) populations in northeastern Wisconsin. Environ. Entomol. 2008, 37, 1174–1184. [**Google Scholar**] [**CrossRef**]

35. Elkinton, J.S.; Liebhold, A.M. Population-dynamics of gypsy moth in North America. Annu. Rev. Entomol. 1990, 35, 571–596. **[Google Scholar]** **[CrossRef]**

36. Herrick, O.W.; Gansner, D.A. Gypsy moth on a new frontier: Forest tree defoliation and mortality. North. J. Appl. For. 1987, 4, 128–133. **[Google Scholar]** **[CrossRef]**

37. Lentini, A.; Mannu, R.; Cocco, A.; Ruiu, P.A.; Cerboneschi, A.; Luciano, P. Long-term monitoring and microbiological control programs against lepidopteran defoliators in Sardinian cork oak forests (Italy). Ann. Silvic. Res. 2020, 45, 21–30. **[Google Scholar]**

38. Allee, W.C. Animal Aggregations: A Study in General Sociology; University of Chicago Press: Chicago, IL, USA, 1932. **[Google Scholar]**

39. Whitmire, S.L.; Tobin, P.C. Persistence of invading gypsy moth populations in the United States. Oecologia 2006, 147, 230–237. **[Google Scholar]** **[CrossRef]** **[PubMed]**

40. Taylor, C.M.; Hastings, A. Allee effects in biological invasions. Ecol. Lett. 2005, 8, 895–908. **[Google Scholar]** **[CrossRef]**

41. Contarini, M.; Onufrieva, K.S.; Thorpe, K.W.; Raffa, K.F.; Tobin, P.C. Mate-finding failure as an important cause of allee effects along the leading edge of an invading insect population. Entomol. Exp. Appl. 2009, 133, 307–314. **[Google Scholar]** **[CrossRef]**

42. Tobin, P.C.; Robinet, C.; Johnson, D.M.; Whitmire, S.L.; Bjornstad, O.N.; Liebhold, A.M. The role of allee effects in gypsy moth, Lymantria dispar (L.), invasions. Popul. Ecol. 2009, 51, 373–384. **[Google Scholar]** **[CrossRef]** **[Green Version]**

43. Jankovic, M.; Petrovskii, S. Gypsy moth invasion in North America: A simulation study of the spatial pattern and the rate of spread. Ecol. Complex. 2013, 14, 132–144. **[Google Scholar]** **[CrossRef]**

44. Sharov, A.A.; Roberts, E.A.; Liebhold, A.M.; Ravlin, F.W. Gypsy moth (Lepidoptera: Lymantriidae) spread in the central Appalachians: Three methods for species boundary estimation. Environ. Entomol. 1995, 24, 1529–1538. **[Google Scholar]** **[CrossRef]**

45. Tobin, P.C.; Bai, B.B.; Eggen, D.A.; Leonard, D.S. The ecology, geopolitics, and economics of managing Lymantria dispar in the United States. Int. J. Pest Manag. 2012, 58, 195–210. **[Google Scholar]** **[CrossRef]**

46. Keena, M.A.; Wallner, W.E.; Grinberg, P.S.; Carde, R.T. Female flight propensity and capability in Lymantria dispar (Lepidoptera: Lymantriidae) from Russia, North America, and their reciprocal f-1 hybrids. Environ. Entomol. 2001, 30, 380–387. **[Google Scholar]** **[CrossRef]**

47. Gray, D.R. Unwanted spatial bias in predicting establishment of an invasive insect based on simulated demographics. Int. J. Biometeorol. 2014, 58, 949–961. **[Google Scholar]** **[CrossRef]**

48. Gray, D.R. The gypsy moth life stage model: Landscape-wide estimates of gypsy moth establishment using a multi-generational phenology model. Ecol. Model. 2004, 176, 155–171. **[Google Scholar]** **[CrossRef]**

49. Soukhovolsky, V.G.; Ponomarev, V.I.; Sokolov, G.I.; Tarasova, O.V.; Krasnoperova, P.A. Gypsy moth Lymantria dispar L. in the southern Urals: Patterns in population dynamics and modeling. Biol. Bull. Rev. 2016, 6, 57–69. **[Google Scholar]** **[CrossRef]**

50. May, C.; Hillerbrand, N.; Thompson, L.M.; Faske, T.M.; Martinez, E.; Parry, D.; Agosta, S.J.; Grayson, K.L. Geographic variation in larval metabolic rate between northern and southern populations of the invasive gypsy moth. J. Insect Sci. 2018, 18. **[Google Scholar]** **[CrossRef]** **[PubMed]**

51. Tobin, P.C.; Gray, D.R.; Liebhold, A.M. Supraoptimal temperatures influence the range dynamics of a non-native insect. Divers. Distrib. 2014, 20, 813–823. **[Google Scholar] [CrossRef]**

52. Faske, T.M.; Thompson, L.M.; Banahene, N.; Levorse, A.; Herrera, M.Q.; Sherman, K.; Timko, S.E.; Yang, B.R.; Gray, D.R.; Parry, D.; et al. Can gypsy moth stand the heat? A reciprocal transplant experiment with an invasive forest pest across its southern range margin. Biol. Invasions 2019, 21, 1365–1378. **[Google Scholar] [CrossRef]**

53. Banahene, N.; Salem, S.K.; Faske, T.M.; Byrne, H.M.; Glackin, M.; Agosta, S.J.; Eckert, A.J.; Grayson, K.L.; Thompson, L.M. Thermal sensitivity of gypsy moth (Lepidoptera: Erebidae) during larval and pupal development. Environ. Entomol. 2018, 47, 1623–1631. **[Google Scholar] [CrossRef]**

54. Peterson, A.T.; Williams, R.; Chen, G. Modeled global invasive potential of Asian gypsy moths, Lymantria dispar. Entomol. Exp. Appl. 2007, 125, 39–44. **[Google Scholar] [CrossRef]**

55. Charlton, R.E.; Carde, R.T.; Wallner, W.E. Synchronous crepuscular flight of female Asian gypsy moths: Relationships of light intensity and ambient and body temperatures. J. Insect Behav. 1999, 12, 517–531. **[Google Scholar] [CrossRef]**

56. Iwaizumi, R.; Arakawa, K.; Koshio, C. Nocturnal flight activities of the female Asian gypsy moth, Lymantria dispar (Linnaeus) (Lepidoptera: Lymantriidae). Appl. Entomol. Zool. 2010, 45, 121–128. **[Google Scholar] [CrossRef] [Green Version]**

57. Chen, F.; Shi, J.; Keena, M. Evaluation of the effects of light intensity and time interval after the start of scotophase on the female flight propensity of Asian gypsy moth (Lepidoptera: Erebidae). Environ. Entomol. 2016, 45, 404–409. **[Google Scholar] [CrossRef] [Green Version]**

58. Keena, M.A.; Grinberg, P.S.; Wallner, W.E. Comparison of female flight capability of Lymantria dispar L. reared on artificial diet versus foliage. In US Department of Agriculture Interagency Gypsy Moth Research Forum 1997; Fosbroke, S.L.C., Gottschalk, K., Eds.; Department of Agriculture, Forest Service, Northeastern Forest Experiment Station: Radnor, PA, USA, 1997; p. 58. **[Google Scholar] [CrossRef]**

59. Liebhold, A.M.; Turcani, M.; Kamata, N. Inference of adult female dispersal from the distribution of gypsy moth egg masses in a Japanese city. Agric. For. Entomol. 2008, 10, 69–73. **[Google Scholar] [CrossRef]**

60. Yang, F.; Luo, Y.; Shi, J. The influence of geographic population, age, and mating status on the flight activity of the Asian gypsy moth Lymantria dispar (Lepidoptera: Erebidae) in China. Appl. Entomol. Zool. 2017, 52, 265–270. **[Google Scholar] [CrossRef]**

61. Robinet, C.; Lance, D.R.; Thorpe, K.W.; Onufrieva, K.S.; Tobin, P.C.; Liebhold, A.M. Dispersion in time and space affect mating success and Allee effects in invading gypsy moth populations. J. Anim. Ecol. 2008, 77, 966–973. **[Google Scholar] [CrossRef] [PubMed]**

62. Munson, A.S.; Leonard, D.; Mastro, T.; McGovern, T.; Levy, J.; Kucera, D. Russian Far East Lymantrid Monitoring Project: Project Summary 1993–1994; United Sates Department of Agriculture Forest Service Intermountain Region: Ogden, UT, USA, 1995.

63. Trotter, R.T., III; Limbu, S.; Hoover, K.; Nadel, H.; Keena, M.A. Comparing Asian gypsy moth [Lymantria dispar asiatica (Lepidoptera: Erebidae) and L. dispar japonica] trap data from East Asian ports with lab parameterized phenology models: New tools and questions. Ann. Entomol. Soc. Am. 2020, 113, 125–138. **[Google Scholar] [CrossRef] [Green Version]**

64. Garner, K.J.; Slavicek, J.M. Identification and characterization of a RAPD-PCR marker for distinguishing Asian and North American gypsy moths. Insect Mol. Biol. 1996, 5, 81–91. **[Google Scholar]** **[CrossRef]** **[PubMed]** **[Green Version]**

65. Stewart, D.; Zahiri, R.; Djoumad, A.; Freschi, L.; Lamarche, J.; Holden, D.; Cervantes, S.; Ojeda, D.I.; Potvin, A.; Nisole, A.; et al. A multi-species TaqMan PCR assay for the identification of Asian gypsy moths (Lymantria spp.) and other invasive lymantriines of biosecurity concern to North America. PLoS ONE 2016, 11. **[Google Scholar]** **[CrossRef]**

66. Stewart, D.; Nisole, A.; Djoumad, A.; Zahiri, R.; Lamarche, J.; Levesque, R.C.; Hamelin, R.C.; Cusson, M. A needle in a haystack: A multigene TaqMan assay for the detection of Asian gypsy moths in bulk pheromone trap samples. Biol. Invasions 2019, 21, 1843–1856. **[Google Scholar]** **[CrossRef]** **[Green Version]**

67. Djoumad, A.; Nisole, A.; Stewart, D.; Holden, D.; Zahiri, R.; Inoue, M.N.; Martemyanov, V.V.; Levesque, R.C.; Hamelin, R.C.; Cusson, M. Reassessment of the status of Lymantria albescens and Lymantria postalba (Lepidoptera: Erebidae: Lymantriinae) as distinct 'Asian gypsy moth' species, using both mitochondrial and nuclear sequence data. Syst. Entomol. 2020, 45, 493–504. **[Google Scholar]** **[CrossRef]** **[Green Version]**

68. Wu, Y.; Molongoski, J.J.; Winograd, D.F.; Bogdanowicz, S.M.; Louyakis, A.S.; Lance, D.R.; Mastro, V.C.; Harrison, R.G. Genetic structure, admixture and invasion success in a holarctic defoliator, the gypsy moth (Lymantria dispar, Lepidoptera: Erebidae). Mol. Ecol. 2015, 24, 1275–1291. **[Google Scholar]** **[CrossRef]**

69. Picq, S.; Keena, M.; Havill, N.; Stewart, D.; Pouliot, E.; Boyle, B.; Levesque, R.C.; Hamelin, R.C.; Cusson, M. Assessing the potential of genotyping-by-sequencing-derived single nucleotide polymorphisms to identify the geographic origins of intercepted gypsy moth (Lymantria dispar) specimens: A proof-of-concept study. Evol. Appl. 2017. **[Google Scholar]** **[CrossRef]**

70. Hamelin, R.C.; Roe, A.D. Genomic biosurveillance of forest invasive alien enemies: A story written in code. Evol. Appl. 2020, 13, 95–115. **[Google Scholar]** **[CrossRef]** **[PubMed]** **[Green Version]**

71. Getahun, M.N.; Thoma, M.; Lavista-Llanos, S.; Keesey, I.; Fandino, R.A.; Knaden, M.; Wicher, D.; Olsson, S.B.; Hansson, B.S. Intracellular regulation of the insect chemoreceptor complex impacts odour localization in flying insects. J. Exp. Biol. 2016, 219, 3428–3438. **[Google Scholar]** **[CrossRef]** **[PubMed]** **[Green Version]**

72. McCormick, A.C.; Grosse-Wilde, E.; Wheeler, D.; Mescher, M.C.; Hansson, B.S.; De Moraes, C.M. Comparing the expression of olfaction-related genes in gypsy moth (Lymantria dispar) adult females and larvae from one flightless and two flight-capable populations. Front. Ecol. Evol. 2017, 5. **[Google Scholar]** **[CrossRef]** **[Green Version]**

73. Hebert, F.O.; Freschi, L.; Blackburn, G.; Beliveau, C.; Dewar, K.; Boyle, B.; Gundersen-Rindal, D.E.; Sparks, M.E.; Cusson, M.; Hamelin, R.C.; et al. Expansion of lines and species-specific DNA repeats drives genome expansion in Asian gypsy moths. Sci. Rep. UK 2019, 9. **[Google Scholar]** **[CrossRef]** **[Green Version]**

74. Zhang, J.; Cong, Q.; Rex, E.A.; Hallwachs, W.; Janzen, D.H.; Grishin, N.V.; Gammon, D.B. Gypsy moth genome provides insights into flight capability and virus-host interactions. Proc. Natl. Acad. Sci. USA 2019, 116, 1669–1678. **[Google Scholar]** **[CrossRef]** **[Green Version]**

75. Marden, J.H. Maximum lift production during takeoff in flying animals. J. Exp. Biol. 1987, 130, 235–258. **[Google Scholar]**

76. Shi, J.; Chen, F.; Keena, M.A. Differences in wing morphometrics of Lymantria dispar (Lepidoptera: Erebidae) between populations that vary in female flight capability. Ann. Entomol. Soc. Am. 2015, 108, 528–535. [**Google Scholar**] [**CrossRef**] [**Green Version**]

77. Bierl, B.A.; Beroza, M.; Collier, C.W. Potent sex attractant of gypsy moth—Its isolation, identification, and synthesis. Science 1970, 170. [**Google Scholar**] [**CrossRef**]

78. Roe, A.D.; Torson, A.S.; Bilodeau, G.; Bilodeau, P.; Blackburn, G.S.; Cui, M.M.; Cusson, M.; Doucet, D.; Griess, V.C.; Lafond, V.; et al. Biosurveillance of forest insects: Part integration and application of genomic tools to the surveillance of non-native forest insects. J. Pest Sci. 2019, 92, 51–70. [**Google Scholar**] [**CrossRef**]

08 COLLECTION OF ADULT EMERALD ASH BORER (AGRILUS PLANIPENNIS) SPECIMENS AT WHITE ASH SITES AFTER INVASION, WITH DIFFERENT LEVELS OF LIVE PHLOEM

1. INTRODUCTION: AN OVERVIEW

The Emerald ash borer (EAB), scientifically known as Agrilus planipennis Fairmaire, is a type of beetle that feeds on the phloem of trees. It is originally from Asia and was first discovered in North America in 2002. The beetles were found on ash trees in the Detroit metropolitan area in southeast Michigan, USA, which were already dying [1,2]. A comprehensive dendrochronological investigation covering an area of 1.5 million hectares revealed that the Emerald Ash Borer (EAB) has caused the death of ash trees in southeast Michigan by 1998, suggesting that it probably established itself in this region by the early 1990s [3]. The infestations grew larger when Emerald Ash Borer (EAB) adults spread out and as separate populations, formed by the unintentional movement of infested ash nursery trees, logs, firewood, and similar items, came together. By 2003, the mortality of ash trees caused by the Emerald Ash Borer (EAB) had spread throughout a significant portion of a five-county region in southeastern Michigan [3]. By February 2021, populations of the Emerald Ash Borer (EAB) had been detected in a minimum of 35 states in the United States and five provinces in Canada [4].

The emerald ash borer (EAB) has caused the death of a vast number of ash trees in forests, rural areas, and urban areas. It is now recognized as the most damaging and expensive invasive forest pest in North America [5,6,7]. Within the original core of the EAB invasion in southeast Michigan, from 80 to 99% of green ash (F.

pennsylvanica Marsh.), black ash (F. nigra Marsh.), and white ash (F. americana L.) trees in the forest overstories had perished by the early- to mid-2000s [8,9,10]. During their whole lifespan, fully grown beetles consume ash leaves, while mature females deposit eggs beneath bark flakes or in bark cracks [1,11]. The larvae consume the phloem and cambium in narrow tunnels, usually damaging the exterior sapwood. This disrupts the movement of nutrients and water [1,6]. As the number of larvae increases, damage to the vascular system intensifies, resulting in the thinning of the canopy, dieback, and ultimately the death of the tree [1,2,6].

Interspecific variations among North American ash species in their preference for and/or ability to fight the Emerald Ash Borer (EAB) might impact the mortality rates of ash trees. This, in turn, can possibly influence the persistence or dynamics of EAB at a local level and ultimately affect the composition of forests. Multiple studies have repeatedly demonstrated that green ash and black ash are highly favored and susceptible hosts for the Emerald Ash Borer (EAB). On the other hand, blue ash (F. quadrangulata Michx.) is significantly less favored and/or comparatively resistant to EAB. White ash, however, seems to be a host that falls in between, displaying intermediate susceptibility to EAB. These findings are supported by references [7,12,13]. While white ash trees in certain locations in southeast Michigan have experienced significant mortality [9,10,14], a substantial number of white ash trees in other forested areas within the same region have managed to maintain their health despite the presence of the Emerald Ash Borer (EAB) for over 12 years. In a recent survey of 28 forested sites in southeast and south-central Michigan, the mortality rate of white ash trees varied from over 90% to under 10% each site, in terms of both the number of trees and the basal area [15]. Within a plantation with 21 randomly arranged sections, each consisting of five types of ash trees, both green ash and black ash were significantly infested. The infestation was severe, with over 200 larvae of the Emerald Ash Borer (EAB) per square meter of phloem area. However, the majority of these larvae perished due to competition among individuals of the same species for phloem resources. The larval densities on blue ash and Manchurican ash (F. mandshurica), which is an Asian species, were equal to or less than 2.0 per square meter. Additionally, the majority of trees were not affected [13]. The average number of larvae per square meter for white ash trees was 41.0 ± 12. Although some trees were extensively infested, seven out of the 21 white ash trees had between zero and ten larvae per square meter [13].

The potential density of Emerald Ash Borer (EAB) within a specific tree or area is determined by the presence of live ash phloem, which is necessary for larval growth [16,17]. McCullough and Siegert [18] established a relationship between the diameter of ash trees at breast height (DBH) and the amount of phloem, as well as the generation of emerald ash borer (EAB). By utilizing data obtained from 148 green ash and white ash trees with diameters at breast height (DBH) ranging from 5.2 to 65.0 cm, it was established that, on average, there is a possibility for 89 ± 4.6 emerald ash borer (EAB) adults to develop per square meter of phloem. Empirical evidence obtained from debarked trees at various locations demonstrated that each individual EAB larva necessitates roughly 10 cm2 of phloem for its growth and development [16,17,18]. Therefore, the quantity of EAB adults that may be generated in a certain region is contingent upon the prevalence and dimensions of ash trees.

The carrying capacity for Emerald Ash Borer (EAB) might be significantly lower in regions that have been invaded by the pest, compared to areas before the invasion. This is particularly true in places where there were high infestations of EAB several years ago, resulting in the death of most of the mature ash trees. This difference in carrying capacity has been seen in previous studies [19,20,21]. Ash regeneration, which includes young trees with a diameter at breast height (DBH) of 2.5 to 10 cm, saplings, or stump sprouts, can be plentiful in certain post-invasion stands. However, this may not be the case for all stands [8,15,22,23,24]. While it is possible for trees with a diameter as small as 2.5 cm to be infested and killed by EAB, they tend to generate a limited number of EAB adults. Additionally, EAB larvae that develop in small trees are more susceptible to being killed by parasitoids compared to larvae in larger trees with thicker bark. The extent to which little ash trees in post-invasion settings can supply sufficient phloem to sustain a population of EAB over an extended period, at densities that can cause tree mortality, is uncertain.

Our hypothesis is that the population levels of EAB in a specific location will fluctuate based on the presence of live ash phloem, which is determined by the quantity and size of live white ash trees. Multiple trap designs and lures have been created and assessed for the purpose of detecting and monitoring populations of the Emerald Ash Borer (EAB). Please refer to the reviews cited in References [29,30,31,32] for more information. Green or purple double-decker traps, which include upper and lower sticky prisms connected to a 3-meter tall PVC pipe and baited with ash volatiles, reliably catch a significant number of EAB and exhibit high rates of detection at field sites ranging from minimal to severe EAB infestation levels [31,33,34,35,36]. In a prior investigation carried out in 24 green ash locations throughout 2010 and 2011, the populations of the emerald ash borer (EAB) were observed using dark purple double-decker traps. These traps were lured with cis-3-hexenol on the upper prism and Manuka oil on the lower prism. Sticky bands were also used on both girdled and ungirdled trap trees [19]. The captures of adult EAB insects mirrored the gradual spread of EAB invasion in southern Michigan over time, from east to west. These captures were found to be strongly associated with the density of EAB larvae attacking trees. at 2010, 13% of Emerald Ash Borer (EAB) adults were taken at post-invasion Core sites in the southeast, where the majority of overstory ash trees were dead. Additionally, 12% of adults were captured in recently infested Cusp sites in the southwest. 19% of the EAB adults were found in south central Crest locations, where EAB populations were at their highest levels and a combination of healthy, declining, and dying ash trees were observed. Poland and McCullough [36] found that heavily infested sites with more than 35% dieback had a higher number of beetles (>100 per trap) captured on baited DD traps compared to sites with moderate infestations and less than 35% dieback (60-100 beetles per trap), or sites with low EAB population densities and no visible signs of tree infestation (less than 25 beetles per trap). At a recently invaded location with extremely low levels of emerald ash borer infestations, only 1-4 beetles were caught on green or purple DD traps that were baited [35].

Our hypothesis is that the living white ash phloem is a crucial yet scarce resource for EAB populations in places that have been invaded by the species. The main goals of our study were to: (1) quantify the extent of live white ash phloem that could be colonized by the Emerald Ash Borer (EAB) in areas that were heavily infested by EAB and had significant overstory mortality at least 10 years ago; (2) evaluate the presence and abundance of EAB at these sites using baited DD traps; and (3) investigate the potential correlation between EAB trap catches and the availability of white ash phloem.

2. METHODS

2.1. Site Selection

We used a Michigan county atlas in April 2014 to identify 70 large, forested areas on state, county, or municipal property in southeast and south central Lower Michigan where EAB infestations had peaked at least 8 to 10 years earlier, and nearly all overstory green ash had died. We queried managers of each area to assess white ash presence and abundance. We visited 40 post-invasion sites where managers confirmed the presence of white ash trees and identified 30 areas where white ash trees, whether dead or alive, appeared to be relatively abundant (**Figure 1**). We established a center point and recorded GPS coordinates in the midst of the white ash trees within each site. Stand conditions, including species composition of overstory trees and regeneration, are presented in detail in Robinett and McCullough [**15**].

Briefly, white ash trees ranged from 10.0 to 44.0 cm DBH (diameter at breast height), and survival varied substantially among sites, ranging from 0 to 100%. Across all sites, however, 75% of white ash trees, and 66% of the white ash basal area, ranging from 0.3 to 3.5 $m^2 \cdot ha^{-1}$, were alive. Nearly all live white ash had signs of previous EAB colonization, but 83% of trees in 2016 had healthy canopies (≤10%

canopy dieback). Green ash trees were recorded in 27 sites, but 92% had been killed by EAB in previous years.

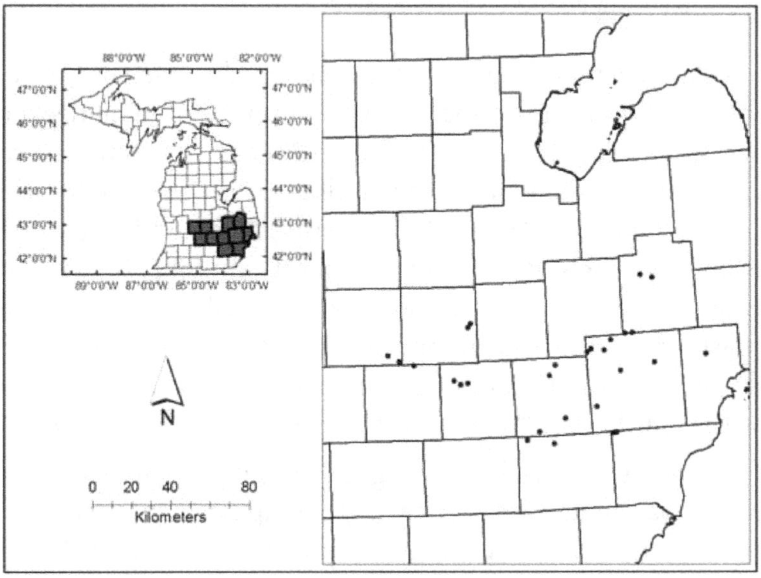

Figure 1. Distribution of 30 white ash sites in southeast and south central Lower Michigan.

2.2. Fixed Radius Plots

We centered a circular plot with an 18 m radius around each of the two DD traps in each site and recorded DBH of live and dead white ash trees (≥6 cm DBH) during the summer in 2014 and again in 2016. We visually estimated canopy dieback in increments of 10%, where <10% indicated a full, healthy canopy, and ≥80% indicated severe decline [**37**]. Ash trees were grouped by dieback class (0–20%, 21–40%, 41–60%, 61–80%, 81–100%) for some analyses. Each ash tree was examined for signs of EAB infestation, including holes left by woodpeckers preying on EAB larvae, D-shaped adult EAB emergence holes, and bark cracks over old larval galleries. Trees were re-surveyed from early July to mid-August in 2016.

Density of white ash stems and white ash basal area ($m^2 \cdot ha^{-1}$) were calculated for each plot and summed for each site in 2014 and 2016. White ash trees were assigned to four DBH classes (6.0–13.0, 13.1–25.0, 25.1–42.0, and 42.1 to 60.0 cm). Area of phloem (m^2) in live and dead white ash trees was calculated using DBH (cm) from trees recorded in 2014 and 2016, following methods of McCullough and Siegert [**18**]. We also adjusted area of live phloem in individual trees to account for dead portions of the canopy. When dieback estimates were >10%, the estimated live phloem area was reduced by the proportion of the canopy that was dead. Previous studies have shown correlations between canopy decline and larval densities or the portion of phloem covered by larval galleries [**12,13,38,39**]. We also estimated live phloem area by DBH and canopy dieback classes.

2.3. Trapping

Presence and relative abundance of EAB was assessed at each of the 30 sites over a period of 3 years based on trap captures. We set up two double decker (DD) traps, 36 to 50 m apart, between 24 May and 12 June 2014 in each site. We recorded GPS coordinates at each trap, allowing us to install traps in the same locations between mid-May and mid-June in 2015 and 2016. Traps remained in place through mid-September each year. Each trap was placed in full sunlight, 5–10 m from the edge of wooded areas, or in open areas among scattered ash trees. A DD trap consisted of two coroplast prisms (36 × 60 cm on each side; Great Lakes IPM, Vestaburg, MI, USA) fastened to a 3-m-tall PVC pipe (10 cm diam at 3.0 m and 1.8 m above-ground and coated with clear Pestick™ (Hummert International, Earth City, MO, USA).

The PVC pipe was supported by sliding it over a T-post embedded in the ground (34). One DD trap was comprised of a dark green upper prism (Sabic green, λ = 530 nm, 48% relative reflectance) and a light purple lower prism (Sabic purple, λ = 420 nm 21.7% reflectance; 670 nm 13.6% reflectance [**40,41**]. Two cis-3-hexenol bubble caps (release rate of 3.7 mg/day per bubble cap for a combined release rate of 7.4 mg/day; Contech Enterprises Inc., Delta, BC, Canada) were attached to the lower edge of both prisms. The second DD trap was comprised of two dark purple prisms (λ = 430 nm 20% reflectance, 600 nm 6% reflectance; 670 nm 13.5% reflectance (56). The top prism was baited with two cis-3-hexenol bubble caps, while the lower prism was baited with a Manuka oil pouch (release rate of 50 mg/day, Synergy Semiochemicals, Corp., Delta, BC, Canada). Lures were replaced after 6 weeks to ensure consistent volatile release, at which time Pestick™ was also re-applied. The same trap designs were placed in the same locations in all three years.

Captured EAB adult beetles were collected at two- to three-week intervals beginning on 10 June and continuing through mid-September each year. Beetles were typically collected from multiple sites on a given day, but it required approximately a week to check traps at all 30 sites. For each collection period, sites were visited in the same order as when they were set up to ensure uniform time between collection periods. Beetles on each prism of the two traps were bagged separately and returned to the Michigan State University (MSU) Forest Entomology Laboratory, East Lansing, MI USA. Beetles were soaked in Histoclear II (National Diagnostics, Atlanta, GA, USA) to remove Pestick™ and then examined under a microscope to confirm species and sex. We acquired cumulative growing degree days (GDD) (base 10 °C; beginning 1 January annually) corresponding to each collection period from the nearest MSU Enviro-weather station (**Enviroweather.**

msu.edu) at each site. Euclidean distances between each site and the nearest weather station were determined using ArcMap 10.3 and ranged from 5 to 43 km.

2.4. Statistical Analysis

Adult male and female beetles captured per trap (total on both prisms) and per prism were tallied for each collection period and summed by year. Differences in EAB captures were compared among years and between the two trap designs, and among prism colors or lures within trap types, using generalized linear mixed models (PROC GLIMMIX). Main effects of trap type, year, prism color, position (upper or lower), and lure were tested as fixed effects in separate models. Site was a random effect in all models. To account for collinearity among years, an autoregressive order one covariance structure was used for all models, with the Kenward-Rogers modification of the denominator degrees of freedom [**42**]. Because the response variable was count data, a negative binomial distribution was used with the log link function [**43**]. Differences among years, trap types, prism color within trap types, and lure within trap types for the separate models were tested with the Tukey-Kramer means comparison procedure. Residuals were tested for homogeneity of variance assumption using Levene's test, and models were adjusted if needed.

Simple linear regression was used to assess whether the area of live white ash phloem (m^2) in the 18 m radius around each trap significantly predicted EAB adult captures within sites (PROC REG). Captures of EAB in 2014, 2015, and 2016 were compared to total area of live phloem measured in 2014, and EAB captures in 2016 were regressed on live phloem area measured in 2016. All analyses were conducted using the SAS 9.4 for Windows statistical package [**44**] with an α-level of 0.05.

3. RESULTS

3.1. Captures of EAB on Traps

Overall, 580, 585, and 932 EAB adults were captured on the 60 DD traps in 2014, 2015, and 2016, respectively. In 2014, no EAB adults were captured in two sites, while one to 61 beetles were captured in the other 28 sites, with an average of 19 ± 2.8 beetles captured per site (**Figure 2**). In 2015, EAB adults were captured in 29 sites; the site with no EAB captures was one of the sites with no EAB captures the previous year. Total beetles captured per site in 2015 ranged from one to 64, with an average of 19 ± 3.0 per site (**Figure 2**). Between 2014 and 2015, EAB captures increased by ≥50% in 12 sites, remained consistent in four sites, and decreased by ≥50% in five sites. In 2016, EAB adults were captured in all 30 sites. Total captures ranged from one to 127 EAB adults per site and averaged 31 ± 5.3 beetles per site (**Figure 2**). Between 2014 and 2016, the number of captured beetles increased by approximately 60% in 14 of the 30 sites.

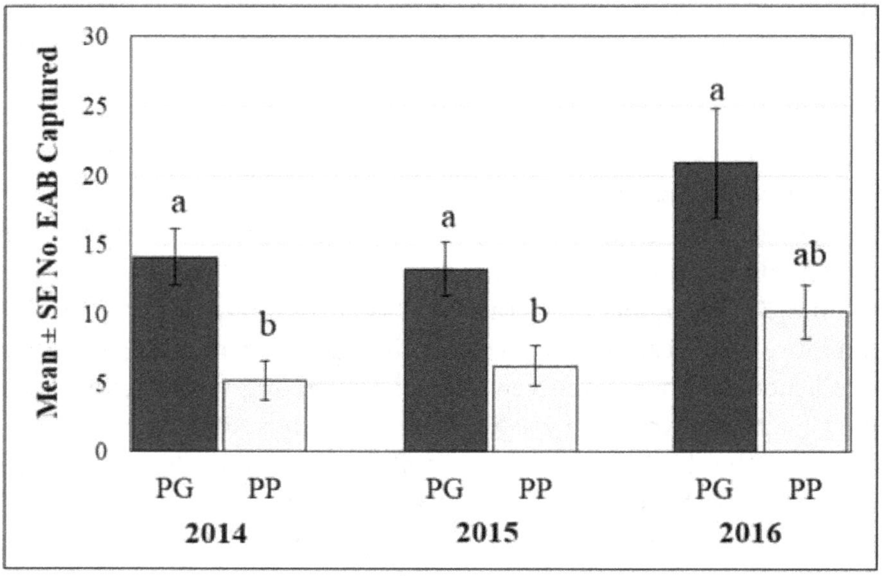

Figure 2. Mean (+SE) number of emerald ash borer adults captured per trap in 2014, 2015, and 2016 on green-purple (GP) double decker (DD) traps and purple-purple (PP) DD traps. Bars topped by the same letter are not significantly different; Tukey-Kramer means comparison procedure, p > 0.05.

Cumulative growing degree days (GDD) (base 10 °C/50 °F) associated with the first trap checks in June in each site ranged from a minimum of 347 to a maximum of 538 $GDD_{10\ °C}$ in 2014–2016. Peak EAB activity occurred in late June to early July each year, corresponding to a minimum of 521 to a maximum of 789 $GDD_{10\ °C}$ in 2014–2016, when 54%, 52%, and 62% of the total EAB adults were captured, respectively. Following this peak, EAB captures decreased by 48% and 39% in 2014 and 2015, respectively, and by 85% in 2016. Traps captured ≤21 EAB total during our final checks, which occurred between 17 and 23 September each year.

Cumulative number of EAB adults captured throughout the season differed between the two trap designs (F = 8.86; degrees of freedom [df] = 1.138; p < 0.003) and among years (F = 6.08; df = 2.42; p = 0.005), but the interaction was not significant (F = 0.46; df = 2.42; p = 0.6). Overall, for all three years combined, traps with green upper and light purple lower prisms (GP) caught significantly more EAB adults (mean = 15.75 ± 3.0) than traps with two dark purple prisms (PP) (mean = 6.85 ± 1.4), accounting for 73%, 68%, and 67% of total captures in 2014, 2015, and 2016, respectively. Significantly more EAB adults were captured in 2016 for both trap types combined (mean = 14.54 ± 2.4) than in 2014 (mean = 8.5 ± 1.4) or in 2015 (mean = 9.06 ± 1.5) (**Figure 2**).

Prism color and lures contributed to differences in trap captures between the two DD trap designs; however, prism color, position, and lures were confounded, making it difficult to fully determine their relative importance for EAB attraction. Overall, green or light purple prisms baited with cis-3-hexenol lures captured more EAB adults than dark purple prisms baited with either cis-3-hexenol or Manuka oil (F = 15.50; df = 3.87; p < 0.001). Prism position (upper or lower) did not have a significant effect on trap captures (F = 3.34; df = 1.87; p = 07).

Male beetles dominated captures on both trap designs in all three years. Overall, 70%, 69%, and 77% of the EAB adults captured in 2014, 2015, and 2016, respectively, were males. Across all years, green prisms caught significantly more EAB males than either upper or lower dark purple prisms, while captures on light purple prisms were intermediate (F = 14.4; df = 3.48; p < 0.0001). Light purple prisms on the GP traps caught significantly more EAB females than either of the dark purple prisms on the PP traps or the upper green prisms on the GP traps (F = 13.3; df = 3.348; p < 0.0001). Significantly more EAB males (F = 7.50; df = 2.348; p = 0.0006) and females (F = 6.06; df = 3.348; p = 0.003) were captured in 2016 than in 2015 or 2014 (**Table 1**).

Table 1. Mean (± SE) number of emerald ash borer (EAB) adult males and females captured per prism on double-decker (DD) traps with green upper and light purple lower prisms (GP) or dark purple upper and lower prisms (PP) in 2014, 2015, and 2016 in 30 sites located in southeast and south central Lower Michigan (N = 30).

		Double Decker (GP) Trap		Double Decker (PP) Trap		*F; df; p*
		Green Upper	Light Purple Lower	Dark Purple Upper	Dark Purple Lower	
2014	Males	7.1 ± 1.2a [1]	4.5 ± 0.9ab	2.2 ± 0.5b	1.6 ± 0.6b	6.21; 3.116; 0.0006
	Females	0.7 ± 0.2ab	1.9 ± 0.4a	0.6 ± 0.2b	0.7 ± 0.3b	8.73; 3.116; <0.0001
	Total	7.8 ± 1.3a	6.4 ± 1.1a	2.9 ± 0.7b	2.3 ± 0.9b	8.76; 3.116; <0.0001
2015	Males	5.5 ± 1.1a	4.1 ± 0.8a	2.3 ± 0.6ab	1.3 ± 0.4b	4.72; 3.116; 0.004
	Females	1.4 ± 0.3ab	2.3 ±0.4a	1.4 ±0.3ab	1.1 ± 0.3b	2.56; 3.116; 0.05
	Total	6.9 ± 1.3a	6.4 ± 1.1ab	3.8 ± 0.9ab	2.4 ± 0.6b	4.40; 3.116; 0.006
2016	Males	8.7 ± 2.0a	7.9 ± 1.6a	4.6 ± 0.9ab	2.8 ± 0.8b	4.11; 3.116; 0.008
	Females	1.0 ± 0.2b	3.3 ±0.7a	3.9 ± 0.8ab	1.0 ± 0.3b	5.81; 3.116; 0.001
	Total	9.8 ± 2.1a	11.2 ± 2.2a	6.3 ± 1.2ab	3.9 ± 0.9b	4.49; 3.116; 0.005

		Double Decker (GP) Trap		Double Decker (PP) Trap		F; df; p
		Green Upper	Light Purple Lower	Dark Purple Upper	Dark Purple Lower	
All years	Males	7.1 ± 0.8a	5.5 ± 0.7ab	3.1 ± 0.4b	1.9 ± 0.3b	14.4; 3.348; <0.0001
	Females	1.0 ± 0.1b	2.5 ± 0.3a	1.2 ± 0.2b	0.9 ± 0.3b	13.3; 3.348; <0.0001
	Total	8.1 ± 0.9a	7.8 ± 0.9a	4.3 ± 0.6b	2.9 ± 0.5b	16.3; 3.348; <0.0001

[1] Means within the same year and sex followed by the same letter are not statistically different; Tukey-Kramer means comparison procedure ($p > 0.05$).

3.2. White Ash Phloem

In 2014, we recorded a total of 1109 white ash trees (≥6 cm DBH) across the 30 sites, including 736 live and 373 dead trees. Most (87%) of the live white ash had <50% canopy dieback and 477 trees had <20% dieback (**Table 2**). Conversely, 44 trees were severely declining with ≥80% dieback (**Table 2**). External signs of EAB infestation were apparent on 58% of the live white ash, including 78% of trees with <50% dieback and 43 of the 44 trees with ≥80% dieback. Nearly all (98%) of the dead trees had clearly been killed by EAB, as evidenced by abundant galleries and sloughing bark.

Table 2. Number of live and dead white ash (F. americana) trees and estimated area of phloem (m²) recorded for trees in fixed radius plots (18-m radius) in 30 sites in Michigan in 2014 and 2016. Phloem area was calculated following methods of McCullough and Siegert [18] and adjusted to account for the proportion of the canopy that was dead.

Percent Canopy Dieback Class	No. Trees	Total Phloem (m²)	Adjusted Live Phloem (m²)
2014			
<10.0%	353	829	829
10.0 to 19.9%	124	248	233
20.0 to 39.9%	131	307	232
40.0 to 59.9%	49	151	80
60.0 to 79.9%	35	106	37
80.0 to 99.0%	44	154	20
Dead	373	1833	0
2016			
<10.0%	608	1300	1300

Percent Canopy Dieback Class	No. Trees	2014 Total Phloem (m^2)	Adjusted Live Phloem (m^2)
10.1 to 19.9%	54	138	124
20.0 to 39.9%	50	119	90
40.0 to 59.9%	14	72	39
60.0 to 79.9%	17	48	16
80.0 to 99.9%	14	58	8
Dead	430	1837	0

Overall, white ash trees across the 30 sites represented an estimated 3627 m^2 of phloem area, which included 1794 m^2 (49.5% of the total) in the live trees tallied in 2014. Within sites, phloem area in live white ash trees averaged (± SE) 120.9 ± 17.51 m^2 and ranged from 24.4 to 421.0 m^2. Small ash (6.0 to 13.0 cm DBH) contributed approximately 685 m^2 of phloem. Live trees in the two largest size classes (25.1 to 42.0 and 42.1 to 60.0 cm DBH) contributed 279 m^2 of phloem, while pole-sized trees (13.1 to 25.0 cm DBH) contributed the remainder. On average, an estimated 10,761 ± 1558 EAB adult beetles per site could have potentially developed if all the phloem available in 2014 was colonized, ranging from a minimum of 1673 beetles at the site with the least phloem to a maximum of 37,466 beetles at the site with the highest amount of white ash phloem. When we adjusted phloem area to account for dead portions of the canopy, total live phloem decreased to 1427 m^2. This included 557 m^2 of live phloem within trees in the smallest DBH class (6.0 to 13.0 cm), 675 m^{2i}n pole-sized trees (13.1 to 25.0 cm), and 195 m^2 of live phloem in trees in the two largest DBH classes (25.1 to 42.0 and 42.1 to 60.0 cm).

In 2016, we again established plots (18 m radius) and measured DBH of white ash trees surrounding the traps. A total of 1187 white ash trees were tallied across the 30 sites, including 743 live and 430 dead trees (**Table 2**). The slight increase in number of trees between 2014 and 2016 reflected growth of young trees that were > 6 cm DBH in 2016 but not in 2014, as well as minor differences in plot boundaries between the two years. Overall, 96% of the live trees had < 50% canopy dieback and 662 trees, including most of the small trees, had ≤ 10% dieback (**Table 2**). Fourteen trees were heavily infested and severely declining with ≥ 80% canopy dieback (**Table 2**). External signs of previous EAB infestation were apparent on 71% of the live white ash and 99% of the dead trees. All of the severely declining trees were or had been heavily infested and 94% of the 118 trees with 10 to 50% dieback had external signs of past EAB infestation.

In 2016, white ash trees (live and dead) represented a total of 3596 m^2 of phloem area, including 1759 m^2 (48.5%) of phloem in the live trees. Phloem area of live white ash within sites averaged (±SE) 119.0 ± 16.44 m^2 and ranged from 24.4 to

410.9 m^2. Live trees in the smallest DBH class (6.0 to 13.0 cm) accounted for 729 m^2 of the phloem recorded, and trees in the two largest DBH classes (25.1 to 60.0 cm) accounted for 259 m^2 of phloem, while pole-sized trees (13.1 to 25.0 cm DBH) provided the remainder. Live phloem area decreased between 2014 and 2016 in 13 sites, increased in 11 sites, and differed by < 1.5 m^2 in six sites. On average, an estimated 10,881 ± 1434.9 EAB adults could have developed if all phloem available in 2016 were fully colonized, ranging from a minimum of 2095 at the site with the lowest amount of white ash phloem to a maximum of 36,567 beetles at the site with the most phloem available. After adjusting phloem to account for dead portions of the canopy, total live phloem decreased to 1457 m^2. Trees in the smallest DBH class (6.0 to 13.0 cm) provided 636 m^2 of live phloem, and pole-sized trees (13.1 to 25.0 cm) contributed another 620 m^2 of live phloem, while trees in the two largest DBH classes (25.1 to 42.0 and 42.1 to 60.0 cm) contributed 201 m^2 of live phloem.

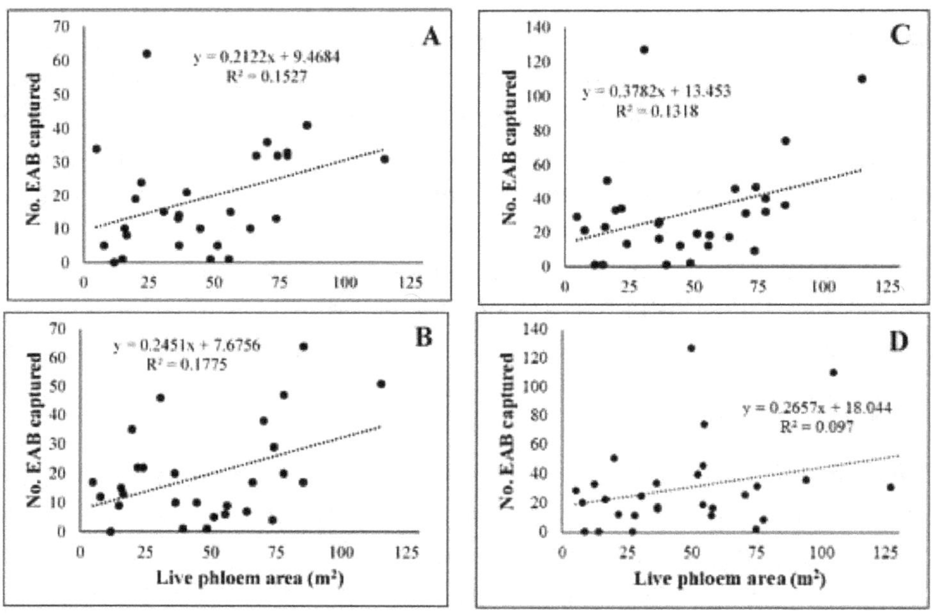

Figure 3. Linear relationship between emerald ash borer (EAB) adult captures in (A) 2014, (B) 2015, and (C) 2016 and adjusted area of live white ash phloem (m^2) surrounding traps (18 m radius) measured in 2014 and between EAB adult captures in 2016 (D) and adjusted area of live white ash phloem (m^2) measured in 2016. Phloem area (m^2) of tallied trees was estimated following methods of McCullough and Siegert [18]; then, it was reduced by the proportion of the canopy that was dead.

Levels of EAB infestation and tree condition ranged from sites in which nearly all trees were alive and appeared relatively healthy to sites where ≥90% of the white ash were dead [15]. Linear regressions between live phloem area, adjusted for dieback, and numbers of EAB captured on traps were run after excluding one site with an unusually high amount of live phloem. Linear relationships between area of live white ash phloem derived from the 2014 surveys and the number of EAB adults captured on traps were significant in 2014 ($F = 4.86$; $p = 0.0361$) and 2015 ($F = 5.83$; $p = 0.0229$) but explained less than 20% of the variation in EAB captures (**Figure 3**). Captures of EAB adults increased substantially in 2016 but were not significantly related to live phloem area measured in 2014 ($F = 4.10$; $p = 0.0529$) nor measured in 2016 ($F = 2.90$; $p = 0.1001$) (**Figure 3**).

4. DISCUSSION

Populations of both EAB and white ash trees continue to co-exist in southeast and south central Lower Michigan more than a decade after these forests were invaded. Across the 30 sites, the level of EAB infestation and condition of white ash trees varied from locations where nearly all trees were alive and in relatively good condition, to areas where ≥90% of the white ash were dead [**19**]. Even in areas where ash mortality was high, however, a substantial amount of live phloem remained available for EAB colonization; estimated area of live phloem per site ranged from 24.4 to 421.0 m^2 in 2014 and from 24.4 to 410.9 m^2 in 2016. Between 2014 and 2016, 57 trees that were seriously declining in 2014 had died, but, overall, live phloem area was only 3.5% lower in 2016 than in 2014. Moreover, across sites, canopy condition of live trees generally improved. For example, 65% and 89% of the live trees had healthy canopies with <20% dieback in 2014 and 2016, respectively.

We originally hypothesized that if availability of live phloem was a limiting resource for larval development, fewer EAB adults would be captured on traps in sites where ash mortality was high than in sites where live ash phloem was abundant. Relationships between EAB captures and live phloem area, adjusted to account for canopy dieback, were significant, but live phloem area explained less than 20% of the variation in captures in any year. Overall, results indicate that phloem availability was not a limiting factor driving EAB population levels and trap captures at our sites. Similarly, McCullough et al. [**35**] found that abundance of ash phloem near traps had no discernable effect on EAB captured on traps at sites with low population densities. They reported that position of traps and sun exposure had only weak effects, while trap type had a significant effect on the number of EAB captured. Additionally, Mercader et al. [**45**] found little relationship between EAB captures on baited purple prism traps suspended from branches of ash trees and local EAB densities across a large, recently infested area with low EAB populations.

While availability of live phloem was nearly the same in our sites in 2014 and 2016, captures of EAB adults increased significantly in 2016 compared to the previous two years. More dead trees were tallied in 2016 than in 2014, reflecting mortality of the 57 trees that were seriously declining in 2014. However, canopy condition of live trees generally improved; the number of trees tallied in each dieback class (>10% to >80% dieback) decreased between 2014 and 2016. Therefore, increased trap catches in 2016 did not appear to be related to live phloem available or increased levels of canopy dieback within the sites. It is possible that some portion of the EAB adults captured on our traps developed on trees beyond our plots and immigrated into our sites. While most EAB colonize ash trees within 100 to 200 m of their point of emergence when phloem is abundant [**46,47**], mature adult females are physiologically capable of dispersing several kilometers [**48,49**]. Ash trees up to 800 m from the emergence point of adults have been colonized [**46,47,48,50**], and adults have been captured on traps at least 300 m away from the nearest ash trees [**33**].

An alternative explanation is that larvae from eggs laid by beetles in 2014 required two years to complete development on the relatively healthy trees in our plots and were captured as adults in 2016. Previous studies have shown that, on recently infested, healthy trees where larval densities remain low, a high proportion of larvae overwinter as early instars the first year, feed and overwinter as prepupae the second winter, and then emerge as adults the following summer [**1,48,51**]. If most beetles captured in 2016 developed on trees within the plots, EAB densities may have been building, although signs of infestation and canopy decline did not increase between 2014 and 2016.

Despite the increase in EAB adult captures in 2016, the number of beetles captured per trap was still relatively low and indicative of very low population levels. For example, McCullough et al. [**35**] captured an average of approximately 1 to 4 EAB on baited green or purple DD traps in a newly infested forest where the EAB population was ultra-low, trees had no visible signs of EAB infestation, and mean larval densities were approximately 2 larvae per m^2 and 12 larvae per m^2 in non-girdled and girdled trees, respectively. Similarly, <25 beetles were captured on baited DD traps at sites with low EAB population densities and no visible signs of tree infestation, while approximately 60–100 beetles were captured at sites with moderate infestations and <35% canopy dieback, and >100 EAB were captured at heavily infested sites with >35% dieback [**36**]. In heavily infested sites where canopy dieback ranged from 35–50%, green or purple prism traps hung 1.5 m or 13 m above ground captured >60 and >100 beetles per trap, respectively [**52,53**]. In our sites, trap catches averaging <20 beetles per trap indicate population levels were very low, and few visible signs of infestation and little canopy decline would be expected.

Estimating the area of phloem available for EAB development provides a means to estimate potential EAB production for a specific location and to assess pre- and post-invasion dynamics of a local EAB population. McCullough and Siegert [**18**] found that, on average, roughly 89 EAB could successfully develop and emerge per m^2 of phloem surface area, and that EAB production averaged 68.8 ± 5.9 beetles per m^2 on small trees (2.5–13 cm DBH) and 105.3 ± 5.7 beetles per m^2 on larger trees (DBH > 13 cm). We estimated that approximately 10,761 ± 1558.4 and 10,881 ± 1434.9 EAB adults could have potentially emerged across the sites based on the trees recorded and phloem area calculated in 2014 and 2016, respectively. However, low trap catches even in 2016, along with relatively stable canopy conditions and continued abundance of live white ash, indicate that EAB populations have not come close to achieving projected levels in most of the sites.

Persistence of both EAB populations and live white ash in the post-invasion sites we monitored may reflect multiple factors, including host preference and oviposition behavior of adult females. Previous research with ash trees in common gardens or plantations, landscapes, and forests has consistently shown that, when green ash and white ash tree co-occur, EAB females will preferentially oviposit on green ash trees [**7,12,13,38**]. In the large forested parks and game areas where our study was conducted, most overstory green ash were killed by EAB several years ago, but green ash regeneration, including recruits, saplings, and stump sprouts, remains abundant in many areas [**8,10,22,23,24**]. Some portion of female beetles that developed on white ash trees in our plots may have dispersed and oviposited on young green ash. Additionally, while DBH of live white ash trees in our sites ranged from 6.0 to 40.6 cm, many trees were small, with a DBH of ≤ 13 cm, and some were likely too small to have been colonized when local EAB populations were peaking in the early and mid-2000s [**3,8,10,19**]. When young white ash trees are colonized by EAB, females typically lay most eggs on the trunk or around branch junctions where rough-textured bark is present, avoiding the relatively smooth bark on branches and upper portions of the trunk [**54**]. If larvae feed predominantly on the trunk of small trees, mortality typically occurs rapidly compared to the gradual decline and dieback characteristic of larger trees where feeding begins on canopy branches. However, with the exception of 57 severely declining trees that died, most white ash trees tallied in our sites appeared healthy in 2014 and remained healthy in 2016 [**15**]. Further, more than half of the white ash trees we tallied, including many small trees, had holes left by woodpeckers preying on late instar EAB larvae, exit holes left by EAB adults, or old larval galleries, indicating these trees were suitable hosts.

Natural enemies of EAB larvae, including woodpeckers [**55,56,57**], native parasitoids [**54,57**], and introduced parasitoids native to Asia [**58,59**], could be playing a role in keeping EAB populations below carrying capacity within these sites. In study sites in southern Michigan where two introduced parasitoids, T. planipennisi and O. agrili, have been established since 2008, healthy ash saplings (400–1600 per ha) and young trees (200–900 per ha) are persisting despite formerly high EAB populations that killed most overstory ash [**28**]. Life table analyses of EAB population dynamics at these sites indicate T. planipennisi killed 36%–85% of the late instar larvae in saplings, contributing to net population growth rate of EAB that was near or below replacement levels. Another study in southeast Michigan indicated ash mortality was lower, while ash regeneration and diameter growth were generally higher in post-invasion sites with parasitoid releases compared to control sites [**60**]. Based on the increasing trend in EAB trap catches between 2014/2015 and 2016, it is difficult to predict whether populations will continue to build in our study sites. Additional years of monitoring to track both EAB dynamics and tree condition in these sites would be helpful to project the long term outlook for white ash in these sites.

The two double-decker (DD) trap designs we used to assess EAB presence and relative infestation levels in our sites also provided useful information on beetle attraction to the green and purple prisms. Overall, the GP (green-light purple) DD traps baited with cis-3-hexenol on both prisms captured significantly more EAB beetles than the PP (dark purple-dark purple) traps baited with cis-3-hexenol on the upper prism and Manuka oil on the lower prism. Males were captured in significantly higher numbers on the upper green panels, while the lower light purple panels captured significantly more females. In another study with DD traps, Poland et al. [**31**] similarly reported that GP traps captured significantly more females, males, and total EAB than did PP traps, but differences were minimal between traps with cis-3-hexenol lures on both prisms compared to traps with cis-3-hexenol on the upper prism and Manuka oil on the lower prism. Previous studies have shown that males were more attracted than females to green prisms, while females were more attracted than males to purple prisms [**31,41,53,61,62,63,64**]. Males, which tend to hover around the canopies of ash trees and then land on leaves or approach females feeding in the canopy [**65,66**], are presumably attracted to colors and volatiles associated with ash leaves. In contrast, purple wavelengths, which are similar to the spectra associated with tree bark, attract females who likely use visual cues to locate suitable oviposition sites on branches and trunks of ash trees [**66**].

Multiple studies across dozens of sites have shown detection rates (e.g., capture of ≥one EAB adult) or total EAB adult captures (standardized by trapping surface area) are higher on DD traps baited with cis-3-hexenol or a combination of cis-3-hexenol and Manuka oil than on baited purple or green prisms or multiple funnel traps hung from ash branches [**33,34,35,36,40,67**]. Furthermore, DD traps tend to have lower variation in trap catches within sites and among sites with different levels of EAB infestation [**31,40**]. In contrast, the number of EAB captured in traps hung in the canopy of ash trees is substantially more variable and may reflect the infestation level of the individual trap-bearing trees or nearby trees [**31,36,45**].

Emergence of EAB adults typically begins in mid-May in southern Michigan when accumulated growing degree days (GDD) (base 10 °C) reach approximately $250_{10\ °C}$ **[68]**. As in previous studies, EAB adult captures peaked in late June to early July in all three years **[33,68,69]**, but, in 2016, the proportion of EAB adults (approximately 29%) captured during our first trap check beginning June 10 was twice as high as in the previous two years (approximately 13%). We suspect this pattern was driven by early spring weather. For example, on 1 June, a total of 215, 254, and 248 growing degree days (base 10 °C) had accumulated in Commerce Township in southeast Michigan in 2014, 2015, and 2016, respectively. On 1 April, however, at this same location, a total of 1.9 and 4.6 growing degree days (base 10 °C) had accumulated in 2014 and 2015, respectively, compared to 29.7 degree days in 2016. A period of unusually warm weather in March 2016 likely enabled many larvae to begin pupating, complete development, and emerge as adults relatively early in the summer.

Our results highlight the need for long-term monitoring to track persistence and dynamics of native ash species and EAB populations in post-invasion sites in North America. While nearly complete mortality of overstory white ash trees has certainly been documented in areas of southeast Michigan **[9,10,70]**, in many sites we evaluated, EAB populations appear to be persisting at densities below carrying capacity, and white ash are growing and presumably reproducing. Whether this will continue over time remains to be seen, especially if trees become stressed by severe drought or other unfavorable conditions that would increase their attractiveness and vulnerability to EAB **[2,68,69,70,71]**.

If EAB populations build and white ash dieback and mortality begin to increase, releasing parasitoids for EAB biological control could perhaps help protect remaining trees and regeneration. Research on individual "lingering" green ash trees that survived longer than surrounding trees or resistance mechanisms of Asian ash species is ongoing **[72]**, but there has been comparatively little effort to address white ash resistance to EAB. Furthermore, practical aspects of establishing resistant ash in forested settings, such as the need for herbicide applications to reduce competing vegetation, protection from deer browse, and probably irrigation, may limit the viability of this approach in many areas. A comprehensive, area-wide integrated strategy that could include periodically treating a portion of trees with a highly effective systemic insecticide, biological control releases, and perhaps use of girdled ash "trap trees" would reduce EAB density, population growth, and ash mortality **[73]**, preserving white ash as a functional component in these forests **[7]**.

5. CONCISE SUMMARY

The findings of our study indicate that white ash trees are enduring in multiple forested locations inside the center area of the emerald ash borer (EAB) infestation in southeast and south central Michigan, even after more than ten years since the first invasion. Around 60% to 70% of the living white ash trees showed signs of previous Emerald Ash Borer (EAB) infestations, and several trees that had significant canopy dieback in 2014 were dead by 2016. However, the majority of trees displayed canopies that were rather healthy, with less than 20% dieback. Overall, the condition of the canopies across the 30 sites showed improvement from 2014 to 2016. The low catches of EAB traps, together with the high number of live white ash trees and consistent canopy conditions, suggest that the EAB populations are currently below the maximum capacity of the locations. Additionally, the availability of ash phloem does not seem to be a limiting factor for the abundance of EAB. Additional surveillance is required to monitor the fluctuation of Emerald Ash Borer (EAB) populations and assess the health of white ash trees in these areas in order to ascertain their future prospects.

6. AT A GLANCE

The Emerald ash borer (EAB), scientifically known as Agrilus planipennis Fairmaire, was initially discovered in 2002 in southeast Michigan. Since then, it has caused extensive death of ash trees (Fraxinus spp.) in the central area of its invasion. The infestation has already spread to 35 states in the United States and five provinces in Canada, resulting in significant forest mortality. There is limited knowledge regarding the endurance and concentrations of EAB populations in areas that have been invaded, once the majority of ash trees have perished. We conducted surveillance of EAB populations from 2014 to 2016 by employing double decker (DD) traps placed among white ash (F. americana) trees in 30 locations in southeast and south-central Michigan that were affected by the invasion. Each site was equipped with two DD traps. One trap was equipped with a dark green upper prism and a light purple lower prism, both of which were baited with cis-3-hexenol lures. The second prism was equipped with two dark purple prisms, one coated with cis-3-hexenol on the upper side and the other coated with Manuka oil on the lower side. The size and condition of ash trees were documented in 2014 and 2016. Additionally, the area of live white ash phloem was assessed within an 18-meter radius plot centered around each DD trap. The area of living white ash phloem per site varied from around 24 to 421 m2 in 2014 and from 24 to 411 m2 in 2016. The overall health of living white ash trees showed improvement over time, with 65% and 89% of the trees having healthy canopies (less than 20% dieback) in 2014 and 2016, respectively. The traps located in sites 28, 29, and 30 captured a combined total of 580, 585, and 932 Emerald Ash Borer (EAB) adults throughout the period of 2014 to 2016. The area of live ash phloem had a relatively small impact on the variation in total EAB catches in all three years. The low number of captured EAB insects, together with consistently stable canopy conditions and an ongoing abundance of live white ash trees, suggest that the EAB populations are currently below the maximum capacity of the locations. Additionally, the availability of ash phloem is not a restricting factor for the abundance of EAB. Additional surveillance is required to observe the fluctuations of the Emerald Ash Borer (EAB) population and the health of the trees in order to assess the future prospects of white ash in these locations.

REFERENCES

1. Cappaert, D.L.; McCullough, D.G.; Poland, T.M.; Siegert, N.W. Emerald ash borer in North America: A research and regulatory challenge. Am. Entomol. 2005, 51, 152–165. [**Google Scholar**] [**CrossRef**] [**Green Version**]
2. Poland, T.M.; McCullough, D.G. Emerald ash borer: Invasion of the urban forest and the threat to North America's ash resource. J. For. 2006, 104, 118–124. [**Google Scholar**]
3. Siegert, N.W.; McCullough, D.G.; Liebhold, A.M.; Telewski, F.W. Dendrochronological reconstruction of the epicentre and early spread of emerald ash borer in North America. Divers. Distrib. 2014, 20, 847–858. [**Google Scholar**] [**CrossRef**]
4. Emerald Ash Borer Information Network (EAB.info). Available online: **http://www.emeraldashborer.info/index.cfm** (accessed on 30 October 2020).
5. Aukema, J.E.; Leung, B.; Kovacs, K.; Chivers, C.; Britton, K.O.; Englin, J.; Frankel, S.J.; Haight, R.G.; Holmes, T.P.; Liebhold, A.; et al. Economic impacts of non-native forest insects in the continental United States. PLoS ONE 2011, 6, e24587. [**Google Scholar**] [**CrossRef**]
6. Herms, D.A.; McCullough, D.G. Emerald ash borer invasion of North America: History, biology, ecology, impacts, and management. Ann. Rev. Entomol. 2014, 59, 13–30. [**Google Scholar**] [**CrossRef**] [**PubMed**] [**Green Version**]
7. McCullough, D.G. Challenges, tactics and integrated management of emerald ash borer in North America. For. Int. J. For. Res. 2020, 93, 197–211. [**Google Scholar**] [**CrossRef**]
8. Burr, S.J.; McCullough, D.G. Condition of green ash (Fraxinus pennsylvanica) overstory and regeneration at three stages of the emerald ash borer invasion wave. Can. J. For. Res. 2014, 44, 768–776. [**Google Scholar**] [**CrossRef**]
9. Klooster, W.S.; Herms, D.A.; Knight, K.S.; Herms, C.P.; McCullough, D.G.; Smith, A.; Gandhi, K.J.K.; Cardina, J. Ash (Fraxinus spp.) mortality, regeneration, and seed bank dynamics in mixed hardwood forests following invasion by emerald ash borer (Agrilus planipennis). Biol. Invas. 2014, 16, 859–873. [**Google Scholar**] [**CrossRef**]
10. Smith, A.; Herms, D.A.; Long, R.P.; Gandhi, K.J. Community composition and structure had no effect on forest susceptibility to invasion by the emerald ash borer (Coleoptera: Buprestidae). Can. Entomol. 2015, 147, 318–328. [**Google Scholar**] [**CrossRef**]
11. Wang, X.Y.; Yang, Z.Q.; Gould, J.R.; Zhang, Y.N.; Liu, G.J.; Liu, E.S. The biology and ecology of the emerald ash borer Agrilus planipennis, in China. J. Insect Sci. 2010, 10, 128. [**Google Scholar**] [**CrossRef**] [**Green Version**]
12. Anulewicz, A.C.; McCullough, D.G.; Cappaert, D.L. Emerald ash borer (Agrilus planipennis) density and canopy dieback in three North American ash species. Arboricult. Urban For. 2007, 33, 338–349. [**Google Scholar**]

13. Tanis, S.R.; McCullough, D.G. Host resistance of five Fraxinus species to Agrilus planipennis (Coleoptera: Buprestidae) and effects of paclobutrazol and fertilization. Environ. Entomol. 2015, 44, 287–299. [**Google Scholar**] [**CrossRef**]

14. Tanis, S.R.; McCullough, D.G. Differential persistence of blue ash and white ash following emerald ash borer invasion. Can. J. For. Res. 2012, 42, 1542–1550. [**Google Scholar**] [**CrossRef**]

15. Robinett, M.A.; McCullough, D.G. White ash (Fraxinus americana) survival in the core of the emerald ash borer (Agrilus planipennis) invasion. Can. J. For. Res. 2019, 49, 510–520. [**Google Scholar**] [**CrossRef**]

16. Mercader, R.J.; Siegert, N.W.; Liebhold, A.M.; McCullough, D.G. Estimating the effectiveness of three potential management options to slow the spread of emerald ash borer populations in localized outlier sites. Can. J. For. Res. 2011, 41, 254–264. [**Google Scholar**] [**CrossRef**]

17. Mercader, R.J.; Siegert, N.W.; Liebhold, A.M.; McCullough, D.G. Simulating the influence of the spatial distribution of host trees on the spread of the emerald ash borer, Agrilus planipennis, in recently colonized sites. Popul. Biol. 2011, 53, 271–285. [**Google Scholar**]

18. McCullough, D.G.; Siegert, N.W. Estimating potential emerald ash borer (Coleoptera: Buprestidae) populations using ash inventory data. J. Econ. Entomol. 2007, 100, 1577–1586. [**Google Scholar**] [**CrossRef**]

19. Burr, S.J.; McCullough, D.G.; Poland, T.M. Density of emerald ash borer (Coleoptera: Buprestidae) adults and larvae at three stages of the invasion wave. Environ. Entomol. 2018, 47, 121–132. [**Google Scholar**] [**CrossRef**]

20. McCullough, D.G.; Poland, T.M.; Tluczek, A.R.; Anulewicz, A.C.; Wieferich, J.; Siegert, N.W. Emerald ash borer (Coleoptera: Buprestidae) densities over a 6-yr period on untreated trees and trees treated with systemic insecticides at 1-, 2-, and 3-yr intervals in a central Michigan forest. J. Econ. Entomol. 2019, 112, 201–212. [**Google Scholar**] [**CrossRef**]

21. Siegert, N.W.; Engelken, P.; McCullough, D.G. Changes in demography and carrying capacity of green ash and black ash ten years after emerald ash borer invasion of two ash-dominant forests. For. Ecol. Manage. 2021. submitted. [**Google Scholar**]

22. Engelken, P.; McCullough, D.G. Riparian forest conditions along three northern Michigan rivers following emerald ash borer invasion. Can. J. For. Res. 2020, 50, 800–810. [**Google Scholar**] [**CrossRef**]

23. Engelken, P.; Benbow, M.E.; McCullough, D.G. Legacy effects of emerald ash borer on riparian forest vegetation and structure. For. Ecol. Manage. 2020, 457, 117684. [**Google Scholar**] [**CrossRef**]

24. Kashian, D.M. Sprouting and seed production may promote persistence of green ash in the presence of the emerald ash borer. Ecosphere 2016, 7, e01332. [**Google Scholar**] [**CrossRef**] [**Green Version**]

25. Aubin, I.; Cardou, F.; Ryall, K.; Kreutzweiser, D.; Scarr, T. Ash regeneration capacity after emerald ash borer (EAB) outbreaks: Some early results. For. Chron. 2015, 91, 291–298. [**Google Scholar**] [**CrossRef**]

26. Siegert, N.W.; McCullough, D.G.; Poland, T.M.; Heyd, R.L. Optimizing use of girdled ash trees for management of low-density emerald ash borer populations. J. Econ. Entomol. 2017, 110, 1096–1106. [**Google Scholar**] [**CrossRef**] [**PubMed**]

27. Duan, J.J.; Abell, K.J.; Bauer, L.S.; Gould, J.; Van Driesche, R.G. Natural enemies implicated in the regulation of an invasive pest: A life table analysis of the population dynamics of the emerald ash borer. Agric. For. Entomol. 2014, 16, 406–416. **[Google Scholar]** **[CrossRef]**

28. Duan, J.J.; Bauer, L.S.; Van Driesche, R.G. Emerald ash borer biocontrol in ash saplings: The potential for early stage recovery of North American ash trees. For. Ecol. Manag. 2017, 394, 64–72. **[Google Scholar]** **[CrossRef]** **[Green Version]**

29. Crook, D.J.; Mastro, V.C. Chemical ecology of the emerald ash borer Agrilus planipennis. J. Chem. Ecol. 2010, 36, 101–112. **[Google Scholar]** **[CrossRef]**

30. Ryall, K. Detection and sampling of emerald ash borer (Coleoptera: Buprestidae) infestations. Can. Entomol. 2015, 147, 290–299. **[Google Scholar]** **[CrossRef]**

31. Poland, T.M.; Petrice, T.R.; Ciaramitaro, T.M. Trap designs, colors, and lures for emerald ash borer detection. Front. For. Glob. Chang. 2019, 2, 80. **[Google Scholar]** **[CrossRef]** **[Green Version]**

32. Poland, T.M.; McCullough, D.G.; Anulewicz, A.C. Evaluation of double-decker traps for emerald ash borer (Coleoptera: Buprestidae). J. Econ. Entomol. 2011, 104, 517–531. **[Google Scholar]** **[CrossRef]**

33. Silk, P.J.; Mayo, P.; Ryall, K.; Roscoe, L. Semiochemical and communication ecology of the emerald ash borer, Agrilus planipennis (Coleoptera: Buprestidae). Insects 2019, 10, 323. **[Google Scholar]** **[CrossRef]** **[PubMed]** **[Green Version]**

34. McCullough, D.G.; Poland, T.M. Building double-decker traps for early detection of emerald ash borer. J. Vis. Exp. 2017, 128, e55252. **[Google Scholar]** **[CrossRef]** **[PubMed]**

35. McCullough, D.G.; Siegert, N.W.; Poland, T.M.; Pierce, S.J.; Ahn, S.Z. Effects of trap type, placement and ash distribution on emerald ash borer captures in a low density site. Environ. Entomol. 2011, 40, 1239–1252. **[Google Scholar]** **[CrossRef]** **[PubMed]** **[Green Version]**

36. Poland, T.M.; McCullough, D.G. Comparison of trap types and colors for capturing emerald ash borer adults at different population densities. Environ. Entomol. 2014, 43, 157–179. **[Google Scholar]** **[CrossRef]** **[PubMed]** **[Green Version]**

37. Zarnoch, S.J.; Bechtold, W.A.; Stolte, K.W. Using crown condition variables as indicators of forest health. Can. J. For. Res. 2004, 34, 1057–1070. **[Google Scholar]** **[CrossRef]**

38. Limback, C.K. Tree Vigor and Its Relation to Emerald Ash Borer (Agrilus planipennis Fairmarie) Adult Host Preference and Larval Development on Green and White Ash Trees. Master's Thesis, Department of Entomology, Michigan State University, East Lansing, MI, USA, 2010; p. 98. **[Google Scholar]**

39. Flower, C.E.; Knight, K.S.; Rebbeck, J.; Gonzalez-Meler, M.A. The relationship between the emerald ash borer (Agrilus planipennis) and ash (Fraxinus spp.) tree decline: Using visual canopy condition assessments and leaf isotope measurements to assess pest damage. For. Ecol. Manag. 2013, 303, 143–147. **[Google Scholar]** **[CrossRef]**

40. Tobin, P.C.; Strom, B.L.; Francese, J.A.; Herms, D.A.; McCullough, D.G.; Poland, T.M.; Ryall, K.L.; Scarr, T.S.; Silk, P.J.; Thistle, H.W. Evaluation of trapping schemes to detect emerald ash borer. Environ. Entomol. 2021. Accepted. **[Google Scholar]**

41. Francese, J.; Rietz, M.L.; Crook, D.J.; Fraser, I.; Lance, D.R.; Mastro, V.C. Improving detection tools for the emerald ash borer (Coleoptera: Buprestidae): Comparison of prism and multifunnel traps at varying population densities. J. Econ. Entomol. 2013, 106, 2407–2414. **[Google Scholar]** **[CrossRef]**

42. Alnosaier, W.S. Kenward-Roger Approximate F Test for Fixed Effects in Mixed Linear Models. Ph.D. Dissertation, Department of Statistics, Oregon State University, Corvallis, OR, USA, 2007; p. 131. [**Google Scholar**]

43. Hilbe, J.M. Negative Binomial Regression, 2nd ed.; Cambridge University Press: Cambridge, UK, 2011. [**Google Scholar**]

44. SAS Institute Inc. Base SAS 9.4 Procedures Guide: Statistical Procedures, 2nd ed.; SAS Institute Inc.: Cary, NC, USA, 2012. [**Google Scholar**]

45. Mercader, R.J.; McCullough, D.G.; Bedford, J.M. A comparison of girdled ash detection trees and baited artificial traps for emerald ash borer (Agrilus planipennis Fairmaire) detection. Environ. Entomol. 2013, 42, 1027–1039. [**Google Scholar**] [**CrossRef**] [**Green Version**]

46. Mercader, R.J.; Siegert, N.W.; Liebhold, A.M.; McCullough, D.G. Dispersal of the emerald ash borer, Agrilus planipennis, in newly-colonized sites. Agric. For. Entomol. 2009, 11, 421–424. [**Google Scholar**] [**CrossRef**]

47. Mercader, R.J.; Siegert, N.W.; McCullough, D.G. Estimating the influence of population density and dispersal behavior on the ability to detect and monitor Agrilus planipennis (Coleoptera: Buprestidae) populations. J. Econ. Entomol. 2012, 105, 272–281. [**Google Scholar**] [**CrossRef**]

48. Siegert, N.W.; McCullough, D.G.; Williams, D.W.; Fraser, I.; Poland, T.M. Dispersal of Agrilus planipennis (Coleoptera: Buprestidae) from discrete epicenters in two outlier sites. Environ. Entomol. 2010, 39, 253–265. [**Google Scholar**] [**CrossRef**]

49. Taylor, R.A.J.; Bauer, L.S.; Poland, T.M.; Windell, K.N. Flight performance of Agrilus planipennis (Coleoptera: Buprestidae) on a flight mill and in free flight. J. Insect Behav. 2010, 23, 128–148. [**Google Scholar**] [**CrossRef**]

50. Siegert, N.W.; Mercader, R.J.; McCullough, D.G. Spread and dispersal of emerald ash borer (Coleoptera: Buprestidae): Estimating the spatial dynamics of a difficult-to-detect invasive forest pest. Can. Entomol. 2015, 147, 338–348. [**Google Scholar**] [**CrossRef**]

51. Tluczek, A.R.; McCullough, D.G.; Poland, T.M. Influence of host stress on emerald ash borer (Coleoptera: Buprestidae) adult density, development, and distribution in Fraxinus pennsylvanica trees. Environ. Entomol. 2011, 40, 357–366. [**Google Scholar**] [**CrossRef**] [**Green Version**]

52. Crook, D.J.; Khrimian, A.; Francese, J.A.; Fraser, I.; Poland, T.M.; Sawyer, A.J.; Mastro, V.C. Development of a host-based semiochemical lure for trapping emerald ash borer Agrilus planipennis (Coleoptera: Buprestidae). Environ. Entomol. 2008, 37, 356–365. [**Google Scholar**] [**CrossRef**] [**PubMed**] [**Green Version**]

53. Crook, D.J.; Francese, J.A.; Zylstra, K.E.; Fraser, I.; Sawyer, A.J.; Bartels, D.W.; Lance, D.R.; Mastro, V.C. Laboratory and field response of the emerald ash borer (Coleoptera: Buprestidae), to selected regions of the electromagnetic spectrum. J. Econ. Entomol. 2009, 102, 2160–2169. [**Google Scholar**] [**CrossRef**]

54. Anulewicz, A.C.; McCullough, D.G.; Cappaert, D.L.; Poland, T.M. Host range of the emerald ash borer (Agrilus planipennis Fairmaire) (Coleoptera:Buprestidae) in North America: Results of multiple-choice field experiments. Environ. Entomol. 2008, 37, 230–241. [**Google Scholar**] [**CrossRef**] [**Green Version**]

55. Cappaert, D.; McCullough, D.G. Phenology of Atanycolus cappaerti (Hymenoptera: Braconidae), a native parasitoid of emerald ash borer, Agrilus planipennis (Coleoptera: Buprestidae). Great Lakes Entomol. 2009, 41, 141–154. [**Google Scholar**]

56. Lindell, C.A.; McCullough, D.G.; Cappaert, D.; Apostolou, N.M.; Roth, M.B. Factors influencing woodpecker predation on emerald ash borer. Am. Midl. Nat. 2008, 159, 434–444. [**Google Scholar**] [**CrossRef**]

57. Duan, J.J.; Ulyshen, M.D.; Bauer, L.S.; Gould, J.; Van Driesche, R. Measuring the impact of biotic factors on populations of immature emerald ash borers (Coleoptera: Buprestidae). Environ. Entomol. 2010, 39, 1513–1522. [**Google Scholar**] [**CrossRef**] [**PubMed**] [**Green Version**]

58. Duan, J.J.; Bauer, L.S.; Abell, K.J.; Van Driesche, R. Population responses of hymenopteran parasitoids to the emerald ash borer (Coleoptera: Buprestidae) in recently invaded areas in north central United States. BioControl 2012, 57, 199–209. [**Google Scholar**] [**CrossRef**]

59. Duan, J.J.; Bauer, L.S.; Van Driesche, R.G.; Gould, J.R. Progress and challenges of protecting North American ash trees from the emerald ash borer using biological control. Forests 2018, 9, 142. [**Google Scholar**] [**CrossRef**] [**Green Version**]

60. Kashian, D.M.; Bauer, L.S.; Spei, B.A.; Duan, J.J.; Gould, J.R. Potential impacts of emerald ash borer biocontrol on ash health and recovery in southern Michigan. Forests 2018, 9, 296. [**Google Scholar**] [**CrossRef**] [**Green Version**]

61. Crook, D.J.; Khrimian, A.; Cossé, A.A.; Fraser, I.; Mastro, V.C. Influence of trap color and host volatiles on capture of the emerald ash borer (Coleoptera: Buprestidae). J. Econ. Entomol. 2012, 105, 429–437. [**Google Scholar**] [**CrossRef**] [**PubMed**] [**Green Version**]

62. Francese, J.A.; Crook, D.J.; Fraser, I.; Lance, D.R.; Sawyer, A.J.; Mastro, V.C. Optimization of trap color for emerald ash borer (Coleoptera: Buprestidae). J. Econ. Entomol. 2010, 103, 1235–1241. [**Google Scholar**] [**CrossRef**]

63. Grant, G.G.; Poland, T.M.; Ciaramitaro, T.; Lyons, D.B.; Jones, G.C. Comparison of male and female emerald ash borer (Coleoptera: Buprestidae) responses to phoebe oil and (Z)-3-hexenol lures in light green prism traps. J. Econ. Entomol. 2011, 104, 173–179. [**Google Scholar**] [**CrossRef**] [**PubMed**]

64. Ryall, K.L.; Silk, P.J.; Mayo, P.; Crook, D.; Khrimian, A.; Cossé, A.A.; Sweeney, J.; Scarr, T.A. Attraction of Agrilus planipennis (Coleoptera: Buprestidae) to a volatile pheromone: Effects of release rate, host volatile, and trap placement. Environ. Entomol. 2012, 41, 648–656. [**Google Scholar**] [**CrossRef**] [**PubMed**]

65. Lelito, J.P.; Fraser, I.; Mastro, V.C.; Tumlinson, J.H.; Böröczky, K.; Baker, T.C. Visually mediated 'paratrooper copulations' in the mating behavior of Agrilus planipennis (Coleoptera: Buprestidae), a highly destructive invasive pest of North American ash trees. Insect Behav. 2007, 20, 537–552. [**Google Scholar**] [**CrossRef**] [**Green Version**]

66. Rodriguez-Saona, C.R.; Miller, J.R.; Poland, T.M.; Kuhn, T.M.; Otis, G.W.; Turk, T.; Ward, D.L. Behaviors of adult Agrilus planipennis (Coleoptera: Buprestidae). Great Lakes Entomol. 2007, 40, 1–16. [**Google Scholar**]

67. Marshall, J.M.; Storer, A.J.; Fraser, I.; Mastro, V.C. Multi-state comparison of detection tools at low emerald ash borer densities. In Proceedings of the Emerald Ash Borer and Asian Longhorned Beetle Research and Technology Development Meeting, Pittsburgh, PA, USA, 19–21 October 2009; Lance, D., Buck, J., Binion, D., Reardon, R., Mastro, V.U.S., Eds.; Forest Service Publication FHTET 2010-01; United States Department of Agriculture: Morgantown, WV, USA, 2010; pp. 124–125. [**Google Scholar**]

68. McCullough, D.G.; Poland, T.M.; Cappaert, D.; Anulewicz, A.C. Attraction of the emerald ash borer to ash trees stressed by girdling, herbicide and wounding. Can. J. For. Res. 2009, 39, 1331–1345. [**Google Scholar**] [**CrossRef**]

69. McCullough, D.G.; Poland, T.M.; Anulewicz, A.C.; Cappaert, D. Emerald ash borer (Coleoptera: Buprestidae) attraction to stressed or baited ash trees. Environ. Entomol. 2009, 38, 1668–1679. [**Google Scholar**] [**CrossRef**] [**PubMed**] [**Green Version**]

70. Knight, K.S.; Brown, J.P.; Long, R.P. Factors affecting the survival of ash (Fraxinus spp.) trees infested by emerald ash borer (Agrilus planipennis). Biol. Invas. 2013, 15, 371–383. [**Google Scholar**] [**CrossRef**]

71. Koch, J.L.; Carey, D.W.; Mason, M.E.; Poland, T.M.; Knight, K.S. Intraspecific variation in Fraxinus pennsylvanica responses to emerald ash borer (Agrilus planipennis). New For. 2015, 46, 995–1011. [**Google Scholar**] [**CrossRef**]

72. Mercader, R.J.; McCullough, D.G.; Storer, A.J.; Bedford, J.M.; Heyd, R.; Poland, T.M.; Katovich, S. Evaluation of the potential use of a systemic insecticide and girdled trees in area wide management of the emerald ash borer. For. Ecol. Manag. 2015, 350, 70–80. [**Google Scholar**] [**CrossRef**]

73. Wei, X.; Reardon, R.; Wu, Y.; Sun, J.H. Emerald ash borer, Agrilus planipennis Fairmaire (Coleoptera: Buprestidae), in China: A review and distribution survey. Acta Entomol. Sin. 2004, 47, 679–685. [**Google Scholar**]

09 PHYTOPHTHORA COMMUNITIES EXHIBIT VARIATION AMONG DIFFERENT TYPES OF MEDITERRANEAN VEGETATION WITHIN A NATURE RESERVE AREA

1. INTRODUCTION: AN OVERVIEW

The genus Phytophthora belongs to the Pythiaceae family within the Peronosporales order of the Oomycota class, which is part of the Chromista kingdom. Currently, there are more than 180 known species within this genus. However, it is believed that the real number of species is at least two or three times higher than the described ones, based on a conservative estimate [1,2]. Several Phytophthora species, including Phytophthora infestans, P. cinnamomi, and P. ramorum, are highly damaging plant diseases that result in significant agricultural losses and the decrease of trees globally [3,4,5,6,7,8,9,10]. The majority of Phytophthora species that cause plant diseases are able to infect a wide variety of plant species from different families [11,12,13,14,15]. These species are usually found in the soil, but some of them have adapted to an aerial lifestyle by producing sporangia that can be shed or temporarily live in the air [16]. Species within the phylogenetic ITS clade 6 that are found in water courses, lakes, and irrigation basins have a higher level of adaptation to aquatic habitats. This adaptation is observed in a limited number of species [17,18,19,20,21]. Typically, these aquatic Phytophthora species exhibit low levels of aggressiveness as plant infections. As a result, it is commonly believed that they function as saprotrophs in decomposing plant material in water and as opportunistic pathogens in environments near rivers. Nevertheless, the complete comprehension of their ecological function in ecosystems remains elusive.

The transportation of nursery plants by humans has been identified as a significant means by which non-native Phytophthora species are introduced into new places [22,23,24,25,26]. Research has shown that there is a direct connection between the ornamental plant industry and the spread of the harmful oak disease P. ramorum in natural areas in North America [27]. Utilizing nursery plants for forest restoration and afforestation can inadvertently introduce and propagate exotic Phytophthora species in natural habitats and forests [28,29,30]. Out of the 20 Phytophthora species found in a survey of protected natural areas in Sicily, 13 were identified as potentially non-native, whereas only seven were determined to be native to Europe [31]. Phytophthora species commonly found in forests and natural or naturalized habitats are often aggressive plant diseases that can infect cultivated plants. This suggests that these environments may serve as potential sources and reservoirs of Phytophthora inoculum for agricultural crops [32]. Therefore, it is imperative to incorporate the monitoring of forest and natural habitats into Phytophthora species surveillance and biosecurity strategies. Alien invasive Phytophthora species disrupt the stability and resilience of natural ecosystems. The intrusion of these viruses into natural and semi-natural environments poses a risk to the survival of native and uncommon plant species, as well as the overall diversity of plant communities [27,32,33,34]. Therefore, understanding the resident Phytophthora populations is essential for developing an effective management strategy for protected natural areas (PNAs).

The improvement of baiting and sampling techniques, along with rapid advancements in molecular diagnostics and DNA-sequencing technology, has made it easier to detect Phytophthora in environmental samples. This has also encouraged the investigation of Phytophthora communities in forests and natural ecosystems worldwide, including previously unexplored areas and watercourses. The surveys uncovered the extensive variety of Phytophthora species in these environments and resulted in the identification of a significant number of previously unknown species within this genus. Their contributions enhanced our comprehension of the worldwide variety of Phytophthora, the geographical spread patterns of individual Phytophthora species and groups from their original locations, their manner of life, and specifically, their reproductive conduct and ability to adapt to various conditions [2]. The investigation of the environmental elements that shape Phytophthora populations and influence their compositional variations in the wild has been lacking despite the existence of multiple surveys. Nevertheless, Protected Natural Areas (PNAs), because to their little interference from human activity, provide the optimal environment for investigating the ecology of Phytophthora species. The primary aim of this study was to examine the potential correlation between the diversity, frequency, and spatial distribution of Phytophthora species in a small nature reserve in Sicily, Southern Italy, and the type of plant and preferred habitat.

2. MATERIALS AND METHODS

2.1. Sampling Area

Sampling activities were carried out in the "Complesso Speleologico Villasmundo—S. Alfio" regional nature reserve (strict nature reserve), part of the Special Areas of Conservation (SAC) (ITA090024) "Cozzo Ogliastri" in the municipality of Melilli (Sicily, Italy) (**Figure 1**). The reserve is managed by the CUTGANA (University of Catania) and was established in 1998 in order to protect one of the most important karst systems of the Hyblean area. The reserve covers a surface of 72 hectares in the north-eastern sector of the Climiti Mountains, between the "Belluzza" and "Cugno di Rio" streams. Along the river "Cugno di Rio", there are cave entrances (A zone, the core zone of the reserve). The buffer zone of the reserve (B zone) hosts several natural ecosystems of great importance.

Figure 1. Geographical location of "Complesso Speleologico Villasmundo—S. Alfio" regional nature reserve (RNR) and plant communities within the reserve.

The plant communities and their respective habitats (according to the Habitats of Directive 92/43/EEC) where samples were collected are located in the B zone of the reserve and are described herein. (i) Platano-Salicetum pedicellatae Barbagallo, Brullo and Fagotto (Nature CODE 92C0 I): forests and woods, prevalently riparian, dominated by oriental plane (Platanus orientalis L.)

and willow (Salix pedicellata Desf.), probably a relic of a more extended plane tree wood. (ii) Sarcopoterium spinosum community (Nature CODE 5420): low, thorny formations of hemispherical shrubs of the coastal thermo-Mediterranean zone of Aegean islands, of mainland Greece and the Ionian islands, of coastal Anatolia, much more widespread and diverse than the Western Mediterranean formations. (iii) Myrto communis-Pistacietum lentisci (Molinier) Rivas-Martínez (Nature CODE 9320): thermo-Mediterranean woodland dominated by arborescent mastic (Pistacia lentiscus L.) and myrtle (Myrtus communis L.). (iv) Pistacio-Quercetum ilicis Brullo and Marcenò (Nature CODE 9340): Mediterranean oak stand characterized by holm oak (Quercus ilex L.). (v) Oleo-Quercetum virgilianae Brullo (Nature CODE: 91AA): mature climax community typified by southern live oak (Quercus virgiliana (Ten.) Ten.) in association with cork oak (Q. suber L.), holm oak (Q. ilex L.), carob (Ceratonia siliqua L.), wild olive (Olea europaea L. subsp. sylvestris (Mill.) Hegi)) and mastic (P. lentiscus L.). (vi) Gallery forest dominated by oleander (Nerium oleander L.) (Nature CODE 92CO I): thermo-Mediterranean community dominated by oleander (N. oleander L.) in association with willows (Salix spp.) and poplars (Populus spp.). Sampling activities were carried out during the autumn of 2015/2016 and 2017/2018. Plant nomenclature follows Pignatti [**68**], while the syntaxa classification follows Biondi et al. [**69**]. For the correlation between plant communities and habitat types, we referred to the Italian Interpretation Manual for the Habitats of Directive 92/43/EEC [**70**].

2.2. Sampling and Phytophthora Isolation

Twenty rhizosphere soil samples, including fine roots, were collected randomly from 20 mature trees and shrubs growing in all six plant communities.

Soil sampling and isolation were performed in accordance with Jung et al. [**31**]: four soil cores were collected under each tree or shrub, 50–150 cm away from the stem base, and rhizosphere soil from all four cores was bulked together (about 1 L).

For each sample, subsamples of 400 mL were used for baiting tests that were performed in a walk-in growth chamber with 12 h natural daylight at 20 °C. Young leaves of C. siliqua and Quercus spp. floated over flooded soil were used as baits. After 24–48 h incubation, necrotic segments (2 × 2 mm) from symptomatic leaves were plated in Petri dishes onto selective PARPNH agar medium which consisted of 100 mL V8 juice (Campbell Grocery Products Ltd., Ashford, UK), 15 g agar, 3 g CaCO3, 200 mg ampicillin, 10 mg rifampicin, 25 mg pentachloronitrobenzene (PCNB), 50 mg nystatin, 50 mg hymexazol, and 1 L of deionised water [**71**]. Petri dishes were incubated at 20 °C in the dark. Outgrowing Phytophthora hyphae were transferred onto V8 juice agar (V8A) under the stereomicroscope. All the Phytophthora isolates were maintained on V8 agar in the dark at a temperature of 6 °C.

Additional isolations were performed directly from river water by using an in situ baiting technique. To this end, 10 non-wounded young leaves of C. siliqua and Quercus spp. were arranged in a mesh-bag styrofoam raft (25 × 30 cm) fixed to float on the water surface (**Figure 2**). In total, five mesh-bag-styrofoam rafts were placed: two on the surface of the Belluzza stream and three on the Cugno di Rio river, the two water courses crossing the reserve. The rafts were collected after 3 days. All obtained isolates were maintained on V8A and stored at 6 °C in the dark.

Figure 2. Mesh-bag styrofoam raft: (A) raft placed in Cugno di Rio river; (B) raft floating on the water surface of the Belluzza river.

2.3. Morphological Characterization of Isolates

Cultures of seven days, grown on V8A at 20 °C in the dark, were used to group all isolates into morphotypes on the basis of their colony growth patterns. For each host plant and plant community, the different morphological types have been labeled with progressive numbering (Roman numbering); then, isolates belonging to the same sampling hosts have been tagged with the relative type number.

Moreover, morphological features of chlamydospores, sporangia, oogonia, antheridia and hyphal swellings were carefully analyzed and compared with species descriptions in the literature [**17,72**].

2.4. Molecular Identification of Isolates

The DNA of the pure cultures of isolates obtained from soil and rafts was extracted by using PowerPlant® Pro DNA isolation Kit (MO BIO Laboratories, Inc., Carlsbad, CA, USA), following the manufacturer's protocol. The DNA was preserved at −20 °C. The identification of Phytophthora species was performed by the analysis of internal transcribed spacer (ITS) regions of ribosomal DNA (rDNA). DNA was amplified using forward primers ITS6 (5′-GAAGGTGAAGTCGTAACAAGG-3′) [**73**] and reverse primer ITS4 (5′-TCCTCCGCTTATTGATATGC-3′) [**74**]. The PCR amplification mix and thermocycler conditions were in accordance with Cooke et al. [**73**]. All PCRs were carried out in a 25 μL reaction mix containing PCR Buffer (1×), dNTP mix (0.2 mM), $MgCl_2$ (1.5 mM), forward and reverse primers (0.5 mM each), Taq DNA Polymerase

(1 U) and 100 ng of DNA. The thermocycler conditions were as follows: 94 °C for 3 min; followed by 35 cycles of 94 °C for 30 s, 55 °C for 30 s, 72 °C for 30 s and then 72 °C for 10 min.

Amplicons were detected in 1% agarose gel and sequenced in both directions by an external service (Amsterdam, The Netherlands). Derived sequences were analyzed using FinchTV v.1.4.0 (Geospiza Inc., Seattle, WA, United States) [**75**]. For species identification, blast searches [**76**] in the Phytophthora Database [**77**], GenBank [**78**] and in a local database containing sequences of ex-type or key isolates from published studies were performed. Isolates were assigned to a species when their sequences were at least 99–100% identical to a reference isolate.

2.5. Analysis of Phytophthora Diversity

The Phytophthora diversity of soil samples sourced from the six plant communities was assessed by using the Shannon diversity (H = -Σpiln(pi)), the Pielou evenness (J = H/lnS) and the Simpson dominance (λ = 1/Σpi2) indices, where pi represents the frequency of each species and S the number of different species per plant community. Since the assumption of normal distribution was violated (the Shapiro–Wilk test was applied), the statistical differences in the diversity among sampling areas were assessed by the chi-square non-parametric test of Kruskal–Wallis followed by Dunn's multiple comparison post-hoc test (the R software [**79**] was used).

2.6. Soil Analysis and USDA Classification

An additional twenty rhizosphere soil samples were collected from plants within the selected sampling areas. The soil analysis was performed by a private laboratory (Progetto Ambiente & C. s.a.s., Catania, Italy) following the "official method of soil chemical analysis", in accordance with standard protocols defined by D.M. 13/09/1999, G.U. n°248, 21/10/99 and D.M. 25/03/2002, G.U. n°84, 10/04/2002. The following characteristics of the soil were determined: pH-H_2O, electrical conductivity at 25 °C, active limestone, organic matter content, nitrates and soil texture.

To define the soil texture of each sample, the USDA classification method [**80**] was used. The percentage of each soil component (sand, clay and silt) has been used in order to assign each sample to a textural class.

3. RESULTS

Molecular analyses were performed on 228 isolates, of which there were 45 from rivers and 183 from the soil of the reserve. These isolates represented all morphotypes recovered in soil samples and baiting rafts. Morphological and ITS sequence analyses revealed the occurrence of multiple Phytophthora species in each type of plant community with the only exception of the Sarcopoterium spinosum vegetation, where no Phytophthora species was detected. ITS sequence analyses showed that all 228 isolates (65.9%) matched with 99–100% identity reference sequences of 11 known Phytophthora species belonging to five different ITS clades. The ITS sequences of isolates of this study were deposited at GenBank, and, since isolates of the same species were all the same, only some sequences were deposited in GenBank. The Phytophthora species recovered in the reserve, the host species and the accession numbers.

Among the isolates, 100 belonged to species in ITS clade 6 (i.e., P. bilorbang, P. asparagi, P. lacustris and P. gonapodydes), 72 to species in ITS clade 2 (i.e., P. multivora, P. oleae and P. plurivora), 20 to species in ITS clade 8 (i.e., P. cryptogea and P. syringae) and 36 to P. parvispora and P. nicotianae, in ITS clades 7 and 1, respectively. The distribution of each Phytophthora species in each plant community type is shown in Table 1.

Table 1. Diagrammatic representation of the diversity and distribution of Phytophthora species recovered from six plant communities in the "Complesso Speleologico Villasmundo—S. Alfio" RNR. The proportion of isolates of each Phytophthora species recovered from each plant community is reported on the same row. The color intensity indicates the frequency of each species in each plant community (see legend).

Plant Community	*Phytophthora* Species Recovered in the Reserve										
	Clade 1	Clade 2			Clade 6				Clade 7	Clade 8	
	NIC	MUL	OLE	PLU	ASP	LAC	GON	BIL	PAR	CRY	SYR
Platano-Salicetum pedicellatae	16.7%	6.7%		3.3%		36.7%	3.3%	3.3%	6.7%	23.3%	
Sarcopoterium spinosum comm.											
Myrto communis-Pistacietum lentisci				34.4%	59.4%						6.25%
Pistacio-Quercetum ilicis				81.8%	4.5%	13.6%					
Oleo-Quercetum virgilianae	48%		8.7%	4.3%	35%					4.3%	
Gallery forest dominated by *N. oleander*				14.3%		71.4%		14.3%			

ASP = *P. asparagi*; BIL = *P. bilorbang*; CRY = *P. cryptogea*; GON = *P. gonapodyides*; LAC = *P. lacustris*; MUL = *P. multivora*; NIC = *P. nicotianae*; OLE = *P. oleae*; PAR = *P. parvispora*; PLU = *P. plurivora*; SYR = *P. syringae*. Underlined, the species that are considered native to Europe; in bold, species regarded as exotic [20,25,31]; the origin of *P. oleae* has not been established yet. % isolated species in each plant community: <25% 25–50% 50–75% >75%.

3.1. Phytophthora Diversity and Distribution in Different Plant Communities

With the only exception of the Sarcopoterium spinosum community, Phytophthora species from ITS clade 6 were isolated from all the plant communities (**Figure 3**). The most common Phytophthora species were aquatic species from ITS clade 6, such as P. lacustris, which were recovered from all river and riparian systems. P. gonapodyides and P. bilorbang were isolated from rhizosphere soil of willow while P. lacustris was recovered from river water and rhizosphere soil of mastic. With regard to species from clade 2, P. oleae, P. multivora and P. plurivora, they were detected in five out of six surveyed plant communities. In particular, P. oleae was recovered from rhizosphere soil of southern live oak, P. multivora from river water and P. plurivora from river water and rhizosphere soil of mastic, cork oak, southern live oak and holm oak. The clade 8 species P. cryptogea and P. syringae were found in three out of six plant communities. P. cryptogea was recovered from rhizosphere soil of willow, plane tree and southern live oak, while P. syringae was recovered only from rhizosphere soil of mastic. Phytophtora parvispora (ITS clade 7) was isolated exclusively from rhizosphere soil of mature willow trees in the Platano-Salicetum pedicellatae plant community. Finally, the ITS clade 1 species P. nicotianae was found in the Platano-Salicetum pedicellatae and Oleo-Quercetum virigilianae plant communities and was isolated from rhizosphere soil of plane, willow, southern live oak and cork oak (**Table 1**). Phytophthora plurivora, P. asparagi, P. lacustris and P. nicotianae were the prevalent species, accounting for 28%, 25%, 17% and 14% of all the isolates, respectively. Conversely, P. syringae, P. gonapodyides, P. bilorbang, P. multivora, P. oleae and P. parvispora were represented by less than 3% of isolates.

Figure 3. Plant communities: (A) Platano-Salicetum pedicellatae; (B) gallery forest dominated by Nerium oleander; (C) Sarcopoterium spinosum community; (D) Myrto communis-Pistacietum lentisci; (E) Pistacio-Quercetum ilicis; (F) Oleo-Quercetum virgilianae.

The structure of the community of Phytophthora species in soil samples differed significantly between the vegetation types. Among 11 Phytophthora species detected, only P. plurivora was found in all five plant communities. By contrast, P. oleae was detected exclusively in the Oleo-Quercetum virgilianae, P. syringae in the Myrto Communis-Pistacietum lentisci; P. gonapodyides, P. bilorbang and P. parvispora in the Platanum-Salicetum pedicellatae.

Overall, the sets of Phytophthora species from different plant communities strongly overlapped (**Figure 4**).

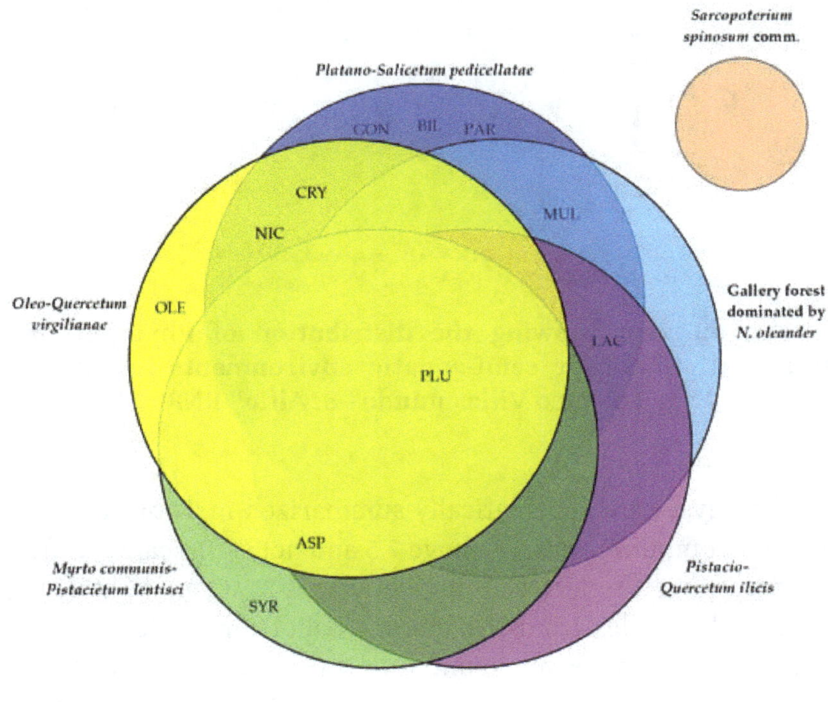

ASP= *P. asparagi*; BIL= *P. bilorbang*; CRY= *P. cryptogea*; GON= *P. gonapodyides*; LAC= *P. lacustris*; MUL= *P. multivora*; NIC= *P. nicotianae*; OLE= *P. oleae*; PAR= P. *parvispora*; PLU= *P. plurivora*; SYR= *P. syringae*

Figure 4. Venn diagram showing the distribution of Phytophthora species in six different plant communities in the "Complesso Speleologico Villasmundo—S. Alfio" RNR.

Comparing the different environments from which the species have been isolated, it can be observed that P. cryptogea, P. lacustris, P. nicotianae and P. plurivora were recovered from both terrestrial and aquatic or semi-aquatic environments, while P. asparagi, P. oleae and P. syringae were found exclusively in terrestrial environments and P. bilorbang, P. gonapodyides, P. multivora and P. parvispora only in aquatic or semi-aquatic environments. Overall, the sets of species from terrestrial habitats and from aquatic or semi-aquatic (riparian) habitats were distinct (**Figure 5**).

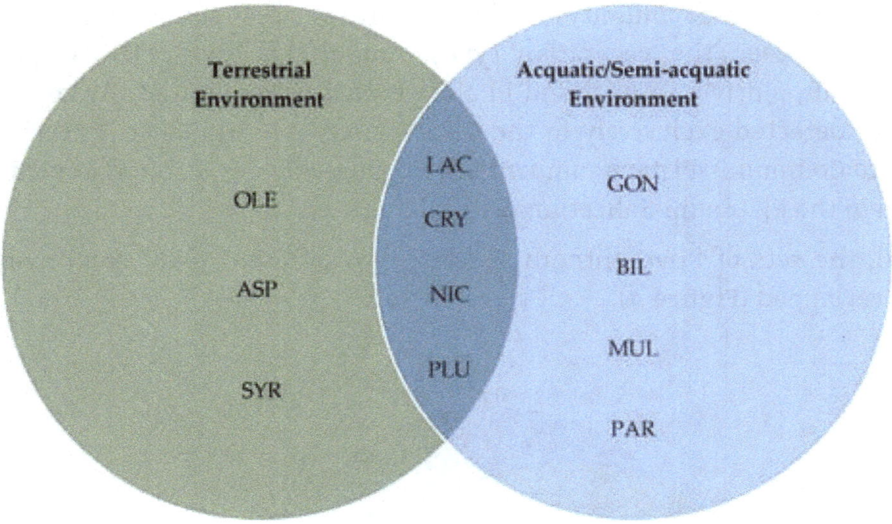

ASP= *P. asparagi*; BIL= *P. bilorbang*; CRY= *P. cryptogea*; GON= *P. gonapodyides*; LAC= *P. lacustris*; MUL= *P. multivora*; NIC= *P. nicotianae*; OLE= *P. oleae*; PAR= P. *parvispora*; PLU= *P. plurivora*; SYR= *P. syringae*

Figure 5. Venn diagram showing the distribution of Phytophthora species in terrestrial and aquatic or semi-aquatic environments in the "Complesso Speleologico Villasmundo—S. Alfio" RNR.

3.2. Analysis of Soil

Results of soil analyses are schematically summarized in **Table 2**. Values of soil pH from all plant communities were above 7.5 and not significantly different from each other. Concerning the electrical conductivity at 25 °C, moderately high values were found in soil samples from the Platano-Salicetum pedicellatae (1100 ± 48 µS/cm), Sarcopoterium spinosum comm. (973 ± 45 µS/cm) and Myrto communis-Pistacietum lentisci (894 ± 43 µS/cm) plant communities, high values in soil of the Pistacio-Quercetum ilicis (1414 ± 63 µS/cm) and significantly lower values in soil from the Oleo-Quercetum virgilianae (439 ± 36 µS/cm) plant community. As far as the active limestone is concerned, a high value was found in soil of the Oleo-Quercetum virgilianae (127 ± 5 g/Kg), while a relatively low value was recorded in soil from Pistacio-Quercetum ilicis (31 ± 2 g/Kg). The amount of nitrates was relatively high in soil from Pistacio-Quercetum ilicis and Oleo-Quercetum virgilianae plant communities (12.8 ± 1 and 11.59 ± 1 mg/Kg, respectively) and significantly lower in soil taken from Platano-Salicetum pedicellatae, Sarcopoterium spinosum comm. and Myrto communis-Pistacietum lentisci. Soils from all plant communities were rich in organic matter. The highest content of organic matter was found in soil from Pistacio-Quercetum ilicis (15%).

Table 2. Soil characteristics in each plant community type of the "Complesso Speleologico Villasmundo—S. Alfio" RNR.

Soil Properties	Plant Community [a]				
	Platano-Salicetum pedicellatae	*Sarcopoterium spinosum* **Comm.**	*Myrto communis-Pistacietum lentisci*	*Pistacio-Quercetum ilicis*	*Oleo-Quercetum virgilianae*
pH	7.5 ± 0.1	7.4 ± 0.1	7.3 ± 0.1	7.4 ± 0.1	7.6 ± 0.1
Electrical conductivity at 25 °C (μS/cm)	1.100 ± 48	973 ± 45	894 ± 43	1.414 ± 63	439 ± 36
Active limestone (g/Kg)	69 ± 3	86 ± 4	81 ± 4	31 ± 2	127 ± 5
Soil texture	Sandy clay loam	Sandy clay loam	Sandy clay	Clay loam	Sandy clay loam
Nitrates (mg/Kg)	6.7 ± 0.7	6.8 ± 0.6	4 ± 0.5	12.8 ± 1	11.59 ± 1
Organic matter (%)	5 ± 0.3	6 ± 0.5	8.2 ± 0.7	15 ± 2	6.1 ± 0.5

[a] Only rafts were placed in the gallery forest dominated by Nerium oleander and no soil sample was collected from this vegetation community

According to the USDA soil textural classification, soils from Platano-Salicetum pedicellatae, Sarcopoterium spinosum comm. and Oleo-Quercetum virgilianae were sandy clay loam; soil from Pistacio-Quercetum ilicis was clay loam and soil from the Myrto communis-Pistacietum lentisci was sandy clay. No obvious correlation was found between soil characteristics and Phytophthora species diversity.

3.3. Analysis of Phytophthora Diversity

The analysis of the diversity of Phytophthora populations from different plant communities showed a high variability of evenness (**Table 3**). Significantly higher values of Shannon and Pielou evenness diversity indices were observed in the Phytophthora populations from the Platano-Salicetum pedicellatae and Oleo-Quercetum virgilianae plant communities, while the evenness was moderate to low in Phytophthora populations from Myrto communis-Pistacietum lentisci, Pistacio-Quercetum ilici and the gallery forest dominated by N. oleander. In contrast, values of the Simpson dominance index were significantly higher in Phytophthora populations from the gallery forest dominated by N. oleander and Pistacio-Quercetum ilicis, intermediate in the Phytophthora population from Myrto communis-Pistacietum lentisci and significantly lower in the Phytophthora populations from Platano-Salicetum pedicellatae and Oleo-Quercetum virgilianae plant communities.

Table 3. Values of the diversity indices, Shannon diversity, evenness and Simpson dominance of Phytophthora populations from six different plant communities in the "Complesso Speleologico Villasmundo—S. Alfio" RNR. Data were analyzed with the Kruskal-Wallis test. Different letters indicate significant differences according to Dunn's multiple comparison test ($p \leq 0.01$).

Plant Communities	Diversity Indexes					
	Shannon Index		Pielou Evenness		Simpson Dominance	
Platano-Salicetum pedicellatae	1.707	a	0.821	a	0.229	c
Sarcopoterium spinosum comm.	-		-		-	
Myrto communis- Pistacietum lentisci	0.760	b	0.692	b	0.563	b
Pistacio-Quercetum ilicis	0.652	bc	0.593	c	0.649	a
Oleo-Quercetum virgilianae	1.205	a	0.749	ab	0.361	c
Gallery forest dominated by *N. oleander*	0.451	c	0.650	b	0.722	a

4. DISCUSSION

As many as 11 Phytophthora species, including putatively endemic and exotic species as well as pathogens associated prevalently to agriculture, such as P. nicotianae [**14,81**], were found in the "Complesso Speleologico Villasmundo—S. Alfio" RNR, a relatively high number compared to the limited extension of the reserve (0.72 km^2). This can be explained with the quite recent establishment of the reserve (D. ARTA n° 616 04/11/1998), which, until 22 years ago, also comprised tree crops whose relics still survive and a high variability of plant communities included within the boundaries of the reserve, each occupying a different ecological niche and constituting a distinct ecosystem. The variability of ecosystems within the reserve is further increased by the presence of two streams, which favor the settlement of species with aquatic or semi-aquatic habitats along the banks and in the rhizosphere soil of the riparian vegetation.

Phytophthora plurivora, P. asparagi and P. lacustris were the most widespread species in the "Complesso Speleologico Villasmundo—S. Alfio" RNR, whereas the other species had a scattered distribution. Among the species recovered from the RNR, some, such as P. cryptogea, P. gonapodyides, P. plurivora and P. syringae, are common in natural and forest ecosystems throughout Europe [**35,60,61,82**], while others, like P. bilorbang, P. oleae and P. parvispora, occur more sporadically. Both P. plurivora and P. gonapodyides were reported in a previous survey of protected natural areas in Sicily [**31**]. In this study, P. oleae was isolated only from the rhizosphere soil of southern live oak (Q. virgilianae (Ten.) Ten.) in the Oleo-Quecetum virgilianae plant community. This was the first time that this recently described species was reported from a host plant other than olive (O. europaea L.) [**83,84**]. Phytophthora bilorbang is a prevalently aquatic species, but it has been occasionally reported as an opportunistic, aggressive plant pathogen [**85**]. Phytophthora parvispora (formerly P. cinnamomi var. parvispora), a species in clade 7a, is an aggressive plant pathogen with a prevalently terrestrial habitat. In this survey, it was recovered from the rhizosphere of willow (S. pedicellata Desf.) in a riparian semi-aquatic environment. In a previous study, it had been reported as a pathogen of strawberry tree (Arbutus unedo) in Sardinia (central Italy) and described as a new species distinct from P. cinnamomi [**86**]. The first record of this species in Italy dates back to 2010 and was from ornamental plants in nursery [**87**].

Phytophthora gonapodyides is a species with a prevalently aquatic lifestyle but may be an aggressive opportunistic plant pathogen [**88**]. It inhibits seed germination and causes root rot and stem lesions in Quercus robur L. and Q. ilex L.; also, in association with other species of Phytophthora, it was recovered from declining oak stands even in xerophytic environments [**71,89,90**]. According to Erwin and Ribeiro [**91**], the damage caused by this species is underestimated and is overshadowed by that of other Phytophthora species, since it is traditionally regarded a minor pathogen. Presumably, the role of this and other Phytophthora species with a prevalently aquatic lifestyle in natural ecosystems is more complex than that of mere plant pathogens and deserves to be further investigated.

Phytophthora plurivora is a very polyphagous pathogen whose host range encompasses more than 80 woody host species including oaks (Quercus spp.), willows (Salix spp.), oleander (N. oleander) and oriental plane (P. orientalis L.) [**31,92**]. In the “Complesso Speleologico Villasmundo—S. Alfio” RNR, it was the species with the widest distribution and was isolated from the rhizosphere soil of mastic (P. lentiscus L.), southern live oak (Q. virgilianae (Ten.) Ten.), holm oak (Q. ilex L.) and cork oak (Q. suber L.), as well as from river water of Platano-Salicetum pedicellatae. Its polyphagia and the ability to produce resting spores, such as thick-walled oospores, might explain the cosmopolitan attitude of this species and its widespread occurrence in Mediterranean natural ecosystems. Interestingly, both P. plurivora and the other three species found exclusively in terrestrial habitats, i.e., P. asaparagi, P. oleae and P. syringae, are homothallic and produce oospores with thick walls. This could be a common adaptive strategy of these species to cope with adverse environmental conditions typical of many ecosystems in Mediterranean climate, such as long periods of drought alternating with intense rainfall, mild wet winters, high temperatures in summer, wide excursion of daily temperature and rapid fluctuation of air and soil humidity. In general, the ability to produce resting structures increases the competitiveness of Phytophthora spp. in terrestrial environments [**93**]. All the species isolated from the rhizosphere soil samples collected from diverse plant communities and streams in the "Complesso Speleologico Villasmundo—S. Alfio" RNR, with the only exception of P. lacustris, produce at least one kind of resting structure. However, only P. asparagi, P. bilorbang, P. multivora, P. plurivora and P. syringae are homothallic. In addition, P. nicotianae, P. oleae, P. parvispora and possibly P. syringae are able to produce chlamydospores.

Phytophthora lacustris is a ubiquitous species in riparian ecosystems, such as reed belts and riparian alder stands, throughout Europe and North America [**20,94**]. Like other species in clade 6, this species tolerates high temperatures, is sexually sterile and produces a great amount of zoospores [**20**]. For the clade as a whole, this combination of characters has been interpreted as an adaptation to riparian environments [**17,95**]. However, although P. lacustris is significantly less aggressive than other Phytophthora spp. with a terrestrial habitat, it can seriously infect fine roots of trees stressed by episodes of flooding or drought [**20**]. Accordingly, in the present study, P. lacustris was isolated prevalently from the stream crossing the gallery forest plant community, which is dominated by oleander (N. oleander), and from the rhizosphere soil of willow (S. pedicellata) trees in the Platano-Salicetum pedicellate riparian plant community, but P. lacustris was recovered even from the rhizosphere soil of southern live oak (Q. virgilianae) in the xerophytic Pistacio-Quercetum ilicis plant community.

Phytophthora asparagi, another species included in clade 6 but in a separate subclade [**96**], was recovered exclusively from terrestrial environments and was common in different plant communities, including Myrto communis-Pistacietum lentisci, Pistacio-Quercetum ilicis and Oleo-Quercetum virgilianae. In a previous survey of the National Park of La Maddalena archipelago (Sardinia), this species was isolated frequently from the rhizosphere soil of typical plants of the Mediterranean maquis vegetation, showing symptoms of decline, including white asparagus (Asparagus albus L.), Phoenician juniper (Juniperus turbinata Guss.) and mastic (P. lentiscus L.) [**65**]. In the "Complesso Speleologico Villasmundo—S. Alfio" RNR, P. asparagi was recovered from the rhizosphere soil of mastic as well as from the rhizosphere soil of evergreen oaks, including southern live oak (Q. virginiana (Ten.) Ten.), holm oak (Q. ilex L.) and cork oak (Q. suber L.). As a result, the list of known host plants with which this species is associated has expanded.

In the present survey, two other species from clade 6, P. gonapodiydes and P. bilorbang, were detected. They were found exclusively in the riparian ecosystem and rivers, confirming a prevalently aquatic lifestyle [**20,65,95,97**]. Very probably, the high number of isolates of clade 6 species obtained in this study and their widespread occurrence is related to the high adaptability of these species to different environmental conditions and their ability to produce numerous zoospores, which in natural ecosystems are easily transported and spread by water courses.

The diversity and species richness of Phytophthora populations associated with different plant communities—as measured using three distinct indexes, Shannon diversity, Pielou evenness and Simpson dominance—were significantly higher in Platano-Salicetum pedicellatae and Oleo-Quercetum virgilianae, very probably reflecting the richness of host plant species and the complexity of these termo-mesophilous communities. Moreover, the higher richness and diversity of species in the Platano-Salicetum pedicellatae might be explained by the presence of both aquatic and terrestrial environments within this plant community. The Pielou evenness index value also suggests that the Phytophthora community in the Platano-Salicetum pedicellatae was more balanced than in the Oleo-Quercetum virgilianae, due to the dominance of P. nicotianae, an aggressive and polyphagous plant pathogen, in the latter community. Considering the proportions of isolates recovered from each of the abovementioned plant communities regarded together as an ecological succession, it can be supposed that the progressive decrease in complexity causes both the reduction in the diversity of Phytophthora communities and the progressive unbalance in their composition, resulting in the dominance of the most aggressive Phytophthora species. In accordance with this hypothesis, the very low complexity of the garrigue of Sarcopoterium spinosum (L.) Spach is the extreme ecological limit for the establishment of Phytophthora communities.

Nevertheless, the sets of Phytophthora species in each plant community, as represented by the Venn diagram, showed a tendency to cluster together, probably due to both the dominant presence of two invasive species (i.e., P. plurivora and P. asparagi) and the proximity of different ecosystems. The strong overlapping of Phytophthora species sets also indicates that environmental conditions are conducive to these oomycetes in all types of plant communities examined in this study. In particular, no obvious correlation was observed between the diversity of distribution of Phytophthora populations and some physicochemical soil properties that may influence the ecology of Phytophthora species, their aggressiveness and ability to survive, such as pH, salinity, textural class, active limestone, nitrates and organic matter content [**98,99,100,101,102,103,104**]. In all plant community types, the content of organic matter was relatively high, but this was particularly true for the Pistacio-Quercetum ilicis. This might explain the occurrence of P. lacustris in the rhizosphere soil of trees in this vegetation community. Phytophthora species, in fact, including several aggressive plant pathogens, have the ability to either survive or complete their lifecycle as saprobes, despite their poor ability to compete with other saprophytic organisms. This saprophytic attitude is more pronounced in clade 6 species, like P. lacustris [**20**]. Therefore, higher levels of organic matter favor the establishment and survival of Phytophthora spp., in general, and of clade 6 species, in particular.

Moreover, other soil parameters, with the only exception of the high level of active limestone in the Oleo-Quercetum virgilianae, which could be a limiting factor, can be considered within the optimum range in all plant community types, including the Sarcopoterium spinosum comm., where no Phytophthora species was found. This indicates that the aforementioned physicochemical soil properties were not a limiting factor for the settlement and survival of Phytophthora spp. in this xerophytic plant community type. Other major environmental factors conditioning the ecology of soilborne Phytophthora species are soil moisture and temperature [**91**]. The effects of soil water status, generally expressed in terms of matric potential, on the ability of Phytophthora species to sporulate and cause disease have been extensively investigated in agricultural systems [**91,105**]. The geographical range of many soilborne Phytophthora species or their ability to thrive and survive at high altitudes depend on their extreme and optimum temperatures for growth [**7,57,106**] and many aquatic Phytophthora species are thermophilic [**19,20,107**], suggesting that this is an adaptive functional trait of this group of species. However, the values of these two soil parameters, matric potential and temperature, are not constant and, in the Mediterranean climate, they vary considerably and suddenly across the seasons or even the same day, so they were not considered in this study.

Overall, comparing Myrto communis- Pistacietum lentisci, Pistacio-Quercetum ilicis and Sarcopoterium spinosum comm., with Platano-Salicetum pedicellatae and Oleo-Quercetum virgilianae, the Phytophthora diversity shows a trend that could be related to the degree of maturity of plant communities. Results of this study are in agreement with those of a similar study carried out in forests of South Africa [**108**]. Conversely, the geographic distribution of P. cinnamomi in the Iberian Peninsula turned out to be influenced primarily by abiotic factors, including soil texture and climate, followed by land use and lastly by the presence of main host plant species [**109**]. However, comparisons with the results of this last study are impaired by the scale of the survey, the heterogeneity of environments investigated and the fact that only one Phytophthora species was involved.

When the Phytophthora species found in the "Complesso Speleologico Villasmundo—S. Alfio" RNR were grouped on the basis of the type of environment from which they were recovered, the Venn diagram clearly separated them into two distinct sets, including species with a prevalent or exclusive terrestrial habitat, such as P. asparagi, P. oleae and P. syringae, and species with a prevalent or exclusive aquatic or semi-aquatic habitat, such as P. bilorbang and P. gonapodyides, respectively. The first group was characterized by homothallic species, able to produce thick-walled oospores allowing them to survive adverse soil conditions; the second group comprised species that have adapted to thrive in aquatic and semi-aquatic environments. The presence of prevalently aquatic species in the first set and the presence of typically terrestrial species in the second set may be explained by the proximity of vegetation to water streams, flooding events and runoff of rain water. Consistently with our results, in a very recent study a correlation was found between the Phytophthora community and both the type of vegetation and environment in the French Guiana rainforest, which, like other neotropical forests, is considered a major plant diversity hotspot [**110**]. However, unexpectedly, the Phytophthora community in this forest showed a low diversity compared to the richness of species recovered from the "Complesso Speleologico Villasmundo—S. Alfio" RNR.

5. CONCISE SUMMARY

This study showed a correlation between the vegetation type and the diversity and distribution of Phytophthora species in a small natural reserve where different types of plant communities typical of the Mediterranean macroregion were represented. Moreover, it confirmed that Phytophthora species may be grouped according to an ecological criterion on the basis of their prevalently terrestrial or aquatic lifestyle. The ecology of these two distinct groups is conditioned by their aggressiveness as well as by some other biological characteristics, such as high temperature requirement, the ability to produce resting structures, the prevalence of asexual reproduction and the amount of zoospores that they produce. The presence of different ecosystems in a restricted area, i.e., in comparable environmental and climatic conditions, was a unique opportunity to highlight this correlation between the type of vegetation and the diversity of Phytophthora communities. However, only a more extensive survey of the same plant community's types in distant geographic areas would definitely demonstrate that the type of vegetation is a major driving factor shaping the Phytophthora communities in natural ecosystems. Other ecological and epidemiological aspects that deserve further attention are the seasonal pattern of Phytophthora populations in these ecosystems, the selective pressure exerted by invasive Phytophthora species on natural vegetation and their impact on spontaneous regeneration.

6. AT A GLANCE

Protected natural areas serve as a repository for Phytophthora species and offer ideal conditions for studying their ecology, as they are less affected by human activity compared to other settings. Context and Goals: The main aim of this study was to establish a connection between the variety and spread of Phytophthora species and the plant life in aquatic, riparian, and terrestrial environments inside a protected region in Eastern Sicily, Southern Italy. Experimental procedures: Water and soil samples were collected from two streams within the reserve. Additionally, samples were taken from six different types of vegetation, namely Platano-Salicetum pedicellatae, the Sarcopoterium spinosum community, Myrto communis-Pistacietum lentisci, Pistacio-Quercetum ilicis, Oleo-Quercetum virgilianae, and a gallery forest primarily consisting of Nerium oleander (classified as habitats according to the Natura 2000 classification). Phytophthora species were isolated from samples by employing leaf baiting and were categorized based on their physical properties and the sequencing of internal transcribed spacer (ITS) regions of ribosomal DNA (rDNA). A total of 11 Phytophthora species belonging to five distinct ITS clades were found. These species include P. asparagi, P. bilorbang, P. cryptogea, P. gonapodyides, P. lacustris, P. multivora, P. nicotianae, P. oleae, P. parvispora, P. plurivora, and P. syringae. No Phytophthora species were detected in the Sarcopoterium spinosum community. Phytophthora asparagi, P. lacustris, and P. plurivora were the most common species in the other five plant communities, but only P. plurivora was found in all of them. In general, aquatic species from clade 6 (100 out of 228 isolates) were the most prevalent. They were found in all five types of vegetation, as well as in streams and riparian environments. The Phytophthora populations discovered in the Platano-Salicetum pedicellatae and Oleo-Quercetum virgilianae exhibit the greatest diversity, but no connection was observed with the physicochemical properties of the soil. Findings: The diversity of Phytophthora communities in this reserve is mostly influenced by the kind of flora and whether the habitat is aquatic or terrestrial.

REFERENCES

1. Scott, P.; Bader, M.K.F.; Burgess, T.; Hardy, G.; Williams, N. Global biogeography and invasion risk of the plant pathogen genus Phytophthora. Environ. Sci. Policy 2019, 101, 175–182. [**Google Scholar**] [**CrossRef**]

2. Jung, T.; Scanu, B.; Brasier, C.M.; Webber, J.; Milenković, I.; Corcobado, T.; Tomšovský, M.; Pánek, M.; Bakonyi, J.; Maia, C.; et al. A survey in natural forest ecosystems of Vietnam reveals high diversity of both new and described phytophthora taxa including P. ramorum. Forests 2020, 11, 93. [**Google Scholar**] [**CrossRef**] [**Green Version**]

3. Rizzo, D.M.; Garbelotto, M.; Hansen, E.M. Phytophthora ramorum: Integrative research and management of an emerging pathogen in California and Oregon forests. Annu. Rev. Phytopathol. 2005, 43, 309–335. [**Google Scholar**] [**CrossRef**] [**PubMed**] [**Green Version**]

4. Haas, B.J.; Kamoun, S.; Zody, M.C.; Jiang, R.H.Y.; Handsaker, R.E.; Cano, L.M.; Grabherr, M.; Kodira, C.D.; Raffaele, S.; Torto-Alalibo, T.; et al. Genome sequence and analysis of the Irish potato famine pathogen Phytophthora infestans. Nature 2009, 461, 393–398. [**Google Scholar**] [**CrossRef**] [**PubMed**]

5. Grünwald, N.J.; Garbelotto, M.; Goss, E.M.; Heungens, K.; Prospero, S. Emergence of the sudden oak death pathogen Phytophthora ramorum. Trends Microbiol. 2012, 20, 131–138. [**Google Scholar**] [**CrossRef**]

6. Kamoun, S.; Furzer, O.; Jones, J.D.G.; Judelson, H.S.; Ali, G.S.; Dalio, R.J.D.; Roy, S.G.; Schena, L.; Zambounis, A.; Panabières, F.; et al. The Top 10 oomycete pathogens in molecular plant pathology. Mol. Plant Pathol. 2015, 16, 413–434. **[Google Scholar]** **[CrossRef]**

7. Burgess, T.I.; Scott, J.K.; Mcdougall, K.L.; Stukely, M.J.C.; Crane, C.; Dunstan, W.A.; Brigg, F.; Andjic, V.; White, D.; Rudman, T.; et al. Current and projected global distribution of Phytophthora cinnamomi, one of the world's worst plant pathogens. Glob. Chang. Biol. 2017, 23, 1661–1674. **[Google Scholar]** **[CrossRef]** **[Green Version]**

8. Moricca, S.; Linaldeddu, B.T.; Ginetti, B.; Scanu, B.; Franceschini, A.; Ragazzi, A. Endemic and emerging pathogens threatening cork oak trees: Management options for conserving a unique forest ecosystem. Plant Dis. 2016, 100, 2184–2193. **[Google Scholar]** **[CrossRef]** **[Green Version]**

9. Frisullo, S.; Lima, G.; Lio, G.M.D.S.; Camele, I.; Melissano, L.; Puglisi, I.; Pane, A.; Agosteo, G.E.; Prudente, L.; Cacciola, S.O. Phytophthora cinnamomi involved in the decline of holm oak (Quercus ilex) stands in southern Italy. For. Sci. 2018, 64, 290–298. **[Google Scholar]** **[CrossRef]**

10. Cacciola, S.O.; Gullino, M.L. Emerging and re-emerging fungus and oomycete soil-borne plant diseases in Italy. Phytopathol. Mediterr. 2019, 58, 451–472. **[Google Scholar]**

11. Hardham, A.R. Phytophthora cinnamomi. Mol. Plant Pathol. 2005, 6, 589–604. **[Google Scholar]** **[CrossRef]** **[PubMed]**

12. Grünwald, N.J.; Goss, E.M.; Press, C.M. Phytophthora ramorum: A pathogen with a remarkably wide host range causing sudden oak death on oaks and ramorum blight on woody ornamentals. Mol. Plant Pathol. 2008, 9, 729–740. **[Google Scholar]** **[CrossRef]** **[PubMed]**

13. Abad, Z.G.; Abad, J.A.; Cacciola, S.O.; Pane, A.; Faedda, R.; Moralejo, E.; Pérez-Sierra, A.; Abad-Campos, P.; Alvarez-Bernaola, L.A.; Bakonyi, J.; et al. Phytophthora niederhauserii sp. nov., a polyphagous species associated with ornamentals, fruit trees and native plants in 13 countries. Mycologia 2014, 106, 431–447. **[Google Scholar]** **[CrossRef]** **[PubMed]**

14. Panabières, F.; Ali, G.S.; Allagui, M.B.; Dalio, R.J.D.; Gudmestad, N.C.; Kuhn, M.L.; Guha Roy, S.; Schena, L.; Zampounis, A. Phytophthora nicotianae diseases worldwide: New knowledge of a long-recognised pathogen. Phytopathol. Mediterr. 2016, 55, 20–40. **[Google Scholar]**

15. Migliorini, D.; Khdiar, M.Y.; Padrón, C.R.; Vivas, M.; Barber, P.A.; Hardy, G.E.S.J.; Burgess, T.I. Extending the host range of Phytophthora multivora, a pathogen of woody plants in horticulture, nurseries, urban environments and natural ecosystems. Urban For. Urban Green. 2019, 46, 126460. **[Google Scholar]** **[CrossRef]**

16. Jung, T.; Pérez-Sierra, A.; Durán, A.; Jung, M.H.; Balci, Y.; Scanu, B. Canker and decline diseases caused by soil- and airborne Phytophthora species in forests and woodlands. Persoonia-Mol. Phylogeny Evol. Fungi 2018, 40, 182–220. **[Google Scholar]** **[CrossRef]** **[Green Version]**

17. Jung, T.; Stukely, M.J.C.; Hardy, G.E.S.J.; White, D.; Paap, T.; Dunstan, W.A.; Burgess, T.I. Multiple new Phytophthora species from ITS Clade 6 associated with natural ecosystems in Australia: Evolutionary and ecological implications. Persoonia Mol. Phylogeny Evol. Fungi 2011, 26, 13–39. **[Google Scholar]** **[CrossRef]** **[Green Version]**

18. Nagel, J.H.; Gryzenhout, M.; Slippers, B.; Wingfield, M.J.; Hardy, G.E.S.J.; Stukely, M.J.C.; Burgess, T.I. Characterization of Phytophthora hybrids from ITS clade 6 associated with

riparian ecosystems in South Africa and Australia. Fungal Biol. 2013, 117, 329–347. [**Google Scholar**] [**CrossRef**] [**Green Version**]

19. Copes, W.E.; Yang, X.; Hong, C. Phytophthora species recovered from irrigation reservoirs in Mississippi and Alabama nurseries and pathogenicity of three new species. Plant Dis. 2015, 99, 1390–1395. [**Google Scholar**] [**CrossRef**] [**Green Version**]
20. Nechwatal, J.; Bakonyi, J.; Cacciola, S.O.; Cooke, D.E.L.; Jung, T.; Nagy, Z.Á.; Vannini, A.; Vettraino, A.M.; Brasier, C.M. The morphology, behaviour and molecular phylogeny of Phytophthora taxon Salixsoil and its redesignation as Phytophthora lacustris sp. nov. Plant Pathol. 2013, 62, 355–369. [**Google Scholar**] [**CrossRef**] [**Green Version**]
21. Redondo, M.A.; Boberg, J.; Stenlid, J.; Oliva, J. Contrasting distribution patterns between aquatic and terrestrial Phytophthora species along a climatic gradient are linked to functional traits. ISME J. 2018, 12, 2967–2980. [**Google Scholar**] [**CrossRef**] [**Green Version**]
22. Mammella, M.A.; Martin, F.N.; Cacciola, S.O.; Coffey, M.D.; Faedda, R.; Schena, L. Analyses of the population structure in a global collection of Phytophthora nicotianae isolates inferred from mitochondrial and nuclear DNA sequences. Phytopathology 2013, 103, 610–622. [**Google Scholar**] [**CrossRef**] [**Green Version**]
23. Migliorini, D.; Ghelardini, L.; Tondini, E.; Luchi, N.; Santini, A. The potential of symptomless potted plants for carrying invasive soilborne plant pathogens. Divers. Distrib. 2015, 21, 1218–1229. [**Google Scholar**] [**CrossRef**]
24. Biasi, A.; Martin, F.N.; Cacciola, S.O.; Lio, G.M.D.S.; Grünwald, N.J.; Schena, L. Genetic analysis of Phytophthora nicotianae populations from different hosts using microsatellite markers. Phytopathology 2016, 106, 1006–1014. [**Google Scholar**] [**CrossRef**] [**PubMed**] [**Green Version**]
25. Jung, T.; Orlikowski, L.; Henricot, B.; Abad-Campos, P.; Aday, A.G.; Casal, O.A.; Bakonyi, J.; Cacciola, S.O.; Cech, T.; Chavarriaga, D.; et al. Widespread Phytophthora infestations in European nurseries put forest, semi-natural and horticultural ecosystems at high risk of Phytophthora diseases. For. Pathol. 2016, 46, 134–163. [**Google Scholar**] [**CrossRef**] [**Green Version**]
26. Garbelotto, M.; Frankel, S.J.; Scanu, B. Soil- and waterborne Phytophthora species linked to recent outbreaks in Northern California restoration sites. Calif. Agric. 2018, 72, 208–216. [**Google Scholar**] [**CrossRef**] [**Green Version**]
27. Garbelotto, M.; Hayden, K.J. Sudden oak death: Interactions of the exotic oomycete Phyotphthora ramorum with native North American hosts. Eukaryot. Cell 2012, 11, 131–1323. [**Google Scholar**] [**CrossRef**] [**PubMed**] [**Green Version**]
28. Simamora, A.V.; Paap, T.; Howard, K.; Stukely, M.J.C.; Hardy, G.E.S.J.; Burgess, T.I. Phytophthora contamination in a nursery and its potential dispersal into the natural environment. Plant Dis. 2018, 102, 132–139. [**Google Scholar**] [**CrossRef**] [**Green Version**]
29. Sims, L.; Tjosvold, S.; Chambers, D.; Garbelotto, M. Control of Phytophthora species in plant stock for habitat restoration through best management practices. Plant Pathol. 2019, 68, 196–204. [**Google Scholar**] [**CrossRef**] [**Green Version**]
30. Sims, L.L.; Chee, C.; Bourret, T.; Hunter, S.; Garbelotto, M. Genetic and phenotypic variation of Phytophthora crassamura isolates from California nurseries and restoration sites. Fungal Biol. 2019, 123, 159–169. [**Google Scholar**] [**CrossRef**]

31. Jung, T.; La Spada, F.; Pane, A.; Aloi, F.; Evoli, M.; Jung, M.H.; Scanu, B.; Faedda, R.; Rizza, C.; Puglisi, I.; et al. Diversity and distribution of Phytophthora species in protected natural areas in Sicily. Forests 2019, 10, 259. [**Google Scholar**] [**CrossRef**] [**Green Version**]

32. Rizzo, D.M.; Garbelotto, M.; Davidson, J.M.; Slaughter, G.W.; Koike, S.T. Phytophthora ramorum as the cause of extensive mortality of Quercus spp. and Lithocarpus densiflorus in California. Plant Dis. 2002, 86, 205–214. [**Google Scholar**] [**CrossRef**] [**PubMed**] [**Green Version**]

33. Cahill, D.M.; Rookes, J.E.; Wilson, B.A.; Gibson, L.; McDougall, K.L. Phytophthora cinnamomi and Australia's biodiversity: Impacts, predictions and progress towards control. Aust. J. Bot. 2008, 56, 279–310. [**Google Scholar**] [**CrossRef**]

34. Jung, T.; Vettraino, A.M.; Cech, T.; Vannini, A. The impact of invasive Phytophthora species on European forests. In Phytophthora: A Global Perspective; CABI: Wallingford, UK, 2013. [**Google Scholar**]

35. Bonants, P.J.M.; Van Gent-Pelzer, M.P.E.; Hooftman, R.; Cooke, D.E.L.; Guy, D.C.; Duncan, J.M. A combination of baiting and different PCR formats, including measurement of real-time quantitative fluorescence, for the detection of Phytophthora fragariae in strawberry plants. Eur. J. Plant Pathol. 2004, 110, 689–702. [**Google Scholar**] [**CrossRef**]

36. Kong, P.; Hong, C.X.; Tooley, P.W.; Ivors, K.; Garbelotto, M.; Richardson, P.A. Rapid identification of Phytophthora ramorum using PCR-SSCP analysis of ribosomal DNA ITS-1. Lett. Appl. Microbiol. 2004, 38, 433–439. [**Google Scholar**] [**CrossRef**]

37. Khaliq, I.; St. Hardy, G.E.J.; White, D.; Burgess, T.I. eDNA from roots: A robust tool for determining Phytophthora communities in natural ecosystems. FEMS Microbiol. Ecol. 2018, 94, fiy048. [**Google Scholar**] [**CrossRef**] [**Green Version**]

38. Kunadiya, M.B.; Dunstan, W.D.; White, D.; Hardy, G.E.S.J.; Grigg, A.H.; Burgess, T.I. A qPCR Assay for the Detection of Phytophthora cinnamomi including an mRNA protocol designed to establish propagule viability in environmental samples. Plant Dis. 2019, 103, 2443–2450. [**Google Scholar**] [**CrossRef**]

39. Ioos, R.; Laugustin, L.; Schenck, N.; Rose, S.; Husson, C.; Frey, P. Usefulness of single copy genes containing introns in Phytophthora for the development of detection tools for the regulated species P. ramorum and P. fragariae. Eur. J. Plant Pathol. 2006, 116, 171–176. [**Google Scholar**] [**CrossRef**]

40. Bilodeau, G.J.; Lévesque, C.A.; De Cock, A.W.A.M.; Duchaine, C.; Brière, S.; Uribe, P.; Martin, F.N.; Hamelin, R.C. Molecular detection of Phytophthora ramorum by real-time polymerase chain reaction using TaqMan, SYBR Green, and molecular beacons. Phytopathology 2007, 97, 632–642. [**Google Scholar**] [**CrossRef**] [**Green Version**]

41. Cooke, D.E.L.; Schena, L.; Cacciola, S.O. Tools to detect, identify and monitor Phytophthora species in natural ecosystems. J. Plant Pathol. 2007, 89, 13–28. [**Google Scholar**]

42. Hwang, J.; Oak, S.W.; Jeffers, S.N. Detecting Phytophthora ramorum and other species of Phytophthora in streams in natural ecosystems using baiting and filtration methods. In Proceedings of the Sudden Oak Death Third Science Symposium, Albany, CA, USA, 5–9 March 2007. [**Google Scholar**]

43. Ghimire, S.R.; Richardson, P.A.; Moorman, G.W.; Lea-Cox, J.D.; Ross, D.S.; Hong, C.X. An in-situ baiting bioassay for detecting Phytophthora species in irrigation runoff containment basins. Plant Pathol. 2009, 58, 577–583. [**Google Scholar**] [**CrossRef**]

44. Martin, F.N.; Coffey, M.D.; Zeller, K.; Hamelin, R.C.; Tooley, P.; Garbelotto, M.; Hughes, K.J.D.; Kubisiak, T.; Bilodeau, G.J.; Levy, L.; et al. Evaluation of molecular markers for Phytophthora ramorum detection and identification: Testing for specificity using a standardized library of isolates. Phytopathology 2009, 99, 390–403. **[Google Scholar]** **[CrossRef]** **[PubMed]** **[Green Version]**

45. O'Brien, P.A.; Williams, N.; Hardy, G.E.S. Detecting Phytophthora. Crit. Rev. Microbiol. 2009, 35, 169–181. **[Google Scholar]** **[CrossRef]**

46. Sutton, W.; Hansen, E.M.; Reeser, P.W.; Kanaskie, A. Stream monitoring for detection of Phytophthora ramorum in Oregon tanoak forests. Plant Dis. 2009, 93, 1182–1186. **[Google Scholar]** **[CrossRef]** **[PubMed]** **[Green Version]**

47. Williams, N.; Hardy, G.E.S.J.; O'Brien, P.A. Analysis of the distribution of Phytophthora cinnamomi in soil at a disease site in Western Australia using nested PCR. For. Pathol. 2009, 39, 95–109. **[Google Scholar]** **[CrossRef]**

48. Tomlinson, J.A.; Dickinson, M.J.; Boonham, N. Rapid detection of Phytophthora ramorum and P. kernoviae by two-minute DNA extraction followed by isothermal amplification and amplicon detection by generic lateral flow device. Phytopathology 2010, 100, 143–149. **[Google Scholar]** **[CrossRef]** **[PubMed]** **[Green Version]**

49. Scibetta, S.; Schena, L.; Chimento, A.; Cacciola, S.O.; Cooke, D.E.L. A molecular method to assess Phytophthora diversity in environmental samples. J. Microbiol. Methods 2012, 88, 356–368. **[Google Scholar]** **[CrossRef]**

50. Schena, L.; Nicosia, M.G.L.D.; Sanzani, S.M.; Faedda, R.; Ippolito, A.; Cacciola, S.O. Development of quantitative PCR detection methods for phytopathogenic fungi and oomycetes. J. Plant Pathol. 2013, 95, 7–24. **[Google Scholar]**

51. Bilodeau, G.J.; Martin, F.N.; Coffey, M.D.; Blomquist, C.L. Development of a multiplex assay for genus- and species-specific detection of Phytophthora based on differences in mitochondrial gene order. Phytopathology 2014, 104, 733–748. **[Google Scholar]** **[CrossRef]** **[Green Version]**

52. Miles, T.D.; Martin, F.N.; Coffey, M.D. Development of rapid isothermal amplification assays for detection of Phytophthora spp. in plant tissue. Phytopathology 2015, 105, 265–278. **[Google Scholar]** **[CrossRef]** **[Green Version]**

53. Prigigallo, M.I.; Abdelfattah, A.; Cacciola, S.O.; Faedda, R.; Sanzani, S.M.; Cooke, D.E.L.; Schena, L. Metabarcoding analysis of Phytophthora diversity using genus-specific primers and 454 pyrosequencing. Phytopathology 2016, 106, 305–313. **[Google Scholar]** **[CrossRef]** **[PubMed]** **[Green Version]**

54. Junker, C.; Pfaff, A.; Werres, S. Validation of the bait test with rhododendron leaves for Phytophthora ramorum. EPPO Bull. 2018, 48, 595–608. **[Google Scholar]** **[CrossRef]**

55. Hüberli, D.; Hardy, G.E.S.J.; White, D.; Williams, N.; Burgess, T.I. Fishing for Phytophthora from Western Australia's waterways: A distribution and diversity survey. Australas. Plant Pathol. 2013, 42, 251–260. **[Google Scholar]** **[CrossRef]**

56. Oh, E.; Gryzenhout, M.; Wingfield, B.D.; Wingfield, M.J.; Burgess, T.I. Surveys of soil and water reveal a goldmine of Phytophthora diversity in South African natural ecosystems. IMA Fungus 2013, 4, 123–131. **[Google Scholar]** **[CrossRef]**

57. Jung, T.; Jung, M.H.; Cacciola, S.O.; Cech, T.; Bakonyi, J.; Seress, D.; Mosca, S.; Schena, L.; Seddaiu, S.; Pane, A.; et al. Multiple new cryptic pathogenic Phytophthora species from

Fagaceae forests in Austria, Italy and Portugal. IMA Fungus 2017, 8, 219–244. [**Google Scholar**] [**CrossRef**] [**Green Version**]

58. Jung, T.; Jung, M.H.; Scanu, B.; Seress, D.; Kovács, G.M.; Maia, C.; Pérez-Sierra, A.; Chang, T.T.; Chandelier, A.; Heungens, K.; et al. Six new Phytophthora species from ITS Clade 7a including two sexually functional heterothallic hybrid species detected in natural ecosystems in Taiwan. Persoonia Mol. Phylogeny Evol. Fungi 2017, 38, 100–135. [**Google Scholar**] [**CrossRef**] [**Green Version**]
59. Jung, T.; Scanu, B.; Bakonyi, J.; Seress, D.; Kovács, G.M.; Durán, A.; Von Stowasser, E.S.; Schena, L.; Mosca, S.; Thu, P.Q.; et al. Nothophytophthora gen. nov., a new sister genus of Phytophthora from natural and semi-natural ecosystems. Persoonia Mol. Phylogeny Evol. Fungi 2017, 39, 143–174. [**Google Scholar**] [**CrossRef**] [**Green Version**]
60. Milenković, I.; Keča, N.; Karadžić, D.; Radulović, Z.; Nowakowska, J.A.; Oszako, T.; Sikora, K.; Corcobado, T.; Jung, T. Isolation and pathogenicity of Phytophthora species from poplar plantations in Serbia. Forests 2018, 9, 330. [**Google Scholar**] [**CrossRef**] [**Green Version**]
61. Mora-Sala, B.; Gramaje, D.; Abad-Campos, P.; Berbegal, M. Diversity of Phytophthora species associated with Quercus ilex L. in three Spanish regions evaluated by NGS. Forests 2019, 10, 979. [**Google Scholar**] [**CrossRef**] [**Green Version**]
62. Riddell, C.E.; Frederickson-Matika, D.; Armstrong, A.C.; Elliot, M.; Forster, J.; Hedley, P.E.; Morris, J.; Thorpe, P.; Cooke, D.E.L.; Pritchard, L.; et al. Metabarcoding reveals a high diversity of woody host-associated Phytophthora spp. in soils at public gardens and amenity woodlands in Britain. PeerJ 2019, 7, e6931. [**Google Scholar**] [**CrossRef**] [**Green Version**]
63. Vannini, A.; Bruni, N.; Tomassini, A.; Franceschini, S.; Vettraino, A.M. Pyrosequencing of environmental soil samples reveals biodiversity of the Phytophthora resident community in chestnut forests. FEMS Microbiol. Ecol. 2013, 85, 433–442. [**Google Scholar**] [**CrossRef**] [**PubMed**] [**Green Version**]
64. Català, S.; Pérez-Sierra, A.; Abad-Campos, P. The use of genus-specific amplicon pyrosequencing to assess Phytophthora species diversity using eDNA from soil and water in northern spain. PLoS ONE 2015, 10, e0119311. [**Google Scholar**] [**CrossRef**] [**PubMed**] [**Green Version**]
65. Scanu, B.; Linaldeddu, B.T.; Deidda, A.; Jung, T. Diversity of Phytophthora species from declining mediterranean maquis vegetation, including two new species, Phytophthora crassamura and P. ornamentata sp. nov. PLoS ONE 2015, 10, e0143234. [**Google Scholar**] [**CrossRef**] [**Green Version**]
66. Sims, L.L.; Sutton, W.; Reeser, P.; Hansen, E.M. The Phytophthora species assemblage and diversity in riparian alder ecosystems of Western Oregon, USA. Mycologia 2015, 107, 889–902. [**Google Scholar**] [**CrossRef**]
67. Dunstan, W.A.; Howard, K.; Hardy, G.E.S.J.; Burgess, T.I. An overview of Australia's Phytophthora species assemblage in natural ecosystems recovered from a survey in Victoria. IMA Fungus 2016, 7, 47–58. [**Google Scholar**] [**CrossRef**] [**PubMed**] [**Green Version**]
68. Pignatti, S.; Guarino, R.; La Rosa, M. (Eds.) Flora d'Italia, 2nd ed.; Edagricole: Milano, Italy, 2018; Volume 4. [**Google Scholar**]
69. Biondi, E.; Blasi, C.; Allegrezza, M.; Anzellotti, I.; Azzella, M.M.; Carli, E.; Casavecchia, S.; Copiz, R.; Del Vico, E.; Facioni, L.; et al. Plant communities of Italy: The vegetation prodrome. Plant Biosyst. 2014, 148, 728–814. [**Google Scholar**] [**CrossRef**] [**Green Version**]

70. Manuale Italiano di Interpretazione Degli Habitat Della Direttiva 92/43/CEE. Available online: **http://vnr.unipg.it/habitat/** (accessed on 15 June 2017).

71. Jung, T.; Blaschke, H.; Neumann, P. Isolation, identification and pathogenicity of Phytophthora species from declining oak stands. Eur. J. For. Pathol. 1996, 26, 253–272. [**Google Scholar**] [**CrossRef**]

72. Jung, T.; Nechwatal, J.; Cooke, D.E.L.; Hartmann, G.; Blaschke, M.; Oßwald, W.F.; Duncan, J.M.; Delatour, C. Phytophthora pseudosyringae sp. nov., a new species causing root and collar rot of deciduous tree species in Europe. Mycol. Res. 2003, 107, 772–789. [**Google Scholar**] [**CrossRef**]

73. Cooke, D.E.; Drenth, A.; Duncan, J.M.; Wagels, G.; Brasier, C.M. A molecular phylogeny of Phytophthora and related oomycetes. Fungal Genet. Biol. 2000, 30, 17–32. [**Google Scholar**] [**CrossRef**]

74. White, T.J.; Bruns, T.; Lee, S.; Taylor, J.W. Amplification and direct sequencing of fungal ribosomal RNA genes for phylogenetics. In PCR Protocols: A Guide to Methods and Applications; Innis, M.A., Gelfand, D.H., Sninsky, J.J., White, T.J., Eds.; Academic Press Inc.: San Diego, CA, USA, 1990; Volume 18, pp. 315–322. [**Google Scholar**]

75. FinchTV v.1.4.0. Available online: **https://digitalworldbiology.com/FinchTV** (accessed on 18 May 2020).

76. BLAST Searches. Available online: **https://blast.ncbi.nlm.nih.gov/Blast.cgi** (accessed on 18 May 2020).

77. Phytophthora Database. Available online: **http://www.phytophthoradb.org/** (accessed on 18 May 2020).

78. GenBank. Available online: **https://www.ncbi.nlm.nih.gov/genbank/** (accessed on 18 May 2020).

79. R: The R Project for Statistical Computing. Available online: **https://www.r-project.org/** (accessed on 18 May 2020).

80. Soil Science Division Staff. Soil Survey Manual; United States Department of Agriculture: Washington, DC, USA, 2017.

81. Meng, Y.; Zhang, Q.; Ding, W.; Shan, W. Phytophthora parasitica: A model oomycete plant pathogen. Mycology 2014, 5, 43–51. [**Google Scholar**] [**CrossRef**]

82. Hansen, E.; Delatour, C. Phytophthora species in oak forests of north-east France. Ann. For. Sci. 1999, 56, 539–547. [**Google Scholar**] [**CrossRef**]

83. Ruano-Rosa, D.; Schena, L.; Agosteo, G.E.; Lio, G.W.D.S.; Cacciola, S.O. Phytophthora oleae sp. nov. causing fruit rot of olive in southern Italy. Plant Pathol. 2018, 67, 1362–1373. [**Google Scholar**] [**CrossRef**]

84. González, M.; Pérez-Sierra, A.; Sánchez, M.E. Phytophthora oleae, a new root pathogen of wild olives. Plant Pathol. 2019, 68, 901–907. [**Google Scholar**] [**CrossRef**]

85. Santilli, E.; Riolo, M.; La Spada, F.; Pane, A.; Cacciola, S.O. First report of root rot of Olea europaea caused by Phytophthora bilorbang in Italy. Plants 2020, 9, 826. [**Google Scholar**] [**CrossRef**]

86. Scanu, B.; Hunter, G.C.; Linaldeddu, B.T.; Franceschini, A.; Maddau, L.; Jung, T.; Denman, S. A taxonomic re-evaluation reveals that Phytophthora cinnamomi and P. cinnamomi var. parvispora are separate species. For. Pathol. 2014, 44, 1–20. [**Google Scholar**]

87. Pane, A.; Faedda, R.; Cacciola, S.O.; Rizza, C.; Scibetta, S.; Lio, G.M.D.S. Root and basal stem rot of mandevillas caused by Phytophthora spp. in eastern Sicily. Plant Dis. 2010, 94, 1374. [**Google Scholar**] [**CrossRef**]

88. Cleary, M.; Ghasemkhani, M.; Blomquist, M.; Witzell, J. First report of Phytophthora gonapodyides causing stem canker on European beech (Fagus sylvatica) in Southern Sweden. Plant Dis. 2016, 100, 2174. [**Google Scholar**] [**CrossRef**]

89. Balci, Y.; Halmschlager, E. Incidence of Phytophthora species in oak forests in Austria and their possible involvement in oak decline. For. Pathol. 2003, 33, 157–174. [**Google Scholar**] [**CrossRef**]

90. Corcobado, T.; Cubera, E.; Pérez-Sierra, A.; Jung, T.; Solla, A. First report of Phytophthora gonapodyides involved in the decline of Quercus ilex in xeric conditions in Spain. New Dis. Rep. 2010, 22, 33. [**Google Scholar**] [**CrossRef**] [**Green Version**]

91. Erwin, D.C.; Ribeiro, O.K. Phytophthora Diseases Worldwide; American Phytopathological Society (APS Press): St. Paul, MN, USA, 1996. [**Google Scholar**]

92. Jung, T.; Burgess, T.I. Re-evaluation of Phytophthora citricola isolates from multiple woody hosts in Europe and North America reveals a new species, Phytophthora plurivora sp. nov. Persoonia 2009, 22, 95–110. [**Google Scholar**] [**CrossRef**] [**Green Version**]

93. Redondo, M.A.; Boberg, J.; Stenlid, J.; Oliva, J. Functional traits associated with the establishment of introduced Phytophthora spp. in Swedish forests. J. Appl. Ecol. 2018, 55, 1538–1552. [**Google Scholar**] [**CrossRef**]

94. Reeser, P.W.; Sutton, W.; Hansen, E.M.; Remigi, P.; Adams, G.C. Phytophthora species in forest streams in Oregon and Alaska. Mycologia 2011, 103, 22–35. [**Google Scholar**] [**CrossRef**]

95. Brasier, C.M.; Cooke, D.E.L.; Duncan, J.M.; Hansen, E.M. Multiple new phenotypic taxa from trees and riparian ecosystems in Phytophthora gonapodyides-P. megasperma ITS Clade 6, which tend to be high-temperature tolerant and either inbreeding or sterile. Mycol. Res. 2003, 107, 277–290. [**Google Scholar**] [**CrossRef**] [**PubMed**]

96. Yang, X.; Tyler, B.M.; Hong, C. An expanded phylogeny for the genus Phytophthora. IMA Fungus 2017, 8, 355–384. [**Google Scholar**] [**CrossRef**] [**PubMed**] [**Green Version**]

97. Aghighi, S.; Hardy, G.E.S.J.; Scott, J.K.; Burgess, T.I. Phytophthora bilorbang sp. nov., a new species associated with the decline of Rubus anglocandicans (European blackberry) in Western Australia. Eur. J. Plant Pathol. 2012, 133, 841–855. [**Google Scholar**] [**CrossRef**] [**Green Version**]

98. Nesbitt, H.J.; Malajczu, N.; Glenn, A.R. Effect of organic matter on the survival of Phytophthora cinnamomi rands in soil. Soil Biol. Biochem. 1978, 11, 133–136. [**Google Scholar**] [**CrossRef**]

99. Allmaras, R.R.; Kraft, J.M.; Miller, D.E. Effects of soil compaction and incorporated crop residue on root health. Annu. Rev. Phytopathol. 1988, 26, 219–243. [**Google Scholar**] [**CrossRef**]

100. Whalley, W.R.; Dumitru, E.; Dexter, A.R. Biological effects of soil compaction. Soil Tillage Res. 1995, 35, 53–68. [**Google Scholar**] [**CrossRef**]

101. Jung, T.; Blaschke, H.; Osswald, W. Involvement of soilborne Phytophthora species in Central European oak decline and the effect of site factors on the disease. Plant Pathol. 2000, 49, 706–718. [**Google Scholar**] [**CrossRef**]

102. Kong, P.; Lea-Cox, J.D.; Moorman, G.W.; Hong, C. Survival of Phytophthora alni, Phytophthora kernoviae, and Phytophthora ramorum in a simulated aquatic environment at different levels of pH. FEMS Microbiol. Lett. 2012, 332, 54–60. **[Google Scholar]** **[CrossRef]** **[Green Version]**

103. Corcobado, T.; Solla, A.; Madeira, M.A.; Moreno, G. Combined effects of soil properties and Phytophthora cinnamomi infections on Quercus ilex decline. Plant Soil 2013, 373, 403–413. **[Google Scholar]** **[CrossRef]**

104. Martin, D.A.N.; Rahmat, A. Relationship of soil physicochemical properties and existence of Phytophthora sp. in pineapple plantations. Indones. J. Sci. Technol. 2017, 2, 81–86. **[Google Scholar]** **[CrossRef]** **[Green Version]**

105. Cacciola, S.O.; Magnano di San Lio, G. Management of citrus diseases caused by Phytophthora spp. In Integrated Management of Plant Diseases Caused by Fungi, Phytoplasma and Bacteria, 1st ed.; Ciancio, A., Mukerji, K.G., Eds.; Springer Science + Business Media B.V.: Berlin, Germany, 2008; Volume 3, pp. 61–84. **[Google Scholar]**

106. Vettraino, A.M.; Morel, O.; Perlerou, C.; Robin, C.; Diamandis, S.; Vannini, A. Occurrence and distribution of Phytophthora species in European chestnut stands and their association with Ink Disease and crown decline. Eur. J. Plant Pathol. 2005, 111, 169–180. **[Google Scholar]** **[CrossRef]**

107. Yang, X.; Copes, W.E.; Hong, C. Phytophthora mississippiae sp. nov., a new species recovered from irrigation reservoirs at a plant nursery in Mississippi. J. Plant Pathol. Microbiol. 2013, 4, 5. **[Google Scholar]** **[CrossRef]** **[Green Version]**

108. Bose, T.; Wingfield, M.J.; Roux, J.; Vivas, M.; Burgess, T.I. Community composition and distribution of Phytophthora species across adjacent native and non-native forests of South Africa. Fungal Ecol. 2018, 36, 17–25. **[Google Scholar]** **[CrossRef]** **[Green Version]**

109. Hernández-Lambraño, R.E.; González-Moreno, P.; Sánchez-Agudo, J.Á. Environmental factors associated with the spatial distribution of invasive plant pathogens in the Iberian peninsula: The case of Phytophthora cinnamomi Rands. For. Ecol. Manag. 2018, 419, 101–109. **[Google Scholar]** **[CrossRef]**

110. Legeay, J.; Husson, C.; Boudier, B.; Louisanna, E.; Baraloto, C.; Schimann, H.; Marcais, B.; Buée, M. Surprising low diversity of the plant pathogen Phytophthora in Amazonian forests. Environ. Microbiol. 2020. **[Google Scholar]** **[CrossRef]**

10 ASSESSMENT OF CHARCOAL CANKER ON OAK TREES AT A SITE OF COMMUNITY IMPORTANCE: A CASE STUDY OF THE RELICT CASTELFIDARDO FOREST (SIC AREA IT520008, CASTELFIDARDO, AN, ITALY)

1. INTRODUCTION: AN OVERVIEW

Forest decline is a very concerning and intricate illness that impacts vast wooded regions in many ecosystems worldwide [1,2,3,4,5]. When addressing the issue of forest loss in the Mediterranean region, it is crucial to give particular consideration to oak forests. This is because oaks play a significant role in Mediterranean forest ecosystems and are widely distributed. Additionally, the reduction in oak formations is severe and extensive [6].

Oak decline encompasses multiple species of the Quercus genus and is characterized as a widespread and gradual deterioration followed by a variety of symptoms. The most prevalent symptoms include: leaf chlorosis and yellowing; microphyllia; premature leaf abscission and crown transparency; branch dieback; exudation of black substances through the trunk; and formation of epicormic shoots along the major branches and stems [7,8]. Oak decline is a well-known illness that is characterized by its complexity and involvement of multiple causes, including predisposing, instigating, and contributing factors [1,9].

The primary predisposing factors are those that diminish the inherent host defenses, leading to chronic metabolic problems that weaken trees and increase their susceptibility to long-term impacts from harmful substances. These factors encompass aspects pertaining to climate, site characteristics, and human influence, such as water scarcity due to prolonged drought, unfavorable soil conditions, excessive water accumulation, improper forest management, and disruption of land use. Inciting causes can manifest as either physical or biological agents, and they can have intermittent and brief effects, such as hail, frost, acute water scarcity, and defoliating insects. These variables cause immediate and temporary harm, so the signs of the decline can improve, and the plants can recover. Nevertheless, if the harm persists and recurs for numerous years, it can permanently undermine the strength of the trees and potentially lead to their demise. Contributing factors are additional elements that weaken plants, and when they affect trees that are already physiologically compromised, they can cause death. Among these, there are secondary biotic agents, including debilitating diseases, burrowing insects, and bark beetles [1,10,11,12,13,14,15].

Each of these detrimental causes can individually or collectively contribute to the decline of oak trees, with the specific activities having variable levels of intensity and hence causing varying degrees of damage. The combined or individual exposure to these stressors might lead to various symptoms that differ in each location. This has led to the development of several and occasionally contradictory hypotheses regarding the cause of these symptoms, resulting in the classification of oak decline as a "complex of different diseases" [1,10].

Temperature and rainfall patterns have become increasingly influential in oak loss in recent years [16,17,18,19]. The climate in the Mediterranean region has exhibited a consistent trend towards increasing warmth and aridity [20,21]. The prolonged water shortage and drought stress can weaken the trees, making them more susceptible to infection by secondary diseases that take advantage of the weakened state [22,23].

The species Quercus is extensively distributed throughout the Italian peninsula, making it a defining feature of both the national and local landscape [24,25]. Castelfidardo Forest, also known as Selva di Castelfidardo, is an exemplary forest that possesses significant beautification and ecological worth. The primary tree species found in this forest are Quercus spp. This forest has been designated as a 'Site of Community Importance' and is part of the European "Natura 2000" network. This network aims to protect European ecosystems and biodiversity by establishing a system of protected areas. Given that Castelfidardo Forest is a designated protected area of significant ecological value, we conducted an investigation into the underlying factors contributing to the widespread and gradual decline and death of numerous oak trees inside this forest.

The Castelfidardo Forest is a type of deciduous forest that has characteristics similar to those found in the Mediterranean region. Currently, it occupies a relatively small area of 36 hectares. However, historical records indicate that in the seventeenth century, it was much larger, stretching all the way to the sea and believed to be around 350 hectares in size. Over time, the original area has experienced substantial reduction, to the extent that Castelfidardo Forest is now regarded as a remnant lowland forest surrounded by a predominantly farmed mountainous terrain [26].

This area is characterized by a moderate bioclimate, with annual rainfall ranging from 300 mm to 600 mm. The majority of the rainfall occurs during the autumn and winter months, leading to lengthy periods of dry summers. The unique combination of the northern exposure, geological characteristics of the site, and the relatively humid soil caused by the superficial aquifer, particularly in the lower part of the forest, has resulted in a microclimate that is distinct from the surrounding hills and coastal areas. As a result, the vegetation in this area exhibits sub-Mediterranean phytoclimatic conditions [27].

Figure 1. Map of Castelfidardo Forest (Selva di Castelfidardo; Site of Community Importance; Castelfidardo, AN, Italy) instituted by Ministerial Decree 30 March 2009.

The formation of Castelfidardo Forest has been influenced by the characteristics of the geological substrate, the morphological structure of the region, and human intervention. The primary components of the forest consist of deciduous Quercus species, including Quercus cerris, Quercus robur, and Quercus pubescens. However, certain sections along the forest's boundaries have been replanted with Pinus spp.

and Cupressus spp. as per the Ministerial Decree issued on 30 March 2009 (Figure 1). Nevertheless, the Castelfidardo Forest has been experiencing widespread degradation, resulting in a decrease in biodiversity and overall deterioration. The presence of a monoplane structure in Castelfidardo Forest, characterized by a dominant layer of trees of similar age and a lack of silvicultural management, high stand density, intense competition among individuals, and specific soil and climate conditions, is responsible for this phenomenon. Moreover, it signifies a diminutive and secluded fragment of a forest from the late Holocene era that has been separated from uninterrupted woods for a considerable period of time.

The objectives of this study were to: (i) assess the health status of the oak trees in the Castelfidardo Forest; (ii) document the prevalence of the charcoal canker pathogen Biscogniauxia mediterranea on the declining oaks; (iii) quantify the amount of deadwood and survey the associated Mycota; and (iv) investigate the potential cause-effect relationship between climatic conditions and the decline of the oak species.

2. MATERIALS AND METHODS

2.1. Study Site, Plot Design, and Tree Health Assessment

An examination of the phytosanitary status of the oak trees was carried out in the Castelfidardo Forest (43°27′59″ N, 13°35′30″ E) from May to July 2015. For this purpose, five plots were located in the upper, middle, and lower areas that were representative of the soil types and vegetation that characterize the forest (**Table 1**). Each plot was circular, with an area ranging from 0.6 ha to 0.9 ha. The plots were of variable sizes and were designed to include all of the heterogeneity conditions in terms of soil type, vegetation, and intensity of decline.

Table 1. Main characteristics of plots studied in Castelfidardo Forest (Castelfidardo, AN, Italy).

Plots	Area (m^2)	Tree Species	Stand Density (Trees/100 m^2)	Trees Monitored (n)
#1	6480.4	*Quercus pubescens* + *Fraxinus ornus*	11	51
#2	8569.6	*Quercus pubescens* + *Fraxinus ornus*	9	49
#3	6870	*Quercus robur* + *Carpinus betulus*	17	95
#4	7700	*Quercus cerris* + *Carpinus orientalis*	15	162
#5	8890	*Quercus cerris* + *Carpinus orientalis*	19	114

In the upper layer, plots #1 and #2 were characterized by the arenaceous substrate where Q. pubescens was predominant, and was associated with sporadic presence of Fraxinus ornus trees. In plot #3, in the middle layer, the dominant species was Q. robur, with rare Carpinus betulus found. In the lower layer, where plots #4 and #5 were located, the arenaceous component definitively disappeared, and the dominant forest vegetation was Q. cerris, with Carpinus orientalis occurring sporadically. In all of these plots, Ruscus aculeatus was the predominant understory species, and this covered large proportions of the plots, strongly limiting the natural regeneration of the oaks.

As summarized in **Table 1**, the main parameters recorded in each plot were: surface area; tree species; tree densities; and total numbers of trees inspected. A number of trees were selected according to the expression of symptoms (i.e., production of tarry exudates, epicormic shoot formation, and eruption of charcoal cankers through the bark), from 49 to 162 trees in each plot. For each tree selected, the diameter at chest height (i.e., 1.3 m above the ground) and its sanitary status were assessed and recorded, noting in particular the presence and frequency of: early symptoms of decline (e.g., emission of exudates, production of epicormic branches); advanced symptoms and/or signs of decline (e.g., dieback, black carbonaceous stromata indicative of charcoal cankers); and tree mortality. The position of each tree was recorded using the Global Positioning System (GPS; e-Trex 30, Garmin), and metallic labels with progressive identification numbers were affixed to the trunks. The data obtained relating to individual trees were managed through the Quantum Geographic Information System (GIS) open source software, and displayed using the Google Earth software.

The occurrence of saprotrophic and facultative necrotrophic fungi (e.g., Armillaria spp.; able to decompose deadwood) was also determined. Fruit bodies of these fungi were photographed, collected, and later identified through conventional morphological keys [**28**,**29**,**30**].

2.2. Sample Collection, Mycological Analysis, and Morphological Identification

Bark tissue with charcoal canker (mean dimension, 4 × 4 cm) was collected from 20 Q. pubescens trees in plots #1 and #2, from 10 Q. robur trees in plot #3, and from 20 Q. cerris trees in plots #4 and #5. These samples were incubated in a humidity chamber with five layers of tissue paper that had previously been soaked in sterile distilled water, and placed in glass Petri dishes (diameter, 14 cm), at 25 °C, in the dark, for 10 days. The samples were then observed under a stereo microscope (MZ125; Leica) to document the sporulation. Various samples were photographed using a digital camera (FinePix S1 Pro; Fujifilm).

The fungal agents were isolated from the bark after the incubation in the humidity chamber, from differentiated fruiting structures. Under aseptic conditions, portions of the mycelia were transferred onto Petri dishes with potato dextrose agar (PDA) with added antibiotics (150 mg/L streptomycin, 150 mg/L ampicillin; to prevent bacterial contamination). Five original mycelial samples were put into each plate, and five replicates were prepared from each sample. Each plate was labeled, sealed with parafilm, and incubated at 25 °C in the dark. After 5 days, the developed fungal colonies were subcultured onto fresh PDA plates, then transferred into glass test tubes, and finally stored at 4 °C. For 50 isolates (10 for each plot), identification was carried out by recording the characteristics of 50–100 units of spores for each isolate, by light microscopy (Eclipse E600; Nikon).

2.3. Molecular Identification of Fungal Colonies

Mycelia were collected from 25 representative fungal colonies grown on PDA, transferred into 1.5 mL microcentrifuge tubes and pulverized, with the addition of 600 µL extraction buffer (20 mM EDTA, 0.1 M Tris-HCl, pH 8.0, 1.4 M NaCl, 2% cetyltrimethylammonium bromide, 4% polyvinylpyrrolidone, 0.1% sodium metabisulfite [added just before use]). The extraction of total DNA was carried out using the CTAB protocol [**31**]. The quality and quantity of the extracted DNA were determined directly on 1% agarose gels, with evaluation using a biophotometer (Eppendorf, Hamburg, Germany). The DNA was finally diluted to 10 ng/µL for further amplification. Nucleic acids were stored at -20 °C, and later analyzed by polymerase chain reaction (PCR). Amplification of the 5.8S ribosomal DNA (rDNA) and flanking internal transcribed spacer (ITS) region was performed in 20 µL PCR reactions that contained 2 µL genomic DNA (at ~20 ng/µL) from the fungal isolate, 10 µL EmeraldAmp GT PCR Master mix 2× (Takara, Madison, WI, USA), and 0.5 µL of each primer (10 µM), in a thermal cycler (MyCycler; Bio-Rad Laboratories, Hercules, CA, USA). MED1/MED2 primers were used, which are specific for B. mediterranea [**32**]. The PCR products (9 µL per sample) were separated by electrophoresis in 1.5% agarose gels stained with Red Gel (Biotium, Hayward, CA, USA), then visualized and captured using an imagining system (Gel Doc XR; BioRad, Hercules, CA, USA). The same ITS region was also amplified from fungal isolates with universal primers ITS1/ITS4 [**33**]. Amplicons from three isolates (WP101, WP20, WP376), were purified using Wizard SV gels and PCR clean-up kits (Promega Corporation, Madison, WI, USA), and quantified using the biophotometer (Eppendorf, Hamburg, Germany). Sequencing was carried out by Genewiz UK. The Bioedit software, v. 7.0.0 (**http://www.mbio.ncsu.edu/Bioedit/bioedit.html**, accessed on 12 April 2021) was used to cut off 20 bp to 30 bp of the terminal end sequence. All of the sequences were used as the query sequences in BLAST searches (**http://blast.ncbi.nlm.nih.gov/Blast.cgi**, accessed on 26 April 2021), to determine the nucleotide identities available in

GeneBank NCBI. The nucleotide sequences were deposited in the NCBI database (MZ407593-MZ407595).

2.4. Deadwood and Associated Lignicolous Macrofungi

As deadwood and deadwood-dwelling macrofungi are fundamental for the maintenance of the biodiversity of forest ecosystems, the quantity and quality of the deadwood served as a proxy indicator for the level of biodiversity of Castelfidardo Forest. In particular, the deadwood was classified into standing dead trees (snags) and lying dead woody debris (logs). The quantities were expressed as percentages, considering the snags and logs with respect to the total number of oak trees in each plot.

The associated lignicolous macrofungal assemblage was analyzed concurrently. The macrofungi were collected gently and taken to the laboratory for detailed mycological examination.

2.5. Climatic Data

Data for the minimum, maximum, and average temperatures, and for the average annual and seasonal rainfall for the period 1951-2015 were obtained from the Regional Informative System located at Ancona Torrette 2944 (RT 701) (13°26′, 43°36′).

3. RESULTS

3.1. Phytosanitary Monitoring

A total of 471 oak trees were assessed from May to July 2015 for their health conditions in the five plots (**Table 1**). Many of the trees showed an advanced decline. The patterns of symptom appearance showed a succession of symptoms of increasing severity. Oaks in the initial stages of decline showed some nonspecific symptoms, such as leaf yellowing, early leaf browning, and thinning of canopy cover. Those in advanced states of decline showed foliar rolling and leaf wilting, production of epicormic shoots on the main branches and stems, and loss of branches, which eventually lead to tree mortality.

The emission of dark-brownish mucilaginous exudates through the bark (**Figure 2**A,B) occurred mainly during the growing season, and this was recorded for 7.2% (34/471) of the trees. Debarking the trees at the points of emission of exudates showed clearly visible necrotic areas of the underbark tissue. Furthermore, the trees with these symptoms often showed marked transparency of the canopy. Trees with production of exudates through the bark were more frequently recorded for plots #5 and #2, at 9.7% (11/114 trees) and 8.2% (4/49), respectively (**Figure 3**A). Epicormic shoot formation along the main trunk was recorded for 42.7% (201/471) of the trees (**Figure** 2C,D), and it occurred equally in all of the five plots (**Figure 3**B). Charcoal cankers erupted through the bark in about 27.6% (130/471) of the trees (**Figure** 2E).

In cross-sections of the trunks, necrosis of the xylem and wood decay were observed in correspondence with the black carbonaceous stromata that had developed externally (**Figure 2**F). The charcoal-black cankers were more frequently recorded for plot #3, at 40.0% (38/95), followed by plots #4 and #5, at 30.3% (49/162) and 25.4% (29/114), respectively; in plots #1 and #2, occurrence of the typical charcoal-black surface was lower (**Figure** 3C). Declining trees generally died in the growing season following the appearance of these erumpent, coriaceous black stromata. In plot #4, there were sporadic fallen trees that showed root rot and thick, black, string-like structures, which were usually seen under the bark or in the soil near the roots, as typical structures of Armillaria sp. (**Figure** 2G,H).

Figure 2. Representative photographs of main symptoms observed on the stems of declining oak trees in Castelfidardo Forest (Castelfidardo, AN, Italy). (A,B) Dark-brownish exudates. (C,D) Epicormic shoot proliferation. (E) Charcoal cankers erupting through the bark. (F) Cross-section of a tree with discoloration and decay of the outermost portions of the woody cylinder in correspondence with areas of external development of B. mediterranea carbonaceous stromata; (G) Lying tree with clear symptoms of root rot. (H) Dark rhizomorphs of Armillaria spp. under the oak bark.

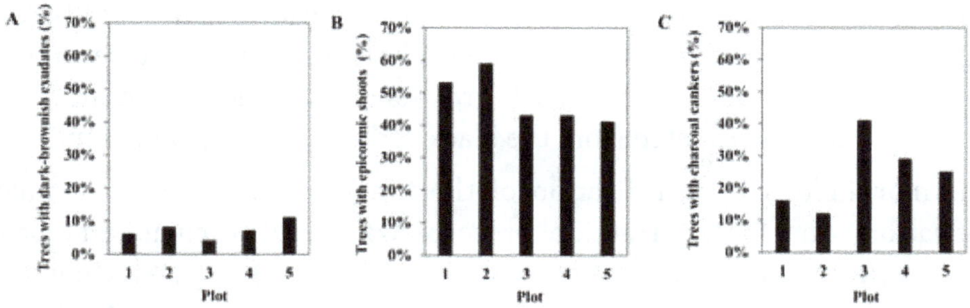

Figure 3. Quantification of the symptoms of oak decline recorded for the five experimental plots (#1–#5) in Castelfidardo Forest. (A) Dark-brownish exudates. (B) Epicormic shoot proliferation. (C) Charcoal cankers.

3.2. Mycological and Molecular Identification

We sporadically isolated from the bark tissue of oaks some fungi belonging to the genera Botryosphaeria, Diplodia, Neofusicoccum, Cytospora, Discula (Apiognomonia). However, we focused our attention on the agent of charcoal canker B. mediterranea as the incidence of this fungus was, in terms of isolation frequencies, extraordinarily higher.

Stereoscopic observations revealed carbonaceous, perithecial stromata in all of the 50 samples. The stromata appeared slightly convex, ellipsoid, and elongated, 7.2 to 20.5 × 3.5 to 4.2 cm. Perithecia were ovoid to tubular, 0.74 to 0.80 × 0.12 to 0.15 mm. A total of 152 fungal colonies were obtained in purity, most of which (94.1%; 143) were identified as B. mediterranea by mycological examination of their distinguishing micromorphological traits (**Figure 4**). The colonies completely filled the agar plates (diameter, 90 mm) after 7 days of incubation at 25 °C. The colonies were white-grey in color and velvety (viewed from the top). Short-stipitate, amyloid asci that were dark brown were observed; these produced ellipsoid ascospores, 14 µm to 19 × 7 to 9 µm, which perfectly matched the description by Mirabolfathy et al. [**34**].

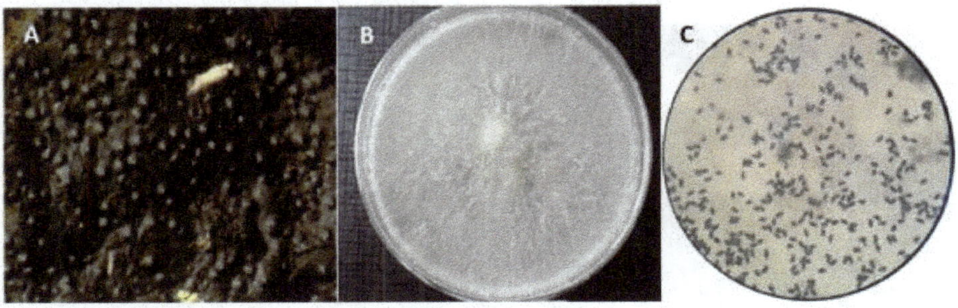

Figure 4. Representative images of carbonaceous stromata of B. mediterranea harboring perithecia, with a clearly visible layer of ostioles on its surface. (A) As observed under the stereo microscope. (B) A colony of B. mediterranea after 5 days on a PDA plate at 25 °C. (C) Ascospores of B. mediterranea as observed under the microscope.

No reproductive structures were observed in cultures that could be used to identify the microorganisms on a micromorphological basis. The reverse of the colonies were darker, tending to dark grey-black. Colonies of Trichoderma sp. were also sporadically isolated from the black carbonaceous stroma (four isolates).

The morphological identification of the fungi was supported by molecular identification. The DNA extraction protocol was very efficient, and provided ~210 ng/μL, with high quality (A260/280 > 1.87; A260/230 > 1.92). Only a few of the extracted samples (n = 6) showed low quality DNA values (A260/280 < 1.80; A260/230 < 1.80), and these were not processed further.

Amplification with the specific MED1/MED2 primers gave fragments of the expected length (377 bp) in all 25 of the samples analyzed. No ambiguous bands were generated, and the lack of bands in the water control test confirmed the absence of contamination. A portion of the ITS region (approximately 580 bp) was sequenced for additional confirmation of the identity of the fungus. Three nucleotide sequences selected revealed 100% identity with reference sequence MT819849 in the NCBI database, as B. mediterranea isolated from Quercus suber in Portugal.

3.3. Deadwood and Deadwood-Dwelling Mycobiota

The ground within Castelfidardo Forest was characterized by high amounts of deadwood, for all of the plots investigated. Plot #2 showed the lowest abundance of snags (12.2%, versus total number of oak trees) and logs (4.0%), while the highest abundance was for plot #5 (39.5%, 18.4%, respectively) (**Figure 5**).

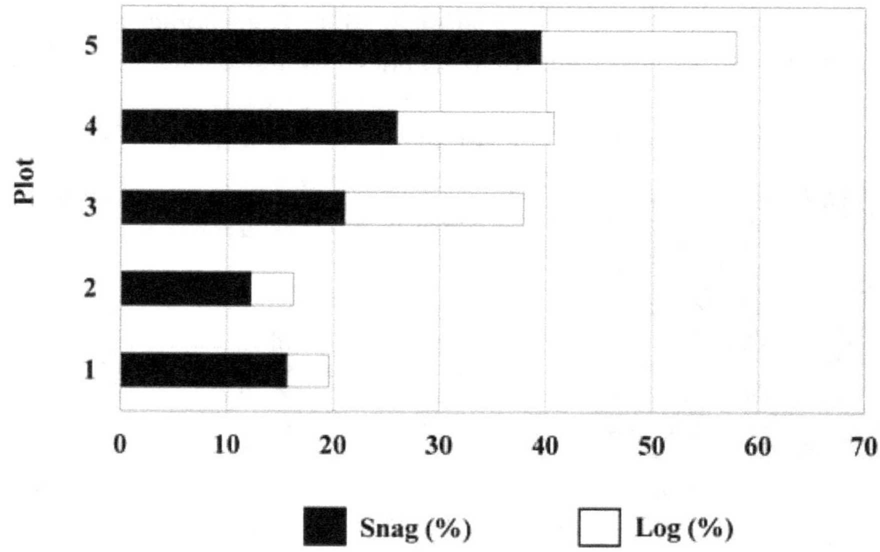

Figure 5. Abundance and type of deadwood (% of snags and logs) recorded in the five plots in Castelfidardo Forest (Castelfidardo, AN, Italy).

The deadwood was colonized by various fungi and other saprophytic microorganisms that thrive on dead and/or decaying trees. Most of these fungi belonged to Basidiomycetes. The most frequently found in plots #4 and #5 were Armillaria mellea, Macrolepiota spp., Trametes spp., and Artomyces spp., while in plots #1, #2, and #3 they were mainly Pluteus spp., Hypholoma spp., Macrolepiota spp., Armillaria mellea and Armillaria ostoyae (**Figure 6**).

Figure 6. Representative photographs of the main macrofungi associated with the deadwood in Castelfidardo Forest (Castelfidardo, AN, Italy): (A) Trametes spp. (B) Cyathus spp. (C) Pluteus spp. (D) Postia spp. (E) Hypholoma spp. (F) Macrolepiota spp. (G) Armillaria ostoyae. (H) Armillaria mellea. (I) Clathrus spp. (J) Tapinella spp. (K) Macrolepiota spp. (L) Mycena spp. (M) Micromphale spp. (N) Coprinus spp. (O) Artomyces spp.

3.4. Climatic Data

Analysis of climatic data related to Castelfidardo Forest revealed that the average annual temperature (17 °C) recorded over the period from 2001 to 2015 has increased by 2 °C with respect to the historical series (1957–2000) (**Figure 7**). For rainfall, there were no evident differences, with two negative peaks recorded in 1983 (376 mm) and 2014 (488 mm). A more detailed analysis of the climatic data relating to the growing seasons from 2001 to 2015 revealed that in the spring and summer periods for that time interval, the temperatures increased by 2 °C and 2.5 °C respectively, compared to the former period of 1957 to 2000.

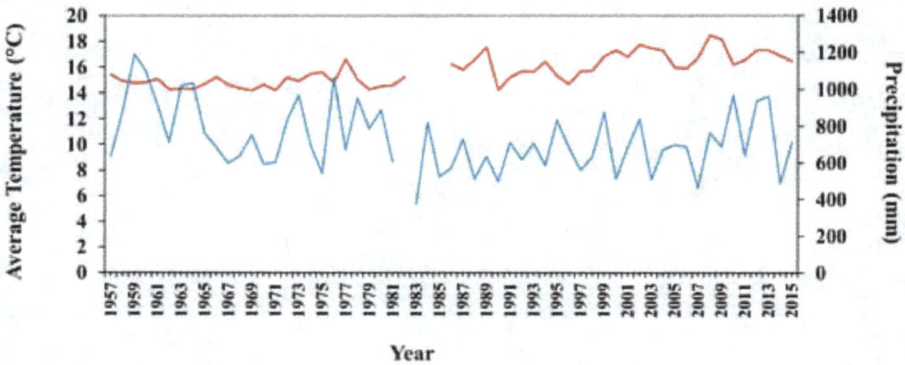

Figure 7. Graph of mean annual temperature (red lines) and rainfall (blue lines) cycles recorded from 1957 to 2015 at the Ancona Torrette 2944 weather station (RT 701) (13°26′, 43°36′) (Regional Informative System).

4. DISCUSSION

Oak forest ecosystems are of paramount importance in the Mediterranean basin due to the multiplicity of functions and services they provide in this region. Indeed, this region is at risk of desertification due to climatic and socio-economic driving forces, including rainfall variability, droughts, changes in land-use patterns, overgrazing, and demographic saturation of rural areas [**35**]. Oak formations have crucial roles here for the prevention of land degradation and for climate mitigation, carbon storage, and the safeguarding of the environment; they also generate economic, esthetic and landscaping benefits [**25**].

Unfortunately, Mediterranean oak forests have been plagued by oak decline for decades now, which is a phenomenon that appears strongly related to climate anomalies, as mainly heat stress, but also water scarcity and droughts [**18,36**]. Indeed, the Mediterranean region is a hotspot for climate change, as it is characterized by a negative water balance and pronounced warming [**37,38**]. This region is currently experiencing high water stress conditions and prolonged droughts, and these are worsening the effects of climate change, with a strongly negative impact on plant biomass production. Our analysis of climatic data showed increased temperatures during spring and summer for the period from 2001 to 2015 in the Castelfidardo area, which confirms the general warming trend that is constraining the whole Mediterranean region.

The stressful conditions determined by the water deficit, high temperatures, and droughts have created conditions conducive to the pervasive spread of the thermophilic fungus B. mediterranea in oak forests [**39**]. Several fungal species have also been isolated from trunk and branch cankers as well as from twigs and foliage of declining oaks, including Botryosphaeria dothidea, Diplodia corticola, Diplodia seriata, Neofusicoccum parvum, Cytospora spp., Discula quercina, and Neocucurbitaria cava [**22,23,40,41**]. The occurrence of these fungi was relatively erratic in the plots investigated, with lower isolation frequency compared to B. mediterranea. However, in particular here, we focused our attention on trees with evident symptoms of decline that were accompanied by the specific presence of B. mediterranea stromata. The relationship between oak decline, climate change, and the proliferation of B. mediterranea has been well documented in Mediterranean countries [**12,42,43,44,45,46**]. Indeed, recent outbreaks of B. mediterranea in countries where it had not been reported to cause harm, like in Turkey, Iran, Slovenia [**47,48**], and Croatia [**49**], confirm that this fungus is expanding its range and is accentuating its role as a contributing factor to the decline of oak formations.

Members of the Xylariaceae fungi are widely known for surviving across a range of lifestyles. They live as saprotrophs, as they can decompose many natural substrates, and they can also be pathogenic, promoting disease on living hosts, and especially those that are water-stressed. The Xylariaceae fungi can persist within trees in a latent stage, due to their endophytic aptitude [**50**]. B. mediterranea is a weakness pathogen that can live as an endophyte inside healthy plant tissues, thus causing extensive, although symptomless, infections. Such a cryptic existence can last for undefined periods, as long as the host does not become weakened by adverse environmental conditions. When stressful conditions that alter the carbohydrate physiology in trees occur, B. mediterranea switches to a pathogenic behavior to extensively colonize its debilitated host. This pathogenic endophyte then moves externally to sporulate profusely over the plant surfaces, a mechanism used to escape from a dying tree and colonize other trees that are already severely weakened, and possibly injured [**11,23**].

The aggressive colonization of oaks by B. mediterranea in the Castelfidardo Forest must thus be mainly ascribed to the chronic debilitation of these trees by hot temperatures, recurrent water deficit and droughts. Exposure to these environmental adversities will have triggered tree infection by this opportunistic fungus whereby its pathogenic action will also be facilitated by the intrinsic, low resilience of this residual forest. As a relict population, Castelfidardo Forest is more prone to any disturbance. Indeed, the extent of oak decline in Castelfidardo Forest appears greater than for other oak stands of the Italian Peninsula. Millennia of anthropogenic pressure, which was mainly represented by agricultural encroachment and overgrazing, led to the isolation of this residual portion of primeval forest. Such forest fragmentation can result in long-lasting and complex changes in biodiversity that can go beyond the loss of species, and thus include alterations to the functional diversity of the remaining communities, with isolated forests being more fragile and vulnerable [**51**].

The saprophytic stage is of fundamental importance in the epidemiology of B. mediterranea in Mediterranean oak stands. Through its extensive production of stromatal charcoal cankers, this xylariaceous fungus can introduce massive amounts of spores into the oak forest ecosystem, enormously increasing its biomass. The number of individuals introduced at a given location is a primary determinant of the successful establishment of an invasive organism into new environments [**52**]. On the other hand, the importance of propagule pressure in the spread of fungal pathogens and in the colonization of new hosts in oak stands has already been established [**53**].

Recent studies have shown how some functional traits can exacerbate the negative effects of drought, such as stand density and intraspecific and interspecific competition, which can then predispose trees to decline [**54,55**]. The Castelfidardo Forest was subjected to intense cutting in 1970, which has not been followed by any other interventions to date. This lack of management for half a century has had repercussions on the tree physiology and vigor; in particular, the high density of the stands causes excessive competition among the trees for the already limited resources (i.e., water, nutrients, light). This has led to depletion of the plant carbon reserves, and therefore to less resistance to colonization by the weakness pathogen B. mediterranea. Indeed, higher incidence of charcoal canker and higher tree mortality were recorded for Q. robur in plot #3 and for Q. cerris in plots #4 and #5. In these plots, tree densities were much higher than in plots #1 and #2. Differential susceptibility to decline across oak species was shown in previous studies. In particular, in France, Q. robur was more susceptible than Quercus petraea [**56**]. In North America, species of the red oak group were more susceptible than those of the white oak group [**57**]. In central and southern Italy, Q. cerris is more susceptible to decline than Q. pubescens [**58**].

In the present study, the disease caused by B. mediterranea was also tree-density dependent, and it represented only the terminal stage of infection. In this context, the fungus would function as an ecological factor that regulates the structure and composition of forest stands, to make them better suited to changing environmental conditions [**59**].

We recorded an abundance of deadwood that was equally distributed across all of the plots. The role of deadwood in forest ecosystems is linked to various aspects, which include improvement of natural diversity [**60**], storage and slow release of carbon and nutrients [**61,62**], maintenance of soil fertility, creation of pedoclimatic conditions conducive to natural regeneration of the forest, conservation of the soil, and improvement of the stability of slopes from hydrogeological risk [**63**]. The deadwood richness of the Castelfidardo Forest suggests that the carbon reserves in the stand are substantial and can therefore be used by the trees. However, it must be considered that when trees are chronically impaired by abiotic (i.e., environmental) and biotic (mainly anthropogenic) stressors beyond a threshold of tolerance, their energies become so depleted that they can no longer benefit from the availability of substrate. This happens especially to physiologically overmature and veteran trees, like the majority of oaks in Castelfidardo Forest [**64,65**]. Armillaria mellea and A. ostoyae were also associated with the deadwood. These two Agaricales deserve special attention because they play a central role in the dynamics of numerous woody ecosystems [**66**]. They are ubiquitous species, more commonly found in mature stands with an abundance of old trees, where they play a valuable ecological role in the recycling of dead wood. However, A. ostoyae and A. mellea are also polyphagous, facultative parasites with parasitic ability, causing white rot to a number of host species [**67**]. In Castelfidardo Forest these two fungi live preferentially as saprotrophs on dead wood and only sporadically are they found to attack pathogenically live but visibly weakened trees [**65**].

5. CONCISE SUMMARY

The current investigation centered on the decrease of oak trees in Castelfidardo Forest, a forest of significant ecological significance. According to Eugenio Paoloni, there has been a significant rise in the attacks of B. mediterranea on oak trees in this forest in recent years. This increase is likely due to high temperatures and frequent droughts, which consistently damage the trees. In various oak woodlands in central Italy, the life history strategy of B. mediterranea has been studied. It has been found, as in our study, that the agent of charcoal canker exploits environmental pressures to aggressively colonize oaks that are physiologically damaged [68]. Given our inability to intervene in climatic conditions, it is recommended to lessen the source of inoculum and decrease tree density as a means to alleviate the consequences of the decline. This fungus, which takes advantage of opportunities, produces a large number of spores on the trunks and branches of trees that are decaying or already dead, as well as on deadwood found on the ground. Hence, it is advisable to implement the following control measures: (i) eradicate all deceased trees with cankers that produce spores; (ii) conduct thinning operations to decrease the number of trees, especially in plots #3, #4, and #5; and (iii) investigate the potential of employing antagonistic microorganisms (such as Trichoderma spp.) to suppress this disease caused by stress [69].

6. AT A GLANCE

Oaks are the primary and essential tree species in Mediterranean woodland habitats. Nevertheless, oak forests have seen significant negative effects from oak decline, a global occurrence that has been worsened by climate change in the past few decades. Biscogniauxia mediterranea, a pathogen, contributes to the deterioration of Mediterranean oak ecosystems in various situations. In this study, we examined the influence and significance of B. mediterranea in the decrease of oak trees in Castelfidardo Forest, which is a remaining woodland from the late Holocene and designated as a Site of Community Importance. We designated five areas where we documented the locations of trees, any indications of decline, the presence of B. mediterranea on declining trees, and the presence of deadwood and related fungi. Out of the 471 oak trees that were examined, 7.0% displayed brownish exudates on their stems, 46.9% had epicormic shoots growing along the main trunk, and 24.4% exhibited black carbonaceous stromata on sick branches and trunks. The reduction was most pronounced for Quercus cerris, with plots #4 and #5 experiencing a decrease of 50.0% (81 out of 162 trees) and 29.0% (33 out of 114 trees) respectively. The loss was also significant for Quercus robur in plot #3, with a decrease of 40.0% (38 out of 95 trees). Lastly, Quercus pubescens in plots #1 and #2 experienced a decline of 13.7% (7 out of 51 trees) and 12.3% (6 out of 49 trees) respectively. Tree bark samples were obtained from trees exhibiting charcoal cankers and transported to the laboratory for microscopic analysis and identification using mycological and molecular techniques. This analysis uncovered a strong correlation between oak trees exhibiting significant decline in vitality and the occurrence of B. mediterranea. The deadwood was evenly distributed across the five plots and was extensively colonized by Basidiomycota. The prevalence of the charcoal canker pathogen B. mediterranea seemed to be associated with environmental stressors. Nevertheless, the lack of silvicultural management, intense competition among fully developed trees, and the geographical isolation of this remaining forest may have made oaks more susceptible to decline.

REFERENCES

1. Manion, P.D. Tree Disease Concepts; Prentice-Hall N.J.: Englewood Cliffs, NJ, USA, 1991. [**Google Scholar**]
2. Hennon, P.E.; D'Amore, D.; Wittwer, D.; Johnson, A.; Schaberg, P.; Hawley, G.; Beier, C.; Sink, S.; Juday, G. Climate warming, reduced snow, and freezing injury could explain the demise of yellow-cedar in southeast Alaska, USA. World Resour. Rev. 2006, 18, 427–450. [**Google Scholar**]

3. Worrall, J.J.; Rehfeldt, G.E.; Hamann, A.; Hogg, E.H.; Marchetti, S.B.; Michaelian, M.; Gray, L.K. Recent declines of Populus tremuloides in North America linked to climate. For. Ecol. Manag. 2013, 299, 35–51. [**Google Scholar**] [**CrossRef**]

4. Bendixsen, D.P.; Hallgren, S.W.; Frazier, A.E. Stress factors associated with forest decline in xeric oak forests of south-central United States. For. Ecol. Manag. 2015, 347, 40–48. [**Google Scholar**] [**CrossRef**]

5. Gaylord, M.L.; Kolb, T.E.; McDowell, N.G. Mechanisms of piñon pine mortality after severe drought: A retrospective study of mature trees. Tree Physiol. 2015, 35, 806–816. [**Google Scholar**] [**CrossRef**] [**PubMed**] [**Green Version**]

6. Ragazzi, A.; Vagniluca, S.; Moricca, S. European expansion of oak decline: Involved microorganisms and methodological approaches. Phytopathol. Mediterr. 1995, 34, 207–226. [**Google Scholar**]

7. Henriques, J.; Inacho, M.L.; Lima, A.; Sousa, E. New outbreaks of charcoal canker on young cork oak trees in Portugal. IOBC-WPRS Bull. 2012, 76, 85–88. [**Google Scholar**]

8. Conte, A.L.; Di Pietro, R.; Iamonico, D.; Di Marzio, P.; Cillis, G.; Lucia, D.; Fortini, P. Oak decline in the Mediterranean basin: A study case from the southern Apennines (Italy). Plant Sociol. 2019, 56, 69–80. [**Google Scholar**]

9. Ciesla, W.M.; Donaubauer, E. Decline and Dieback of Trees and Forests: A Global Overview (No. 120); Food and Agriculture Org.: Rome, Italy, 1994. [**Google Scholar**]

10. Führer, E. Oak decline in central Europe: A synopsis of hypotheses. In Proceedings: Population Dynamics, Impacts, and Integrated Management of Forest Defoliating Insects; General Technical Report NE-247; Kulhavy, D., Ross, W.G., Cahal, R.R., Eds.; USDA For. Serv: Radnor, PA, USA, 1998; pp. 7–24. [**Google Scholar**]

11. Moricca, S.; Ragazzi, A. Fungal endophytes in Mediterranean oak forests: A lesson from Discula quercina. Phytopathology 2008, 98, 380–386. [**Google Scholar**] [**CrossRef**] [**PubMed**] [**Green Version**]

12. Vannini, A.; Lucero, G.; Anselmi, N.; Vettraino, A. Response of endophytic Biscogniauxia mediterranea to variation in leaf water potential of Quercus cerris. For. Pathol. 2009, 39, 8–14. [**Google Scholar**] [**CrossRef**]

13. Ghanbary, E.; Fathizadeh, O.; Pazhouhan, I.; Zarafshar, M.; Tabari, M.; Jafarnia, S.; Parad, G.A.; Bader, M.K.-F. Drought and pathogen effects on survival, leaf physiology, oxidative damage, and defense in two Middle Eastern oak species. Forests 2021, 12, 247. [**Google Scholar**] [**CrossRef**]

14. Keča, N.; Koufakis, I.; Dietershagen, J.; Nowakowska, J.A.; Oszako, T. European oak decline phenomenon in relation to climatic changes. Folia For. Polon. Ser. A For. 2016, 58, 170–177. [**Google Scholar**] [**CrossRef**] [**Green Version**]

15. Panzavolta, T.; Panichi, A.; Bracalini, M.; Croci, F.; Benigno, A.; Ragazzi, A.; Tiberi, R.; Moricca, S. Tree pathogens and their insect-mediated transport: Implications for oak tree die-off in a natural park area. Glob. Ecol. Conserv. 2018, 15, e00437. [**Google Scholar**] [**CrossRef**]

16. Breshears, D.D.; Cobb, N.S.; Rich, P.M.; Price, K.P.; Allen, C.D.; Balice, R.G.; Romme, W.H.; Kastens, J.H.; Floyd, M.L.; Belnap, J.; et al. Regional vegetation die-off in response to global-change-type drought. Proc. Natl. Acad. Sci. USA 2005, 102, 15144–15148. [**Google Scholar**] [**CrossRef**] [**Green Version**]

17. van Mantgem, P.J.; Stephenson, N.L.; Byrne, J.C.; Daniels, L.D.; Franklin, J.F.; Fulé, P.Z.; Harmon, M.E.; Larson, A.J.; Smith, J.M.; Taylor, A.H.; et al. Widespread increase of tree mortality rates in the western United States. Science 2009, 323, 521–524. [**Google Scholar**] [**CrossRef**] [**PubMed**] [**Green Version**]

18. Allen, C.D.; Macalady, A.K.; Chenchouni, H.; Bachelet, D.; McDowell, N.; Vennetier, M.; Kitzberger, T.; Rigling, A.; Breshears, D.D.; Hogg, E.T.; et al. A global overview of drought and heat-induced tree mortality reveals emerging climate change risks for forests. For. Ecol. Manag. 2010, 259, 660–684. [**Google Scholar**] [**CrossRef**] [**Green Version**]

19. Millar, C.I.; Stephenson, N.L. Temperate forest health in an era of emerging megadisturbance. Science 2015, 349, 823–826. [**Google Scholar**] [**CrossRef**] [**PubMed**]

20. Mariotti, A.; Pan, Y.; Zeng, N.; Alessandri, A. Long-term climate change in the Mediterranean region in the midst of decadal variability. Clim. Dyn. 2015, 44, 1437–1456. [**Google Scholar**] [**CrossRef**]

21. Polade, S.D.; Gershunov, A.; Cayan, D.R.; Dettinger, M.D.; Pierce, D.W. Precipitation in a warming world: Assessing projected hydro-climate changes in California and other Mediterranean climate regions. Sci. Rep. 2017, 7, 1–10. [**Google Scholar**] [**CrossRef**] [**PubMed**] [**Green Version**]

22. Linaldeddu, B.T.; Sirca, C.; Spano, D.; Franceschini, A. Variation of endophytic cork oak-associated fungal communities in relation to plant health and water stress. For. Pathol. 2011, 41, 193–201. [**Google Scholar**] [**CrossRef**] [**Green Version**]

23. Moricca, S.; Linaldeddu, B.T.; Ginetti, B.; Scanu, B.; Franceschini, A.; Ragazzi, A. Endemic and emerging pathogens threatening cork oak trees: Management options for conserving a unique forest ecosystem. Plant Dis. 2016, 100, 2184–2193. [**Google Scholar**] [**CrossRef**] [**Green Version**]

24. Brewer, S.; Cheddadi, R.; Beaulieu, J.L.; Reille, M. The migration of deciduous Quercus throughout Europe since the last glacial period. For. Ecol. Manag. 2002, 156, 27–48. [**Google Scholar**] [**CrossRef**]

25. Bussotti, F.; Cenni, E.; Bettini, D.; Sarti, C.; Stergulc, F.; Feducci, M.; Capretti, P. Le condizioni dei boschi in Italia. Risultati dalle indagini estensive di Livello I (1997–2010). For. J. Silvic. For. Ecol. 2014, 11, 8–12. [**Google Scholar**]

26. Taffetani, F. Boschi residui in Italia tra paesaggio rurale e conservazione. In Proceedings of the Atti del Terzo Congresso Nazionale di Selvicoltura per il Miglioramento e la Conservazione dei Boschi Italiani, Sicilia, Italy, 16–19 October 2008; Ciancio, O., Ed.; Accademia di Scienze Forestali: Florence, Italy, 2009. [**Google Scholar**]

27. Aleffi, M.; Allegrini, M.C.; Fuffa, X.; Mucchichini, S.; Tacchi, R. Briofite e Licheni delle Selva di Castelfidardo (Marche); Università degli Studi di Camerino: Camerino, Italy, 2005. [**Google Scholar**]

28. Bernicchia, A.; Gorjòn, S.P. Polypores of the Mediterranean Region; Romar SRL: Verona, Italy, 2020. [**Google Scholar**]

29. Consiglio, G.; Papetti, C. Funghi d'Italia Vol. 1-2-3 Ed; AMB: Trento, Italy, 2009. [**Google Scholar**]

30. Volk, T.J.; Burdsall, H.H. A Nomenclatural Study of Armillaria and Armillariella Species (Basidiomycotina, Tricholomataceae); Fungiflora: Førde, Norway, 1995; p. 121. [**Google Scholar**]

31. Doyle, J.J.; Doyle, J.L. Isolation of plant DNA from fresh tissue. Focus 1990, 12, 13–15. [**Google Scholar**]

32. Mazzaglia, A.; Anselmi, N.; Vicario, S.; Vannini, A. Sequence analysis of the 5.8S rDNA and ITS regions in evaluating genetic relationship among some species of Hypoxylon and related genera. Mycol. Res. 2001, 105, 670–675. [**Google Scholar**] [**CrossRef**]

33. White, T.; Bruns, T.; Lee, S.; Taylor, J. Amplification and direct sequencing of fungal ribosomal RNA genes for phylogenetics. In PCR Protocols: A Guide to Methods and Applications; Innis, M.A.G.D.H., Sninsky, J.J., White, T.J., Eds.; Academic Press, Inc.: New York, NY, USA, 1990; pp. 315–322. [**Google Scholar**]

34. Mirabolfathy, M.; Groenewald, J.Z.; Crous, P.W. The occurrence of charcoal disease caused by Biscogniauxia mediterranea on chestnut-leaved oak (Quercus castaneifolia) in the Golestan Forests of Iran. Plant Dis. 2011, 95, 876. [**Google Scholar**] [**CrossRef**]

35. Puigdefábregas, J.; Mendizabal, T. Perspectives on desertification: Western Mediterranean. J. Arid Environ. 1998, 39, 209–224. [**Google Scholar**] [**CrossRef**]

36. Desprez-Loustau, M.-L.; Marcais, B.; Nageleisen, L.-M.; Piou, D.; Vannini, A. Interactive effects of drought and pathogens in forest trees. Ann. Forensic Sci. 2006, 63, 597–612. [**Google Scholar**] [**CrossRef**] [**Green Version**]

37. Diffenbaugh, N.S.; Pal, J.S.; Giorgi, F.; Gao, X. Heat stress intensification in the Mediterranean climate change hotspot. Geophys. Res. Lett. 2007, 34, 1–6. [**Google Scholar**] [**CrossRef**] [**Green Version**]

38. Leite, C.; Oliveira, V.; Miranda, I.; Pereira, E. Cork oak and climate change: Disentangling drought effects on cork chemical composition. Sci. Rep. 2020, 10, 1–8. [**Google Scholar**]

39. Henriques, J.; Nóbrega, F.; Sousa, E.; Lima, A. Morphological and genetic diversity of Biscogniauxia mediterranea associated to Quercus suber in the Mediterranean Basin. Rev. Ciências Agrárias 2015, 38, 166–175. [**Google Scholar**]

40. Linaldeddu, B.T.; Maddau, L.; Franceschini, A. First report of Diplodia corticola causing canker and dieback of Quercus ilex, Q. petraea, and Q. suber in Corsica (France). Plant Dis. 2017, 101, 256. [**Google Scholar**] [**CrossRef**]

41. Anselmi, N.; Cellerino, G.P.; Franceschini, A.; Granata, G.; Luisi, N.; Marras, F.; Mazzaglia, A.; Mutto Accordi, S.; Ragazzi, A. Geographic distribution of fungal endophytes of Quercus sp. in Italy. In Endophytism in Forest Trees; Ragazzi, A., Moricca, S., Dellavalle, I., Eds.; Accademia Italiana di Scienze Forestali: Firenze, Italy, 2004; pp. 73–89. [**Google Scholar**]

42. Vannini, A.; Valentini, R.; Luisi, N. Impact of drought and Hypoxylon mediterraneum on oak decline in the Mediterranean region. Ann. Sci. For. 1996, 53, 753–760. [**Google Scholar**] [**CrossRef**]

43. Collado, J.; Platas, G.; Pelaez, F. Identification of an endophytic Nodulisporium sp. from Quercus ilex in central Spain as the anamorph of Biscogniauxia mediterranea by rDNA sequence analysis and effect of different ecological factors on distribution of the fungus. Mycologia 2001, 93, 875–886. [**Google Scholar**] [**CrossRef**]

44. Moricca, S.; Capretti, P.; Dellavalle, I.; Ragazzi, A.; Turco, E. Differences in composition of endophytic mycobiota in twigs and leaves of healthy and declining Quercus species in Italy. For. Pathol. 2012, 33, 31–38. [**Google Scholar**]

45. Sánchez-Salguero, R.; Colangelo, M.; Matías, L.; Ripullone, F.; Camarero, J.J. Shifts in growth responses to climate and exceeded drought-vulnerability thresholds characterize dieback in two Mediterranean deciduous oaks. Forests 2020, 11, 714. [**Google Scholar**] [**CrossRef**]

46. Yangui, I.; Jamaâ, M.L.B.; Boutiti, M.Z.; Vettraino, A.M.; Vannini, A.; Messaoud, C. Occurrence of Biscogniauxia mediterranea in cork oak stands in Tunisia. Phytoparasitica 2021, 49, 131–141. [**Google Scholar**] [**CrossRef**]
47. Jurc, D.; Ogris, N. First report outbreak of charcoal disease caused by Biscognauxia mediterranea on Turkey oak in Slovenia. Plant Pathol. 2006, 55, 299. [**Google Scholar**]
48. Alidadi, A.; Kowsari, M.; Javan-Nikkhah, M.; Salehi Jouzani, G.R.; Ebrahimi Rastaghi, M. New pathogenic and endophytic fungal species associated with Persian oak in Iran. Eur. J. Plant Pathol. 2019, 155, 1017–1032. [**Google Scholar**] [**CrossRef**]
49. Diminić, D.; Kranjec Orlović, J.; Lukić, I.; Ježić, M.; Ćurković Perica, M.; Pernek, M. First report of charcoaldisease of oak (Biscogniauxia mediterranea) on Quercus spp. in Croatia. Plant Dis. 2019, 103, 1–4. [**Google Scholar**] [**CrossRef**]
50. Whalley, A.J.S. The xylariaceaous way of life. Mycol. Res. 1996, 100, 897–922. [**Google Scholar**] [**CrossRef**]
51. Zambrano, J.; Cordeiro, N.J.; Garzon-Lopez, C.; Yeager, L.; Fortunel, C.; Ndangalasi, H.J.; Beckman, N.G. Investigating the direct and indirect effects of forest fragmentation on plant functional diversity. PLoS ONE 2020, 15, e0235210. [**Google Scholar**] [**CrossRef**] [**PubMed**]
52. Lockwood, J.L.; Cassey, P.; Blackburn, T.M. The more you introduce the more you get: The role of colonization pressure and propagule pressure in invasion ecology. Divers. Distrib. 2009, 15, 904–910. [**Google Scholar**] [**CrossRef**]
53. Panzavolta, T.; Panichi, A.; Bracalini, M.; Croci, F.; Ginetti, B.; Ragazzi, A.; Tiberi, R.; Moricca, S. Dispersal and propagule pressure of Botryosphaeriaceae species in a declining oak stand is affected by insect vectors. Forests 2017, 8, 228. [**Google Scholar**] [**CrossRef**] [**Green Version**]
54. Greenwood, S.; Ruiz-Benito, P.; Martínez-Vilalta, J.; Lloret, F.; Kitzberger, T.; Allen, C.D.; Fensham, R.; Laughlin, D.C.; Kattge, J.; Bönisch, G.; et al. Tree mortality across biomes is promoted by drought intensity, lower wood density and higher specific leaf area. Ecol. Lett. 2017, 20, 539–553. [**Google Scholar**] [**CrossRef**] [**PubMed**]
55. Martínez-Vilalta, J.; Lloret, F.; Breshears, D. Drought-induced forest decline: Causes, scope and implications. Biol. Lett. 2012, 8, 689–691. [**Google Scholar**] [**CrossRef**] [**PubMed**]
56. Landmann, G.; Becker, M.; Delatour, C.; Dreyer, E.; Dupouey, J.L. Oak dieback in France: Historical and recent records, possible causes, current investigations. Zustand Gefährdung Laubwäder 1993, 5, 97–114. [**Google Scholar**]
57. Oak, S.W.; Starkey, D.A.; Dabney, J.M. Oak decline alters habitat in southern upland forest. In Proceedings of the Annual Conference of the Southeast Association of the Fish and Wildlife Agencies, Hilton Head, SC, USA, 7–10 November 1988; Volume 42, pp. 491–501. [**Google Scholar**]
58. Vannini, A.; Valentini, R. Influence of water relations in Quercus cerris-Hypoxylon mediterraneum interaction: A model of drought induced susceptibility to a weakness parasite. Tree Physiol. 1994, 14, 129–139. [**Google Scholar**] [**CrossRef**]
59. Vannini, A.; Scarascia Mugnozza, G. Water stress: A predisposing factor in the pathogenesis of Hypoxylon mediterraneum on Quercus cerris. Eur. J. For. Path. 1991, 21, 193–202. [**Google Scholar**] [**CrossRef**]
60. Nordén, B.; Ryberg, M.; Götmark, F.; Olausson, B. Relative importance of coarse and fine woody debris for the diversity of wood-inhabiting fungi in temperate broadleaf forests. Biol. Conserv. 2004, 117, 1–10. [**Google Scholar**] [**CrossRef**]

61. Holub, S.M.; Spears, J.D.H.; Lajtha, K. A reanalysis of nutrient dynamics in coniferous coarse woody debris. Can. J. For. Res. 2001, 31, 1894–1902. [**Google Scholar**] [**CrossRef**]

62. Laiho, R.; Prescott, C.E. The contribution of coarse woody debris to carbon, nitrogen, and phosphorus cycles in three Rocky Mountain coniferous forests. Can. J. For. Res. 1999, 29, 1592–1603. [**Google Scholar**] [**CrossRef**]

63. Hagan, J.M.; Grove, S.L. Coarse woody debris. J. For. 1999, 97, 6–11. [**Google Scholar**]

64. Allen, C.D.; Breshears, D.D.; McDowell, N.G. On underestimation of global vulnerability to tree mortality and forest die-off from hotter drought in the Anthropocene. Ecosphere 2015, 6, 129. [**Google Scholar**] [**CrossRef**]

65. Holuša, J.; Lubojacký, J.; Čurn, V.; Tonka, T.; Lukášová, K.; Horák, J. Combined effects of drought stress and Armillaria infection on tree mortality in Norway spruce plantations. For. Ecol. Manag. 2018, 427, 434–445. [**Google Scholar**] [**CrossRef**]

66. Shaw, C.G., III; Kile, G.A. (Eds.) Armillaria Root Disease; Agricultural Handbook No. 691; Forest Service, US Department of Agriculture: Washington, DC, USA, 1991.

67. Heinzelmann, R.; Dutech, C.; Tsykun, T.; Labbé, F.; Soularue, J.P.; Prospero, S. Latest advances and future perspectives in Armillaria research. Can. J. Plant Path. 2019, 41, 1–23. [**Google Scholar**] [**CrossRef**]

68. Moricca, S.; Bracalini, M.; Croci, F.; Corsinovi, S.; Tiberi, R.; Ragazzi, A.; Panzavolta, T. Biotic Factors Affecting Ecosystem Services in Urban and Peri-Urban Forests in Italy: The Role of Introduced and Impending Pathogens and Pests. Forests 2018, 9, 65. [**Google Scholar**] [**CrossRef**] [**Green Version**]

69. Costa, D.; Tavares, R.M.; Baptista, P.; Lino-Neto, T. Cork Oak endophytic fungi as potential biocontrol agents against Biscogniauxia mediterranea and Diplodia corticola. J. Fungi 2020, 6, 287. [**Google Scholar**] [**CrossRef**] [**PubMed**]

11 INFLUENCE OF PLANT DIVERSITY ON THE INCIDENCE OF DIPLODIA MANAGEMENT OF TIP BLIGHT IN NATURAL SECONDARY JAPANESE RED PINE FORESTS

1. INTRODUCTION: AN OVERVIEW

Sphaeropsis sapinea (Fr.) Dyko and B. Sutton is a highly significant pathogen that affects over 60 conifer species, including Abies, Cedrus, Juniperus, Larix, Picea, Pinus, Pseudotsuga, and Thuja [1] (p. 186). It induces tip blight, stem canker, rhizome decay, and sapstain, posing significant risks to forest health in over 40 nations including New Zealand, Australia, South Africa, the United States of America, Finland, Germany, and China [2,3,4,5,6,7,8] (pp. 224–227). As an endophyte, it can exist as a latent pathogen and does not cause any obvious symptoms in the hosts [9]. Sphaeropsis sapinea, a pathogen that affects coniferous trees of all ages, can cause illness when the trees experience physiological stress or physical damage [10]. An elevated prevalence of the illness leads to widespread forest decay, which negatively impacts the integrity and stability of forest ecosystems [11,12,13,14].

The Kunyu Mountains have the primary natural secondary forests of Pinus densiflora Sieb. et Zucc., often known as Japanese red pine. These forests serve as the original habitats and distribution centers for Japanese red pine in China and even Northeast Asia. They possess a plethora of species, genetic variations, and variety within their ecosystems. Their germplasms are regarded as possessing significant scientific research and practical worth [15,16]. Diplodia tip blight, caused by the fungus S. sapinea, is a significant disease that impacts the natural secondary forests of Japanese red pine in the Kunyu Mountains. It poses a threat to the robust and sustainable growth of forests. However, the reason behind why this disease has caused an epidemic in this location remains unexplained.

Research on the correlation between Diplodia tip blight in pine trees and environmental factors revealed that S. sapinea, acting as an endophyte, can turn into a pathogen when its hosts are weakened by abiotic stresses such as reduced rainfall, higher temperatures, drought, and hailstorms [6,7,17,18,19].

Research has shown that the diversity of tree species has a substantial influence on the occurrence of diseases. The risk of Picea abies (L.) H. Karsten root rot caused by Heterobasidion annosum (Fr.) Bref. is reduced in mixed forests of P. abies compared to pure forests of P. abies [20]. Hantsch et al. [21] discovered an inverse relationship between the disease index of oak powdery mildew (Erysiphe alphitoides (Griffon and Maubl.) U. Braun and S. Takam. and Erysiphe hypophylla (Nevod.) U. Braun and Cunningt.) and the variety of tree species. In contrast, Menges and Loucks [22] proposed that the mortality of specific host tree species, such as Quercus velutina Lam., Quercus borealis Michx., and Quercus ellipsoidalis E. J. Hill, infested with Ceratocystis fagacearum (Bretz.) Hunt, increases as the diversity of Quercus species increases. Moreover, the correlation between the variety of plants and the occurrence of diseases is quite intricate in forest ecosystems with multiple hosts [23]. For example, the majority of plants in forest habitats serve as hosts for Phytophthora ramorum Werres, De Cock and Man in't Veld [24,25].

Nevertheless, the presence of alternative hosts can hinder the transmission of diseases to hosts that are extremely vulnerable, hence mitigating the impact of the pathogen on the plants they infect [26]. Xie and Liang (27) discovered that the disease index of shoot blight (Ceratosphaeria phyllostachydis Zhang) in moso bamboo (Phyllostachys edulis (Carriere) J. Houzeau) is highest in the moso bamboo–conifer mixed forest, moderate in the moso bamboo–hardwood mixed forest, and lowest in the moso bamboo pure forest. The shrub-herb plants play a significant role in the forest community. Their diversity can modify the levels of humidity, soil composition, light availability, and other elements inside the forest. This, in turn, can impact the growth of trees and the overall forest environment. These changes have a direct influence on the presence and progression of illnesses [28,29]. The presence of various plant species in the shrub and herb layers of different moso bamboo stands has distinct impacts on the growth of moso bamboo shoot blight [27]. The species richness and Shannon-Wiener index showed an increase, while the Simpson index showed a drop in the shrub-herb layers of the mixed forest of Dalbergia odorifera T. Chen. Additionally, there was a significant decrease in the disease index of black scurf (Phyllachora dalbergiicola Henn.) in D. odorifera [30]. These studies provide evidence of clear connections between disease and environmental variables. However, the correlation between plant diversity and illness is intricate. The variations in plant variety across various stand types and vertical structure layers exert distinct impacts on disease occurrence.

The objective of our study was to investigate the relationship between plant diversity and the occurrence of Diplodia tip blight in P. densiflora. This study examined the relationship between plant diversity and Diplodia tip blight in P. densiflora in different stand types and vertical structure layers of natural secondary Japanese red pine forests in the Kunyu Mountains in China. Throughout this analysis, the environmental parameters were maintained at a reasonably constant level. It was assumed that the diversity of plants has distinct effects on the development of diseases in various types of forests. This study aims to determine the potential of a way to manipulate plant diversity structure in order to regulate the occurrence of Diplodia tip blight in P. densiflora, using the observed impact as evidence.

2. MATERIALS AND METHODS

2.1. Study Area and Plot Layout

The Kunyu Mountains (121°41′34″–121°48′04″ E; 37°11′50″–37°17′22″ N) are located on the Jiaodong Peninsula of Eastern China. The region has a warm temperate monsoon climate with mean annual temperature of 12.3 °C, mean annual precipitation of 800–1200 mm, mean annual relative humidity of 62.6%, and frost-free period of 200–220 d. The forest soil is mainly brown sandy loam [**31**,**32**]. Pinus densiflora is the main constructive species. Between June and October 2020, a total of 120 temporary sample plots, each with an area of 20 m × 20 m, were set up in natural secondary Japanese red pine forests of the Kunyu Mountains. All stand plots distributed evenly and had similar conditions of altitude 200 ± 25 m, middle slope position, slope 30° ± 3°, sunny slope aspect, brown sandy loam, and forest age 40 ± 4 years. There were 40 plots of P. densiflora–hardwood mixed forest, 40 plots of P. densiflora–conifer mixed forest and 40 plots of P. densiflora pure forest.

Plots were considered to contain pure forest if their ratios of cross-sectional area at breast height of P. densiflora to the cross-sectional area at breast height of all tree species were ≥70%. Otherwise, the plots were deemed mixed forests. Shrub plots of 5 m × 5 m were established by the five-point method at each 20 m × 20 m tree plot. In this method, the center and four corners of each plots were selected. Herb plots of 2 m × 2 m were established in each shrub plot.

2.2. Plant Diversity Analysis

Plant species and quantity were counted in the tree, shrub, and herb layers in the plots of P. densiflora–hardwood mixed forest, P. densiflora–conifer mixed forest and P. densiflora pure forest. For the tree layers, the diameter at breast height (DBH) of living trees was ≥ 5 cm. Plant species richness, and the Shannon–Wiener, Simpson, and Pielou indices were calculated. Each plant diversity index has its own unique meaning. Plant species richness refers to the total number of species in plant communities. All plant species are treated equally and only the presence of the plant species is considered, and relative species abundance is not assessed here. The Pielou index measures the distribution of individuals of plant species in plant communities. The Shannon–Wiener index predicts that plant individuals randomly selected within plant communities belong to the uncertainty of a certain plant species. It considers the abundance and evenness of plant species. The Simpson index indicates the probability that two plant species selected randomly within the plant community are the same plant species [**33**] (pp. 7–10), [**34,35**]. It also considers both plant species abundance and evenness. However, the Simpson index is more strongly affected by evenness than the Shannon–Wiener index. The formulae for the above four indices are as follows [**35,36,37**]:

$$\text{Species richness } (S) = \text{Total plant species per plot} \tag{1}$$

$$\text{Shannon} - \text{Wiener index } (H') = -\sum_{i=1}^{s} \frac{n_i}{N} \ln\left(\frac{n_i}{N}\right) \tag{2}$$

$$\text{Simpson index } (D) = 1 - \sum_{i=1}^{s} \left(\frac{n_i}{N}\right)^2 \tag{3}$$

$$\text{Pielou index } (J) = -\sum_{i=1}^{s} \frac{n_i}{N} \ln\left(\frac{n_i}{N}\right) / \ln S \tag{4}$$

where ni is the total abundance of a single plant species per plot and N is the total abundance of all plant species per plot.

2.3. Disease Index

The disease index of Diplodia tip blight in P. dentiflora was measured by the five-point method, in that six trees were taken from each of the four corners and the center of each plot. These 30 P. densiflora were divided into upper, middle, and lower layers, for which one branch from each direction (east, south, west, and north) were selected [**34**]. The needles are treated as cylinders with equal radii. Hence, the ratios of the lesion areas on the needles with tip blight symptoms (**Figure 1**) to the total needle areas could be transformed into the length ratios. Total needle lesion lengths and total needle lengths for all sampled branches per tree were measured with a steel tape. The ratios of the former to the latter were calculated and then the representative values per tree were determined according to the plant disease classification standard in **Table 1**. The representative values and the corresponding total numbers of P. densiflora were substituted into the formula to calculate the sample plot disease index. The formulae are as follows [**38,39**]:

$$DITP = \frac{\sum NDPd \times RV}{TPd \times MRV} \times 100 \quad (5)$$

$$DISP = \frac{TDISP}{TNSP} \quad (6)$$

where DITP is the Diplodia tip blight disease index in each temporary sample plot, NDPd is the number of diseased P. densiflora at a certain disease grade, TPd is the total number of P. densiflora (30), RV is the representative value of the corresponding grade, and MRV is the maximum representative value of the corresponding grade (4). DISP is the disease index of a certain stand type, TDISP is the sum of all DITP values in a single stand type, and TNSP is the sum of temporary sample plots in a single stand type (40).

Figure 1. Symptoms of local (A) and overall (B) tip blight in Pinus densiflora caused by Sphaeropsis sapinea.

Table 1. Plant disease classification standard of Diplodia tip blight in Pinus dentiflora.

Disease Grade	Representative Value	Plant Disease Classification Standard
I	0	No disease symptoms
II	1	The lesion area is less than 25%
III	2	The lesion area is 25–50%
IV	3	The lesion area is 50–75%
V	4	The lesion area is more than 75%

2.4. Data Analysis

The variables of disease indices and plant diversity indices of each P. densiflora stand conformed to the normal distribution according to quantile–quantile (Q–Q) plot tests. Brown-Forsythe variance analysis followed by Tamhane's T2 test at $p < 0.05$ level [**40**] (pp. 114–115) was used to compare the means of the Diplodia tip blight disease indices in all plots of the three Japanese red pine stand types and the plant diversity indices in all plots of various Japanese red pine stand types and vertical structure layers. The scatter plots had linear trends between the plant diversity indices and the disease indices of Diplodia tip blight in P. densiflora, thus Pearson correlation coefficients based on these two normally distributed sets of variables were calculated to determine the correlations [**41,42**].

Generalized linear model (GLM) analysis for all test factors (namely plant species richness and the Shannon–Wiener, Simpson, and Pielou indices among various vertical structure layers and the same stand type) was performed to examine plant diversity indices that had significant impacts on the disease indices of Diplodia tip blight in P. densiflora. The optimal model for each stand was selected by Akaike information criterion (AIC) and Bayesian information criterion (BIC). Plant species richness and the Shannon–Wiener, Simpson, and Pielou indices for each stand were replaced with x_1–x_4 for the tree layer, x_5–x_8 for the shrub layer, and x_9–x_{12} for the herb layer, respectively. In this manner, the prediction models for the Diplodia tip blight disease indices in each stand were respectively acquired. Data processing and statistical analyses were performed by R v4.0.2 (R Core Team, Vienna, Austria), Microsoft Excel 2016 (Microsoft Corp., Redmond, WA, USA), and SPSS v23.0 (IBM Corp., Armonk, NY, USA).

3. RESULTS

3.1. Disease Indices in Three Different Japanese Red Pine Stand Types

The disease indices of Diplodia tip blight in P. densiflora in three Japanese red pine stand types significantly differed ($F(2, 117) = 415.197$; $p < 0.01$) as follows: P. densiflora–conifer mixed forest (44.215 ± 2.288), P. densiflora pure forest (36.202 ± 2.444), and P. densiflora–hardwood mixed forest (29.875 ± 1.930).

3.2. Plant Species Composition and Diversity

3.2.1. Plant Species Composition

The sum of the tree species in the tree layers of the three Japanese red pine stands was in the range of 13–19 (**Table 2**). In the P. densiflora–hardwood mixed forest, a total of 19 tree species in 15 genera and 11 families were identified. The highest relative abundance was of P. densiflora (47.85%), followed by Quercus variabilis

(22.30%), Q. acutissima (15.96%), etc. In the P. densiflora–conifer mixed forest, one tree variety and 16 tree species in 12 genera and eight families were found, the highest relative abundance was of P. densiflora (52.42%), followed by P. thunbergii (29.94%), etc. In the P. densiflora pure forest, a total of 13 tree species in nine genera and six families were detected. The highest relative abundance was of P. densiflora (71.04%), followed by Q. acutissima (14.52%), etc. (**Table 2**).

Table 2. Relative abundance of tree species in the three stands.

Tree Species	Relative Abundance (%)		
	PHF	PCF	PPF
Rhus chinensis Mill.	1.41	0.40	0.33
Catalpa bungei C. A. Mey	0.31	0.07	—
Cryptomeria japonica var. *sinensis* Miquel	—	0.26	—
Cunninghamia lanceolata (Lamb.) Hook.	0.38	0.70	0.41
Metasequoia glyptostroboides Hu & W. C. Cheng	—	—	0.04
Albizia julibrissin Durazz.	—	—	0.07
Albizia kalkora (Roxb.) Prain	2.13	0.40	0.96
Robinia pseudoacacia L.	1.13	1.46	1.15
Castanea mollissima Bl.	—	0.18	—
Quercus acutissima carr.	15.96	5.56	14.52
Quercus serrata Murray	0.34	0.51	—
Quercus variabilis Bl.	22.30	1.87	5.32
Larix kaempferi (Lamb.) Carr.	0.09	0.33	0.07
Pinus armandii Franch.	0.66	4.39	0.55
Pinus densiflora Sieb. et Zucc.	47.85	52.42	71.04
Pinus taeda L.	—	1.32	—
Pinus thunbergii Parl.	6.77	29.94	5.5
Rhamnella franguloides (Maxim.) Weberb.	0.09	—	—
Cerasus serrulata (Lindl.) G. Don ex London	0.06	—	—
Pyrus calleryana Dcne.	0.03	—	—
Sorbus alnifolia (Sieb. et Zucc.) K. Koch	0.19	0.04	—
Ailanthus altissima (Mill.) Swingle	0.22	0.15	0.04
Symplocos paniculata (Thunb.) Miq.	0.03	—	—
Ulmus pumila L.	0.03	—	—

Note: *Pinus densiflora*–hardwood mixed forest, *P. densiflora*–conifer mixed forest, and *P. densiflora* pure forest are denoted as PHF, PCF, and PPF, respectively.

The shrub layers of the three Japanese red pine stands consisted of 29–35 plant species (**Table 3**). In the P. densiflora–hardwood mixed forest, one shrub variety and 34 shrub species in 31 genera and 20 families were found. The highest relative abundance was of Rhus chinensis (29.89%), followed by Indigofera kirilowii (29.38%), etc. In the P. densiflora–conifer mixed forest, one shrub variety and 34 shrub species in 31 genera and 20 families were identified. The highest relative abundance was of I. kirilowii (33.87%), followed by R. chinensis (26.71%), Lespedeza bicolor (12.09%), etc. In the P. densiflora pure forest, one shrub variety and 28 shrub species in 25 genera and 17 families were detected. The highest relative abundance was of I. kirilowii (39.41%), followed by R. chinensis (30.02%), L. bicolor (11.12%), etc. (**Table 3**).

Table 3. Relative abundance of shrub species in the three stands.

Shrub Species	Relative Abundance (%)		
	PHF	PCF	PPF
Rhus chinensis Mill.	29.89	26.71	30.02
Eleutherococcus senticosus (Rupr. Maxim.) Maxim.	—	0.03	—
Blumea balsamifera (L.) DC.	0.03	—	—
Lonicera japonica Thunb.	0.14	0.30	—
Viburnum erosum Thunb.	0.20	0.28	1.45
Celastrus orbiculatus Thunb.	0.61	2.81	0.27
Euonymus oxyphyllus Miq.	0.06	0.03	0.04
Rhododendron simsii Planch.	0.06	—	—
Albizia kalkora (Roxb.) Prain	6.94	3.26	2.97
Amorpha fruticosa L.	0.25	1.87	2.07
Campylotropis macrocarpa (Bge.) Rehd.	—	0.51	0.20
Cercis chinensis Bunge	—	—	0.04
Indigofera kirilowii Maxim. ex Palibin	29.38	33.87	39.41
Lespedeza bicolor Turcz.	5.94	12.09	11.12
Pueraria montana (Lour.) Merr.	0.03	0.05	—
Deutzia grandiflora Bunge	5.69	2.73	3.56
Callicarpa bodinieri Levl.	—	0.13	—
Lindera glauca (Sieb. et Zucc.) Bl.	0.08	0.15	0.39
Lindera obtusiloba Bl.	0.98	0.61	0.39
Yulania denudata (Desr.) D. L. Fu	—	0.13	0.08
Grewia biloba var. *parviflora* (Bunge) Hand.-Mazz.	3.29	3.31	3.48
Broussonetia papyrifera (Linn.) L'Her.r ex Vent.	—	—	0.08
Morus alba L.	—	—	0.04
Morus mongolica (Bur.) Schneid.	—	0.05	0.04

Shrub Species	Relative Abundance (%)		
	PHF	PCF	PPF
Jasminum nudiflorum Lindl.	–	0.03	–
Ligustrum lucidum Ait.	0.11	0.86	–
Flueggea suffruticosa (Pall.) Baill.	0.11	–	0.12
Rhamnella franguloides (Maxim.) Weberb.	0.53	–	–
Rhamnus davurica Pall.	0.06	0.18	0.16
Armeniaca sibirica (L.) Lam.	0.06	0.03	–
Cerasus japonica (Thunb.) Lois.	5.52	2.61	1.84
Cerasus tomentosa (Thunb.) Wall.	0.11	0.03	0.12
Rosa multiflora Thunb.	0.56	0.18	0.08
Rubus parvifolius L.	0.70	0.81	0.12
Rubus phoenicolasius Maxim.	0.45	0.91	–
Zanthoxylum schinifolium Sieb. et Zucc.	2.65	1.67	0.74
Koelreuteria paniculata Laxm.	0.03	–	–
Smilax china L.	0.98	0.76	0.23
Smilax sieboldii Miq.	1.14	2.02	0.35
Symplocos paniculata (Thunb.) Miq.	2.45	0.30	0.31
Ulmus macrocarpa Hance	0.03	0.03	–
Boehmeria japonica (Linnaeus f.) Miq.	0.11	0.53	–
Ampelopsis humulifolia Bge.	0.70	0.15	0.27
Vitis amurensis Rupr.	0.17	–	–
Note: The meanings of PHF, PCF, and PPF are the same as those in Table 2.			

The herb layers of the three Japanese red pine stands comprised of 49–53 plant species (**Table 4**). In the P. densiflora–hardwood mixed forest, one herb subspecies, two herb varieties, and 49 herb species in 45 genera and 24 families were found. The highest relative abundance was of Carex lanceolata (23.90%), followed by Sanguisorba applanata (12.81%), Pteridium aquilinum var. latiusculum (10.76%), Commelina communis (10.39%), etc. In the P. densiflora–conifer mixed forest, two herb subspecies, two herb varieties, and 49 herb species in 47 genera and 21 families were identified. The highest relative abundance was of Imperata cylindrica (18.25%), followed by C. lanceolata (16.28%), S. applanata (12.04%), etc. In the P. densiflora pure forest, one herb subspecies, two herb varieties, and 46 herb species in 44 genera and 23 families were detected. The highest relative abundance was of C. lanceolata (20.37%), followed by S. applanata (18.49%), I. cylindrica (14.50%),

Miscanthus sinensis (10.77%), etc. (**Table 4**).

Table 4. Relative abundance of herb species in the three stands.

Herb Species	Relative Abundance (%)		
	PHF	PCF	PPF
Amaranthus tricolor L.	—	0.04	—
Bupleurum chinense DC.	0.10	—	—
Saposhnikovia divaricata (Turcz.) Schischk.	0.03	—	—
Cynanchum chinense R. Br.	0.20	0.04	—
Cynanchum paniculatum (Bunge) Kitag.	0.79	0.31	0.13
Metaplexis japonica (Thunb.) Makino	—	0.10	0.10
Anaphalis sinica Hance	0.13	—	—
Artemisia argyi Lévl. et Van.	0.03	0.16	0.03
Artemisia japonica Thunb.	—	0.24	0.10
Artemisia stechmanniana Bess.	0.13	0.53	—
Aster hispidus Thunb.	—	0.06	—
Aster indicus L.	0.07	—	—
Atractylodes Lancea (Thunb.) DC.	0.10	0.10	—
Bidens pilosa L.	1.39	0.01	0.06
Carpesium cernuum L.	—	—	0.03
Chrysanthemum chanetii H. Lév.	—	0.27	0.19
Chrysanthemum indicum L.	0.50	3.24	0.77
Cirsium arvense var. *integrifolium* C. Wimm. et Grabowski	—	0.07	—
Crepidiastrum denticulatum (Hout.) Pak & Kawano	0.13	0.01	0.22
Echinops grijsii Hance	—	0.01	—
Elephantopus scaber L.	0.03	—	0.10
Erigeron bonariensis L.	0.03	—	—
Erigeron canadensis L.	—	0.09	0.86
Leibnitzia anandria (L.) Turcz.	0.07	—	—
Senecio scandens Buch.-Ham. ex D. Don	—	0.03	—
Sonchus wightianus DC.	—	0.24	0.48
Taraxacum mongolicum Hand.-Mazz.	—	0.03	—
Brassica juncea (L.) Czern.	0.03	—	—
Adenophora petiolata subsp. *Hunanensis* (Nannfeldt) D. Y. Hong & S. Ge	2.52	1.18	2.04
Adenophora stricta Miq.	0.26	0.10	0.10
Codonopsis lanceolata (Sieb. et Zucc.) Trautv.	—	0.19	—
Platycodon grandiflorus (Jacq.) A. DC.	0.60	0.74	0.54
Dianthus chinensis L.	0.07	0.26	0.03
Gypsophila oldhamiana Miq.	—	0.13	0.41

Herb Species	Relative Abundance (%)		
	PHF	PCF	PPF
Pseudostellaria heterophylla (Miq.) Pax	0.26	—	—
Silene conoidea L.	—	0.03	—
Commelina communis L.	10.39	6.16	0.26
Hylotelephium erythrostictum (Miq.) H. Ohba	—	0.03	0.10
Orostachys fimbriata (Turcz.) A. Berger	—	0.23	0.67
Phedimus aizoon (L.) 't Hart	0.66	0.01	0.03
Trichosanthes kirilowii Maxim.	—	—	0.03
Carex lanceolata Boott	23.90	16.28	20.37
Discorea nipponica Makino	—	—	0.10
Swertia bimaculata (Sieb. et Zucc.) Hook. f. et Thoms. ex C. B. Clark	0.03	—	—
Belamcanda chinensis (L.) Redouté	—	0.33	0.19
Juncus effusus L.	9.63	—	5.71
Agastache rugosa (Fisch. et Mey.) O. Ktze.	0.10	—	—
Isodon amethystoides (Benth.) H. Hara	0.50	—	—
Asparagus cochinchinensis (Lour.) Merr.	0.07	0.01	—
Convallaria majalis L.	—	—	0.29
Lilium concolor var. *pulchellum* (Fisch.) Regel	0.03	—	0.10
Ophiopogon japonicus (L. f.) Ker-Gawl.	0.60	0.23	0.29
Polygonatum odoratum (Mill.) Druce	—	0.01	0.06
Cocculus orbiculatus (L.) DC.	0.53	2.33	1.47
Phryma leptostachya subsp. *Asiatica* (Hara) Kitamura	—	0.09	—
Phytolacca acinosa Roxb.	0.10	—	—
Arthraxon prionodes (Steud.) Dandy	0.03	—	—
Arundinella hirta (Thunb.) Tanaka	1.36	—	0.41
Deyeuxia pyramidalis (Host) Veldkamp	0.33	5.15	1.88
Imperata cylindrica (L.) Beauv.	7.08	18.25	14.50
Miscanthus sacchariflorus (Maxim.) Hackel	—	—	0.19
Miscanthus sinensis Anderss.	7.85	6.65	10.77
Oplismenus undulatifolius (Ard.) Beauv.	0.86	3.92	1.79
Themeda triandra Forsk.	—	0.51	0.92
Fallopia multiflora (Thunb.) Harald.	0.10	—	—
Polygonum divaricatum L.	3.21	0.67	1.34
Androsace umbellata (Lour.) Merr.	—	0.03	—
Lysimachia barystachys Bunge	0.13	—	0.10
Lysimachia fortunei Maxim.	0.03	—	—
Lysimachia pentapetala Bunge	0.10	2.85	1.31
Pteridium aquilinum var. *latiusculum* (Desv.) Underw. ex Heller	10.76	9.19	9.56

Herb Species	Relative Abundance (%)		
	PHF	PCF	PPF
Clematis florida Thunb.	0.03	0.17	0.03
Fragaria vesca L.	—	0.03	—
Sanguisorba applanata Yü et Li	12.81	12.04	18.49
Galium odoratum (L.) Scop.	0.07	—	—
Rubia cordifolia L.	—	0.37	0.03
Rubia yunnanensis Diels	0.17	—	—
Ostericum sieboldii (Miq.) Nakai	0.07	—	—
Patrinia scabiosifolia Link	0.03	5.57	1.66
Viola arcuata Bl.	0.96	0.19	0.51
Viola philippica Cav.	—	0.47	0.22
Alpinia japonica (Thunb.) Miq.	—	—	0.45
Note: The meanings of PHF, PCF, and PPF are the same as those in Table 2.			

3.2.2. Comparison of Plant Diversity Indices

Plant Diversity Indices among Various Stand Types and the Same Vertical Structure Layer Comparison of the diversity indices among various stand types **(Table 5)** showed that the cumulative values for the species richness, the Shannon–Wiener, and Simpson indices in the tree–shrub–herb layers were, in descending order, P. densiflora–conifer mixed forest > P. densiflora–hardwood mixed forest > P. densiflora pure forest. The differences among groups in terms of the species richness ($F (2, 117) = 90.606$; $p < 0.01$) and the Shannon–Wiener ($F (2, 117) = 59.715$; $p < 0.01$) and Simpson ($F (2, 117) = 62.622$; $p < 0.01$) indices were all significant. The cumulative values of the Pielou index in the tree–shrub–herb layers were highest for the P. densiflora–hardwood mixed forest, moderate for the P. densiflora–conifer mixed forest, and lowest for the P. densiflora pure forest. There were no significant differences among groups in terms of this index ($F (2, 117) = 2.651$; $p > 0.05$).

Table 5. The total of each plant's diversity indices for the three stands.

Stand Type	*S*	*H'*	*D*	*J*
PHF	18.917 ± 1.682^{b}	3.792 ± 0.205^{b}	1.924 ± 0.070^{a}	2.221 ± 0.137
PCF	22.797 ± 3.205^{a}	4.083 ± 0.404^{a}	1.975 ± 0.119^{a}	2.171 ± 0.109
PPF	16.066 ± 1.419^{c}	3.379 ± 0.216^{c}	1.743 ± 0.097^{b}	2.152 ± 0.164
Note: The species richness and the Shannon–Wiener, Simpson, and Pielou indices are denoted as *S*, *H'*, *D*, and *J*, respectively. The meanings of PHF, PCF, and PPF are the same as those in Table 2. The lowercase letters (a, b, c) show the significant difference between various stand types and the same vertical structure layer ($p < 0.05$). For non-significant differences among groups, no letters are appended to the values. All values are means ± standard deviation.				

In the tree layers of the three stands, the differences in the species richness (F

(2, 117) = 17.361; p < 0.01) and the Shannon–Wiener (F (2, 117)' = 70.503; p < 0.01), Simpson (F (2, 117) = 127.614; p < 0.01), and Pielou (F (2, 117) = 36.521; p < 0.01) indices among groups were all significant. Multiple comparisons showed that the species richness and the Simpson and Pielou indices of the tree layers in the P. densiflora–hardwood mixed forest and the P. densiflora–conifer mixed forest were significantly different from those of the tree layers in the P. densiflora pure forest (p < 0.01) (**Table 6**).

Table 6. Plant diversity indices among various stand types and vertical structure layers.

Stand Type	Layer	*S*	*H'*	*D*	*J*
PHF	Tree	$5.388 \pm 0.881^{a,B}$	$1.202 \pm 0.099^{a,B}$	$0.624 \pm 0.037^{a,B}$	$0.719 \pm 0.055^{a,B}$
	Shrub	$6.964 \pm 1.182^{a,A}$	1.229 ± 0.131^{B}	0.626 ± 0.048^{B}	$0.719 \pm 0.070^{a,b,B}$
	Herb	$6.566 \pm 0.964^{b,A}$	$1.360 \pm 0.144^{b,A}$	$0.675 \pm 0.051^{b,A}$	0.783 ± 0.079^{A}
PCF	Tree	$5.153 \pm 1.424^{a,C}$	$1.105 \pm 0.218^{b,B}$	$0.598 \pm 0.071^{a,B}$	$0.701 \pm 0.071^{a,B}$
	Shrub	$7.366 \pm 1.680^{a,B}$	1.185 ± 0.141^{B}	0.596 ± 0.059^{B}	$0.693 \pm 0.075^{b,B}$
	Herb	$10.278 \pm 1.584^{a,A}$	$1.794 \pm 0.169^{a,A}$	$0.781 \pm 0.035^{a,A}$	0.777 ± 0.047^{A}
PPF	Tree	$4.075 \pm 0.764^{b,C}$	$0.820 \pm 0.099^{c,C}$	$0.436 \pm 0.057^{b,C}$	$0.592 \pm 0.086^{b,C}$
	Shrub	$5.347 \pm 0.936^{b,B}$	1.163 ± 0.176^{B}	0.614 ± 0.076^{B}	$0.745 \pm 0.100^{a,B}$
	Herb	$6.644 \pm 0.893^{b,A}$	$1.395 \pm 0.158^{b,A}$	$0.693 \pm 0.059^{b,A}$	0.815 ± 0.094^{A}

Note: The meanings of *S*, *H'*, *D*, and *J* are the same as those in Table 5. The meanings of PHF, PCF, and PPF are the same as those in Table 2. The lowercase letters (a, b, c) show the significant difference between various stand types and the same vertical structure layer (*p* < 0.05). The uppercase letters (A, B, C) show the significant difference between various vertical structure layers and the same stand type (*p* < 0.05). For non-significant differences among groups, no letters are appended to the values. All values are means ± standard deviation (SD).

In the shrub layer of the three stands, the differences in the species richness (F (2, 117) = 26.891; p < 0.01;) and the Pielou index (F (2, 117) = 3.904; p < 0.05) among groups were significant. Multiple comparisons revealed that the species richness of the P. densiflora pure mixed forest significantly differed from those of the other two stands (p < 0.01). The Pielou indices significantly differed between the P. densiflora pure forest and the P. densiflora–conifer mixed forest (p < 0.05). The differences among stands in terms of the Shannon–Wiener (F (2, 117) = 1.964; p > 0.05) and Simpson (F (2, 117) = 2.408; p > 0.05) indices were not significant (**Table 6**).

In the herb layers of the three stands, the differences in the species richness (F (2, 117) = 127.444; p < 0.01) and the Shannon–Wiener (F (2, 117) = 93.937; p < 0.01) and Simpson (F (2, 117) = 53.329; p < 0.01) indices between groups were significant. Multiple comparisons showed that these indices significantly differed between the P. densiflora–conifer mixed forest and the other two stands ($p < 0.01$). The differences among the herb layers of the three stands in terms of the Pielou indices ($F (2, 117) = 2.910$; $p > 0.05$) were not significant (**Table 6**).

Plant Diversity Indices among Various Vertical Structure Layers and the Same Stand Type Overall, the species richness and the Shannon–Wiener, Simpson, and Pielou indices were higher for the herb layers than the shrub and tree layers of natural secondary Japanese red pine forests. In the P. densiflora–hardwood mixed forest, the differences in the species richness ($F (2, 117) = 26.001$; $p < 0.01$) and the Shannon–Wiener ($F (2, 117) = 18.011$; $p < 0.01$), Simpson ($F (2, 117) = 15.598$; $p < 0.01$), and Pielou ($F (2, 117) = 11.601$; $p < 0.01$) indices among groups were all significant. Multiple comparisons revealed that the latter three indices of the herb layer significantly differed from those of the shrub and tree layers ($p < 0.01$). The species richness significantly differed between the tree layer and the other two layers ($p < 0.01$). In the P. densiflora–conifer mixed forest, the differences in the species richness ($F (2, 117) = 107.719$; $p < 0.01$) and Shannon–Wiener ($F (2, 117) = 177.587$; $p < 0.01$), Simpson ($F (2, 117) = 139.297$; $p < 0.01$), and Pielou ($F (2, 117) = 20.258$; $p < 0.01$) indices among groups were all significant. Multiple comparisons revealed that the latter three indices of herb layer were significantly differed from those of the shrub and tree layers ($p < 0.01$). In the P. densiflora pure mixed forest, the differences in the species richness ($F (2, 117) = 87.678$; $p < 0.01$) and the Shannon–Wiener ($F (2, 117) = 153.491$; $p < 0.01$), Simpson ($F (2, 117) = 166.290$; $p < 0.01$), and Pielou ($F (2, 117) = 59.575$; $p < 0.01$) indices among groups were all significant (**Table 6**).

3.3. Correlation between Plant Diversity Indices and Disease Index of Diplodia Tip Blight in P. densiflora

3.3.1. Tree Layer

In the tree layer of the P. densiflora–hardwood mixed forest, the disease index of Diplodia tip blight in P. densiflora was significantly negatively correlated with the species richness and the Shannon–Wiener, Simpson, and Pielou indices ($p < 0.05$). In the tree layer of the P. densiflora–conifer mixed forest, the disease index of Diplodia tip blight in P. densiflora was significantly positively correlated with the Shannon–Wiener, Simpson, and Pielou indices ($p < 0.05$). However, the disease index of Diplodia tip blight in P. densiflora was not significantly correlated with the species richness. In the tree layer of the P. densiflora pure mixed forest, the disease index of Diplodia tip blight in P. densiflora was significantly negatively correlated with the Shannon–Wiener and Simpson indices ($p < 0.05$). However, there was no significant correlation between the disease index of Diplodia tip blight in P.

densiflora and the species richness or the Pielou index (**Table** 7).

Table 7. The correlation coefficients between plant diversity indices and disease index of Diplodia tip blight in Pinus densiflora.

Layer	Stand Type	*S*-DI	*H'*-DI	*D*-DI	*J*-DI
	PHF	-0.430 **	-0.883 **	-0.796 **	-0.367 *
Tree	PCF	0.218	0.667 **	0.672 **	0.385 *
	PPF	-0.221	-0.471 **	-0.392 *	-0.139
	PHF	-0.024	0.640 **	0.655 **	0.259
Shrub	PCF	0.114	0.471 **	0.463 **	0.196
	PPF	-0.122	0.522 **	0.444 **	0.228
	PHF	0.118	0.431 **	0.397 *	-0.062
Herb	PCF	0.204	0.637 **	0.660 **	0.433 **
	PPF	-0.059	-0.006	0.023	0.086

Note: * means $p < 0.05$, and ** means $p < 0.01$. '-' before the correlation coefficient means negative correlation, and no '-' means positive correlation. *S*-DI, *H'*-DI, *D*-DI and *J*-DI represent the correlation coefficients between the species richness, the Shannon–Wiener index, the Simpson index, and the Pielou index and Diplodia tip blight in *Pinus densiflora*, respectively.

3.3.2. Shrub Layer

In the shrub layers of all three stands, the disease index of Diplodia tip blight in P. densiflora was significantly positively correlated with the Shannon–Wiener and Simpson indices ($p < 0.01$). However, the disease index of Diplodia tip blight in P. densiflora was not significantly correlated with the species richness or the Pielou index (**Table** 7).

3.3.3. Herb Layer

In the herb layer, the disease index of Diplodia tip blight in P. densiflora was significantly positively correlated with the Shannon–Wiener and Simpson indices ($p < 0.05$) of the P. densiflora–hardwood mixed forest and the Shannon–Wiener, Simpson, and Pielou indices of the P. densiflora–conifer mixed forest. However, the disease index of Diplodia tip blight in P. densiflora was not significantly correlated with the species richness and the Pielou index of the P. densiflora–hardwood mixed forest, the species richness of the P. densiflora–conifer mixed forest, or the species richness, and the Shannon–Wiener, Simpson, and Pielou indices of the P. densiflora pure forest (**Table** 7).

3.4. Generalized Linear Model Analysis between Plant Diversity Indices and Disease Index of Diplodia Tip Blight in P. densiflora in Three Japanese Red Pine Stand Types

In the P. densiflora–hardwood mixed forest, the Simpson index of the tree layer was the most significant influence on the disease index of Diplodia tip blight in P. densiflora ($p < 0.05$), followed by the Simpson index of the shrub layer ($p < 0.01$) and the Shannon–Wiener index of the tree layer ($p < 0.01$). The values of AIC and BIC were 94.214 and 102.659, respectively. The optimal disease index prediction model in this stand was:

$$y = 43.538 - 11.165x_2 - 11.616x_3 + 11.195x_7 \ (p < 0.01) \quad (7)$$

In the P. densiflora-conifer mixed forest, the Simpson index of the herb layer was the most significant influence on the disease index of Diplodia tip blight in P. densiflora ($p < 0.01$), followed by the Simpson indices of the tree and shrub layers ($p < 0.01$). The values of AIC and BIC were 138.477 and 146.921, respectively. The optimal disease index prediction model in this stand was:

$$y = 7.467 + 14.639x_3 + 10.985x_7 + 27.454x_{11} \ (p < 0.01) \quad (8)$$

In the P. densiflora pure forest, the Shannon–Wiener index of the tree layer was the most significant influence on the disease index of Diplodia tip blight in P. densiflora ($p < 0.01$), followed by the Shannon–Wiener index of the shrub layer ($p < 0.01$). The value of AIC and BIC were 172.441 and 179.197, respectively. The optimal disease index prediction model in this stand was:

$$y = 36.495 - 8.693x_2 + 5.874x_6 \ (p < 0.01) \quad (9)$$

4. DISCUSSION

In the present study, the disease index of Diplodia tip blight in the P. densiflora–conifer mixed forest was larger than the disease indices of Diplodia tip blight in the P. densiflora–hardwood mixed forest and the P. densiflora–pure mixed forest. Moreover, the disease index was positively correlated with the tree diversity indices. Similarly, Gerlach et al. **[43]** reported that the extent of damage caused by root rot (Armillaria sp.) rises with the increase of the proportion of coniferous species in a mixed forest. Our result was in good agreement with the associational susceptibility hypothesis that the disease severity intensifies when multiple host plants are combined **[44,45,46,47]**. In the P. densiflora–conifer mixed forest, there are high relative abundances of both native P. densiflora and introduced P. thunbergii. These species have a close genetic relationship and are hosts of the Diplodia tip blight pathogen. They have similar morphological, physiological, and biochemical characteristics, such as similar nutrients, which could provide enough resources to support pathogen proliferation and increase the risks of host parasitization and Diplodia tip blight epidemics in P. densiflora **[48]**.

The present study also showed that the disease index of Diplodia tip blight in P. densiflora was lowest in the P. densiflora–hardwood mixed forest and negatively correlated with the tree diversity indices. It is in line with Karlman et al.'s **[49]** conclusion that certain fungal diseases in conifer are negatively correlated with the hardwood richness. This phenomenon upholds the associational resistance hypothesis that the disease severity gradually decreases with increasing tree diversity in mixed forest consisting of both host trees and non-host trees [**44,46**]. In the P. densiflora–hardwood mixed forest, the host tree species of S. sapinea, P. densiflora, and its non-host tree species, Q. acutissima and Q. variabilis, all have high relative abundances. Non-host trees may create form physical or chemical barriers that impede the extensive localization, spread, and colonization of the pathogen [**50**]. In this way, non-host trees could lower the incidence of Diplodia tip blight in P. densiflora. In addition, the increase of the relative abundance in hardwood in the P. densiflora–hardwood mixed forest diminishes the amount of resources available to the pathogen, thereby hindering its growth and population establishment [**47,50,51**]. For example, as there are fewer P. abies in the P. abies-Pinus sylvestris L. mixed forest than there are in the P. abies pure forest, the range and quantity of H. annosum infections on adjacent trees are reduced in the former case [**20**].

The disease index of Diplodia tip blight in the P. densiflora pure forest is higher than that in the P. densiflora–hardwood mixed forest and lower than that in the P. densiflora–conifer mixed forest. Furthermore, the differences among stand types were significant. Meanwhile, the disease index was negatively correlated with the tree diversity indices. As the P. densiflora pure forest has a large proportion of host-P. densiflora, the disease index there is significantly larger than that in the P. densiflora–hardwood mixed forest. However, the other tree species with ratios of cross-sectional area at breast height <30% are mainly non-host hardwoods. Therefore, the disease index of Diplodia tip blight is lower in the P. densiflora pure forest than it is in the P. densiflora–conifer mixed forest.

Based on the foregoing findings, it is speculated that there are some promotion or inhibition effects between different tree species and the disease of Diplodia tip blight in P. densiflora that need to be confirmed. Pan [**16**] found that Cephalcia Kunyushanica Xiao have significant positive effects with host P. densiflora and P. thunbergii, and have significant negative effects with Populus ussuriensis Kom. and Hovenia acerba Lindl. The purpose of alleviating the harm of C. Kunyushanica could be achieved by properly matching P. ussuriensis and H. acerba among hosts. Therefore, taking some forest management measures to appropriately increase non-host broad-leaved tree species could mitigate the Diplodia tip blight disease.

There were significant positive correlations between the disease index of Diplodia tip blight in P. densiflora and the plant diversity indices in the shrub–herb layers of all three stands except the understory herb layer of P. densiflora pure forest. Clearly, there may be possible explanations that because of forest resource limitations, competition between trees and understory plants might dramatically increase with shrub–herb diversity. These factors influence the host growth regulation and reduce the host resistance to the pathogen of Diplodia tip blight [**30**]. Another reason may be that as plant diversity increases in the shrub–herb layers, high-humidity microclimates rapidly develop in the understory environments, which favor the proliferation and dispersal of pathogen mycelia and conidia in diseased needles, shoots, and stems. [**1,52**]. In this manner, the infection risk of Diplodia tip blight in P. densiflora increases with decreasing host tree vigor. To control the occurrence and dispersal of Diplodia tip blight in P. densiflora, the diversity structure of the understory vegetation must be modified in natural secondary Japanese red pine forests.

Researchers in China reported that excess precipitation causes root waterlogging and poor ventilation in host trees of S. sapinea, which results in weakening of the host trees and further aggravates the occurrence of Diplodia tip blight [**1,17,53**]. By contrast, the occurrence of Diplodia tip blight caused by S. sapinea is different in countries such as Germany, Italy, the United States of America, and South Africa, where asymptomatic endophytic fungi become pathogen since host resistance is weakened by abiotic stresses such as precipitation deficits, elevated temperatures, drought, and hailstorms [**8**] (pp. 224–227), [**6,54**]. The above results inferred that extreme abiotic stresses increase disease susceptibility of host trees, and then lead to the prevalence of the disease. Therefore, to avoid the influence of the related abiotic factors, the effect of plant diversity on the occurrence of Diplodia tip blight in P. densiflora was analyzed in the case of relatively consistent environmental factors in the research plots.

Generalized linear model analysis revealed that the three prediction models respectively established with the Simpson indices of the tree and shrub layers and the Shannon–Wiener index of the tree layer as independent variables in the P. densiflora–hardwood mixed forest, the Simpson indices of the herb, tree, and shrub layers as independent variables in the P. densiflora–conifer mixed forest, and the Shannon–Wiener indices of the tree and shrub layers as independent variables in the P. densiflora pure mixed forest could better explain the variation of Diplodia tip blight in P. densiflora disease index in three stands. Both the Shannon–Wiener index and Simpson index contain the information of plant species richness and evenness and can reflect characteristics of plant diversity. In the practical application, the models obtained in the three stands were used to predict the disease index of Diplodia tip blight in P. densiflora, which could provide a reference for prevention and control of this disease.

5. CONCISE SUMMARY

The variation in plant variety across different stand types and vertical structure levels has distinct impacts on the prevalence of Diplodia tip blight in P. densiflora in the Kunyu Mountains of China. The presence of Diplodia tip blight in the P. densiflora–hardwood mixed forest and the P. densiflora pure forest confirms the idea that these forests are resistant to the disease. On the other hand, the appearance of Diplodia tip blight in the P. densiflora–conifer mixed forest suggests that this kind of forest is more susceptible to the disease. Increased plant diversity within the forest leads to heightened competition between the pathogen host P. densiflora and shrub-herb plants, ultimately leading to reduced disease resistance in the host. Furthermore, the forest can create and support humid microclimates that promote the transmission of pathogens carried by sick tissues. The presence of Diplodia tip blight disease is a major factor in determining plant diversity. Nevertheless, in order to accurately account for the impact of plant diversity on the disease, it is imperative to take into consideration the specific traits of plant species, particularly the genetic links among trees. Forest management strategies can be used to modify the structure of plant species diversity and reduce the occurrence of this disease.

6. AT A GLANCE

Diplodia tip blight, caused by the fungus Sphaeropsis sapinea, poses a significant danger to the well-being of natural secondary forests of Japanese red pine (Pinus densiflora Sieb. et Zucc.). In order to investigate the impact of plant diversity on the occurrence of Diplodia tip blight disease, we examined the correlation between Diplodia tip blight in P. densiflora and plant diversity in different stand types and vertical structure layers. This analysis was conducted using Pearson correlation coefficients and a generalized linear model, while ensuring that environmental factors remained relatively constant. The disease index of Diplodia tip blight in P. densiflora showed a favorable correlation with the plant diversity indices in the tree layers of P. densiflora–conifer mixed forest. Conversely, it exhibited a negative correlation with the plant diversity indices in the tree layers of both the P. densiflora–hardwood mixed forest and the P. densiflora pure forest. The illness index showed a positive correlation with the plant diversity indices in all stands, except for the P. densiflora pure forest herb layer, in the shrub-herb layers. Prediction models were also developed for the disease index of Diplodia tip blight in the P. densiflora–hardwood mixed forest, P. densiflora–conifer mixed forest, and P. densiflora pure forest. The results indicate that in the Kunyu Mountains of China, the variations in plant variety among different stand types and vertical structure layers have distinct impacts on the presence of Diplodia tip blight in P. densiflora. The presence of Diplodia tip blight disease is a major factor influencing plant diversity. Nevertheless, in order to accurately assess the impact of plant diversity on Diplodia tip blight, it is imperative to take into account the specific traits of plant species, particularly the genetic linkages among trees. Forest management techniques can be employed to reduce the occurrence of Diplodia tip blight and alter the plant species diversity structure in natural secondary Japanese red pine forests.

REFERENCES

1. Song, X.D. Sphaeropsis shoot blight of pines. In Major Forest Diseases and Insect Pests in China; Zhang, X.Y., Luo, Y.Q., Eds.; Chinese Forestry Publishing House: Beijing, China, 2003; p. 186. [**Google Scholar**]
2. Jankovský, L.; Palovčíková, D. Dieback of Austrian pine—The epidemic occurrence of Sphaeropsis sapinea in southern Moravia. J. For. Sci. 2003, 49, 389–394. [**Google Scholar**] [**CrossRef**] [**Green Version**]
3. Iturritxa, E.; Ganley, R.J.; Raposo, R.; García-Serna, I.; Mesanza, N.; Kirkpatrick, S.C.; Gordon, T.R. Resistance levels of Spanish conifers against Fusarium circinatum and Diplodia pinea. For. Pathol. 2013, 43, 488–495. [**Google Scholar**] [**CrossRef**]
4. Zlatkovic, M.; Keča, N.; Wingfield, M.J.; Jami, F.; Slippers, B. New and unexpected host associations for Diplodia sapinea in the Western Balkans. For. Path. 2017, 47, 1–11. [**Google Scholar**] [**CrossRef**] [**Green Version**]
5. Manzanos, T.; Aragonés, A.; Iturritxa, E. Genotypic diversity and distribution of Sphaeropsis sapinea within Pinus radiata trees from northern Spain. For. Path. 2019, 49, 1–9. [**Google Scholar**] [**CrossRef**]
6. Bußkamp, J.; Langer, G.J.; Langer, E.J. Sphaeropsis sapinea and fungal endophyte diversity in twigs of scots pine (Pinus sylvestris) in Germany. Mycol. Prog. 2020, 19, 985–999. [**Google Scholar**] [**CrossRef**]
7. Terhonen, E.; Babalola, J.; Kasanen, R.; Jalkanen, R.; Blumenstein, K. Sphaeropsis sapinea found as symptomless endophyte in Finland. Silva Fenn. 2021, 55, 1–14. [**Google Scholar**] [**CrossRef**]
8. Ye, J.R.; He, W. Forest Pathology; Chinese Forestry Publishing House: Beijing, China, 2011; pp. 224–227. [**Google Scholar**]
9. Bihon, W.; Slippers, B.; Burgess, T.; Wingfield, M.J.; Wingfield, B.D. Sources of Diplodia pinea endophytic infections in Pinus patula and P. radiata seedlings in South Africa. For. Pathol. 2011, 41, 370–375. [**Google Scholar**] [**CrossRef**] [**Green Version**]
10. De Wet, J.; Wingfield, M.J.; Coutinho, T.; Wingfield, B.D. Characterisation of the "C" morphotype of the pine pathogen Sphaeropsis sapinea. For. Ecol. Manag. 2002, 161, 181–188. [**Google Scholar**] [**CrossRef**]
11. Palmer, M.A. Isolate Types of Sphaeropsis sapinea associated with main stem cankers and top-kill of Pinus resinosa in Minnesota and Wisconsin. Plant Dis. 1991, 75, 507–510. [**Google Scholar**] [**CrossRef**]
12. Dijk, H.F.G.V.; Gaag, M.V.D.; Perik, P.J.M.; Roelofs, J.G.M. Nutrient availability in Corsican pine stands in the Netherlands and the occurrence of Sphaeropsis sapinea: A field study. Can. J. Bot. 1992, 70, 870–875. [**Google Scholar**] [**CrossRef**]
13. Wu, X.Q. Occurence and control countermeasure of Sphaeropsis sapinea shoot blight on conifer in the world. World For. Res. 1999, 12, 16–21. [**Google Scholar**]
14. Stanosz, G.R.; Smith, D.R.; Albers, J.S. Surveys for asymptomatic persistence of Sphaeropsis sapinea on or in stems of red pine seedlings from seven Great Lakes region nurseries. For.

Pathol. 2005, 35, 233–244. [**Google Scholar**] [**CrossRef**]

15. Wang, R.Q.; Zhou, G.Y. Natural regeneration and developing prospects Japanese red pine forest in Shandong peninsula. J. Ecol. 1989, 8, 18–22. [**Google Scholar**]
16. Pan, Q. Study on Interspecific Affinity of Cephalcia Kunyushania and Plant Species of Natural Forests in Kunyushan. Ph.D. Thesis, Chinese Academy of Forestry, Beijing, China, 2015. [**Google Scholar**]
17. Zhao, S.P.; Xu, H.X.; Yang, M.M.; Gu, S.Y. Environmental triggers and control countermeasures for the occurrence of shoot blight of exotic pines. For. Pest. Dis. 1993, 3, 45–47. [**Google Scholar**]
18. Paoletti, E.; Danti, R.; Strati, S. Pre- and post-inoculation water stress affects Sphaeropsis sapinea canker length in Pinus halepensis seedlings. For. Path. 2001, 31, 209–218. [**Google Scholar**] [**CrossRef**]
19. Sherwood, P.; Villari, C.; Capretti, P.; Bonello, P. Mechanisms of induced susceptibility to Diplodia tip blight in drought-stressed Austrian pine. Tree Physiol. 2015, 35, 549–562. [**Google Scholar**] [**CrossRef**] [**PubMed**] [**Green Version**]
20. Lindén, M.; Vollbrecht, G. Sensitivity of Picea abies to butt rot in pure stands and in mixed stands with Pinus sylvestris in southern Sweden. Silva. Fenn. 2002, 36, 767–778. [**Google Scholar**] [**CrossRef**] [**Green Version**]
21. Hantsch, L.; Braun, U.; Scherer-Lorenzen, M.; Bruelheide, H. Species richness and species identity effects on occurrence of foliar fungal pathogens in a tree diversity experiment. Ecosphere 2013, 4, 1–12. [**Google Scholar**] [**CrossRef**] [**Green Version**]
22. Menges, E.S.; Loucks, O.L. Modeling a disease caused patch disturbance, Oak wilt in the midwestern United States. Ecology 1984, 65, 487–498. [**Google Scholar**] [**CrossRef**]
23. LoGiudice, K.; Duerr, S.T.K.; Newhouse, M.J.; Schmidt, K.A.; Killilea, M.E.; Ostfeld, R.S. Impact of host community composition on Lyme disease risk. Ecology 2008, 89, 2841–2849. [**Google Scholar**] [**CrossRef**]
24. Anandaraj, M.; Mathew, S.K.; Eapen, S.J.; Cissin, J.; Bhai, R.S. Morphological and molecular intervention in identifying Phytophthora spp. causing leaf and nut fall in nutmeg (Myristica fragrans houtt.). Eur. J. Plant. Pathol. 2019, 156, 373–386. [**Google Scholar**] [**CrossRef**]
25. Garbelotto, M.; Schmidt, D.; Popenuck, T. Pathogenicity and infectivity of Phytophthora ramorum vary depending on host species, infected plant part, inoculum potential, pathogen genotype, and temperature. Plant. Pathol. 2020, 70, 287–304. [**Google Scholar**] [**CrossRef**]
26. Haas, S.E.; Hooten, M.B.; Rizzo, D.M.; Meentemeyer, R.K. Forest species diversity reduces disease risk in a generalist plant pathogen invasion. Ecol. Lett. 2011, 14, 1108–1116. [**Google Scholar**] [**CrossRef**] [**PubMed**]
27. Xie, F.; Liang, J. The relationship between species diversity of mixed bamboo stands and the disease index of Ceratosphaeria phyllostachydis Zhang. J. Hunan Agric. Univ. 2018, 44, 520–523. [**Google Scholar**] [**CrossRef**]
28. Liu, J.; He, W.M.; Fang, Z.L. Spatial characteristics of soil moisture and organic matter and light in Pinus tabulaeformis forest and Quercus liaotungensis forest on Dongling Mountain, Beijing, China. Acta Ecol. Sin. 2005, 25, 2954–2960. [**Google Scholar**]
29. Luo, N. The Impact of the Different Stand Structure and Understory Vegetation Species Diversity on the Disease and Insect Pest of the Aquilaria Sinensis Plantations. Master's Thesis, Central South University of Forestry and Technology, Changsha, China, 2016.

[**Google Scholar**]

30. Wang, X.X. The Effect of Plantations Landscape Vertical Stratification Structure on Phyllachora dalbergiicola in Dalbergia odorifera. Master's Thesis, Central South University of Forestry and Technology, Changsha, China, 2016. [**Google Scholar**]
31. Hu, R.R.; Liang, J.; Xie, X.; Zhang, Y.J.; Zhang, X.Y. Stand characteristics and soil properties in Japanese red pine (Pinus densiflora) pure forests with different disease severity index in Kunyushan Mountains region. China Int. J. Agric. Biol. 2019, 21, 726–732. [**Google Scholar**] [**CrossRef**]
32. Xie, X.; Liang, J.; Zhang, M.; Hu, R.R.; Cheng, Y.; Zhang, X.Y. Endophytic fungal community of Pinus densiflora infected by different incidences of Sphaeropsis sapinea in a mixed coniferous forest. Int. J. Agric. Biol. 2020, 23, 1044–1050. [**Google Scholar**] [**CrossRef**]
33. Magurran, A.E. Diversity Indices and Species Abundance Model. Ecological Diversity and Its Measurement; Magurran, A.E., Ed.; Princeton University Press: Princeton, NJ, USA, 1988; pp. 7–10. [**Google Scholar**]
34. Wu, H. Comparative study of species diversity indices of different type communities. J. Cent. South. Cent. Univ. For. Technol. 2015, 35, 84–89. [**Google Scholar**] [**CrossRef**]
35. Wang, D.B.; Ji, S.Y.; Chen, F.P. A review on the species diversity of plant community. Chin. J. Ecol. 2001, 20, 55–60. [**Google Scholar**]
36. Ma, K.P.; Liu, Y.M. Methods to measure the diversity of biological community-I α diversity measurement methods (II). Chin. Biodivesity 1994, 2, 231–239. [**Google Scholar**]
37. Gustavsson, E.; Lennartsson, T.; Emanuelsson, M. Land use more than 200 years ago explains current grassland plant diversity in a Swedish agricultural landscape. Biol. Conserv. 2007, 138, 47–59. [**Google Scholar**] [**CrossRef**]
38. Shen, B.K.; Chu, X.R.; Zhang, M.H.; Xue, M.L.; Yao, C.L. The development of dieback of pines by Sphaeropsis sapinea. For. Res. 1993, 6, 157–191. [**Google Scholar**]
39. Hu, R.R. Establishment and Verification of Forest Pest Based Index Model. Ph.D. Thesis, Chinese Academy of Forestry, Beijing, China, 2019. [**Google Scholar**]
40. Li, Z.H.; Luo, P. SPSS Common Statistical Analysis Course, 4th ed.; Publishing House of Electronics Industry: Beijing, China, 2015; pp. 114–115. [**Google Scholar**]
41. Fieller, E.C.; Pearson, E.S. Tests for rank correlation coefficients I. Biometrika 1961, 48, 29–40. [**Google Scholar**] [**CrossRef**]
42. Hauke, J.; Kossowski, T. Comparison of values of Pearson's and Spearman's correlation coefficients on the same sets of data. Quaest. Geogr. 2011, 30, 87–93. [**Google Scholar**] [**CrossRef**] [**Green Version**]
43. Gerlach, J.P.; Reich, P.B.; Puettman, K.; Baker, T. Species, diversity, and density affect tree seedling mortality from Armillaria root rot. Can. J. For. Res. 1997, 27, 1509–1512. [**Google Scholar**] [**CrossRef**]
44. Tahvanainen, J.O.; Root, R.B. The influence of vegetational diversity on the population ecology of a specialized herbivore, Phyllotreta cruciferae (Coleoptera, Chrysomelidae). Oecologia 1972, 10, 321–346. [**Google Scholar**] [**CrossRef**] [**PubMed**]
45. White, J.A.; Whitham, T.G. Associational susceptibility of cottonwood to a box elder herbivore. Ecology 2000, 81, 1795–1803. [**Google Scholar**] [**CrossRef**]

46. Barbosa, P.; Hines, J.; Kaplan, I.; Martinson, H.; Szczepaniec, A.; Szendrei, Z. Associational resistance and associational susceptibility, Having right or wrong neighbors. Annu. Rev. Ecol. Evol. Syst. 2009, 40, 1–20. [**Google Scholar**] [**CrossRef**] [**Green Version**]
47. Liang, J.; Zhu, Y.P.; Sun, X.Q.; Zhang, X.Y. Relationship of forest ecosystem composition and structure with management of pests. For. Pest. Dis. 2012, 31, 7–12. [**Google Scholar**]
48. Zhu, Y.P. The Occurrence Mechanism of Cephalcia Kunyushanica Population. Ph.D. Thesis, Chinese Academy of Forestry, Beijing, China, 2013. [**Google Scholar**]
49. Karlman, M.; Hansson, P.; Witzell, J. Scleroderris canker on lodgepole pine introduced in northern Sweden. Can. J. For. Res. 1994, 24, 1948–1959. [**Google Scholar**] [**CrossRef**]
50. Keesing, F.; Holt, R.D.; Ostfeld, R.S. Effects of species diversity on disease risk. Ecol. Lett. 2006, 9, 485–498. [**Google Scholar**] [**CrossRef**] [**PubMed**]
51. Zeng, F.Y.; Sun, Z.Q. Mechanism, hypothesis and evidence of herbivorous insect-host interactions in forest ecosystem. Acta Ecol. Sin. 2014, 34, 1061–1071. [**Google Scholar**] [**CrossRef**]
52. Lu, X.; Nan, Z.B. Perspectives in effects of grazing on diversity of grassland plant community and forage diseases. Pratacult. Sci. 2015, 32, 1423–1431. [**Google Scholar**]
53. Wu, L.; Xue, L. Studies on cedar blight. For. Pest. Dis. 1995, 1, 1–6. [**Google Scholar**]
54. Luchi, N.; Oliveira Longa, C.M.; Danti, R.; Capretti, P.; Maresi, G. Diplodia sapinea: The main fungal species involved in the colonization of pine shoots in Italy. For. Pathol. 2014, 44, 372–381. [**Google Scholar**] [**CrossRef**]

12 INVESTIGATION OF ROOT ROT INFECTION AND SPREAD CAUSED BY HETEROBASIDION PARVIPORUM IN PICEA ABIES STANDS FOLLOWING THINNING: CASE STUDIES ON FORMER PASTURE AND MEADOW LANDS

1. INTRODUCTION: AN OVERVIEW

Conversion of pastures, meadows, or agricultural fields to forest plantations is a prevalent practice in numerous nations. As an example, throughout the European Union, a total of 12.9 million hectares of unused lands were converted into forests between the years 1995 and 2015 [1]. When comparing agricultural fields to lands that were formerly covered by forests, it is important for forest management methods to take into account the most suitable tree species and associated threats [2,3,4]. Hence, multiple research have conducted a comparison between the pace of stand growth and growth circumstances in forest lands as opposed to previous agricultural lands [5,6].

Norway spruce (Picea abies (L.) Karst.) is a frequently planted conifer tree species in Latvia, as well as other Baltic and Nordic nations, particularly on former agricultural fields [7]. The Norway spruce is extremely vulnerable to root rot caused by Heterobasidion spp. [8,9,10,11,12,13]. The country-wide survey conducted in Latvia found that approximately 22% of spruce trees were affected by heart rot, primarily caused by Heterobasidion parviporum Niemelä & Korhonen. This fungal infection leads to significant economic losses [14,15]. Over 50% of Latvia's land is forested, and the forest industry accounts for nearly 20% of Latvia's total exports [16]. Therefore, it is essential to minimize the economic damages resulting from root rot.

Previous-generation stumps that are infected play a crucial role in transferring Heterobasidion infection to the next generation of trees through root contacts [10,11,17,18,19,20,21]. Hence, implementing forest planting in regions that are free from infection and carrying out afforestation on previously used agricultural lands or pastures are crucial techniques for reducing the risk for infection [22].

However, multiple studies have found that the rate of butt rot occurrence is comparable in former fields or grazing area when compared to former forestlands [23], and in some cases, even significant damage has been observed in previous agricultural lands [24,25,26,27]. Modelling using stump infection rate with Heterobasidion in spruce stands indicated that former pastures and arable lands are more vulnerable to root rot compared to forest stands [28]. A recent study conducted in the Czech Republic revealed variations in infection rates based on the prior land uses. The study found that former arable areas had a higher infection rate compared to former pastures [29].

The higher infection rate on former agricultural lands can be attributed to several factors. These include soil chemical parameters such as high carbonate content, high nitrogen content, or an overall unbalanced nutrient status [4,30,31,32]. Another factor is soil compaction, which stimulates the superficial root system of Norway spruce [26]. Biotic factors, such as the absence of fungi antagonistic to Heterobasidion, also contribute to the higher infection rate [4,26,31,33,34,35,36]. Additionally, an increased percentage of grass cover has been linked to the higher infection rate [30].

The implementation of forest management measures, particularly cuts, has a crucial role in the occurrence of Heterobasidion infections in stands located on previously used agricultural lands. In these areas, the predominant method of disease transmission is through the infection of recently cut conifer stumps by spores. Conversely, in previous forest areas, the spread of diseases might be more substantial due to mycelial infections that originate from old stumps or roots. Consequently, multiple research investigations have specifically examined the occurrence of root rot following the initial thinning process on previously cultivated ground [37, 38, 39]. The occurrence of Heterobasidion root rot is mostly influenced by the extent of stump infection, however the potential for direct root infection cannot be entirely disregarded. For example, Rönnberg et al. [40] demonstrated a significant occurrence of Heterobasidion root rot in unthinned spruce stands that were planted on pastureland. Thor and Stenlid [41] discovered that between 0.5% and 8.1% of trees in unthinned spruce stands on previously used agricultural land were infected. In addition, they observed that an unthinned stand on former pastureland had an infection rate of up to 16.9%.

Secondary infections arise when the hyphae of the pathogens extend to the root systems of neighboring trees through root contacts, resulting in the formation of disease centers that grow in size [42,43]. In order to gain comprehensive understanding of secondary infection, Swedjemark and Stenlid [44] conducted an analysis on the population structures of Heterobasidion annosum s.l. in spruce stands that were previously used for agriculture. Their findings revealed that seven years after thinning, 63% of trees exhibited decay symptoms and were infected by mycelial spread originating from stumps. A separate study conducted in Sweden examined the growth rate of artificially introduced Heterobasidion mycelia in the stem and roots of Norway spruce trees growing in forest and arable land. The study found that, on average, there was no significant difference in the growth rate of Heterobasidion in both the tree and stump roots between the former arable and forest sites [19].

Nevertheless, the aforementioned research were conducted on former agricultural lands in Sweden and Scotland, and it is important to note that the findings may not be directly relevant to various types of soil and forest, as well as other agricultural and forestry management approaches. There is a lack of comparable studies conducted in Norway spruce stands in Latvia or nearby countries, highlighting the necessity for conducting such investigations. A recent study conducted in Latvia by Zaļuma et al. [45] examined the impact of Heterobasidion root rot on Scots pine stands. The researchers found that the expansion rate of Heterobasidion root rot was lower in pine plantations developed on former agricultural land compared to those created on former forestland. Therefore, it is crucial for practical forestry in Latvia to gather precise and extensive information about the dissemination of Heterobasidion root rot in Norway spruce stands located on sites with no past forest activity, in addition to the research stated earlier. The objective of this study was to examine the distribution patterns, both primary and secondary infection, of Heterobasidion spp. in managed Norway spruce plantations that were developed on previously used meadows and pasturelands in the central region of Latvia.

2. MATERIALS AND METHODS

2.1. Study Sites and Experimental Design

The study was established in three Norway spruce stands in Central Latvia, in the municipalities of Babite, Auce, and Olaine (**Figure 1**). In this region, average annual temperature ranges from 6.8 to 7.9 °C and annual precipitation rate is 574 to 636 mm. For Central Latvia, dry pine and spruce forests on sandy or loamy soils, water-saturated areas (peat bogs, wet forest types, flood plain meadows, etc.) with peat soils or drained forests and a large proportion of former agricultural fields on loamy soils are characteristic. The first-rotation spruce plantations in the study represent this variety: a drained forest within a region of peat bogs (Stand 1), a spruce plantation within a region dominated by agricultural fields (Stand 2), and a dry-type (Oxalidosa) stand (Stand 3). Characteristics of the investigated areas are presented in **Table 1**.

Figure 1. Location of the study sites in the central part of Latvia.

Table 1. Description of Norway spruce stands selected for the study: general information, forest characteristics, and details of sampling area and soil parameters.

	Stand 1	Stand 2	Stand 3
General information			
Location	56°84′ N, 23°77′ E	56°38′ N, 23°12′ E	56°82′ N, 24°07′ E
Municipality	Babite	Auce	Olaine
Altitude, m	17	85	13
Annual rainfall, mm	600–650	550–600	650–700
Forest data			

	Stand 1	Stand 2	Stand 3
Total area, ha	8.7	2.75	1.54
Age, years	73	82	52
Average height, m	27.1	28.8	23.5
DBH, cm	35	30	24
Yield per ha	342	387	268
Forest type	*Vacciniosa mel.*	*Myrtillosa mel.*	*Oxalidosa*
Thinning years *	2007, 2015	1991, 2001	1987, 2017
Previous land use	Pastureland	Wet meadow	Meadow
Other tree species (<5% admixture)	*Betula pendula*	*Populus tremula*	*Pinus sylvestris*
Sampling area			
Length of transect, m	500	400	250
Area of transect, m^2	1000	800	500
Total No. of circle sample plots	11	6	5
Total area of circles, m^2	3454	1884	1570
Soil			
pH (KCl)	4.1	5.9	4.6
C, (g/kg)	146	103	106
N, (g/kg)	7.6	8.5	7.5
P, (g/kg)	0.03	0.06	0.03
K, (g/kg)	0.02	0.05	0.02
Ca, (g/kg)	4.4	8.9	6.9
Mg, (g/kg)	0.16	0.88	0.25
* All stands were thinned twice. No data of the exact thinning period and season is available			

All stands had selective or sanitary thinning (**Table 1**); since after thinning, stumps were not treated with any control agents to prevent infection of root rot, at least potentially, stumps were exposed to primary infection of Heterobasidion.

Soil samples for chemical analysis were collected at 0–20 cm depth along transects every 50 m. Chemical analyses of soil were conducted using established standard methods (International Organization for Standardization (ISO) standard) at LSFRI Silava, Laboratory of Forest Environment, Salaspils, Latvia. Samples were prepared for analysis according to ISO 11464:1994. The soil pH was determined in 1 M KCl (soil—extractant mixture 1:2.5). Total carbon content was determined using an elemental analyser ELTRA CS530 (Eltra GmbH, Haan, Germany) according to LVS ISO 10694. The concentration of total N was determined using the Kjeldahl method (ISO 11261:1995) and P was assayed by colourimetry using the ammonium molybdate spectrometric method in 1 M HCl solution (soil/extractant mixture

1:5). Concentration of K, Ca, and Mg was determined by atomic absorption spectrophotometer with an acetylene-air flame in 1 M HCl extract.

To determine average infection rate, we sampled all standing trees within a 2-m-wide transect parallel to the longest edge of each stand with starting points in the middle of the shortest edge. Health status was assigned to each tree in the transect: (i) visually healthy crowns and (ii) trees with severe dieback symptoms, dead and windthrown trees [**9**]. To detect the pathogen under laboratory conditions, wood samples were collected from all trees on the transect with the aid of an increment borer (one sample per tree at root collar height). Additionally, to characterise stand growth rate, we measured the diameter at breast height and, for living trees, the tree height (**Table 1**).

Based on the transect data about the distribution of Heterobasidion infected trees and visual observation of potential infection centers, we established from five to eleven circular sample plots per stand (10 m radius, 314 m^2) where all trees and stumps were sampled and examined for Heterobasidion fruit bodies. In addition, stumps were classified as either stumps from the 1st or 2nd thinning. Broken dead tree trunks of different heights were categorized as stumps. Wood samples from trees were collected with an increment borer (one sample per tree at root collar height) and from stumps—either with the same method or with the aid of an axe if they were heavily decomposed. Wood samples from each tree were placed into individual plastic tubes and stump samples into individual plastic bags. All sampled trees and stumps (both in transects and circular plots) were numbered and their location mapped. Field work was carried out from March 2019 to April 2020.

2.2. Isolation and Identification of Heterobasidion Genotypes

In the laboratory, all wood samples were surface sterilized by flaming and placed on petri dishes containing malt extract agar (MEA) media (for tree samples) or Hagem agar media (for stump samples). After a 5 to 7-day incubation, the plates were examined under a Leica stereomicroscope M205C (20–30× magnification) to detect the presence of Heterobasidion spp., observing its characteristic asexual sporulation (conidiophores and conidia) on wood samples [**10**]. If Heterobasidion was observed, colonies were sampled with sterilised surgical forceps and placed on new MEA media. Outgrowing Heterobasidion mycelia were subcultured to obtain pure fungal cultures.

Isolates originating from the same circular sample plot were subjected to pairwise somatic compatibility tests by confronting their mycelia on Hagem agar media in 9 cm Petri dishes in all combinations. The genets were identified either by recording the line of demarcation in the contact zone of confrontation (demonstrating that the genotypes are different) or by observing free fusion of the mycelia (implying that the genotypes were identical) [**46**].

The results of this test provided an information about the total number and size of Heterobasidion genets: number of genets encompassing only one tree or stump and number of genets encompassing two or more trees and (or) stumps. Based on these data, we estimated the number of trees/stumps infected by Heterobasidion genets per hectare and per disease centre. To estimate the area and maximum width of each genet including two or more trees/stumps, we used the map of each site with all sampled trees and stumps mapped. We used species specific primers (Heterobasidion parviporum specific primers KJ-F and KJ-R and Heterobasidion annosum specific primers MJ-F and MJ-R) to determine the Heterobasidion species of each genotype [**47**]. DNA was extracted from fungal cultures representing each genotype using a CTAB method [**48**]. Two PCR reactions were run with each sample—one with H.parviporum and other with H.annosum species specific primers listed above. PCR conditions, were a described by Hantula and Vainio [**47**]. DNA from H. parviporum and H. annosum was used as a positive control.

2.3. Data Analysis

A chi-square (χ^2) test of independence from the actual number of observations of infected or decaying trees and stumps was used to compare abundance patterns between sites and sample plots [**49**].

Using the geospatial tool QGIS 3.2.0-Bonn (bc43194) (**https://qgis.org/api/3.2/**), tree spatial location was visualised and the width and the area of genets including two or more trees/stumps was measured (tool "MeasureLine" and "MeasureArea"). The maximum expansion rates were calculated as a reference taking years from the first thinning for each stand (2007, 1991 and 1987, respectively, Stand 1 to 3) or from the second thinning (2015, 2010 and 2017, respectively, Stand 1 to 3).

To assess which factors have an impact on the number of trees per genet, we used generalized linear mixed models (GLMM) assuming a Poisson error distribution. The full model included number of trees per genet as a response variable and as fixed effects we tested site, stand age class (older stands (Stand 1 and 2) vs. younger stand (Stand 3), disease centre infection indicators (classes of infection (<40% vs. >40%), stump and dead tree proportion) and as a random factor in all models we included disease centre. Models were calculated using the program package lme4 [**50**].

Area and width of genets including two or more trees/stumps as well as maximum expansion rates (area and width) were tested for normality using the Shapiro and Wilk normality test at significance level $\alpha = 0.001$ [**51**]. If normality was assessed, one-way analysis of variance (ANOVA) with Fisher LSD post-hoc test was calculated [**52**]. If normality was not assessed, nonparametric Wilcoxon test (unpaired samples) was used to compare parameters. For calculations, R version 4.0.2 was used [**53**].

3. RESULTS

3.1. Incidence of Dieback and Heterobasidion Infection

In total, in all three study sites, 851 wood samples were collected for Heterobasidion isolation (397 from living trees, 118 from dead trees and 336 from stumps). The number of sampled trees and stumps are given in **Table 2.** In all stands, stumps from both 1st and 2nd thinning were present and in addition there were some trees broken close to their base (mainly in Stand 2) (**Table 2**).

Table 2. Characteristics and dieback rate in Norway spruce stands on former pastures (Stand 1) and meadows (Stands 2 and 3). (T—transect; C—disease centres (circular sample plots).

	Stand 1		Stand 2		Stand 3	
	T	C	T	C	T	C
Living trees						
Total number *	72	95	25	73	34	153
Number per hectare	720	275	313	387	680	975
Dead trees						
Total number *	8	41	10	37	8	29
Number per hectare	80	119	125	196	160	185
Stumps						
Total number	n.d.	142	n.d.	96	n.d.	98
No. of 1st thinning stumps	n.d.	65	n.d.	45	n.d.	29
No. of 2nd thinning stumps	n.d.	75	n.d.	39	n.d.	69
No. of other stumps **	n.d.	2	n.d.	12	n.d.	-
Number per hectare	n.d.	411	n.d.	510	n.d.	624

* Several trees were present in both transect and disease centre. ** Category "other stumps" includes trees broken at different heights.

In Stand 2 and 3, chi-square test of independence showed that transect and disease center data have similar tendencies regarding the number of living and dead trees ($p > 0.05$). In contrast, chi-square test for Stand 1 data indicated significant differences in living and dead tree abundance in disease centers and the transect ($\chi^2 = 11.65$; df = 1; n = 216) (there were less dead trees in the transect ($\chi^2 = 5.67$) than expected).

Additionally, a chi-square test of independence was performed to examine the numbers of living trees and stumps among sites. The relation between these variables was significant, $\chi^2 = 24.17$ (df = 2, n = 657), $p < 0.05$. The highest chi-square values were observed for both living trees and stumps in Stand 3: more living trees ($\chi^2 = 7.52$) and less stumps ($\chi^2 = 7.18$) than predicted.

The transects of both Stand 2 and 3 had several openings, which explains the lower number of trees per hectare in comparison with disease centres (**Table 2**). In Stand 3, these openings were related to stand planting design (there were several gaps left to facilitate thinning). Stand 3 had higher planting density that reflects in the number of trees per hectare (**Table 2**).

In addition, fruit bodies were recorded as an indicator of disease of trees and stumps. There were more records of fruit bodies in Stand 1 (two dead trees and 10 stumps with fruit bodies) compared to other stands (there were two fruit bodies recorded in Stand 2 and four fruit bodies in Stand 3, found exclusively on stumps in both stands).

After incubation of 851 collected wood samples, we detected pathogen conidiophores with conidia in 302 samples (35%); infection rate varied among stands (30% in Stand 1 and 39% in Stands 2 and 3) (**Table 3**).

Table 3. Frequency (% *) of Norway spruce trees or stumps infected by Heterobasidion parviporum in managed forests on former pastures and meadows.

	Stand 1	Stand 2	Stand 3
Transect; %	16.3	28.6	33.3
Living trees; %	12.5	25.7	23.8
Dead trees; %	3.8	2.9	9.5
Disease Centres; %	33.5	41.3	38.6
Living trees; %	11.5	15.5	18.9
Dead trees; %	8.6	7.3	5.4
Stumps; %	13.3	18.4	14.3
Total; %	29.6	38.5	38.9

* Percentage expresses the proportion of trees/stumps where *Heterobasidion parviporum* was detected of all trees/stumps sampled.

Chi-square test indicated that in transects and disease centres, the incidence of Heterobasidion infection was similar among sites ($p > 0.05$). In addition, the infection incidence of both living and dead trees in disease centres have a similar pattern among sites ($p > 0.05$). In total, we confirmed the presence of the pathogen in 31.8 and 48.3% of all living and dead tree samples, respectively. Infection incidences of living trees among sites was 24.5, 38.6 and 35.2% (Stand 1 to 3, respectively) and of dead trees 58.1, 35.7 and 51.5% (Stand 1 to 3, respectively). In total, we confirmed the presence of the pathogen in 34.2% of all stump samples. Infection incidences varied among stump samples at different sites: 26.1, 39.6 and 40.8% (Stand 1 to 3, respectively). However, the differences were not significant.

3.2. Analysis of Heterobasidion Genets

Pure Heterobasidion sp. cultures were successfully isolated from 90.5% of a total of 286 wood samples producing Heterobasidion conidiophores with conidia collected from disease centers (76.3, 98.8, and 96.3% from Stands 1 to 3, respectively) (**Table 3**). Identification with species specific primers revealed that all obtained isolates belonged to species Heterobasidion parviporum.

In total, 141 Heterobasidion genets were detected resulting in 72 to 484 Heterobasidion genets per hectare among study sites (**Table 4**). Less genets per hectare were detected in Stand 1 (2.9 times and 6.7 times less in comparison with Stand 2 and Stand 3, respectively).

Table 4. Distribution pattern and characteristics of Heterobasidion parviporum genets in disease centres of managed Norway spruce stands on former pasture and meadows.

	Stand 1	Stand 2	Stand 3
Genets including one tree/stump			
Infection per hectare	38	143	376
Trees infected	35	64	274
Stumps infected	3	79	102
Frequency (%)	21%	25%	56%
Genets including two or more trees/stumps			
Number of trees and stumps per genet (min.-max.)	4.8 (2–12)	4.4 (2–11)	2.6 (2–7)
Infection per hectare	170	302	293
Trees infected	120	185	191
Stumps infected	50	117	102
Distance among trees and (or) stumps, m			
Mean ± SE	9.6 ± 1.7	6.9 ±1.0	3.1 ± 0.5
Min.-max.	1.8–18.0	2.1–14.6	1.0–6.9
Area, m^2			
Mean ± SE	56.4 ± 19.7	12.4 ± 4.9	5.9 ± 2.9
Min.-max.	1.2–137.4	0.3–52.2	1.3–17.3
Frequency (%)	79%	75%	44%
Total infection per hectare	208	445	669

The mean number of trees/stumps per genet was 2.8, 2.1 and 1.4 in Stand 1 to 3, respectively. Genets including one tree/stump were the most abundant in all sites: 13, 27, 59 (i.e., 52, 68 and 78% of all genets per stand), in Stand 1 to 3, respectively.

Expanded genets including two or more trees/stumps were less frequent 12, 13 and 17 (Stand 1 to 3 respectively). Chi-square test indicated differences among abundance of genets including one tree/stump and genets including two or more trees/stumps per sites ($\chi^2 = 6.1$, df = 2, n = 141; $p > 0.05$).

Frequency of trees or stumps infected by genets including one tree/stump of all infected trees and stumps was 21, 25 and 56% (Stand 1 to 3, respectively). Chi-square test indicated differences among abundance of trees and stumps infected by genets including one tree/stump per sites ($\chi^2 = 26.3$, df = 2, n = 259; $p > 0.05$), in particular abundance of trees and stumps of genets including one tree/stump in Stand 1 was lower than predicted ($\chi^2 = 7.4$) and in Stand 3 higher than predicted ($\chi^2 = 8.4$).

In comparison with Stand 1 and 2, Stand 3 had more isolates and consequently more genets (in total and both genets including one tree/stump and genets including two or more trees/stumps). Stand 3 had a lower number of trees per genet and less expanded genets including two or more trees/stumps (lower distance among trees and area) (**Table 4**). Modeling the number of trees per genet and including disease centres as a random factor revealed that grouping Stand 1 and 2 together (older stands) vs. Stand 3 identified a significant difference ($p < 0.05$).

The maximum number of trees infected by genets including two or more trees/stumps are presented in **Table 4**. The largest genets included both living and dead trees and stumps. The longest mean distance between trees and (or) stumps with shared genets was about 7–10 m (maximum 15–18 m) (Stands 1 and 2) compared with values of Stand 3 (mean longest distance 3 m; maximum 7 m) (**Table 4**). Analysis of variance showed significant differences and post-hoc tests indicated that the genets including two or more trees/stumps in Stand 3 were significantly ($p < 0.05$) smaller in size than genets in Stand 1 (3.3 times smaller) and in Stand 2 (2.3 times smaller).

The maximum area of genets including two or more trees/stumps in Stand 1 was 2.6 to 8 times greater than in Stand 2 and 3 (**Table 4**). The mean area of genets including two or more trees/stumps was also 5 to 9 time larger in Stand 1. Analysis of variance showed significant differences and post-hoc tests indicated that the genets including two or more trees/stumps in Stand 1 had significantly ($p < 0.05$) larger area than both Stand 2 and 3 ($p < 0.05$).

The maximum expansion length per year calculated from the first thinning was 1.5, 0.5 and 0.2 m (Stand 1 to 3, respectively) and maximum expansion area per year: 11.5, 1.9, and 0.5 m^2 (Stand 1 to 3, respectively). Wilcoxon test showed that also mean expansion indicators (distance and area per year) were higher in Stand 1 in comparison with Stand 2 and Stand 3 ($p < 0.05$).

4. DISCUSSION

4.1. Dieback and Observed Heterobasidion Infection of Norway Spruce on Former Pasture and Meadow Lands

Visual observation of stand health status revealed expansive spread of root rot in the studied Norway spruce stands on former non-forest lands (pastures and meadows) in Latvia. The greater total number of trees and stumps per hectare in Stand 3 is mainly related with greater planting density. The number of stumps indicates intensity of thinning and, therefore, potential infection risk of stumps by Heterobasidion spores. Greater abundance of stumps might also favor the mycelial spread of Heterobasidion since its growth rate in stump roots is significantly faster than in living trees [**19,54**]. Moreover, in densely planted stands, where the trees or stumps are closer to each other, the possibility of root contacts is higher, thus allowing greater probability of infection transmission [**4**].

In Stand 2, there were several trees broken at different heights (**Table 2**), which indirectly indicates reduced tree vitality. In addition, gaps and a low number of trees per transect might indicate not only a reduced number of trees due to the stand age but also poor current and previous pathological conditions within the stand.

Fruit bodies were found in all study sites, which indicates that environmental conditions were suitable for fruiting of H. parviporum. Consequently, the risk of primary infection by H. parviporum spores should be taken into account when planning future thinning in these stands or other similar sites.

The detected H. parviporum infection rate (30% to 39%) was on average in line with data from other studies from both former agricultural and forest lands. It indicates that even though no inoculum is present in former non-forest lands, infection can establish and expand as successfully as in forest lands or even more successfully. For instance, in a Swedish study carried out on a spruce stand on former agricultural land, seven years after thinning, 63% of the trees had decay symptoms [**45**]. In Sweden, modelling of spruce stands including different former land use indicated 2–90% of rotted trees in final cuttings [**55**]. In Denmark, on average, 23% (max. 94%) of Norway spruce individuals growing on former agricultural or grazing fields were found to suffer from butt rot [**23**].

The success rate of H. parviporum isolation from dead spruce individuals was 35.7 to 58.1%. However, since the pathogen was not detected in all dead spruce individuals, it cannot be excluded that either the infection has not yet spread higher in the stem or that causes other than Heterobasidion root rot were responsible for a proportion of declining or dead trees. Other fungal agents as Armillaria sp. can cause root rot in Norway spruce stands [**56**]. In addition, spruce dieback not related to root rot has been observed in several regions in Latvia [**57**].

The frequency of stumps colonized by H. parviporum (34.2% on average) was relatively high in spruce stands planted in former pastures and meadows in Latvia. Viable Heterobasidion mycelium can survive in stumps for decades [**58**]. However, with time, it develops more intensively in deeper layers of stumps [**59**] and isolation can be difficult due to wood decomposition and presence of other stump colonizing fungi. Since the stumps in the studied stands were not recently cut, total stump infection may be even higher. Infected spruce stumps left after thinning transfer root rot to residual trees and serve as a source of infection [**21,54**]. Therefore, winter thinning and use of biological control agents containing spores of Phlebiopsis gigantea or other preventive substances such as urea is recommended to limit the infection of conifer stumps by basidiospores of Heterobasidion spp. [**43,60,61,62**], especially in stands with no previous rot incidences such as in stands established on former agricultural lands [**4,63,64**].

4.2. Heterobasidion Genets

All obtained Heterobasidion isolates were identified as Heterobasidion parviporum, which is the most common Heterobasidion species causing damage in spruce stands in Northern Europe and Latvia [**15,65**]. H. parviporum is adapted to spruce as the host species [**66,67**] and is therefore more often observed in spruce stands, being predominant or exclusive in some cases [**9,68,69,70**], and is able to spread more frequently from stumps to neighbouring trees than H. annosum [**71**]. Korhonen et al. [**72**] reported that Heterobasidion infections in spruce stands on former agricultural lands were mainly caused by H. parviporum and concluded that in such first rotation spruce stands without any history of pine growth and woody inoculum, spores of H. parviporum are the principal infection source. Moreover, Vasiliauskas and Stenlid [**18**] showed that H. parviporum is better adapted to spruce wood, and spreads more extensively in spruce stems compared to H. annosum.

In our study, the number of trees/stumps infected by H. parviporum per hectare was 208, 445 and 669 (Stand 1 to 3, respectively) (**Table 4**). In addition, the total number of Heterobasidion genets per hectare in the present study was high: 72, 212, and 484 (Stand 1 to 3, respectively). In comparison, in a Finnish study on Heterobasidion infections on clear-felled forests on old forest sites, the number of fungal genets per hectare varied from 2 to 158 (mean 28) [**73**]. The high infection rate and high number of Heterobasidion genets in the present study may indicate a higher susceptibility of Norway spruce to Heterobasidion root rot on former agricultural lands compared to old forest lands [**2**].

The number of genets including only one tree or stump was lower in older stands (Stand 1 and 2) than in the youngest one (Stand 3). Time from the first thinning did, however, not differ significantly between the stands (**Table 1**). However, even in final cuttings on old forest sites, genets including one tree/

stump and therefore primary infection can be dominant [**73**]. In general, a lower number of trees infected by the same Heterobasidion genet and occupying a smaller stand area indicate that these clones are relatively younger than those comprised of more trees and covering a larger area [**44**]. However, the age of the genets of Heterobasidion may affect the pathogenicity of the fungus, with younger genets being more pathogenic [**39**].

Development of territorial clones following primary infection of Heterobasidion takes at least 7–8 years [**42,44,74**]. In our study, 22 to 48% of detected genets formed expanding territorial clones, which included 44 to 79% of infected trees and (or) stumps. This indicates that infection in all sites is well established and expanding. Due to well-developed clonality, in Stand 1 and 2, it is most probable that infection was initiated due to the first thinning (12, 28 years ago), and not by the more recent ones (4, 18 years ago) (Stand 1 and 2, respectively). In Stand 3, infection history seems to be much more recent than in other stands and probably more expansive after the last thinning (two-years before sampling) than after the first one (32-years before sampling). In addition, greater planting density in Stand 3 could favour development of territorial clones in a shorter period as indicated previously [**35,42,44**].

Comparing the maximum size of Heterobasidion genets with other studies, we found genets of similar size also reported from stands with a much longer infection history. For instance, Vasiliauskas and Stenlid [**18**] reported territorial clones of H. annosum s.l. up to 20 m in diameter in a 60 year-old spruce stand planted in a previously infected pine site, while in the oldest stands of our study (Stand 1 and 2), territorial clones reached up to 14–17 m. In addition, the largest distance between isolates from the same genet in Finnish study [**69**] in 70 to 100 year-old spruce stands was 13 m, which is less than in our study on former non-forest lands. Previous studies indicate that H. parviporum can form large territorial clones. Stenlid [**46**] studied a 120-year-old and thinned P. abies stand and detected H. parviporum territorial clones reaching up to 30 m in diameter. Similarly, up to 40 m long territorial clones of H. parviporum were observed in an intensively thinned P. abies stands in Serbia [**74**]. In related studies from Finland, the maximal length of territorial clones varied between approximately 10 and 30 m [**9,66**] or reached up to 55 m [**73**]. Our study shows that even 12 to 32 years after thinnings in stands on former pastures and meadows, H. parviporum can form large territorial clones.

The maximum number of trees per genet in our study was 12 and this is lower in comparison to stands on former forest lands (i.e., a Finnish study [**67**] reported 16 trees per genet) and also to stands on former pasturelands, where a greater number of trees (up to 25) forming territorial clones have been reported [**40**]. However, the real number of infected trees and stumps in our study could be greater than detected due to various reasons: (i) stump decomposition and poor isolation success from some samples, (ii) sampling height of trees (root collar) excluding infection in lower zones, (iii) sample number (one sample per tree) not covering all stem inner surface where decay column can be located, and (iv) limited sampling area (territory of circular sample plot). In addition, to estimate the exact borders and size of Heterobasidion genets, a thorough analyses of excavated root systems would be required. Consequently, the size of the genets may actually be bigger.

All the largest H. parviporum genets of our study contained three to nine dead trees or stumps. Stumps colonized by Heterobasidion provide a substrate from which the fungus can spread and cause infection in standing trees [**21**]. A study by Capretti and Mugnai [**26**] indicated that dead wood (buried wood in their study) might be an important factor for extensive spread of the pathogen especially in former pasture soil. Recent studies show that root fragments are important for H. parviporum infection transfer [**75**]. Therefore, both abundance of dead wood and roots could be related with the spread of the pathogen in Stand 1 (former pastureland). The average expansion rate of territorial clones of H. parviporum on Stand 2 and 3 (former meadows) (0.25 and 0.10 m·year^{-1}) corresponds to results obtained by Bendz-Hellgren et al. [**19**], where the average H. annosum s.l. growth rate on former agricultural lands was 0.25 m·year^{-1} in spruce stump roots and 0.09 m·year^{-1} in living tree roots. The average H. parviporum growth rate in the Stand 1 (former pastureland) was high, 0.8 m·year^{-1}, in comparison with results on both forest and arable lands reported by Bendz-Hellgren et al. [**19**]. This might indicate an erroneous assumption of infection year (i.e., infection had occurred before first thinning via wounding). Since no wounds were observed on trunks, current primary infections probably occurred via root injuries. Spores of Heterobasidion are present in soil, where they can maintain their vitality at least for a year and infect injured tree roots [**32**]. Rönnberg et al. [**40**] reported a high incidence of Heterobasidion root rot in unthinned spruce stands planted on pastureland. In addition, Vollbrecht and Agestam [**28**] and Capretti and Mugnai [**26**] indicated that former pastures were highly susceptible to root rot. Capretti and Mugnai [**26**] also concluded that Heterobasidion growth rate in dead wood was particularly high in pasture lands. The increased capacity of H. parviporum to colonise dead wood and form fruitbodies in Stand 1 of our study is a probable scenario, explaining the greater expansion of genets.

In research sites, stumps were not treated against Heterobasidion root rot, which in a short period resulted in severe damage as reported above. Therefore, we suggest that stump treatment should be considered as a general practice on former pastures and meadows to prevent both primary and subsequent infections, as suggested by other authors [**41,43,62,64**]. This is particularly relevant for private forest owners who are often managing forests on former agricultural lands, but often do not utilise preventive stump treatments.

Further studies should focus on the presence of Heterobasidion spores in soils of former agricultural lands in relation to the soil microorganism community and root injuries.

5. CONCISE SUMMARY

This study collected novel data on the ecology and dissemination patterns of Heterobasidion parviporum in Norway spruce forests that were formerly used as pastures and meadows. Based on this new and unique information, we have reached the following conclusions: (i) Typically, primary infection through spores is more common in such stands, but secondary infection becomes more prevalent as the stand ages. (ii) H. parviporum has the ability to create large territorial clones (up to 137 m2) in woods that were formerly used for pasture and meadow, resulting in significant tree dieback and mortality. Hence, in order to minimize financial losses in spruce plantations on previously used pastures or meadows, it is advisable to contemplate stump treatment during thinning operations using biological or chemical control agents that can effectively prevent Heterobasidion infections. Facilitating the implementation of stump treatment for private forest owners, particularly those who own forests on former agricultural lands, is crucial.

6. AT A GLANCE

The reforestation of former agricultural fields is a widespread activity in numerous nations. This practice is advantageous for preventing the transmission of illnesses from earlier generations of forests and for increasing the size of forested regions. Nevertheless, numerous biotic and abiotic hazards have been documented in these stands, notably an increased susceptibility to Heterobasidion root rot following thinning. This study aims to examine the dissemination of Heterobasidion root rot in three Picea abies (L.) Karst. plantations in Latvia that were formerly used as pasture and meadow areas and have undergone forest management activities. At first, in order to calculate the average infection rate, we collected samples from all the trees that were upright (a total of 157) along the designated paths inside the locations where sampling was conducted. The prevalence of Heterobasidion root rot among stands ranged from 16 to 33% on the transects. Using the collected data and noting the observed dieback in all the areas, we created circular sample plots in the centers of the illness where we examined all the trees and stumps. The infection rate in the circular plots ranged from 34% to 41% on average. The Heterobasidion isolates that were obtained were examined using primers specific to the species, and it was established that all of them belonged to Heterobasidion parviporum Niemelä & Korhonen. Isolates were matched in order to determine the quantity and dimensions of genotypes. Out of the 141 genets that were analyzed, 99 were found to be present in only one tree or stump, while 42 genets were found to include two or more trees or stumps, indicating that the infection spread through root connections. The observed stands exhibited a range of 72 to 484 Heterobasidion genets per hectare. The following deductions were made: (i) Areas that were previously non-forest lands, such as pastures and meadows, can be very vulnerable to Heterobasidion infection after thinning. (ii) The pathogen has the ability to create expanding territorial clones in both former pastures and meadows. (iii) It is recommended to treat stumps with biological or chemical control agents in order to prevent Heterobasidion infections.

REFERENCES

1. Forest Europe. State of Europe's Forests 2015. Summary for Policy Makers. In Proceedings of the Ministerial Conference on the Protection of Forests in Europe, Madrid, Spain, 20–21 October 2015. [**Google Scholar**]
2. Korhonen, K.; Stenlid, J. Biology of Heterobasidion annosum. In Heterobasidion annosum: Biology, Ecology, Impact and Control; Woodward, S., Stenlid, J., Karjalainen, R., Hütermann, A., Eds.; CAB International: Wallingford, UK, 1998; pp. 43–70. [**Google Scholar**]
3. Mańka, K.; Mańka, M. The control of fungal disease in forests planted on agricultural lands. In Proceedings of the XIII Czech and Slovak Plant Protection Conference, Prague, Czech Republic, 12–15 September 1994. [**Google Scholar**]
4. Sierota, Z. Heterobasidion root rot in forests on former agricultural lands in Poland: Scale of threat and prevention. Sci. Res. Essays 2013, 8, 2298–2305. [**Google Scholar**] [**CrossRef**]
5. Cukor, J.; Vacek, Z.; Linda, R.; Sharma, R.P.; Vacek, S. Afforested farmland vs. forestland: Effects of bark stripping by Cervus elaphus and climate on production potential and structure of Picea abies forests. PLoS ONE 2019, 14, e0221082. [**Google Scholar**] [**CrossRef**] [**PubMed**] [**Green Version**]
6. Rytter, L.; Lutter, R. Early growth of different tree species on agricultural land along a latitudinal transect in Sweden. For. Int. J. For. Res. 2020, 93, 376–388. [**Google Scholar**] [**CrossRef**]
7. Daugaviete, M.; Lazdina, D.; Bambe, B.; Lazdins, A.; Makovskis, K.; Daugavietis, U. Plantation Forests: A Guarantee of Sustainable Management of Abandoned and Marginal Farmlands. In Energy Efficiency and Sustainable Lighting—A Bet for the Future; Hermoso-Orzaez, M.J., Gago-Calderon, A., Eds.; IntechOpen: London, UK, 2020. [**Google Scholar**] [**CrossRef**]
8. Vollbrecht, G.; Johansson, U.; Eriksson, H.; Stenlid, J. Butt rot incidence, yield and growth pattern in a tree species experiment in south western Sweden. For. Ecol. Manag. 1995, 76, 87–93. [**Google Scholar**] [**CrossRef**]
9. Piri, T. The spreading of the S type of Heterobasidion annosum from Norway spruce stumps to the subsequent tree stand. Eur. J. For. Path. 1996, 26, 193–204. [**Google Scholar**] [**CrossRef**]
10. Woodward, S.; Stenlid, J.; Karjalainen, R.; Huttermann, A. Heterobasidion annosum: Biology, Ecology, Impact and Control; CAB International: Wallingford, UK, 1998. [**Google Scholar**]
11. Greig, B.J.W.; Gibbs, J.N.; Pratt, J.E. Experiments on the susceptibility of conifers to Heterobasidion annosum in Great Britain. For. Pathol. 2001, 31, 219–228. [**Google Scholar**] [**CrossRef**]

12. Gonthier, P.; Brun, F.; Lione, G.; Nicolotti, G. Modelling the incidence of Heterobasidion annosum butt rots and related economic losses in alpine mixed naturally regenerated forests of northern Italy. For. Pathol. 2012, 42, 57–68. [**Google Scholar**] [**CrossRef**]

13. Gonthier, P. Frequency of stump infections by Heterobasidion annosum s.l. and benefits from urea treatments vary with tree species and season in European Alpine forests. For. Ecol. Manag. 2019, 434, 76–86. [**Google Scholar**] [**CrossRef**]

14. Gaitnieks, T.; Arhipova, N.; Donis, J.; Stenlid, J.; Vasaitis, R. Butt rot incidence and related losses in Latvian Picea abies (L.) Karst. stands. In Proceedings of the 12th International Conference on Root and Butt Rot of Forest trees, Berkley, CA, USA, 12–19 August 2007; The University of California: Berkley, CA, USA, 2008; pp. 177–179. [**Google Scholar**]

15. Arhipova, N.; Gaitnieks, T.; Donis, J.; Stenlid, J.; Vasaitis, R. Butt rot incidence, causal fungi, and related yield loss in Picea abies stands of Latvia. Can. J. For. Res. 2011, 41, 2337–2345. [**Google Scholar**] [**CrossRef**]

16. Anonymous. Forest Industry in Numbers and Facts (Meža Nozare Skaitļos un Faktos); Biedrība "Zaļās Mājas": Riga, Latvia, 2020. (In Latvian) [**Google Scholar**]

17. Stenlid, J. Controlling and predicting the spread of Heterobasidion annosum from infected stumps and trees of Picea abies. Scand. J. For. Res. 1987, 2, 187–198. [**Google Scholar**] [**CrossRef**]

18. Vasiliauskas, R.; Stenlid, J. Spread of S and P group isolates of Heterobasidion annosum within and among Picea abies trees in central Lithuania. Can. J. For. Res. 1998, 28, 961–966. [**Google Scholar**] [**CrossRef**]

19. Bendz-Hellgren, M.; Brandtberg, P.-O.; Johansson, M.; Swedjemark, G.; Stenlid, J. Growth rate of Heterobasidion annosum in Picea abies established on forest land and arable land. Scand. J. For. Res. 1999, 14, 402–407. [**Google Scholar**] [**CrossRef**]

20. Lygis, V.; Vasiliauskas, R.; Stenlid, J. Planting Betula pendula on pine sites infested by Heterobasidion annosum: Disease transfer, silvicultural evaluation, and community of wood-inhabiting fungi. Can. J. For. Res. 2004, 34, 120–130. [**Google Scholar**] [**CrossRef**]

21. Piri, T.; Korhonen, K. The effect of winter thinning on the spread of Heterobasidion parviporum in Norway spruce stands. Can. J. For. Res. 2008, 38, 2589–2595. [**Google Scholar**] [**CrossRef**]

22. Allikmäe, E.; Laarmann, D.; Korjus, H. Vitality Assessment of Visually Healthy Trees in Estonia. Forests 2017, 8, 223. [**Google Scholar**] [**CrossRef**] [**Green Version**]

23. Vollbrecht, G.; Bilde Jørgensen, B. Modelling the incidence of butt rot in plantations of Picea abies in Denmark. Can. J. For. Res. 1995, 25, 1887–1896. [**Google Scholar**] [**CrossRef**]

24. Červinkova, H. Problems of decays of Norway spruce stands in the ČSSR. In Proceedings of the VIIIth Czechoslovak plant protection conference, Prague, Czech Republic, 9–11 September 1980; pp. 287–288. [**Google Scholar**]

25. Capretti, P.; Moriondo, F. Danni in alcuni impianti di conifer associate alla presenza di Heterobasidion annosum (Fomes annosus). Phytopath. Medit. 1983, 22, 157–167. [**Google Scholar**]

26. Capretti, P.; Mugnai, L. Saprophytic growth of Heterobasidion annosum on silver-fir logs interred in different types of forest soils. Eur. J. For. Pathol. 1989, 19, 257–262. [**Google Scholar**] [**CrossRef**]

27. Graber, S. Die Kernfäuleschäden an Fichte (Picea abies Karst.) in der Schweiz nördlich der Alpen: Untersuchungen über das Schadenausmass, die ökologischen, waldbaulichen und mykologischen Einflussfaktoren sowie die ökonomischen Auswirkungen. Beih. Schweiz. Z. Forstwes. 1996, 79, 283. **[Google Scholar]**

28. Vollbrecht, G.; Agestam, E. Modelling incidence of root rot in Picea abies plantations in Southern Sweden. Scand. J. For. Res. 1995, 10, 74–81. **[Google Scholar]** **[CrossRef]**

29. Mareš, R. The extent of root rot damage in Norway spruce stands established on fertile sites of former agricultural land. J. For. Sci. 2010, 56, 1–6. **[Google Scholar]** **[CrossRef]** **[Green Version]**

30. Froelich, R.C.; Dell, T.R.; Walkinshaw, C.H. Soil factores associated with Fomes annosus in the Gulf states. For. Sci. 1966, 12, 356–361. **[Google Scholar]**

31. Dmitri, L. Influence of nutrition and application of fertilizers on the resistance of forest plants to fungal diseases. Eur. J. For. Path. 1977, 7, 177–186. **[Google Scholar]** **[CrossRef]**

32. Brūna, L.; Kļaviņa, D.; Korhonen, K.; Zaļuma, A.; Burņeviča, N.; Gaitnieks, T. Effect of soil properties on the spread of Heterobasidion root rot. Proc. Latv. Acad. Sci. Sect. B Nat. Exact Appl. Sci. 2019, 73, 466–471. **[Google Scholar]** **[CrossRef]** **[Green Version]**

33. Manka, K. Some investigations on Fomes annosus Fr. carried out in Poland. In Proceedings of the Third International Conference on Fomes annosus, Aarhus, Denmark, 29 July–3 August 1968; Hodges, C.S., Rishbeth, J., Yde-Andersen, A., Eds.; USDA Forest Service: Asherville, NC, USA, 1970; pp. 83–90. **[Google Scholar]**

34. Gaitnieks, T.; Arhipova, N.; Nikolajeva, V.; Vulfa, L.; Klavina, D. Heterobasidion annosum izraisītā egļu sakņu trupe lauksaimniecības zemēs. (Heterobasidion annosum root rot of Norway spruce in agricultural soils). Mežzinātne 2008, 17, 22–37. (In Latvian) **[Google Scholar]**

35. Vasiliauskas, A. Root Fungus (Heterobasidion annosum) and the Resistance of Coniferous Forest Ecosystems; Mokslas Publishers: Vilnius, Lithuania, 1989. (In Russian) **[Google Scholar]**

36. Pešková, V.; Soukup, F.; Landa, J. Comparison of mycobiota of diverse aged spruce stands on former agricultural soil. J. For. Sci. 2009, 55, 452–460. **[Google Scholar]** **[CrossRef]** **[Green Version]**

37. Pratt, J.E.; Greig, B.J.W. Heterobasidion annosum: Development of butt rot following thinning in two young first rotation stands of Norway spruce. Forestry 1988, 61, 339–347. **[Google Scholar]** **[CrossRef]**

38. Vollbrecht, G.; Gemmel, P.; Pettersson, N. The effect of precommercial thinning on the incidence of Heterobasidion annosum in planted Picea abies. Scand. J. For. Res. 1995, 10, 37–41. **[Google Scholar]** **[CrossRef]**

39. Rönnberg, J.; Jørgensen, B.B. Incidence of root and butt rot in consecutive rotations of Picea abies. Scand. J. For. Res. 2000, 15, 210–217. **[Google Scholar]** **[CrossRef]**

40. Rönnberg, J.; Berglund, M.; Johansson, U.; Cleary, M. Incidence of Heterobasidion spp. following different thinning regimes in Norway spruce in southern Sweden. For. Ecol. Manag. 2013, 289, 409–415. **[Google Scholar]** **[CrossRef]**

41. Thor, M.; Stenlid, J. Heterobasidion annosum infection of Picea abies following manual or mechanized stump treatment. Scand. J. For. Res. 2005, 20, 154–164. **[Google Scholar]** **[CrossRef]**
42. Stenlid, J.; Redfern, D.B. Spread within the tree and stand. In Heterobasidion annosum: Biology, Ecology, Impact and Control; Woodward, S., Stenlid, J., Karjalainen, R., Hütterman, A., Eds.; CAB International: Wallingford, UK, 1998; pp. 125–143. **[Google Scholar]**
43. Garbelotto, M.; Gonthier, P. Biology, epidemiology, and control of Heterobasidion species worldwide. Annu. Rev. Phytopathol. 2013, 51, 39–59. **[Google Scholar]** **[CrossRef]** **[PubMed]** **[Green Version]**
44. Swedjemark, G.; Stenlid, J. Population dynamics of the root rot fungus Heterobasidion annosum following thinning of Picea abies. Oikos 1993, 66, 247–254. **[Google Scholar]** **[CrossRef]**
45. Zaļuma, A.; Muižnieks, I.; Gaitnieks, T.; Burņeviča, N.; Jansons, Ā.; Jansons, J.; Stenlid, J.; Vasaitis, R. Infection and spread of root rot caused by Heterobasidion spp. in Pinus contorta plantations in Northern Europe: Three case studies. Can. J. For. Res. 2019, 48, 969–977. **[Google Scholar]** **[CrossRef]** **[Green Version]**
46. Stenlid, J. Population structure of Heterobasidion annosum as determined by somatic incompatibility, sexual incompatibility and isozyme patterns. Can. J. Bot. 1985, 63, 2268–2273. **[Google Scholar]** **[CrossRef]**
47. Hantula, J.; Vainio, E. Specific primers for the differentiation of Heterobasidion annosum (s.str.) and H. parviporum infected stumps in northern Europe. Silva Fenn. 2003, 37, 181–187. **[Google Scholar]** **[CrossRef]** **[Green Version]**
48. Kåren, O.; Högberg, N.; Dahlberg, A.; Jonsson, L.; Nylund, J.E. Inter- and intraspecific variation in the ITS region of rDNA of ectomycorrhizal fungi in Fennoscandia as detected by endonuclease analysis. New Phytol. 1997, 136, 313–325. **[Google Scholar]** **[CrossRef]**
49. Mead, R.; Curnow, R.N. Statistical Methods in Agriculture and Experimental Biology; Chapman & Hall: London, UK, 1983. **[Google Scholar]**
50. Bates, D.; Maechler, M.; Bolker, B.; Walker, S. Fitting Linear Mixed-Effects Models Using lme4. J. Stat. Softw. 2015, 67, 1–48. **[Google Scholar]** **[CrossRef]**
51. Royston, P. An extension of Shapiro and Wilk's W test for normality to large samples. Appl. Stat. 1982, 31, 115–124. **[Google Scholar]** **[CrossRef]**
52. Labouriau, R. postHoc: Tools for Post-Hoc Analysis. R Package Version 0.1.3. 2020. Available online: **https://CRAN.R-project.org/package=postHoc** (accessed on 10 January 2020).
53. R Core Team. R: A Language and Environment for Statistical Computing; R Foundation for Statistical Computing: Vienna, Austria, 2020; Available online: **https://www.R-project.org/** (accessed on 10 January 2020).
54. Pettersson, M.; Rönnberg, J.; Vollbrecht, G.; Gemmel, P. Effect of thinning and Phlebiopsis gigantea stump treatment on the growth of Heterobasidion parviporum inoculated in Picea abies. Scand. J. For. Res. 2003, 18, 362–367. **[Google Scholar]** **[CrossRef]**
55. Thor, M.; Arlinger, J.; Stenlid, J. Heterobasidion annosum root rot in Picea abies: Modelling economic outcomes of stump treatment in Scandinavian coniferous forests. Scand. J. For. Res. 2006, 21, 414–423. **[Google Scholar]** **[CrossRef]**

56. Korhonen, K.; Piri, T. How cultivate Norway spruce on site infested by Heterobasidion? In Proceedings of the Nordic/Baltic Forest Pathology Meeting, Copenhagen, Denmark, 31 August–4 September 2002; Thomsen, I.M., Ed.; Danish Forest and Landscape Research Institute: Horsholm, Denmark, 2003; pp. 21–29. [**Google Scholar**]

57. Klavina, D.; Menkis, A.; Gaitnieks, T.; Velmala, S.; Lazdins, A.; Rajala, T.; Pennanen, T. Analysis of Norway spruce dieback phenomenon in Latvia—A belowground perspective. Scand. J. For. Res. 2016, 31, 156–165. [**Google Scholar**] [**CrossRef**]

58. Piri, T. Silvicultural Control of Heterobasidion Root Rot in Norway Spruce Forests in Southern Finland. Regeneration and Vitality Fertilization of Infected Stands. Ph.D. Thesis, Finnish Forest Research Institute METLA, University of Helsinki, Helsinki, Finland, 2003. [**Google Scholar**]

59. Varese, G.C.; Gonthier, P.; Nicolotti, G. Long-term effects on other fungi are studied in biological and chemical stump treatments in the fight against Heterobasidion annosum coll. Mycologia 2003, 95, 379–387. [**Google Scholar**] [**CrossRef**] [**PubMed**]

60. Vollbrecht, G.; Bilde Jørgensen, B. The effect of stump treatment on the spread rate of butt rot in Picea abies in Danish permanent forest yield research plots, Scand. J. For. Res. 1995, 10, 271–277. [**Google Scholar**] [**CrossRef**]

61. Oliva, J.; Samils, N.; Johansson, U.; Bendz-Hellgren, M.; Stenlid, J. Urea treatment reduced Heterobasidion annosum s.l. root rot in Picea abies after 15 years. For. Ecol. Manag. 2008, 255, 2876–2882. [**Google Scholar**] [**CrossRef**]

62. Blomquist, M.; Larsson-Herrera, S.; Hofmann, J.; Ceyd Beram, R.; Cleary, M.; Rönnberg, J. Size matters but is big always better? Effectiveness of urea and Phlebiopsis gigantea as treatment against Heterobasidion on Picea abies stumps of variable size. For. Ecol. Manag. 2020, 462, 117998. [**Google Scholar**] [**CrossRef**]

63. Brandtberg, P.-O.; Johansson, M.; Seeger, P. Effects of season and urea treatment on infection of stumps of Picea abies by Heterobasidion annosum in stands on former arable land. Scand. J. For. Res. 1996, 11, 261–268. [**Google Scholar**] [**CrossRef**]

64. Oliva, J.; Thor, M.; Stenlid, J. Long-term effects of mechanized stump treatment against Heterobasidion annosum root rot in Picea abies. Can. J. For. Res. 2010, 40, 1020–1033. [**Google Scholar**] [**CrossRef**]

65. Niemelä, T.; Korhonen, K. Taxonomy of the genus Heterobasidion. In Heterobasidion annosum: Biology, Ecology, Impact and Control; Woodward, S., Stenlid, J., Karjalainen, R., Hütterman, A., Eds.; CAB International: Wallingford, UK, 1998; pp. 27–35. [**Google Scholar**]

66. Hanso, S.; Korhonen, K.; Hanso, M. Attack of spruce and pine by S and P groups of Heterobasidion annosum on forest and former agricultural soils in Estonia. In Root and Butt Rots, Proceedings of the IUFRO Working Party S2.06.01, Wik, Sweden and Haikko, Finland, 9–16 August 1993; Johansson, M., Stenlid, J., Eds.; Swedish University of Agricultural Sciences: Uppsala, Sweden, 1994; pp. 254–259. [**Google Scholar**]

67. Korhonen, K.; Piri, T. The main hosts and distribution of S and P groups of Heterobasidion annosum in Finland. In Root and Butt Rots, Proceedings of the IUFRO Working Party S2.06.01, Wik, Sweden and Haikko, Finland, 9–16 August 1993; Johansson, M., Stenlid, J., Eds.; Swedish University of Agricultural Sciences: Uppsala, Sweden, 1994; pp. 260–267. [**Google Scholar**]

68. Piri, T.; Korhonen, K. Infection of advance regeneration of Norway spruce by Heterobasidion parviporum. Can. J. For. Res. 2001, 31, 937–942. [**Google Scholar**] [**CrossRef**]

69. Piri, T.; Valkonen, S. Incidence and spread of Heterobasidion root rot in uneven-aged Norway spruce stands. Can. J. For. Res. 2013, 43, 872–877. [**Google Scholar**] [**CrossRef**]

70. Piri, T. Early development of root rot in young Norway spruce planted on sites infected by Heterobasidion in southern Finland. Can. J. For. Res. 2003, 33, 604–611, 654. [**Google Scholar**] [**CrossRef**]

71. Oliva, J.; Bendz-Hellgren, M.; Stenlid, J. Spread of Heterobasidion annosum s.s. and Heterobasidion parviporum in Picea abies 15 years after stump inoculation. FEMS Microbiol. Ecol. 2011, 75, 414–429. [**Google Scholar**] [**CrossRef**] [**PubMed**]

72. Korhonen, K.; Bobko, I.; Hanso, S.; Piri, T.; Vasiliauskas, A. Intersterility groups of Heterobasidion annosum in some spruce and pine stands in Byelorussia, Lithuania and Estonia. Eur. J. For. Pathol. 1992, 22, 384–391. [**Google Scholar**] [**CrossRef**]

73. Piri, T.; Korhonen, K.; Sairanen, A. Occurrence of Heterobasidion annosum in pure and mixed spruce stands in Southern Finland. Scand. J. For. Res. 1990, 5, 113–125. [**Google Scholar**] [**CrossRef**]

74. Keca, N.; Keca, L. Distribution of Heterobasidion parviporum genets in Norway spruce forests in Serbia. In Proceedings of the 13th International Conference: Root and Butt Rot of Forest Trees, Firenze, Italy, 4–10 September 2011; Carpetti, P., Comparini, M., La Porta, N., Santini, A., Eds.; Firenze University Press: Firenze, Italy, 2013; pp. 155–158. [**Google Scholar**]

75. Piri, T.; Hamberg, L. Persistence and infectivity of Heterobasidion parviporum in Norway spruce root residuals following stump harvesting. For. Ecol. Manag. 2015, 353, 49–58. [**Google Scholar**] [**CrossRef**]

13 ASSESSING THE VULNERABILITY OF WOOD FROM VARIOUS TREE SPECIES TO INITIAL INFECTIONS BY HETEROBASIDION ANNOSUM SENSU LATO AND THE EXTENT OF PATHOGEN SPORE DISPERSAL

1. INTRODUCTION: AN OVERVIEW

Conversion of pastures, meadows, or agricultural fields into forest plantations is a widespread practice in numerous nations. As an example, throughout the European Union, a total of 12.9 million hectares of unused lands were converted into forests between 1995 and 2015 [1]. Agricultural lands and lands with past forest covers have distinct characteristics, therefore forest management plans should take into account the optimal tree species and associated hazards [2,3,4]. Consequently, multiple research have conducted comparisons between the pace of stand growth and growth circumstances in forest lands and former agricultural lands [5,6].

Norway spruce (Picea abies (L.) Karst.) is a frequently planted conifer tree species in Latvia, as well as other Baltic and Nordic nations, particularly in areas that were previously used for agriculture [7]. The Norway spruce is extremely vulnerable to root rot caused by Heterobasidion spp. [8,9,10,11,12,13]. The country-wide survey conducted in Latvia found that approximately 22% of spruce trees were affected by heart rot, primarily caused by Heterobasidion parviporum Niemelä &

Korhonen. This fungal infection leads to significant economic losses [14,15]. Over 50% of Latvia's land is forested, and the forest industry accounts for nearly 20% of Latvia's total exports [16]. Therefore, it is essential to minimize the economic losses resulting from root rot.

The transmission of Heterobasidion infection to future generations of trees through root contacts is facilitated by the presence of infected stumps from earlier generations [10,11,17,18,19,20,21]. Hence, implementing forest planting in areas free from infection and afforestation on previously used agricultural lands or pastures are crucial tactics to reduce the risk of infection [22].

However, multiple studies have found that the rate of butt rot occurrence is comparable in former fields or grazing area when compared to former forestlands [23], and in some cases, even significant damage has been observed in previous agricultural lands [24,25,26,27]. Modelling using stump infection rate with Heterobasidion in spruce stands indicated that former pastures and arable lands are more vulnerable to root rot compared to forest stands [28]. A recent study conducted in the Czech Republic found variations in infection rates based on the prior land use, with greater infection rates observed in former arable lands compared to former pastures [29].

The higher infection rate on former agricultural lands can be attributed to several factors. These include soil chemical parameters such as high carbonate content, high nitrogen content, or an overall unbalanced nutrient status [4,30,31,32]. Additionally, soil compaction can stimulate the superficial root system of Norway spruce, leading to increased infection rates [26]. Biotic factors, such as the absence of fungi antagonistic to Heterobasidion, also contribute to higher infection rates [4,26,31,33,34,35,36]. Lastly, an increased percentage of grass cover has been linked to higher infection rates [30].

Effective forest management methods, particularly cuts, play a crucial role in the development of Heterobasidion infections in forests that were formerly used for agriculture. In these areas, the predominant method of disease transmission is through the infection of recently cut conifer stumps by spores. Conversely, in previous forest areas, the spread of diseases can be more substantial due to mycelial infections that originate from the ancient stumps or roots. Consequently, multiple research have specifically examined the occurrence of root rot following the initial thinning process on previously used farmland [37, 38, 39]. The occurrence of Heterobasidion root rot is mostly influenced by the extent of stump infection, however the potential for direct root infection cannot be entirely disregarded. For example, Rönnberg et al. [40] demonstrated a significant occurrence of Heterobasidion root rot in unthinned spruce stands that were planted on pastureland. Thor and Stenlid [41] discovered that between 0.5% and 8.1% of trees in unthinned spruce stands on previously used agricultural land were infected. In addition, they observed that up to 16.9% of trees in an unthinned stand on former pastureland were sick.

Secondary infections arise when the hyphae of the pathogens extend to the root systems of neighboring trees through root contacts, resulting in the formation of disease centers that gradually increase [42,43]. In order to gain comprehensive understanding of secondary infection, Swedjemark and Stenlid [44] conducted an analysis on the population structures of Heterobasidion annosum s.l. in spruce stands that were previously used for agriculture. Their findings revealed that, seven years after thinning, 63% of trees exhibited decay symptoms and were infected by mycelial spread from stumps. A separate study conducted in Sweden compared the growth rate of artificially introduced Heterobasidion mycelia in the stem and roots of Norway spruce trees growing in forest land and arable land. The study found that, on average, there was no significant difference in the growth rate of Heterobasidion in both the tree and stump roots between the former arable and forest sites [19].

Nevertheless, the aforementioned research were conducted on former agricultural lands in Sweden and Scotland, and their findings may not be universally relevant to other soil and forest types, as well as diverse agricultural and forestry management approaches. There is a lack of research conducted in Norway spruce stands in Latvia or nearby countries that may be compared to each other, highlighting the necessity for doing such studies. A recent study conducted in Latvia by Zaļuma et al. [45] examined the impact of Heterobasidion root rot on Scots pine stands. The study found that the expansion rate of Heterobasidion root rot was lower in pine plantations developed on former agricultural land compared to those created on former forestland. Therefore, it is crucial for practical forestry in Latvia to gather precise and detailed information about the extent of Heterobasidion root rot in Norway spruce stands located on areas with no prior history of forestation, in addition to the research stated earlier. The objective of this study was to examine the distribution patterns, both primary and secondary infection, of Heterobasidion spp. in managed Norway spruce plantations that were developed on previously used meadows and pasturelands in the central region of Latvia.

2. MATERIALS AND METHODS

2.1. Study Sites and Experimental Design

The study was established in three Norway spruce stands in Central Latvia, in the municipalities of Babite, Auce, and Olaine (**Figure 1**). In this region, average annual temperature ranges from 6.8 to 7.9 °C and annual precipitation rate is 574 to 636 mm. For Central Latvia, dry pine and spruce forests on sandy or loamy soils, water-saturated areas (peat bogs, wet forest types, flood plain meadows, etc.) with peat soils or drained forests and a large proportion of former agricultural fields on loamy soils are characteristic. The first-rotation spruce plantations in the study represent this variety: a drained forest within a region of peat bogs (Stand 1), a spruce plantation within a region dominated by agricultural fields (Stand 2), and a dry-type (Oxalidosa) stand (Stand 3). Characteristics of the investigated areas are presented in **Table 1**.

Figure 1. Location of the study sites in the central part of Latvia.

Table 1. Description of Norway spruce stands selected for the study: general information, forest characteristics, and details of sampling area and soil parameters.

	Stand 1	Stand 2	Stand 3
General information			
Location	56°84′ N, 23°77′ E	56°38′ N, 23°12′ E	56°82′ N, 24°07′ E
Municipality	Babite	Auce	Olaine
Altitude, m	17	85	13
Annual rainfall, mm	600–650	550–600	650–700
Forest data			
Total area, ha	8.7	2.75	1.54

	Stand 1	Stand 2	Stand 3
Age, years	73	82	52
Average height, m	27.1	28.8	23.5
DBH, cm	35	30	24
Yield per ha	342	387	268
Forest type	*Vacciniosa mel.*	*Myrtillosa mel.*	*Oxalidosa*
Thinning years *	2007, 2015	1991, 2001	1987, 2017
Previous land use	Pastureland	Wet meadow	Meadow
Other tree species (<5% admixture)	*Betula pendula*	*Populus tremula*	*Pinus sylvestris*
Sampling area			
Length of transect, m	500	400	250
Area of transect, m^2	1000	800	500
Total No. of circle sample plots	11	6	5
Total area of circles, m^2	3454	1884	1570
Soil			
pH (KCl)	4.1	5.9	4.6
C, (g/kg)	146	103	106
N, (g/kg)	7.6	8.5	7.5
P, (g/kg)	0.03	0.06	0.03
K, (g/kg)	0.02	0.05	0.02
Ca, (g/kg)	4.4	8.9	6.9
Mg, (g/kg)	0.16	0.88	0.25
* All stands were thinned twice. No data of the exact thinning period and season is available.			

All stands had selective or sanitary thinning (**Table 1**); since after thinning, stumps were not treated with any control agents to prevent infection of root rot, at least potentially, stumps were exposed to primary infection of Heterobasidion.

Soil samples for chemical analysis were collected at 0–20 cm depth along transects every 50 m. Chemical analyses of soil were conducted using established standard methods (International Organization for Standardization (ISO) standard) at LSFRI Silava, Laboratory of Forest Environment, Salaspils, Latvia. Samples were prepared for analysis according to ISO 11464:1994. The soil pH was determined in 1 M KCl (soil—extractant mixture 1:2.5). Total carbon content was determined using an elemental analyser ELTRA CS530 (Eltra GmbH, Haan, Germany) according to LVS ISO 10694. The concentration of total N was determined using the Kjeldahl method (ISO 11261:1995) and P was assayed by colourimetry using the ammonium molybdate spectrometric method in 1 M HCl solution (soil/extractant mixture 1:5). Concentration of K, Ca, and Mg was determined by atomic absorption spectrophotometer with an acetylene-air flame in 1 M HCl extract.

To determine average infection rate, we sampled all standing trees within a 2-m-wide transect parallel to the longest edge of each stand with starting points in the middle of the shortest edge. Health status was assigned to each tree in the transect: (i) visually healthy crowns and (ii) trees with severe dieback symptoms, dead and windthrown trees [**9**]. To detect the pathogen under laboratory conditions, wood samples were collected from all trees on the transect with the aid of an increment borer (one sample per tree at root collar height). Additionally, to characterise stand growth rate, we measured the diameter at breast height and, for living trees, the tree height (**Table 1**).

Based on the transect data about the distribution of Heterobasidion infected trees and visual observation of potential infection centers, we established from five to eleven circular sample plots per stand (10 m radius, 314 m^2) where all trees and stumps were sampled and examined for Heterobasidion fruit bodies. In addition, stumps were classified as either stumps from the 1st or 2nd thinning. Broken dead tree trunks of different heights were categorized as stumps. Wood samples from trees were collected with an increment borer (one sample per tree at root collar height) and from stumps—either with the same method or with the aid of an axe if they were heavily decomposed. Wood samples from each tree were placed into individual plastic tubes and stump samples into individual plastic bags. All sampled trees and stumps (both in transects and circular plots) were numbered and their location mapped. Field work was carried out from March 2019 to April 2020.

2.2. Isolation and Identification of Heterobasidion Genotypes

In the laboratory, all wood samples were surface sterilized by flaming and placed on petri dishes containing malt extract agar (MEA) media (for tree samples) or Hagem agar media (for stump samples). After a 5 to 7-day incubation, the plates were examined under a Leica stereomicroscope M205C (20–30× magnification) to detect the presence of Heterobasidion spp., observing its characteristic asexual sporulation (conidiophores and conidia) on wood samples [**10**]. If Heterobasidion was observed, colonies were sampled with sterilised surgical forceps and placed on new MEA media. Outgrowing Heterobasidion mycelia were subcultured to obtain pure fungal cultures.

Isolates originating from the same circular sample plot were subjected to pairwise somatic compatibility tests by confronting their mycelia on Hagem agar media in 9 cm Petri dishes in all combinations. The genets were identified either by recording the line of demarcation in the contact zone of confrontation (demonstrating that the genotypes are different) or by observing free fusion of the mycelia (implying that the genotypes were identical) [**46**].

The results of this test provided an information about the total number and size of Heterobasidion genets: number of genets encompassing only one tree or stump and number of genets encompassing two or more trees and (or) stumps. Based on these data, we estimated the number of trees/stumps infected by Heterobasidion genets per hectare and per disease centre. To estimate the area and maximum width of each genet including two or more trees/stumps, we used the map of each site with all sampled trees and stumps mapped. We used species specific primers (Heterobasidion parviporum specific primers KJ-F and KJ-R and Heterobasidion annosum specific primers MJ-F and MJ-R) to determine the Heterobasidion species of each genotype [**47**]. DNA was extracted from fungal cultures representing each genotype using a CTAB method [**48**]. Two PCR reactions were run with each sample—one with H.parviporum and other with H.annosum species specific primers listed above. PCR conditions, were a described by Hantula and Vainio [**47**]. DNA from H. parviporum and H. annosum was used as a positive control.

2.3. Data Analysis

A chi-square (χ^2) test of independence from the actual number of observations of infected or decaying trees and stumps was used to compare abundance patterns between sites and sample plots [**49**].

Using the geospatial tool QGIS 3.2.0-Bonn (bc43194) (**https://qgis.org/api/3.2/**), tree spatial location was visualised and the width and the area of genets including two or more trees/stumps was measured (tool "MeasureLine" and "MeasureArea"). The maximum expansion rates were calculated as a reference taking years from the first thinning for each stand (2007, 1991 and 1987, respectively, Stand 1 to 3) or from the second thinning (2015, 2010 and 2017, respectively, Stand 1 to 3).

To assess which factors have an impact on the number of trees per genet, we used generalized linear mixed models (GLMM) assuming a Poisson error distribution. The full model included number of trees per genet as a response variable and as fixed effects we tested site, stand age class (older stands (Stand 1 and 2) vs. younger stand (Stand 3), disease centre infection indicators (classes of infection (<40% vs. >40%), stump and dead tree proportion) and as a random factor in all models we included disease centre. Models were calculated using the program package lme4 [**50**].

Area and width of genets including two or more trees/stumps as well as maximum expansion rates (area and width) were tested for normality using the Shapiro and Wilk normality test at significance level □ = 0.001 [**51**]. If normality was assessed, one-way analysis of variance (ANOVA) with Fisher LSD post-hoc test was calculated [**52**]. If normality was not assessed, nonparametric Wilcoxon test (unpaired samples) was used to compare parameters. For calculations, R version 4.0.2 was used [**53**].

3. RESULTS

3.1. Incidence of Dieback and Heterobasidion Infection

In total, in all three study sites, 851 wood samples were collected for Heterobasidion isolation (397 from living trees, 118 from dead trees and 336 from stumps). The number of sampled trees and stumps are given in **Table 2.** In all stands, stumps from both 1st and 2nd thinning were present and in addition there were some trees broken close to their base (mainly in Stand 2) (**Table 2**).

Table 2. Characteristics and dieback rate in Norway spruce stands on former pastures (Stand 1) and meadows (Stands 2 and 3). (T—transect; C—disease centres (circular sample plots).

	Stand 1		Stand 2		Stand 3	
	T	C	T	C	T	C
Living trees						
Total number *	72	95	25	73	34	153
Number per hectare	720	275	313	387	680	975
Dead trees						
Total number *	8	41	10	37	8	29
Number per hectare	80	119	125	196	160	185
Stumps						
Total number	n.d.	142	n.d.	96	n.d.	98
No. of 1st thinning stumps	n.d.	65	n.d.	45	n.d.	29
No. of 2nd thinning stumps	n.d.	75	n.d.	39	n.d.	69
No. of other stumps **	n.d.	2	n.d.	12	n.d.	-
Number per hectare	n.d.	411	n.d.	510	n.d.	624

* Several trees were present in both transect and disease centre. ** Category "other stumps" includes trees broken at different heights.

In Stand 2 and 3, chi-square test of independence showed that transect and disease center data have similar tendencies regarding the number of living and dead trees ($p > 0.05$). In contrast, chi-square test for Stand 1 data indicated significant differences in living and dead tree abundance in disease centers and the transect ($\chi^2 = 11.65$; df = 1; n = 216) (there were less dead trees in the transect ($\chi^2 = 5.67$) than expected).

Additionally, a chi-square test of independence was performed to examine the numbers of living trees and stumps among sites. The relation between these variables was significant, $\chi^2 = 24.17$ (df = 2, n = 657), $p < 0.05$. The highest chi-square values were observed for both living trees and stumps in Stand 3: more living trees ($\chi^2 = 7.52$) and less stumps ($\chi^2 = 7.18$) than predicted.

The transects of both Stand 2 and 3 had several openings, which explains the lower number of trees per hectare in comparison with disease centres (**Table 2**). In Stand 3, these openings were related to stand planting design (there were several gaps left to facilitate thinning). Stand 3 had higher planting density that reflects in the number of trees per hectare (**Table 2**).

In addition, fruit bodies were recorded as an indicator of disease of trees and stumps. There were more records of fruit bodies in Stand 1 (two dead trees and 10 stumps with fruit bodies) compared to other stands (there were two fruit bodies recorded in Stand 2 and four fruit bodies in Stand 3, found exclusively on stumps in both stands).

After incubation of 851 collected wood samples, we detected pathogen conidiophores with conidia in 302 samples (35%); infection rate varied among stands (30% in Stand 1 and 39% in Stands 2 and 3) (**Table 3**).

Table 3. Frequency (% *) of Norway spruce trees or stumps infected by Heterobasidion parviporum in managed forests on former pastures and meadows.

	Stand 1	Stand 2	Stand 3
Transect; %	16.3	28.6	33.3
Living trees; %	12.5	25.7	23.8
Dead trees; %	3.8	2.9	9.5
Disease Centres; %	33.5	41.3	38.6
Living trees; %	11.5	15.5	18.9
Dead trees; %	8.6	7.3	5.4
Stumps; %	13.3	18.4	14.3
Total; %	29.6	38.5	38.9

* Percentage expresses the proportion of trees/stumps where *Heterobasidion parviporum* was detected of all trees/stumps sampled.

Chi-square test indicated that in transects and disease centres, the incidence of Heterobasidion infection was similar among sites ($p > 0.05$). In addition, the infection incidence of both living and dead trees in disease centres have a similar pattern among sites ($p > 0.05$). In total, we confirmed the presence of the pathogen in 31.8 and 48.3% of all living and dead tree samples, respectively. Infection incidences of living trees among sites was 24.5, 38.6 and 35.2% (Stand 1 to 3, respectively) and of dead trees 58.1, 35.7 and 51.5% (Stand 1 to 3, respectively). In total, we confirmed the presence of the pathogen in 34.2% of all stump samples. Infection incidences varied among stump samples at different sites: 26.1, 39.6 and 40.8% (Stand 1 to 3, respectively). However, the differences were not significant.

3.2. Analysis of Heterobasidion Genets

Pure Heterobasidion sp. cultures were successfully isolated from 90.5% of a total of 286 wood samples producing Heterobasidion conidiophores with conidia collected from disease centers (76.3, 98.8, and 96.3% from Stands 1 to 3, respectively) (**Table 3**). Identification with species specific primers revealed that all obtained isolates belonged to species Heterobasidion parviporum.

In total, 141 Heterobasidion genets were detected resulting in 72 to 484 Heterobasidion genets per hectare among study sites (**Table 4**). Less genets per hectare were detected in Stand 1 (2.9 times and 6.7 times less in comparison with Stand 2 and Stand 3, respectively).

Table 4. Distribution pattern and characteristics of Heterobasidion parviporum genets in disease centres of managed Norway spruce stands on former pasture and meadows.

	Stand 1	Stand 2	Stand 3
Genets including one tree/stump			
Infection per hectare	38	143	376
Trees infected	35	64	274
Stumps infected	3	79	102
Frequency (%)	21%	25%	56%
Genets including two or more trees/stumps			
Number of trees and stumps per genet (min.-max.)	4.8 (2–12)	4.4 (2–11)	2.6 (2–7)
Infection per hectare	170	302	293
Trees infected	120	185	191
Stumps infected	50	117	102
Distance among trees and (or) stumps, m			
Mean ± SE	9.6 ± 1.7	6.9 ±1.0	3.1 ± 0.5
Min.-max.	1.8–18.0	2.1–14.6	1.0–6.9
Area, m^2			
Mean ± SE	56.4 ± 19.7	12.4 ± 4.9	5.9 ± 2.9
Min.-max.	1.2–137.4	0.3–52.2	1.3–17.3
Frequency (%)	79%	75%	44%
Total infection per hectare	208	445	669

The mean number of trees/stumps per genet was 2.8, 2.1 and 1.4 in Stand 1 to 3, respectively. Genets including one tree/stump were the most abundant in all sites: 13, 27, 59 (i.e., 52, 68 and 78% of all genets per stand), in Stand 1 to 3, respectively. Expanded genets including two or more trees/stumps were less frequent 12, 13 and 17 (Stand 1 to 3 respectively). Chi-square test indicated differences among abundance of genets including one tree/stump and genets including two or more trees/stumps per sites ($\chi^2 = 6.1$, df = 2, n = 141; $p > 0.05$).

Frequency of trees or stumps infected by genets including one tree/stump of all infected trees and stumps was 21, 25 and 56% (Stand 1 to 3, respectively). Chi-square test indicated differences among abundance of trees and stumps infected by genets including one tree/stump per sites ($\chi^2 = 26.3$, df = 2, n = 259; $p > 0.05$), in particular abundance of trees and stumps of genets including one tree/stump in Stand 1 was lower than predicted ($\chi^2 = 7.4$) and in Stand 3 higher than predicted ($\chi^2 = 8.4$).

In comparison with Stand 1 and 2, Stand 3 had more isolates and consequently more genets (in total and both genets including one tree/stump and genets including two or more trees/stumps). Stand 3 had a lower number of trees per genet and less expanded genets including two or more trees/stumps (lower distance among trees and area) (**Table 4**). Modeling the number of trees per genet and including disease centres as a random factor revealed that grouping Stand 1 and 2 together (older stands) vs. Stand 3 identified a significant difference ($p < 0.05$).

The maximum number of trees infected by genets including two or more trees/stumps are presented in **Table 4**. The largest genets included both living and dead trees and stumps. The longest mean distance between trees and (or) stumps with shared genets was about 7–10 m (maximum 15–18 m) (Stands 1 and 2) compared with values of Stand 3 (mean longest distance 3 m; maximum 7 m) (**Table 4**). Analysis of variance showed significant differences and post-hoc tests indicated that the genets including two or more trees/stumps in Stand 3 were significantly ($p < 0.05$) smaller in size than genets in Stand 1 (3.3 times smaller) and in Stand 2 (2.3 times smaller).

The maximum area of genets including two or more trees/stumps in Stand 1 was 2.6 to 8 times greater than in Stand 2 and 3 (**Table 4**). The mean area of genets including two or more trees/stumps was also 5 to 9 time larger in Stand 1. Analysis of variance showed significant differences and post-hoc tests indicated that the genets including two or more trees/stumps in Stand 1 had significantly ($p < 0.05$) larger area than both Stand 2 and 3 ($p < 0.05$).

The maximum expansion length per year calculated from the first thinning was 1.5, 0.5 and 0.2 m (Stand 1 to 3, respectively) and maximum expansion area per year: 11.5, 1.9, and 0.5 m^2 (Stand 1 to 3, respectively). Wilcoxon test showed that also mean expansion indicators (distance and area per year) were higher in Stand 1 in comparison with Stand 2 and Stand 3 ($p < 0.05$).

4. DISCUSSION

4.1. Dieback and Observed Heterobasidion Infection of Norway Spruce on Former Pasture and Meadow Lands

Visual observation of stand health status revealed expansive spread of root rot in the studied Norway spruce stands on former non-forest lands (pastures and meadows) in Latvia. The greater total number of trees and stumps per hectare in Stand 3 is mainly related with greater planting density. The number of stumps indicates intensity of thinning and, therefore, potential infection risk of stumps by Heterobasidion spores. Greater abundance of stumps might also favor the mycelial spread of Heterobasidion since its growth rate in stump roots is significantly faster than in living trees [**19,54**]. Moreover, in densely planted stands, where the trees or stumps are closer to each other, the possibility of root contacts is higher, thus allowing greater probability of infection transmission [**4**].

In Stand 2, there were several trees broken at different heights (**Table 2**), which indirectly indicates reduced tree vitality. In addition, gaps and a low number of trees per transect might indicate not only a reduced number of trees due to the stand age but also poor current and previous pathological conditions within the stand. Fruit bodies were found in all study sites, which indicates that environmental conditions were suitable for fruiting of H. parviporum. Consequently, the risk of primary infection by H. parviporum spores should be taken into account when planning future thinning in these stands or other similar sites.

The detected H. parviporum infection rate (30% to 39%) was on average in line with data from other studies from both former agricultural and forest lands. It indicates that even though no inoculum is present in former non-forest lands, infection can establish and expand as successfully as in forest lands or even more successfully. For instance, in a Swedish study carried out on a spruce stand on former agricultural land, seven years after thinning, 63% of the trees had decay symptoms [**45**]. In Sweden, modelling of spruce stands including different former land use indicated 2–90% of rotted trees in final cuttings [**55**]. In Denmark, on average, 23% (max. 94%) of Norway spruce individuals growing on former agricultural or grazing fields were found to suffer from butt rot [**23**].

The success rate of H. parviporum isolation from dead spruce individuals was 35.7 to 58.1%. However, since the pathogen was not detected in all dead spruce individuals, it cannot be excluded that either the infection has not yet spread higher in the stem or that causes other than Heterobasidion root rot were responsible for a proportion of declining or dead trees. Other fungal agents as Armillaria sp. can cause root rot in Norway spruce stands [**56**]. In addition, spruce dieback not related to root rot has been observed in several regions in Latvia [**57**].

The frequency of stumps colonized by H. parviporum (34.2% on average) was relatively high in spruce stands planted in former pastures and meadows in Latvia. Viable Heterobasidion mycelium can survive in stumps for decades [**58**]. However, with time, it develops more intensively in deeper layers of stumps [**59**] and isolation can be difficult due to wood decomposition and presence of other stump colonizing fungi. Since the stumps in the studied stands were not recently cut, total stump infection may be even higher. Infected spruce stumps left after thinning transfer root rot to residual trees and serve as a source of infection [**21,54**]. Therefore, winter thinning and use of biological control agents containing spores of Phlebiopsis gigantea or other preventive substances such as urea is recommended to limit the infection of conifer stumps by basidiospores of Heterobasidion spp. [**43,60,61,62**], especially in stands with no previous rot incidences such as in stands established on former agricultural lands [**4,63,64**].

4.2. Heterobasidion Genets

All obtained Heterobasidion isolates were identified as Heterobasidion parviporum, which is the most common Heterobasidion species causing damage in spruce stands in Northern Europe and Latvia [**15,65**]. H. parviporum is adapted to spruce as the host species [**66,67**] and is therefore more often observed in spruce stands, being predominant or exclusive in some cases [**9,68,69,70**], and is able to spread more frequently from stumps to neighbouring trees than H. annosum [**71**]. Korhonen et al. [**72**] reported that Heterobasidion infections in spruce stands on former agricultural lands were mainly caused by H. parviporum and concluded that in such first rotation spruce stands without any history of pine growth and woody inoculum, spores of H. parviporum are the principal infection source. Moreover, Vasiliauskas and Stenlid [**18**] showed that H. parviporum is better adapted to spruce wood, and spreads more extensively in spruce stems compared to H. annosum.

In our study, the number of trees/stumps infected by H. parviporum per hectare was 208, 445 and 669 (Stand 1 to 3, respectively) (**Table 4**). In addition, the total number of Heterobasidion genets per hectare in the present study was high: 72, 212, and 484 (Stand 1 to 3, respectively). In comparison, in a Finnish study on Heterobasidion infections on clear-felled forests on old forest sites, the number of fungal genets per hectare varied from 2 to 158 (mean 28) [**73**]. The high infection rate and high number of Heterobasidion genets in the present study may indicate a higher susceptibility of Norway spruce to Heterobasidion root rot on former agricultural lands compared to old forest lands [**2**].

The number of genets including only one tree or stump was lower in older stands (Stand 1 and 2) than in the youngest one (Stand 3). Time from the first thinning did, however, not differ significantly between the stands (**Table 1**). However, even in final cuttings on old forest sites, genets including one tree/stump and therefore primary infection can be dominant [**73**]. In general, a lower number of trees infected by the same Heterobasidion genet and occupying a smaller stand area indicate that these clones are relatively younger than those comprised of more trees and covering a larger area [**44**]. However, the age of the genets of Heterobasidion may affect the pathogenicity of the fungus, with younger genets being more pathogenic [**39**].

Development of territorial clones following primary infection of Heterobasidion takes at least 7–8 years [**42,44,74**]. In our study, 22 to 48% of detected genets formed expanding territorial clones, which included 44 to 79% of infected trees and (or) stumps. This indicates that infection in all sites is well established and expanding. Due to well-developed clonality, in Stand 1 and 2, it is most probable that infection was initiated due to the first thinning (12, 28 years ago), and not by the more recent ones (4, 18 years ago) (Stand 1 and 2, respectively). In Stand 3, infection history seems to be much more recent than in other stands and probably more expansive after the last thinning (two-years before sampling) than after the first one (32-years before sampling). In addition, greater planting density in Stand 3 could favour development of territorial clones in a shorter period as indicated previously [**35,42,44**].

Comparing the maximum size of Heterobasidion genets with other studies, we found genets of similar size also reported from stands with a much longer infection history. For instance, Vasiliauskas and Stenlid [**18**] reported territorial clones of H. annosum s.l. up to 20 m in diameter in a 60 year-old spruce stand planted in a previously infected pine site, while in the oldest stands of our study (Stand 1 and 2), territorial clones reached up to 14–17 m. In addition, the largest distance between isolates from the same genet in Finnish study [**69**] in 70 to 100 year-old spruce stands was 13 m, which is less than in our study on former non-forest lands. Previous studies indicate that H. parviporum can form large territorial clones. Stenlid [**46**] studied a 120-year-old and thinned P. abies stand and detected H. parviporum territorial clones reaching up to 30 m in diameter. Similarly, up to 40 m long territorial clones of H. parviporum were observed in an intensively thinned P. abies stands in Serbia [**74**]. In related studies from Finland, the maximal length of territorial clones varied between approximately 10 and 30 m [**9,66**] or reached up to 55 m [**73**]. Our study shows that even 12 to 32 years after thinnings in stands on former pastures and meadows, H. parviporum can form large territorial clones.

The maximum number of trees per genet in our study was 12 and this is lower in comparison to stands on former forest lands (i.e., a Finnish study [**67**] reported 16 trees per genet) and also to stands on former pasturelands, where a greater number of trees (up to 25) forming territorial clones have been reported [**40**]. However, the real number of infected trees and stumps in our study could be greater than detected due to various reasons: (i) stump decomposition and poor isolation success from some samples, (ii) sampling height of trees (root collar) excluding infection in lower zones, (iii) sample number (one sample per tree) not covering all stem inner surface where decay column can be located, and (iv) limited sampling area (territory of circular sample plot). In addition, to estimate the exact borders and size of Heterobasidion genets, a thorough analyses of excavated root systems would be required. Consequently, the size of the genets may actually be bigger.

All the largest H. parviporum genets of our study contained three to nine dead trees or stumps. Stumps colonized by Heterobasidion provide a substrate from which the fungus can spread and cause infection in standing trees [**21**]. A study by Capretti and Mugnai [**26**] indicated that dead wood (buried wood in their study) might be an important factor for extensive spread of the pathogen especially in former pasture soil. Recent studies show that root fragments are important for H. parviporum infection transfer [**75**]. Therefore, both abundance of dead wood and roots could be related with the spread of the pathogen in Stand 1 (former pastureland). The average expansion rate of territorial clones of H. parviporum on Stand 2 and 3 (former meadows) (0.25 and 0.10 m·year^{-1}) corresponds to results obtained by Bendz-Hellgren et al. [**19**], where the average H. annosum s.l. growth rate on former agricultural lands was 0.25 m·year^{-1} in spruce stump roots and 0.09 m·year^{-1} in living tree roots. The average H. parviporum growth rate in the Stand 1 (former pastureland) was high, 0.8 m·year^{-1}, in comparison with results on both forest and arable lands reported by Bendz-Hellgren et al. [**19**]. This might indicate an erroneous assumption of infection year (i.e., infection had occurred before first thinning via wounding). Since no wounds were observed on trunks, current primary infections probably occurred via root injuries. Spores of Heterobasidion are present in soil, where they can maintain their vitality at least for a year and infect injured tree roots [**32**]. Rönnberg et al. [**40**] reported a high incidence of Heterobasidion root rot in unthinned spruce stands planted on pastureland. In addition, Vollbrecht and Agestam [**28**] and Capretti and Mugnai [**26**] indicated that former pastures were highly susceptible to root rot. Capretti and Mugnai [**26**] also concluded that Heterobasidion growth rate in dead wood was particularly high in pasture lands. The increased capacity of H. parviporum to colonise dead wood and form fruitbodies in Stand 1 of our study is a probable scenario, explaining the greater expansion of genets.

In research sites, stumps were not treated against Heterobasidion root rot, which in a short period resulted in severe damage as reported above. Therefore, we suggest that stump treatment should be considered as a general practice on former pastures and meadows to prevent both primary and subsequent infections, as suggested by other authors [**41,43,62,64**]. This is particularly relevant for private forest owners who are often managing forests on former agricultural lands, but often do not utilise preventive stump treatments.

Further studies should focus on the presence of Heterobasidion spores in soils of former agricultural lands in relation to the soil microorganism community and root injuries.

5. CONCISE SUMMARY

This study collected novel data on the ecology and dissemination patterns of Heterobasidion parviporum in Norway spruce forests that were formerly used as pastures and meadows. Based on this new facts, we have reached the following conclusions: (i) Typically, primary infection through spores is more common in such stands, but secondary infection becomes more prevalent as the stand ages. (ii) H. parviporum has the ability to create large territorial clones (up to 137 m2) in woods that were previously used as pasture or meadow areas, leading to significant tree dieback and mortality. Thus, in order to minimize financial losses in spruce plantations on previously used pastures or meadows, it is advisable to employ stump treatment using biological or chemical control agents that can effectively prevent Heterobasidion infections during thinning operations. Facilitating the implementation of stump treatment for private forest owners, particularly those who own forests on former agricultural lands, is crucial.

6. AT A GLANCE

The reforestation of previously cultivated fields is a widespread activity in numerous nations. This practice is advantageous for preventing the transmission of illnesses from earlier generations of forests and for increasing the size of forested regions. Nevertheless, numerous biotic and abiotic hazards have been documented in these stands, notably an increased susceptibility to Heterobasidion root rot following thinning. This study aims to examine the dissemination of Heterobasidion root rot in three Picea abies (L.) Karst. plantations in Latvia that were previously used as pastures and meadows and are now managed as forests. At first, we collected samples from all the standing trees (a total of 157) located along transects inside the designated sampling regions in order to calculate the average infection rate. The prevalence of Heterobasidion root rot among stands ranged from 16 to 33% on the transects. Using the collected data and noting the observed dieback in all the areas, we set up circular sample plots in the centers of disease where we thoroughly examined all the trees and stumps. The infection rate in the circular plots ranged from 34% to 41% on average. The Heterobasidion isolates that were obtained were examined using primers specific to the species, and it was found that all of them were Heterobasidion parviporum Niemelä & Korhonen. Isolates were matched to determine the quantity and dimensions of genotypes. Out of the 141 genets that were analyzed, 99 were found to be present in only one tree or stump, while 42 genets were seen to include two or more trees or stumps, indicating the transmission of infection through root connections. The observed stands exhibited a range of 72 to 484 Heterobasidion genets per hectare. The following deductions were made: (i) Areas that were previously non-forest lands, such as pastures and meadows, can be very vulnerable to Heterobasidion infection after thinning. (ii) The pathogen has the ability to create expanding territorial clones in both former pastures and meadows. (iii) To prevent Heterobasidion infections, it is recommended to treat stumps with biological or chemical control agents.

REFERENCES

1. Forest Europe. State of Europe's Forests 2015. Summary for Policy Makers. In Proceedings of the Ministerial Conference on the Protection of Forests in Europe, Madrid, Spain, 20–21 October 2015. [**Google Scholar**]
2. Korhonen, K.; Stenlid, J. Biology of Heterobasidion annosum. In Heterobasidion annosum: Biology, Ecology, Impact and Control; Woodward, S., Stenlid, J., Karjalainen, R., Hütermann, A., Eds.; CAB International: Wallingford, UK, 1998; pp. 43–70. [**Google Scholar**]
3. Mańka, K.; Mańka, M. The control of fungal disease in forests planted on agricultural lands. In Proceedings of the XIII Czech and Slovak Plant Protection Conference, Prague, Czech Republic, 12–15 September 1994. [**Google Scholar**]
4. Sierota, Z. Heterobasidion root rot in forests on former agricultural lands in Poland: Scale of threat and prevention. Sci. Res. Essays 2013, 8, 2298–2305. [**Google Scholar**] [**CrossRef**]
5. Cukor, J.; Vacek, Z.; Linda, R.; Sharma, R.P.; Vacek, S. Afforested farmland vs. forestland: Effects of bark stripping by Cervus elaphus and climate on production potential and structure of Picea abies forests. PLoS ONE 2019, 14, e0221082. [**Google Scholar**] [**CrossRef**] [**PubMed**] [**Green Version**]
6. Rytter, L.; Lutter, R. Early growth of different tree species on agricultural land along a latitudinal transect in Sweden. For. Int. J. For. Res. 2020, 93, 376–388. [**Google Scholar**] [**CrossRef**]
7. Daugaviete, M.; Lazdina, D.; Bambe, B.; Lazdins, A.; Makovskis, K.; Daugavietis, U. Plantation Forests: A Guarantee of Sustainable Management of Abandoned and Marginal Farmlands. In Energy Efficiency and Sustainable Lighting—A Bet for the Future; Hermoso-Orzaez, M.J., Gago-Calderon, A., Eds.; IntechOpen: London, UK, 2020. [**Google Scholar**] [**CrossRef**]
8. Vollbrecht, G.; Johansson, U.; Eriksson, H.; Stenlid, J. Butt rot incidence, yield and growth pattern in a tree species experiment in south western Sweden. For. Ecol. Manag. 1995, 76, 87–93. [**Google Scholar**] [**CrossRef**]
9. Piri, T. The spreading of the S type of Heterobasidion annosum from Norway spruce stumps to the subsequent tree stand. Eur. J. For. Path. 1996, 26, 193–204. [**Google Scholar**] [**CrossRef**]
10. Woodward, S.; Stenlid, J.; Karjalainen, R.; Huttermann, A. Heterobasidion annosum: Biology, Ecology, Impact and Control; CAB International: Wallingford, UK, 1998. [**Google Scholar**]
11. Greig, B.J.W.; Gibbs, J.N.; Pratt, J.E. Experiments on the susceptibility of conifers to Heterobasidion annosum in Great Britain. For. Pathol. 2001, 31, 219–228. [**Google Scholar**] [**CrossRef**]
12. Gonthier, P.; Brun, F.; Lione, G.; Nicolotti, G. Modelling the incidence of Heterobasidion annosum butt rots and related economic losses in alpine mixed naturally regenerated forests of northern Italy. For. Pathol. 2012, 42, 57–68. [**Google Scholar**] [**CrossRef**]
13. Gonthier, P. Frequency of stump infections by Heterobasidion annosum s.l. and benefits from urea treatments vary with tree species and season in European Alpine forests. For. Ecol. Manag. 2019, 434, 76–86. [**Google Scholar**] [**CrossRef**]
14. Gaitnieks, T.; Arhipova, N.; Donis, J.; Stenlid, J.; Vasaitis, R. Butt rot incidence and related losses in Latvian Picea abies (L.) Karst. stands. In Proceedings of the 12th International Conference on Root and Butt Rot of Forest trees, Berkley, CA, USA, 12–19 August 2007; The University of California: Berkley, CA, USA, 2008; pp. 177–179. [**Google Scholar**]

15. Arhipova, N.; Gaitnieks, T.; Donis, J.; Stenlid, J.; Vasaitis, R. Butt rot incidence, causal fungi, and related yield loss in Picea abies stands of Latvia. Can. J. For. Res. 2011, 41, 2337–2345. [**Google Scholar**] [**CrossRef**]

16. Anonymous. Forest Industry in Numbers and Facts (Meža Nozare Skaitļos un Faktos); Biedrība "Zaļās Mājas": Riga, Latvia, 2020. (In Latvian) [**Google Scholar**]

17. Stenlid, J. Controlling and predicting the spread of Heterobasidion annosum from infected stumps and trees of Picea abies. Scand. J. For. Res. 1987, 2, 187–198. [**Google Scholar**] [**CrossRef**]

18. Vasiliauskas, R.; Stenlid, J. Spread of S and P group isolates of Heterobasidion annosum within and among Picea abies trees in central Lithuania. Can. J. For. Res. 1998, 28, 961–966. [**Google Scholar**] [**CrossRef**]

19. Bendz-Hellgren, M.; Brandtberg, P.-O.; Johansson, M.; Swedjemark, G.; Stenlid, J. Growth rate of Heterobasidion annosum in Picea abies established on forest land and arable land. Scand. J. For. Res. 1999, 14, 402–407. [**Google Scholar**] [**CrossRef**]

20. Lygis, V.; Vasiliauskas, R.; Stenlid, J. Planting Betula pendula on pine sites infested by Heterobasidion annosum: Disease transfer, silvicultural evaluation, and community of wood-inhabiting fungi. Can. J. For. Res. 2004, 34, 120–130. [**Google Scholar**] [**CrossRef**]

21. Piri, T.; Korhonen, K. The effect of winter thinning on the spread of Heterobasidion parviporum in Norway spruce stands. Can. J. For. Res. 2008, 38, 2589–2595. [**Google Scholar**] [**CrossRef**]

22. Allikmäe, E.; Laarmann, D.; Korjus, H. Vitality Assessment of Visually Healthy Trees in Estonia. Forests 2017, 8, 223. [**Google Scholar**] [**CrossRef**] [**Green Version**]

23. Vollbrecht, G.; Bilde Jørgensen, B. Modelling the incidence of butt rot in plantations of Picea abies in Denmark. Can. J. For. Res. 1995, 25, 1887–1896. [**Google Scholar**] [**CrossRef**]

24. Červinkova, H. Problems of decays of Norway spruce stands in the ČSSR. In Proceedings of the VIIIth Czechoslovak plant protection conference, Prague, Czech Republic, 9–11 September 1980; pp. 287–288. [**Google Scholar**]

25. Capretti, P.; Moriondo, F. Danni in alcuni impianti di conifer associate alla presenza di Heterobasidion annosum (Fomes annosus). Phytopath. Medit. 1983, 22, 157–167. [**Google Scholar**]

26. Capretti, P.; Mugnai, L. Saprophytic growth of Heterobasidion annosum on silver-fir logs interred in different types of forest soils. Eur. J. For. Pathol. 1989, 19, 257–262. [**Google Scholar**] [**CrossRef**]

27. Graber, S. Die Kernfäuleschäden an Fichte (Picea abies Karst.) in der Schweiz nördlich der Alpen: Untersuchungen über das Schadenausmass, die ökologischen, waldbaulichen und mykologischen Einflussfaktoren sowie die ökonomischen Auswirkungen. Beih. Schweiz. Z. Forstwes. 1996, 79, 283. [**Google Scholar**]

28. Vollbrecht, G.; Agestam, E. Modelling incidence of root rot in Picea abies plantations in Southern Sweden. Scand. J. For. Res. 1995, 10, 74–81. [**Google Scholar**] [**CrossRef**]

29. Mareš, R. The extent of root rot damage in Norway spruce stands established on fertile sites of former agricultural land. J. For. Sci. 2010, 56, 1–6. [**Google Scholar**] [**CrossRef**] [**Green Version**]

30. Froelich, R.C.; Dell, T.R.; Walkinshaw, C.H. Soil factores associated with Fomes annosus in the Gulf states. For. Sci. 1966, 12, 356–361. [**Google Scholar**]

31. Dmitri, L. Influence of nutrition and application of fertilizers on the resistance of forest plants to fungal diseases. Eur. J. For. Path. 1977, 7, 177–186. [**Google Scholar**] [**CrossRef**]

32. Brūna, L.; Kļaviņa, D.; Korhonen, K.; Zaļuma, A.; Burņeviča, N.; Gaitnieks, T. Effect of soil properties on the spread of Heterobasidion root rot. Proc. Latv. Acad. Sci. Sect. B Nat. Exact Appl. Sci. 2019, 73, 466–471. [**Google Scholar**] [**CrossRef**] [**Green Version**]

33. Manka, K. Some investigations on Fomes annosus Fr. carried out in Poland. In Proceedings of the Third International Conference on Fomes annosus, Aarhus, Denmark, 29 July–3 August 1968; Hodges, C.S., Rishbeth, J., Yde-Andersen, A., Eds.; USDA Forest Service: Asherville, NC, USA, 1970; pp. 83–90. [**Google Scholar**]

34. Gaitnieks, T.; Arhipova, N.; Nikolajeva, V.; Vulfa, L.; Klavina, D. Heterobasidion annosum izraisītā egļu sakņu trupe lauksaimniecības zemēs. (Heterobasidion annosum root rot of Norway spruce in agricultural soils). Mežzinātne 2008, 17, 22–37. (In Latvian) [**Google Scholar**]

35. Vasiliauskas, A. Root Fungus (Heterobasidion annosum) and the Resistance of Coniferous Forest Ecosystems; Mokslas Publishers: Vilnius, Lithuania, 1989. (In Russian) [**Google Scholar**]

36. Pešková, V.; Soukup, F.; Landa, J. Comparison of mycobiota of diverse aged spruce stands on former agricultural soil. J. For. Sci. 2009, 55, 452–460. [**Google Scholar**] [**CrossRef**] [**Green Version**]

37. Pratt, J.E.; Greig, B.J.W. Heterobasidion annosum: Development of butt rot following thinning in two young first rotation stands of Norway spruce. Forestry 1988, 61, 339–347. [**Google Scholar**] [**CrossRef**]

38. Vollbrecht, G.; Gemmel, P.; Pettersson, N. The effect of precommercial thinning on the incidence of Heterobasidion annosum in planted Picea abies. Scand. J. For. Res. 1995, 10, 37–41. [**Google Scholar**] [**CrossRef**]

39. Rönnberg, J.; Jørgensen, B.B. Incidence of root and butt rot in consecutive rotations of Picea abies. Scand. J. For. Res. 2000, 15, 210–217. [**Google Scholar**] [**CrossRef**]

40. Rönnberg, J.; Berglund, M.; Johansson, U.; Cleary, M. Incidence of Heterobasidion spp. following different thinning regimes in Norway spruce in southern Sweden. For. Ecol. Manag. 2013, 289, 409–415. [**Google Scholar**] [**CrossRef**]

41. Thor, M.; Stenlid, J. Heterobasidion annosum infection of Picea abies following manual or mechanized stump treatment. Scand. J. For. Res. 2005, 20, 154–164. [**Google Scholar**] [**CrossRef**]

42. Stenlid, J.; Redfern, D.B. Spread within the tree and stand. In Heterobasidion annosum: Biology, Ecology, Impact and Control; Woodward, S., Stenlid, J., Karjalainen, R., Hütterman, A., Eds.; CAB International: Wallingford, UK, 1998; pp. 125–143. [**Google Scholar**]

43. Garbelotto, M.; Gonthier, P. Biology, epidemiology, and control of Heterobasidion species worldwide. Annu. Rev. Phytopathol. 2013, 51, 39–59. [**Google Scholar**] [**CrossRef**] [**PubMed**] [**Green Version**]

44. Swedjemark, G.; Stenlid, J. Population dynamics of the root rot fungus Heterobasidion annosum following thinning of Picea abies. Oikos 1993, 66, 247–254. [**Google Scholar**] [**CrossRef**]

45. Zaļuma, A.; Muižnieks, I.; Gaitnieks, T.; Burņeviča, N.; Jansons, Ā.; Jansons, J.; Stenlid, J.; Vasaitis, R. Infection and spread of root rot caused by Heterobasidion spp. in Pinus contorta plantations in Northern Europe: Three case studies. Can. J. For. Res. 2019, 48, 969–977. **[Google Scholar]** **[CrossRef]** **[Green Version]**

46. Stenlid, J. Population structure of Heterobasidion annosum as determined by somatic incompatibility, sexual incompatibility and isozyme patterns. Can. J. Bot. 1985, 63, 2268–2273. **[Google Scholar]** **[CrossRef]**

47. Hantula, J.; Vainio, E. Specific primers for the differentiation of Heterobasidion annosum (s.str.) and H. parviporum infected stumps in northern Europe. Silva Fenn. 2003, 37, 181–187. **[Google Scholar]** **[CrossRef]** **[Green Version]**

48. Kåren, O.; Högberg, N.; Dahlberg, A.; Jonsson, L.; Nylund, J.E. Inter- and intraspecific variation in the ITS region of rDNA of ectomycorrhizal fungi in Fennoscandia as detected by endonuclease analysis. New Phytol. 1997, 136, 313–325. **[Google Scholar]** **[CrossRef]**

49. Mead, R.; Curnow, R.N. Statistical Methods in Agriculture and Experimental Biology; Chapman & Hall: London, UK, 1983. **[Google Scholar]**

50. Bates, D.; Maechler, M.; Bolker, B.; Walker, S. Fitting Linear Mixed-Effects Models Using lme4. J. Stat. Softw. 2015, 67, 1–48. **[Google Scholar]** **[CrossRef]**

51. Royston, P. An extension of Shapiro and Wilk's W test for normality to large samples. Appl. Stat. 1982, 31, 115–124. **[Google Scholar]** **[CrossRef]**

52. Labouriau, R. postHoc: Tools for Post-Hoc Analysis. R Package Version 0.1.3. 2020. Available online: **https://CRAN.R-project.org/package=postHoc** (accessed on 10 January 2020).

53. R Core Team. R: A Language and Environment for Statistical Computing; R Foundation for Statistical Computing: Vienna, Austria, 2020; Available online: **https://www.R-project.org/** (accessed on 10 January 2020).

54. Pettersson, M.; Rönnberg, J.; Vollbrecht, G.; Gemmel, P. Effect of thinning and Phlebiopsis gigantea stump treatment on the growth of Heterobasidion parviporum inoculated in Picea abies. Scand. J. For. Res. 2003, 18, 362–367. **[Google Scholar]** **[CrossRef]**

55. Thor, M.; Arlinger, J.; Stenlid, J. Heterobasidion annosum root rot in Picea abies: Modelling economic outcomes of stump treatment in Scandinavian coniferous forests. Scand. J. For. Res. 2006, 21, 414–423. **[Google Scholar]** **[CrossRef]**

56. Korhonen, K.; Piri, T. How cultivate Norway spruce on site infested by Heterobasidion? In Proceedings of the Nordic/Baltic Forest Pathology Meeting, Copenhagen, Denmark, 31 August–4 September 2002; Thomsen, I.M., Ed.; Danish Forest and Landscape Research Institute: Horsholm, Denmark, 2003; pp. 21–29. **[Google Scholar]**

57. Klavina, D.; Menkis, A.; Gaitnieks, T.; Velmala, S.; Lazdins, A.; Rajala, T.; Pennanen, T. Analysis of Norway spruce dieback phenomenon in Latvia—A belowground perspective. Scand. J. For. Res. 2016, 31, 156–165. **[Google Scholar]** **[CrossRef]**

58. Piri, T. Silvicultural Control of Heterobasidion Root Rot in Norway Spruce Forests in Southern Finland. Regeneration and Vitality Fertilization of Infected Stands. Ph.D. Thesis, Finnish Forest Research Institute METLA, University of Helsinki, Helsinki, Finland, 2003. **[Google Scholar]**

59. Varese, G.C.; Gonthier, P.; Nicolotti, G. Long-term effects on other fungi are studied in biological and chemical stump treatments in the fight against Heterobasidion annosum coll. Mycologia 2003, 95, 379–387. **[Google Scholar]** **[CrossRef]** **[PubMed]**

60. Vollbrecht, G.; Bilde Jørgensen, B. The effect of stump treatment on the spread rate of butt rot in Picea abies in Danish permanent forest yield research plots, Scand. J. For. Res. 1995, 10, 271–277. **[Google Scholar]** **[CrossRef]**

61. Oliva, J.; Samils, N.; Johansson, U.; Bendz-Hellgren, M.; Stenlid, J. Urea treatment reduced Heterobasidion annosum s.l. root rot in Picea abies after 15 years. For. Ecol. Manag. 2008, 255, 2876–2882. **[Google Scholar]** **[CrossRef]**

62. Blomquist, M.; Larsson-Herrera, S.; Hofmann, J.; Ceyd Beram, R.; Cleary, M.; Rönnberg, J. Size matters but is big always better? Effectiveness of urea and Phlebiopsis gigantea as treatment against Heterobasidion on Picea abies stumps of variable size. For. Ecol. Manag. 2020, 462, 117998. **[Google Scholar]** **[CrossRef]**

63. Brandtberg, P.-O.; Johansson, M.; Seeger, P. Effects of season and urea treatment on infection of stumps of Picea abies by Heterobasidion annosum in stands on former arable land. Scand. J. For. Res. 1996, 11, 261–268. **[Google Scholar]** **[CrossRef]**

64. Oliva, J.; Thor, M.; Stenlid, J. Long-term effects of mechanized stump treatment against Heterobasidion annosum root rot in Picea abies. Can. J. For. Res. 2010, 40, 1020–1033. **[Google Scholar]** **[CrossRef]**

65. Niemelä, T.; Korhonen, K. Taxonomy of the genus Heterobasidion. In Heterobasidion annosum: Biology, Ecology, Impact and Control; Woodward, S., Stenlid, J., Karjalainen, R., Hütterman, A., Eds.; CAB International: Wallingford, UK, 1998; pp. 27–35. **[Google Scholar]**

66. Hanso, S.; Korhonen, K.; Hanso, M. Attack of spruce and pine by S and P groups of Heterobasidion annosum on forest and former agricultural soils in Estonia. In Root and Butt Rots, Proceedings of the IUFRO Working Party S2.06.01, Wik, Sweden and Haikko, Finland, 9–16 August 1993; Johansson, M., Stenlid, J., Eds.; Swedish University of Agricultural Sciences: Uppsala, Sweden, 1994; pp. 254–259. **[Google Scholar]**

67. Korhonen, K.; Piri, T. The main hosts and distribution of S and P groups of Heterobasidion annosum in Finland. In Root and Butt Rots, Proceedings of the IUFRO Working Party S2.06.01, Wik, Sweden and Haikko, Finland, 9–16 August 1993; Johansson, M., Stenlid, J., Eds.; Swedish University of Agricultural Sciences: Uppsala, Sweden, 1994; pp. 260–267. **[Google Scholar]**

68. Piri, T.; Korhonen, K. Infection of advance regeneration of Norway spruce by Heterobasidion parviporum. Can. J. For. Res. 2001, 31, 937–942. **[Google Scholar]** **[CrossRef]**

69. Piri, T.; Valkonen, S. Incidence and spread of Heterobasidion root rot in uneven-aged Norway spruce stands. Can. J. For. Res. 2013, 43, 872–877. **[Google Scholar]** **[CrossRef]**

70. Piri, T. Early development of root rot in young Norway spruce planted on sites infected by Heterobasidion in southern Finland. Can. J. For. Res. 2003, 33, 604–611, 654. **[Google Scholar]** **[CrossRef]**

71. Oliva, J.; Bendz-Hellgren, M.; Stenlid, J. Spread of Heterobasidion annosum s.s. and Heterobasidion parviporum in Picea abies 15 years after stump inoculation. FEMS Microbiol. Ecol. 2011, 75, 414–429. **[Google Scholar]** **[CrossRef]** **[PubMed]**

72. Korhonen, K.; Bobko, I.; Hanso, S.; Piri, T.; Vasiliauskas, A. Intersterility groups of Heterobasidion annosum in some spruce and pine stands in Byelorussia, Lithuania and Estonia. Eur. J. For. Pathol. 1992, 22, 384–391. **[Google Scholar]** **[CrossRef]**

73. Piri, T.; Korhonen, K.; Sairanen, A. Occurrence of Heterobasidion annosum in pure and mixed spruce stands in Southern Finland. Scand. J. For. Res. 1990, 5, 113–125. **[Google Scholar]** **[CrossRef]**

74. Keca, N.; Keca, L. Distribution of Heterobasidion parviporum genets in Norway spruce forests in Serbia. In Proceedings of the 13th International Conference: Root and Butt Rot of Forest Trees, Firenze, Italy, 4–10 September 2011; Carpetti, P., Comparini, M., La Porta, N., Santini, A., Eds.; Firenze University Press: Firenze, Italy, 2013; pp. 155–158. [**Google Scholar**]

75. Piri, T.; Hamberg, L. Persistence and infectivity of Heterobasidion parviporum in Norway spruce root residuals following stump harvesting. For. Ecol. Manag. 2015, 353, 49–58. [**Google Scholar**] [**CrossRef**]

14 PRESENCE OF INVERTEBRATES WITHIN THE FRUITING BODIES OF HETEROBASIDION SPP., INFECTED PICEA ABIES LOGS, AND THE SURROUNDING SOIL

1. INTRODUCTION: AN OVERVIEW

Heterobasidion spp., which are pathogenic fungus, result in significant economic losses in forestry as they infect living trees and cause wood degradation. The disease has a more pronounced effect on Norway spruce (Picea abies) stands due to the ability of the fungal mycelia to extend several meters up the stem from the root system. This leads to rot in the most valuable section of the stem [1].

Approximately 22% of Picea abies stands in Latvia are affected by Heterobasidion spp. Furthermore, the economic loss per hectare at the time of final cutting reaches EUR 4000 [2]. Heterobasidion is disseminated through basidiospores during primary infections and through vegetative mycelia via root contacts during secondary infections [3]. The prevalence of fruitbodies of the pathogen in a specific area affects the initial infection caused by basidiospores. These fruitbodies mostly impact the infection rate within a specific stand, rather than in neighboring stands [4]. Both Heterobasidion parviporum Niemelä & Korhonen and Heterobasidion annosum (Fr.) Bref. are present in Latvia, and both of these species produce fruitbodies on infected P. abies wood at a comparable frequency [5]. Research undertaken in Germany and Finland [6,7] has shown that a significant number of fruiting bodies are produced on 3- to 4-year-old, large-sized P. abies logs that are left in the forest following logging activities. In Latvia, the average area covered by fruitbodies on 3- to 4-year-old P. abies logs is 3759 cm2/m3. However, on logs in stands growing on drained peat soils, this area can surpass 5000 cm2/m3 [5].

Heterobasidion fruitbodies have the capacity to generate a range of 35,000 to 151,000 basidiospores per square centimeter of the fruitbody within a span of one hour [8]. In P. abies stands, the spore count can surpass 1000 per square meter per hour [9]. The high density of these spores poses a significant risk to recently cut P. abies stumps during tree felling, as these new stumps are the main surface for Heterobasidion spores to establish colonies. Thus, in order to minimize the likelihood of infection, various studies have suggested the extraction of sizable P. abies logs that are contaminated by Heterobasidion spp. from forest areas [5,6,7]. Nevertheless, additional research has demonstrated that sizable P. abies logs play a crucial role as a home for numerous fungal and invertebrate species [10,11]. It is well-established that fungal fruitbodies are colonized by many invertebrate species [12,13].

Fruitbodies serve as both a food source and a habitat for invertebrates, impacting the population density of invertebrates in a specific environment [11]. The invertebrate fauna present in the fruitbodies of Fomitopsis pinicola (Fr.) Karst, a fungal species typically found in conifer forests, has been studied [14,15,16,17]. However, in the majority of research examining invertebrate fauna that live in fungal fruitbodies, many fungal species were analyzed without any distinction being made between them. Previous research has primarily focused on larger arthropods that consume fungus, specifically beetles (Coleoptera) and flies (Diptera) [17,18,19]. Additionally, Acari and Collembola are frequently seen living in fungal fruitbodies [20,21]. Several research have also analyzed Heterobasidion spp. fruitbodies, however most of them have focused on a limited number of fruitbodies. These studies only gathered adult Coleoptera, as they are reasonably straightforward to identify. In a study involving many fungal species, adult Coleoptera were discovered in just two fruitbodies of H. parviporum out of a total of 44 collected [22]. In another comparable study, only 50% of the studied H. parviporum fruitbodies were inhabited by Coleoptera [23].

The study demonstrated that Heterobasidion annosum was found to be connected with various species of Coleoptera from the Ciidae family [24]. Additionally, it was established that Cis lineatocribatus reproduces in fruitbodies, with both adults and larvae being identified [18]. Prior studies by Stenlid [25] and Piri et al. [26] have emphasized the potential involvement of insect vectors in spreading Heterobasidion spp. It is well established that the spores of Heterobasidion spp. are spread by the huge pine weevil (Hylobius abietis) [27,28], but the contribution of other insects in this process remains mainly unexplored. In order to assess the impact of removing infected P. abies wood from forest stands on invertebrate diversity, it is necessary to conduct a thorough evaluation of the invertebrate fauna found in Heterobasidion spp. fruitbodies. This examination should specifically concentrate on uncommon and endangered species. This study aimed to evaluate and compare the variety of invertebrates in Heterobasidion spp. fruitbodies, infected P. abies wood, and the surrounding soil. Additionally, the study aimed to design and test techniques for collecting invertebrate samples from Heterobasidion spp. fruitbodies.

2. MATERIALS AND METHODS

2.1. Sample Plots

Samples were collected from two sample plots in the autumn of 2016: (i) a 70-year-old forest stand consisting of 80% P. abies with an admixture of Betula pendula and Pinus sylvestris on peatland (forest type Oxalidosa turf.mel. [**29**]) near the Forest Research Station in the Kalsnava forest region (56°41′40.4″ N, 25°54′21.8″ E), further referred to as Kalsnava; (ii) a 50-year-old forest stand consisting of 90% P. abies with an admixture of B. pendula on mineral soil (forest type Oxalidosa [**29**]) in JSC Latvia's State ForestsCentral Daugava region, Ogre division (56°50′50.8″ N, 24°48′17.8″ E), further referred to as Ogre. Sample plots contained infected P. abies logs of different sizes and decomposition stages, where Heterobasidion spp. fruitbodies of varying sizes and ages were previously identified.

2.2. Field Methods

To sample fruitbodies, 46 infected P. abies logs with large Heterobasidion spp. fruitbodies (**Figure 1**) were randomly selected in each plot. For each log, the root end diameter and length were measured. The length of logs ranged between 107 and 487 cm (average 274 cm); diameters ranged between 15 and 42 cm (average 25 cm). Fruitbody samples (100 cm^2) were collected by cutting a 10 × 10 cm piece from the fruitbody closest to the root end of the log.

During the collection of fruitbody samples, it was noted if the fruitbody was in contact with the soil and if any damage was visible. A similarly sized (100 cm^2) wood sample was collected 20 cm from the fruitbody in the direction away from the root end of the log. One wood sample and at least one fruitbody sample was collected from each log. If multiple and well-separated fruitbodies were found on one log, two to three fruitbody samples were collected. Where the log was in contact with the soil, a 5 cm deep, 100 cm^2 soil sample was collected. If multiple logs were found less than 10 m from each other (in log piles), only one soil sample was collected. After collection, all samples were placed into plastic bags, and kept at +4 °C until processing. A total of 69 Heterobasidion spp. fruitbodies, 46 P. abies wood samples and 19 soil samples were collected. In Kalsnava, 46 fruitbodies, 30 P. abies wood samples and 15 soil samples were collected, while in Ogre, 23 fruitbodies, 16 P. abies wood samples and 4 soil samples were collected.

Figure 1. Large Heterobasidion spp. fruitbody.

2.3. Laboratory Methods

In the laboratory, each sample was weighed using a Kern EMB 200-2 balance (Balingen, Germany) before and after placing samples in Tullgren funnels (**Figure 2**), and macrofauna were collected (if present) using forceps. By weighing the samples before and after placing them in Tullgren funnels, the relative humidity of samples was calculated. The relative humidity of fruitbodies ranged from 40.6% to 85.6% with an average of 70.1%. In Ogre, the relative humidity of fruitbodies was on average 69.4%, but in Kalsnava it was on average 71.6%. The average weight of a fruitbody sample before drying was 68.7 g, maximal and minimal values being 151.0 g and 14.2 g, respectively. After drying, the weight varied between 5.1 g and 42.3 g with an average of 19.9 g.

Figure 2. Tullgren funnels utilized for the collection of invertebrates.

The thickness of fruitbodies was measured with a ruler. Most of the collected fruitbodies were between 0.4 and 2.5 cm thick, but some fruitbodies that were collected close to the root collar were up to 5.0 cm thick. The majority of fruitbodies collected were at least 3 years old with some being between 6 and 7 years old.

After measurements, samples were placed in Tullgren funnels for at least 72 h. Tullgren funnels (**Figure 2**) are commonly utilized to obtain quantitative data about arthropods in soil [**30**], but this method is also effective to obtain data on invertebrates in fruitbodies and wood [**14,31,32**]. A 1:3 mixture of glycerine and 70% ethanol was placed in the collection vials to preserve the collected invertebrate individuals.

The collected invertebrates were examined using Leica NBO MSV266 (Wetzlar, Germany) and Optika B-383PH (Ponteranica, Italy) microscopes. Invertebrates were classified to the lowest possible taxonomic level, most to family level. Springtails (Collembola) were classified according to W. Dunger [**33**], mites (Acari) according to unpublished material by Dr U. Kagainis on the classification of oribatid mites (Oribatida) morphogroups in Latvia. Beetles (Coleoptera) were classified to the family level using the key by D. M. Unwin [**34**]. Taking into account the available information on Coleoptera of the Ciidae family inhabiting Heterobasidion spp. fruitbodies, these were classified to species level using the publicly available key by Lompes [**35**], which was developed from the results of studies by Reibnitz [**24**].

2.4. Statistical Analyses

To analyse invertebrate diversity in each substrate, the invertebrate data (actual observations) were used to construct accumulation curves [**36**], which was conducted in R [**37**] using the vegan package and specaccum function [**38**]. Differences in invertebrate taxa richness in different substrates (study sites as well as adults and larvae combined) were compared by nonparametric chi-square tests, taking into account the Bonferroni correction [**39**]. Shannon index values [**40**] were calculated for each substrate to determine the diversity in invertebrate communities. Values were calculated using invertebrate frequencies by taxa, i.e., the subclass or family taxonomic level. The invertebrate community structure in different substrates (fruitbodies, wood and soil) was analysed using principal component analysis in Canoco v.5.02 (Microcomputer Power, Ithaca, NY, USA).

3. RESULTS

A total of 7311 invertebrate individuals were collected from all substrates and from both sample plots, including 7151 adults and 160 larvae. The number of invertebrates collected per individual fruitbody was between 1 and 396 (average 70). In wood and soil samples, invertebrate numbers ranged from 0 to 316 (average 35) and 0 to 192 (average 41), respectively. There were 55 different invertebrate taxa in 69 fruitbody samples, 47 taxa in 46 wood samples and 36 taxa in 19 soil samples. Consequently, the chi-square test showed that the absolute richness of invertebrate taxa differed significantly among all substrates, being highest in the soil, then in the wood and lowest in fungal fruitbodies ($p < 0.004$). A possible reason for the observed differences may be an uneven number of samples collected of different substrates. Indeed, when the same number of samples was taken from each substrate following the rarefaction analysis, the richness of invertebrate taxa did not differ significantly among different substrates ($p > 0.05$).

The accumulation curves for different substrates did not reach the asymptote (**Figure 3**).

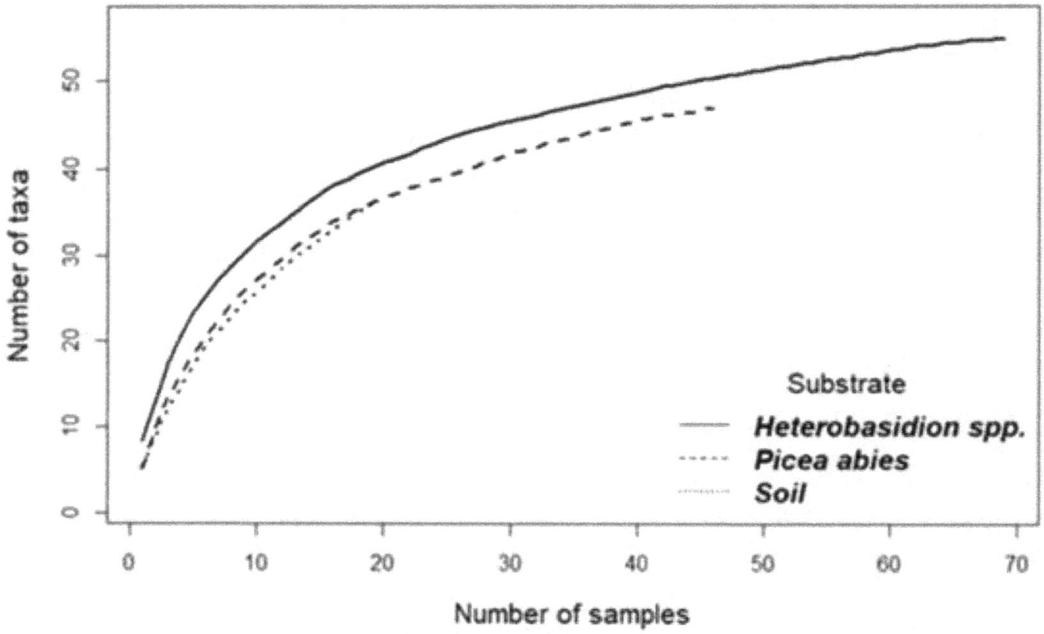

Figure 3. Accumulation curves showing the relationship between the cumulative number of invertebrate taxa and the number of collected samples.

The largest number of invertebrates was collected from Heterobasidion spp. fruitbodies—4886 adults and 101 larvae (**Table 1**). The dominant invertebrate taxa were Acari (82.4%), with the other invertebrates being Collembola (6.2%), Coleoptera (4.7%), Diplopoda (2.5%) and Diptera (1.8%). In addition, Haplotaxida (0.8%), Thysanoptera (0.7%), Hymenoptera (0.3%), Aranea (0.2%), Gastropoda (0.1%) and Chilopoda (0.1%) were found in fruitbodies. Several invertebrate groups were represented by very few individuals in fruitbodies—Lepidoptera, Isopoda and Hemiptera (0.04% each) and Protura and Opiliones (0.02% each).

Table 1. Occurrence and relative abundance (%) of invertebrate taxa identified in Heterobasidion fruitbodies, Picea abies wood and soil samples. Absolute numbers for higher taxonomic groups are shown in parentheses.

Invertebrate Taxa	*Heterobasidion* Fruitbodies		*Picea abies* Wood		Soil	
Subclass/Family	Adults	Larvae	Adults	Larvae	Adults	Larvae
Acari	(4111)		(1277)		(634)	
Hypochthoniidae	0.6	-	2.6	-	1.4	-
Pthiracaridae	0.9	-	4.8	-	2.2	-
Euphthiracaridae	0.2	-	5.9	-	0.3	-
Crotonioidae	0.7	-	0.1	-	1.4	-
Nanhermanniidae	0.1	-	0.3	-	0.4	-
Damaeidae	0.2	-	0.1	-	0.7	-
Eremaeidae	0.1	-	0.1	-	-	-
Tenuialidae	0.04	-	0.1	-	0.3	-
Xenillidae	0.02	-	0.2	-	-	-
Carabodidae	65.9	-	12.4	-	11.4	-
Oppiidae	7.3	-	39.2	-	32.7	-
Scheloribatidae	-	-	-	-	0.6	-
Ceratozetidae	0.8	-	5.3	-	34.9	-
Galumnidae	0.04	-	0.1	-	0.6	-
Other [1]	7.1	-	10.3	-	4.3	-
Collembola	(309)		(156)		(24)	
Hypogastruridae	0.2	-	0.2	-	0.1	-
Neanuridae	4.2	-	6.4	-	2.3	-
Onychiuridae	0.8	-	1.5	-	0.4	-
Entomobryidae	0.1	-	1.2	-	0.3	-
Isotomidae	0.7	-	0.1	-	0.3	-
Tomoceridae	0.2	-	0.4	-	-	-
Sminthuridae	0.02	-	0.1	-	-	-
Coleoptera	(170)	(62)	(12)	(29)	(7)	(3)
Anobiidae	0.02	-	-	-	-	-
Cantharidae	-	17.8	-	16.7	-	40.0
Carabidae	0.1	-	-	-	-	-
Ciidae	3.0	11.9	-	1.9	-	-
Cryptophagidae	-	-	-	-	0.1	-
Elateridae	0.02	23.8	-	33.3	-	-

Invertebrate Taxa	*Heterobasidion* Fruitbodies		*Picea abies* Wood		Soil	
Pselaphidae	-	-	0.1	-	0.1	-
Ptilidae	0.1	-	0.5	-	-	-
Silphidae	0.1	-	0.1	-	-	-
Staphylinidae	0.2	7.9	0.1	1.9	0.7	20.0
Diplopoda	(126)		(50)		(1)	
Blaniulidae	2.3	-	2.9	-	0.1	-
Julidae	0.02	-	-	-	-	-
Polydesmidae	0.3	-	0.3	-	-	-
Diptera	(50)	(39)	(10)	(22)	(3)	(1)
Bibionidae	0.02	0.9	-	-	-	-
Brachycera	0.1	15.8	-	7.4	-	20.0
Cecidomyiidae	0.04	2.9	-	-	-	-
Ceratopogonidae	0.1	-	-	3.7	-	-
Chiromidae	0.4	14.8	0.3	25.9	-	-
Drosophilidae	-	-	0.1	-	-	-
Limoniidae	-	-	-	3.7	-	-
Psychodidae	0.02	-	-	-	-	-
Rhagionidae	-	-	-	-	0.1	-
Sciaridae	0.3	0.9	0.3	-	0.3	-
Stratiomyidae	0.02	2.9	-	-	-	-
Haplotaxida	(41)		(18)		(8)	
Lumbricidae	0.4	-	0.6	-	0.7	-
Enchytraeidae	0.4	-	0.5	-	0.4	-
Thysanoptera	0.7 (35)	-	0.6 (10)	-	0.1 (1)	-
Hymenoptera	(15)		(7)		(8)	
Parasitica	0.3	-	0.4	-	1.2	-
Formicidae	0.02	-	-	-	-	-
Aranea	0.2 (10)	-	0.5 (8)	-	0.4 (3)	-
Gastropoda	(6)		(6)		(1)	
Arionidae	0.02	-	-	-	-	-
Clausiliidae	0.02	-	-	-	0.1	-
Zonitiidae	0.1	-	0.4	-	-	-
Chilopoda	0.1 (5)	-	0.2 (3)	-	0.1 (1)	-
Hemiptera	0.04 (2)	-	0.1 (1)	-	0.3 (2)	-
Lepidoptera	(2)			(3)		(1)

Invertebrate Taxa	*Heterobasidion* Fruitbodies		*Picea abies* Wood		Soil	
Tineidae	0.04	-	-	5.5	-	20.0
Isopoda	(2)		(2)			
Trichoniscidae	0.04	-	0.1	-	-	-
Opiliones	0.02 (1)	-	-		-	-
Protura	0.02 (1)	-	-	-	-	-
Pseudoscorpionida	-	-	0.6 (10)	-	0.3 (2)	-

[1] Acari individuals that were not identified to family level.

Coleoptera of the Ciidae family were most prevalent in fruitbodies (86.8% of all beetles collected): Cis nitidus (27.6% of all beetles collected), C. castaneus (19.4%), C. lineatocribatus (18.8%), C. jacquemartii (14.7%) and C. glabratus (5.8%). Adult Phosphuga atrata, Daliopius marginatus and Oxypselaphus obscurus were observed, as well as larvae of several beetle families, i.e., Elateridae, Cantharidae, Ciidae and Staphylinidae. Larvae of these taxa were also found in wood samples; however, they were in a smaller number.

Three Diplopoda species were found in fruitbodies. The most prevalent species was Proteroiulus fuscus, which was twice as prevalent than in wood samples. One Cylindroiulus caeruleocinctus individual was found only in fruitbodies, as well as thirteen Polydesmus denticalatus individuals. Three earthworm species Lumbricus castaneus, L. rubellus and L. terrestris were found in fruitbodies. These species were also found in wood samples; however, they were in a smaller number.

The second largest number of invertebrates was collected from wood samples, i.e., 1570 adults and 54 larvae (**Table 1**). The dominant invertebrate taxa were Acari (78.6%), with the other invertebrates being Collembola (9.6%), Diplopoda (3.1%), Coleoptera (2.5%), Diptera (1.9%) and Haplotaxida (1.1%). Several invertebrate groups were represented by a few individuals in wood samples, which were Pseudoscorpionida and Thysanoptera (0.6% each), Aranea (0.5%), Hymenoptera (0.4%), Gastropoda (0.4%) Lepidoptera larvae (0.2%), Chilopoda (0.2%), Isopoda (0.1%) and Hemiptera (0.1%).

The smallest number of samples were collected from soil samples, i.e., 695 adults and five larvae (**Table 1**). As in the fruitbody and wood samples, Acari were the dominant invertebrate taxon (90.6%). Other invertebrates were present in small numbers: Collembola (3.4%), Coleoptera (1.4%), Haplotaxida (1.1%) and Hymenoptera (1.1%). Several invertebrate groups were represented by very few individuals in soil samples: Diptera (0.6%), Aranea (0.4%), Pseudoscorpionida (0.3%) and Hemiptera (0.3%), and Chilopoda, Diplopoda, Thysanoptera, Lepidoptera larvae and Gastropoda (0.14% each). Shannon index values were 2.37 in wood samples, 1.63 in fruitbodies and 1.95 in soil samples.

Dominance of one invertebrate taxonomic group was pronounced with Acari dominating in all substrates, and Collembola was the second most common group (**Table 1**). The third most prevalent taxonomic group varied between substrates, i.e., in Heterobasidion spp. samples it was Coleoptera, in wood it was Diplopoda and in soil it was Coleoptera and Haplotaxida. Among the 62 invertebrate taxa found in the present study, eleven (17.7%) were exclusively found in fruitbodies, two (3.2%) in the wood and three (4.8%) in the soil, while thirty (48.4%) were common to all substrates (**Figure 4**). There were 43 shared invertebrate taxa between fruitbodies and wood, 31 between fruitbodies and soil and 32 between wood and soil (**Figure 4**).

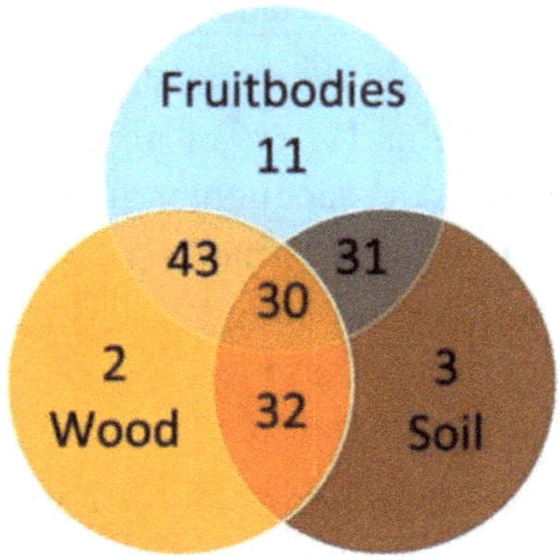

Figure 4. Venn diagram showing the diversity and overlap of invertebrate taxa in different substrates (sites as well as adults and larvae are combined).

In principal component analysis (PCA) of invertebrate communities, axis 1 explained 81.7% of variation and axis 2 explained 18.3%. PCA showed that invertebrate communities from different substrates were separated from each other on axis 1, i.e., on the axis, which explained most of the variation (**Figure 5**).

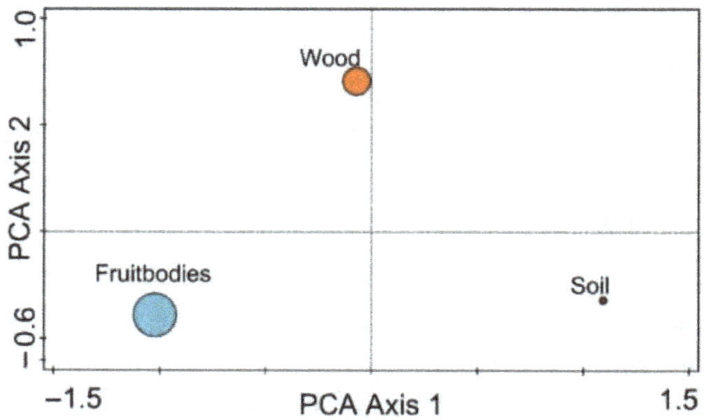

Figure 5. Ordination diagram based on principal component analysis of invertebrate communities from Heterobasidion spp. fruitbodies, Picea abies wood and soil samples. The size of each point reflects the relative richness of invertebrate taxa.

4. DISCUSSION

The taxonomic composition of invertebrates found in Heterobasidion spp. samples reflects the biology of fruitbodies. Fruitbodies develop on the part of the infected substrate that are in contact with the soil [7]. If the wood substrate is embedded in the soil, the fruitbody develops in a narrow line at the point of contact between the soil and the wood [5]. Therefore, the dominant invertebrate taxa in fruitbodies were also present in wood and soil samples (**Table 1**, **Figure 4**). Although the majority of invertebrate taxa were shared among different substrates, the PCA showed that invertebrate communities in different substrates were separated from each other, which was likely due to differences in the abundance data of different taxa (**Table 1**), thereby demonstrating the specificity and substrate preferences of different invertebrates. Our data show that Acari were the predominant invertebrate taxa found in Heterobasidion spp. fruitbodies (**Table 1**). This is consistent with other reports about fauna of invertebrates found in Heterobasidion spp. [**32**] and F. pinicola [**14**] fruitbodies. Some of the Acari that were not identified to family level belong to the order Mesostigmata, which was previously identified in fruitbodies from other species, such as F. pinicola [**41**]. Acari of the genus Carabodes were the most prevalent in fruitbodies. The presence of this genus in fruitbodies has been studied extensively in coniferous forests in Norway [**14,42**], and many taxa inhabiting soil also occupy fruitbodies as an additional food resource. This was also observed for Acari collected from Heterobasidion spp. [**32**].

Several studies have noted that Heterobasidion spp. fruitbody development is promoted by shade and moisture [**5,43,44**]. In this study, invertebrate groups that prefer moist environments, such as Haplotaxida, Diplopoda and Chilopoda, were also frequently detected in fruitbodies. However, these invertebrates are probably incidental and temporary inhabitants of fruitbodies even though their numbers in the analysed samples was relatively high (**Table 1**). Diplopoda and Chilopoda are prone to dehydration, and therefore seek moist places to forage [**45**]. The relative moisture content of the fruitbodies analysed in this study was 70%, and water drained from them when compressed. A portion of the Haplotaxida was collected from the oldest peripheries of the fruitbodies, where other invertebrates had already formed tunnels.

Heterobasidion spp. form perennial fruitbodies with a layered, cork-like structure [**3**]. Several Diptera adults and larvae were identified in fruitbodies, however the number of individuals collected was relatively small in comparison with the total number of invertebrates collected, which could be due the structure of fruitbodies. Diptera most often inhabit softer, fruitbodies of annual fungi that decay faster [**46**]. However, despite the structural differences of various fungal fruitbodies, our study is in agreement with the results from other studies [**47,48**], i.e., showing the low level of specialization between Diptera. Therefore, fungal fruitbodies can be colonized by various groups of Diptera. The number of amassed Diptera could be affected by the time of sampling as these insects are mostly inactive during the dormancy period.

One of the aims of this study was to compare invertebrate fauna inhabiting fruitbodies, rotting wood and soil. While these substrates are in close contact with each other in nature, differences in the invertebrate fauna inhabiting them were observed (**Table 1**, **Figure 4** and **Figure 5**). These differences may be due to the season when samples were collected (October). Assessment of invertebrate fauna during the vegetation period may be more representative. Each substrate had a characteristic invertebrate fauna profile (**Figure 5**), and the number of individuals and taxa collected varied between different substrates. For example, Acari of the genus Carabodes were most prevalent in fruitbodies but were less common in wood and soil samples (**Table 1**). By contrast, some other Acari were most often found in wood and soil samples (Oppiidae and Ceratozetidae, respectively) but were less commonly found in fruitbodies. In total, 90% of invertebrate individuals collected from soil samples were Acari, while in fruitbody and wood samples the proportion of Acari was lower (82% and 78%, respectively). Although the majority of invertebrate taxa were found in all substrates, some characteristic taxa were only found in fruitbodies (e.g., Ciidae), but not in wood or soil samples (**Figure 4**).

In studies of polypore inhabitants, usually the entire fruitbody is collected [**17**,**32**]. Studies in Latvia on the development of Heterobasidion spp. fruitbodies on decaying P. abies wood indicate that the size of fruitbodies can vary widely, and older fruitbodies coalesce with each other [**5**]. Therefore, in this study, only partial fragments of fruitbodies were collected. Although Heterobasidion spp. form large fruitbodies on windthrown trees and large dimension logging debris in favourable conditions [**3**,**5**], collection of a large number of 10 cm × 10 cm fragments of Heterobasidion spp. fruitbodies is difficult. Although the area of collected fruitbodies was uniform (100 cm^2), the volume of fruitbodies varied. This was due to differing thicknesses of the collected fruitbodies, which is influenced by the age of fruitbodies. In addition, the dry weight of collected fruitbodies varied between 5.1 g and 47.3 g, which has also been reported to influence invertebrate numbers and diversity in H. annosum fruitbodies [**32**]. The obtained data indicate that in further research, a larger number of samples of smaller size could be collected.

The Shannon diversity index values indicated that invertebrate diversity in fruitbodies was moderate. As mentioned before, the results of the study were influenced by the season when samples were collected. Heterobasidion spp. sporulate throughout the year, however, the maximum rate is between June and August [**49**,**50**]. Collection of fruitbodies at this time may identify invertebrate taxa that feed on spores at some developmental stage or taxa that are inactive during colder periods of the year. Analysis of invertebrates present in fruitbodies during peak sporulation times may identify potential vectors that can influence the primary infection pathways of this pathogen [**51**]. Previous studies on the role of different invertebrates as vectors of Heterobasidion have mainly focused on fungal conidia [**3,52**], but further studies are required to better understand the nature of such associations.

In this study, samples were placed in Tullgren funnels to collect invertebrates for three days. In total, 95% of the individuals collected were collected within the first 24 h. Some reports [**45,53**] indicate that in order to identify certain invertebrates, samples need to be held in Tullgren funnels for a longer period, for example, to allow development of larvae. Some Diptera and Coleoptera larvae were collected in this study, however, the proportion of larvae of the total number of individuals collected was low (2% in fruitbodies). To obtain a more comprehensive assessment of invertebrate fauna inhabiting Heterobasidion spp., methods that allow the development of larvae should be considered, by not preserving collected individuals in glycerine and placing collected larvae in rearing chambers, or by placing fruitbodies in vials and collecting emerged insects. In addition, Tullgren funnels are only partially effective for the collection of mobile invertebrates such as Collembola and should be used in conjunction with other collection methods [**54**], for example, by covering funnels. In this study, Collembola were observed on the surface of the samples during both sample sorting and collection in Tullgren funnels.

The stage of wood decay also influences invertebrate numbers and diversity, similarly to the decay stage of fruitbodies. The majority of organisms associated with polypores utilise fruitbodies in the later succession stages [**11**]. During the sterile phase, when the fungal hymenium layer is developing, perennial polypore species produce chemicals that protect them against invertebrate damage, which ceases when the fruitbody starts to decay [**14**]. Invertebrates may only start to inhabit fruitbodies after several years. In addition, fruitbodies on a single log do not start to decay simultaneously, ensuring suitable conditions for invertebrates that colonise decaying fruitbodies, even over several decades. Studies in Latvia have identified Heterobasidion spp. fruitbodies in various stages of decay on more than ten-year-old rotting spruce wood [**5**].

Ciidae was the most prevalent Coleoptera family collected from fruitbodies. This is a functionally important group that inhabits fruitbodies during both larval and adult stages, and several successive generations are often found in a single fruitbody. The diversity of Ciidae species is significantly influenced by host organism incidence within a territory, as well as fruitbody developmental stages [**17,55**]. Cis bidentatus was the only Coleoptera species collected from H. parviporum fruitbodies in Finland, and that colonised decaying fruitbodies or decaying parts of still living fruitbodies [**23**]. However, this species was not identified in this study.

While a comparatively small number of Cis glabratus individuals were identified in this study, this species has been reported to be prevalent in H. annosum [**24**] and F. pinicola fruitbodies [**15,17**]. In addition to species collected in this study (C. lineatocribatus, C. nitidus, C. glabratus and C. castaneus), another study also reported the presence of C. bidentatus and Ennearthron cornutum on Heterobasidion spp. [**24**].

According to the European Union Council directive 92/43/EEK (Protecting Europe's biodiversity (Natura 2000)), no rare or protected species of invertebrates were identified in this study.

Further research is needed to reveal the effect of the decay stage of wood and fruitbodies, as well as the effect of wood dimensions on the invertebrate fauna found in Heterobasidion spp. fruitbodies. Differences in the invertebrate fauna inhabiting decaying wood in areas of mineral and peat soils could be assessed by collecting samples from forest stands of different forest types.

5. CONCISE SUMMARY

The findings revealed the existence of a diverse community of invertebrates living in Heterobasidion fruitbodies, decomposed P. abies logs, and the surrounding soil. While many invertebrates were found across diverse substrates, their abundance varied significantly, indicating distinct preferences of certain species for particular substrates.

6. AT A GLANCE

Heterobasidion spp., which are pathogenic fungus, create noticeable fruitbodies on wood that has been infected. These fruitbodies can serve as a habitat for many creatures, including invertebrates. The objective was to evaluate and contrast the richness of invertebrates in Heterobasidion spp. fruitbodies, infected Picea abies wood, and the surrounding soil. Additionally, methods for collecting invertebrates from Heterobasidion spp. fruitbodies were tested. A grand total of 69 Heterobasidion fruitbody samples, 46 wood samples, and 19 soil samples were gathered from two forest sites in Latvia. A total of 7311 invertebrate individuals were gathered, encompassing 62 distinct taxa. Of them, 55 were found in fruitbodies, 47 in wood, and 36 in soil. The primary invertebrates found in fruitbodies were Acari (82.4%), Collembola (6.2%), and Coleoptera (4.7%). In wood, the leading invertebrates were Acari (78.6%), Collembola (9.6%), and Diplopoda (3.1%). In soil, the prevailing invertebrates were Acari (90.6%), Collembola (3.4%), and Coleoptera (1.4%). Out of all the distinct groups, 17.7% were only found in fruitbodies, 3.2% in wood, 4.8% in soil, and 48.4% were found in several substrates. While there were similarities in the types of invertebrate groups found, their distribution varied significantly across different surfaces. The employed sampling methods yielded useful data on invertebrate communities. Additional refinement of these techniques may be necessary to acquire more intricate and precise data regarding the invertebrates residing in various ecosystems.

REFERENCES

1. Garbelotto, M.; Gonthier, P. Biology, epidemiology, and control of Heterobasidion species worldwide. Annu. Rev. Phytopathol. 2013, 51, 39–59. [**Google Scholar**] [**CrossRef**] [**PubMed**] [**Green Version**]
2. Gaitnieks, T.; Arhipova, N.; Donis, J.; Stendlid, J.; Vasaitis, R. Butt rot incidence and related losses in Latvian Picea abies (L.) Kasrt. stands. In Proceedings of the 12th International Conference on Root and Butt Rots, 12–19 August 2007; The University of California: Berkley, CA, USA, 2008; pp. 177–179. [**Google Scholar**]
3. Korhonen, K.; Stenlid, J. Biology of Heterobasidion annosum. In Heterobasidion Annosum: Biology, Ecology, Impact and Control; Woodward, S., Stenlid, J., Karjalainen, R., Hüttermann, A., Eds.; CAB International: Wallingford, UK, 1998; pp. 43–70. [**Google Scholar**]
4. Stenlid, J. Regional differentiation in Heterobasidion annosum. In Proceedings of the Root and Butt Rots: Proceedings of the International Conference on Root and Butt Rots (IUFRO) Working Party S2.06.01, Wik, Sweden and Haikko, Finland, 9–16 August 1993; Johansson, M., Stenlid, J., Eds.; Swedish University of Agricultural Sciences: Uppsala, Sweden, 1994; pp. 243–248. [**Google Scholar**]
5. Gaitnieks, T.; Bruna, L.; Zaluma, A.; Burnevica, N.; Klavina, D.; Legzdina, L.; Jansons, J.; Piri, T. Development of Heterobasidion spp. fruit bodies on decayed Piecea abies. For. Ecol. Manag. 2021, 482, 118835. [**Google Scholar**] [**CrossRef**]
6. Schütt, P.; Schuck, H.J.; Lautenschlager, K.; Prestle, W.; Stimm, B. Fomes annosus sporocarps—Their abundance on decayed logs left in the forest. Eur. J. For. Pathol. 1979, 9, 57–61. [**Google Scholar**] [**CrossRef**] [**Green Version**]
7. Müller, M.M.; Heinonen, J.; Korhonen, K. Occurrence of Heterobasidion basidiocarps on cull pieces of Norway spruce left on cutting areas and in mature spruce stands. For. Pathol. 2007, 37, 374–386. [**Google Scholar**] [**CrossRef**]
8. Möykkynen, T.; Von Weissenberg, K.; Pappinen, A. Estimation of dispersal gradients of S- and P-type basidiospores of Heterobasidion annosum. Eur. J. For. Pathol. 1997, 27, 291–300. [**Google Scholar**] [**CrossRef**]
9. Gonthier, P.; Garbelotto, M.M.; Nicolotti, G. Seasonal patterns of spore deposition of Heterobasidion species in four forests of the western Alps. Phytopathology 2005, 95, 759–767. [**Google Scholar**] [**CrossRef**] [**Green Version**]
10. Penttilä, R.; Siitonen, J.; Kuusinen, M. Polypore diversity in managed and old-growth boreal Picea abies forests in southern Finland. Biol. Conserv. 2004, 117, 271–283. [**Google Scholar**] [**CrossRef**]
11. Stokland, J.N.; Siitonen, J.; Jonsson, B.G. Biodiversity in Dead Wood; Cambridge University Press: Cambridge, UK, 2012. [**Google Scholar**]

12. Hanski, I. Fungivory: Fungi, Insects and Ecology. In Insect Fungus Interactions, Proceedings of the 14th Symposium of the Royal Entomological Society of London in Collabaration with the British Mycological Society, London, UK, 16–17 September 1989; Wilding, N., Collins, N.M., Hammond, P.M., Webber, J.F., Eds.; Academic Press: London, UK, 1989; pp. 25–61. [**Google Scholar**]
13. Hammond, P.; Lawrence, J.F. Appendix. Mycophagy in insects: A summary. In Insect Fungus Interactions, Proceedings of the 14th Symposium of the Royal Entomological Society of London in Collabaration with the British Mycological Society, London, UK, 16–17 September 1989; Wilding, N., Collins, N.M., Hammond, P.M., Webber, J.F., Eds.; Academic Press: London, UK, 1989; pp. 275–324. [**Google Scholar**]
14. Hågvar, S.; Steen, R. Succession of beetles (genus Cis) and oribatid mites (genus Carabodes) in dead sporocarps of the red-banded polypore fungus Fomitopsis pinicola. Scand. J. For. Res. 2013, 28, 436–444. [**Google Scholar**] [**CrossRef**]
15. Jonsell, M.; Nordlander, G. Insects in polypore fungi as indicator species: A comparison between forest sites differing in amounts and continuity of dead wood. For. Ecol. Manag. 2002, 157, 101–118. [**Google Scholar**] [**CrossRef**]
16. Komonen, A.; Jonsell, M.; Okland, B.; Sverdrup-Thygeson, A.; Thunes, K. Insect assemblage associated with the polypore Fomitopsis pinicola: A comparison across Fennoscandia. Entomol. Fenn. 2004, 15, 102–112. [**Google Scholar**] [**CrossRef**]
17. Thunes, K.H.; Midtgaard, F.; Gjerde, I. Diversity of coleoptera of bracket fungus Fomitopsis pinicola in a Norweigian spruce forest. Biodivers. Conserv. 2000, 9, 833–852. [**Google Scholar**] [**CrossRef**]
18. Benick, L. Pilzkäfer und Käferpilze. Ökologische und statistische Untersuchungen. Acta Zool. Fenn. 1952, 70, 1–250. [**Google Scholar**]
19. Okland, B. Insect fauna compared between six polypore species in a southern Norwegian spruce forest. Fauna Nor. Ser. B 1995, 42, 21–26. [**Google Scholar**]
20. Graves, R.C.; Graves, A.C.F. The Insects and Other Inhabitants of Shelf Fungi in the Southern Blue Ridge Region of Western North Carolina. I. Introduction, Description of the Habitat, and the Host Fungi. Ann. Entomol. Soc. Am. 1966, 59, 381–390. [**Google Scholar**] [**CrossRef**]
21. Takahashi, K.H.; Tuno, N.; Kagaya, T. Abundance of mycophagous arthropods present on different species of fungi in relation to resource abundance at different spatial scales. Eur. J. Entomol. 2005, 102, 39–46. [**Google Scholar**] [**CrossRef**] [**Green Version**]
22. Schigel, D.S.; Niemelä, T.; Kinnunen, J. Polypores of western Finnish Lapland and seasonal dynamics of polypore beetles. Karstenia 2006, 46, 37–64. [**Google Scholar**] [**CrossRef**] [**Green Version**]
23. Schigel, D.S. Polypore–beetle associations in Finland. Fenn. Ann. Zool. Fenn. 2011, 48, 319–348. [**Google Scholar**] [**CrossRef**]
24. Reibnitz, J. Verbreitung und Lebensräume der Baumschwammfresser Südwestdeutschlands (Coleoptera: Cisidae). Mitt. Entomol. Ver. Stuttg. 1999, 34, 1–76. [**Google Scholar**]
25. Stenlid, J. Population structure of Heterobasidion annosum as determined by somatic incompatibility, sexual incompatibility, and isoenzyme patterns. Can. J. Bot. 1985, 63, 2268–2273. [**Google Scholar**] [**CrossRef**]

26. Piri, T.; Korhonen, K.; Sairanen, A. Occurrence of Heterobasidion annosum in Pure and Mixed Spruce Stands in Southern Finland. Scand. J. For. Res. 1990, 5, 113–125. [**Google Scholar**] [**CrossRef**]

27. Kadlec, Z.; Stary, P.; Zumr, V. Field evidence for the large pine weevil, Hylobius abietis as a vector of Heterobasidion annosum. Eur. J. For. Pathol. 1992, 22, 316–318. [**Google Scholar**] [**CrossRef**]

28. Drenkhan, T.; Sibul, I.; Kasanen, R.; Vainio, E.J. Viruses of Heterobasidion parviporum persist within their fungal host during passage through the alimentary tract of Hylobius abietis. For. Pathol. 2013, 43, 317–323. [**Google Scholar**] [**CrossRef**]

29. Bušs, K. Forest ecosystem classification in Latvia. Proc. Latv. Acad. Sci. Sect. B Nat. Exact Appl. Sci. 1997, 51, 204–218. [**Google Scholar**]

30. Crossley, D.A.; Blair, J.M. A high-efficiency, "low-technology" Tullgren-type extractor for soil microarthropods. Agric. Ecosyst. Environ. 1991, 34, 187–192. [**Google Scholar**] [**CrossRef**]

31. Lindquist, E.E. Remarkable convergence between two taxa of ascid mites (Acari: Mesostigmata) adapted to living in pore tubes of bracket fungi in North America, with description of Mycolaelaps new genus. Can. J. Zool. 1995, 73, 104–128. [**Google Scholar**] [**CrossRef**] [**Green Version**]

32. O'Connell, T.; Bolger, T. Fungal fruiting bodies and the structure of fungus-micro-arthropod assemblages. Biol. Environ. 1997, 97, 249–262. [**Google Scholar**]

33. Dunger, W.; Fiedler, H.J. Methoden der Bodenbiologie; Gustav Fischer Verlag: Stuttgart, Germany, 1997. [**Google Scholar**]

34. Unwin, D.M. A key to the families of British Beetles. Field Stud. Counc. 1988, 166, 149–197. [**Google Scholar**]

35. Lompe, A. Käfer Europas. Familir Ciidae. Available online: **http://www.coleonet.de/coleo/texte/ciidae.htm** (accessed on 13 May 2017).

36. Colwell, R.K.; Chang, X.M.; Jing, C. Interpolating, extrapolating and comparing incidence-based species accumulation curves. Ecology 2004, 85, 2717–2727. [**Google Scholar**] [**CrossRef**] [**Green Version**]

37. Team, R.C. R: A Language and Environment for Statistical Computing; R Core Team: Vienna, Austria, 2015. [**Google Scholar**]

38. Oksanen, J.; Blanchet, F.G.; Friendly, M.; Kindt, R.; Legendre, P.; McGlinn, D.; Minchin, P.R.; O'Hara, R.B.; Simpson, G.L.; Solymos, P.; et al. "Package 'Vegan' " Community Ecology Package, Version 2.5-7; 2020. Available online: **http://CRAN.R-project.org/package=vegan** (accessed on 25 January 2021).

39. Magurran, A.E. Ecological Diversity and Its Measurement; Princeton University Press: Princeton, NJ, USA, 1988. [**Google Scholar**]

40. Shannon, C.E. A Mathematical Theory of Communication. Bell Syst. Tech. J. 1948, 27, 379–423. [**Google Scholar**] [**CrossRef**] [**Green Version**]

41. Makarova, O.L. Gamasid mites (Parasitiformes, Mesostigmata) dwelling in bracket fungi at the pechoro-ilychskii reserve (Republic of Komi). Zool. Zhurnal 2004, 83, 1335–1340. [**Google Scholar**]

42. Hågvar, S.; Amundsen, T.; Økland, B. Mites of the genus Carabodes (Acari, Oribatida) in Norwegian coniferous forests: Occurrence in different soils, vegetation types and polypore hosts. Scand. J. For. Res. 2014, 29, 629–638. [**Google Scholar**] [**CrossRef**]
43. Negruckii, S.F. Kornevaya Gubka, 2nd ed.; Agropromizdat: Moscow, Russian, 1986. [**Google Scholar**]
44. Vasiliauskas, A.P. Root Fungus and Resistance of Coniferous Forests Ecosystems; Mokslas Publishers: Vilnius, Lithuania, 1989. [**Google Scholar**]
45. Graves, R.C. Ecological observation on the insects and other inhabinants of woody shelf fungi (Basidiomycetes: Polyporaceae) in the Chichago area. Ann. Entomol. Soc. Amer. 1960, 53, 61–78. [**Google Scholar**] [**CrossRef**]
46. Buxton, P.A. British Diptera associated with fungi. 111: Plies of all families reared from about 150 species of fungi. Entomol. Mon. 1960, 96, 61–94. [**Google Scholar**]
47. Põldmaa, K.; Jürgenstein, S.; Bahram, M.; Teder, T.; Kurina, O. Host diversity and trophic status as determinants of species richness and community composition of fungus gnats. Basic Appl. Ecol. 2015, 16, 46–53. [**Google Scholar**] [**CrossRef**]
48. Põldmaa, K.; Kaasik, A.; Tammaru, T.; Kurina, O.; Jürgenstein, S.; Teder, T. Polyphagy on unpredictable resources does not exclude host specialization: Insects feeding on mushrooms. Ecology 2016, 97, 2824–2833. [**Google Scholar**] [**CrossRef**]
49. Brandtberg, P.; Johansson, M.; Seeger, P. Effects of season and urea treatment on infection of stumps of Picea abies by Heterobasidion annosum in stands on former arable land. Scand. J. For. Res. 1996, 11, 261–268. [**Google Scholar**] [**CrossRef**]
50. Gonthier, P.; Garbelotto, M.; Varese, G.C.; Nicolotti, G. Relative abundance and potential dispersal range of intersterility groups of Heterobasidion annosum in pure and mixed forests. Can. J. Bot. 2001, 79, 1057–1065. [**Google Scholar**] [**CrossRef**]
51. Vollbrecht, G.; Johansson, U.; Eriksson, H.; Stendlid, J. Butt rot incidence, yield and growth pattern in a tree species experiment in southwestern Sweden. For. Ecol. Manag. 1995, 76.1, 87–93. [**Google Scholar**] [**CrossRef**]
52. Greig, B.J.W. Field recognition and diagnosis of Heretobasidion annisum. In Heterobasidion Annosum: Biology, Ecology, Impact and Control; Woodward, S., Stenlid, J., Karjalainen, R., Hüttermann, A., Eds.; CAB International: Wallingford, UK, 1998; pp. 35–92. [**Google Scholar**]
53. Schigel, D.S. Collecting and rearing fungivorous Coleoptera. Rev. d'Ecol. 2008, 63, 15–20. [**Google Scholar**]
54. Querner, P. Combining pitfall traps and soil samples to collect Collembola for site scale biodiversity assessments. Appl. Soil Ecol. 2010, 45, 293–297. [**Google Scholar**] [**CrossRef**]
55. Thorn, S.; Müller, J.; Bässler, C.; Gminder, A.; Brandl, R.; Heibl, C. Host abundance, durability, basidiome form and phylogenetic isolation determine fungivore species richness. Biol. J. Linn. Soc. 2015, 114, 699–708. [**Google Scholar**] [**CrossRef**] [**Green Version**]

15 DIVERSITY PATTERNS IN THE SYMBIOTIC MITE COMMUNITY ASSOCIATED WITH THE MOUNTAIN PINE BEETLE, DENDROCTONUS PONDEROSAE HOPKINS

1. INTRODUCTION: AN OVERVIEW

On a global scale, forests are experiencing heightened levels of stress due to insect pests, specifically bark beetles (Coleoptera: Scolytinae) [1,2,3]. The consequences of bark beetles can be significant as they actively harm the phloem tissue of trees and transmit phytopathogens, leading to widespread tree death [4,5]. In the past ten years, the mountain pine beetle, scientifically known as Dendroctonus ponderosae Hopkins, has caused the death of trees in almost 30 million hectares of forest in western North America [4,5]. The mountain pine beetle has been able to extend its range to higher elevations and move northward due to a changing climate, which was previously limiting its dispersal based on temperature [6,7,8]. The mountain pine beetle infests 13 species of pine, with the most common targets being lodgepole pine (Pinus contorta Doug.), white pine (Pinus strobiformis Englm.), limber pine (Pinus flexilis E. James), ponderosa pine (Pinus ponderosa Doug. ex C. Lawson) [9], and more recently, jack pine (Pinus banksiana Lamb.) [7,10].

The majority of the mountain pine beetle's lifespan is spent inside the phloem layer of its host tree. During this time, the beetles create tunnels, introduce blue-staining fungus, lay eggs, and undergo the process of maturation from larva to adult stage [11,12,13]. The microhabitats formed by the activities of beetles are utilized by many species, known as 'symbiota' or 'symbionts' [14]. The symbiotic organisms associated with the beetle can also affect its fitness through ecological interactions, such as mutualistic relationships involving food and antagonistic relationships affecting fitness [15,16]. For instance, the fungal organisms that are connected with the mountain pine beetle can have a favorable impact on the beetle's ability to reproduce by providing nutritional resources. However, they can also have a detrimental effect through parasitism and competition. Mites, specifically Acari, play significant roles in the galleries of bark beetles. They have multiple direct and indirect functions, including vectoring fungi, preying on each other and the juvenile beetles, and more. Some beetles may focus on one specific host, while others are more adaptable and can survive on different hosts. Additionally, there are groups of hidden species that have different levels of preference for specific hosts. Similar to other closely related bark beetles, the roles of mites have both direct and indirect effects on the success of a beetle [26,27,28,29]. Mites might indirectly contribute to tree mortality by enabling the spread of plant diseases to healthy trees [29,30]. For instance, mites that are connected to the bark beetle, Scolytus multistriatus (Marsham), have a role in spreading the harmful vascular wilt disease agent, Ophiostoma novo-ulmi Brasier, in Ulmus spp. It is worth noting that this fungus can have a major negative impact on the overall health and reproductive success of the beetles [31].

Mites form associations with bark beetles and other insects in the forest through phoresy, which involves using the beetles as a method of passive transportation between trees and under the bark. When there are large numbers of phoretic mites, it can hinder the mobility of the host and have a negative impact on both the host beetle and the mites themselves. High populations of mites in carrion beetles (Coleoptera: Silphidae) had a detrimental impact on the fitness of the host beetles [34]. Phoretic mite burdens in beetle populations can be influenced by various factors, such as forest tree type [16,35], temperature [36,37,38], geographic distance and dispersal (as discussed in [39]), and beetle abundance [29,40]. Due to the mountain pine beetle's changing preferences for specific pine species [12], as well as the varying number of preferred host trees and local climate in different woods [4,5,6], it is anticipated that mite populations will differ among beetle populations. However, this hypothesis is seldom validated.

Ignoring mites in the investigation of bark beetles not only underestimates the overall biodiversity linked to bark beetles, but also diminishes our capacity to analyze fluctuations in beetle population dynamics [16,21,41]. Previous research on mountain pine beetle mites has mostly focused on individual populations, without comparing different populations. This has limited our understanding of how distinct groups of mites are related to the success of their beetle hosts or the stage of the population.

In this study, we investigate, depict, and evaluate the groups of mites that attach themselves to mountain pine beetles in five distinct beetle populations distributed along a latitudinal gradient. These populations differ in terms of the kind of trees they infest, the local climate, and the stage of the beetle population. We employ alpha and beta diversity metrics to characterize the variations in composition and examine potential factors influencing these variations, such as variations in host tree species, climate, beetle population, and mite functional groups. Our prediction is that there is a correlation between increased latitudinal separation and variations in mite assemblages. These variations are likely to be influenced by climatic and environmental factors. Our objective is to provide the groundwork for future research that will investigate large-scale patterns of mite communities linked to mountain pine beetles and other forest insect species that are significant in terms of both economic and ecological impact.

2. MATERIALS AND METHODS

2.1. Mite Collection

Between 2009 to 2017, 2225 mountain pine beetle specimens were collected in Arizona, Colorado, Utah, South Dakota (United States), and Alberta (Canada) during the beetle's main emergence event, which usually occurs in the late summer months between July–August (**Figure 1**; **Table 1**). Beetles were captured using a combination of baited flight traps (Lindgren funnel traps; ref. [**43**]) and infested logs placed inside emergence chambers [**44**] (**Table 1**).

Sampling size of beetles varied due to differences in funding and resources, and the sampling strategies of each author/collaborator. Host tree species, sampling years, and beetle population phase also varied with location (**Table 1**). This broad dataset, therefore, represents nearly all stages and climate types of mountain pine beetle populations across its North American distribution. Differences in sample size of beetles and associated mite loads were accounted for through the use of multi-variate non-parametric statistical analyses. Between 161 and 910 beetles from each population were examined for mites. Mites were removed from each beetle, identified by clearing and mounting on glass slides under a dissecting microscope, and stored using mite collection protocols and identification resources as described in Vissa, et al. [**35**] for Arizona and Utah populations, Mori, et al. [**42**] for Alberta, Reboletti [**45**] for South Dakota, and Mercado, et al. [**21**] for Colorado. These vary only subtly in mite storage (particularly location of specimen), and otherwise converge on the same collection, clearing and mounting protocols used for processing bark-beetle mites.

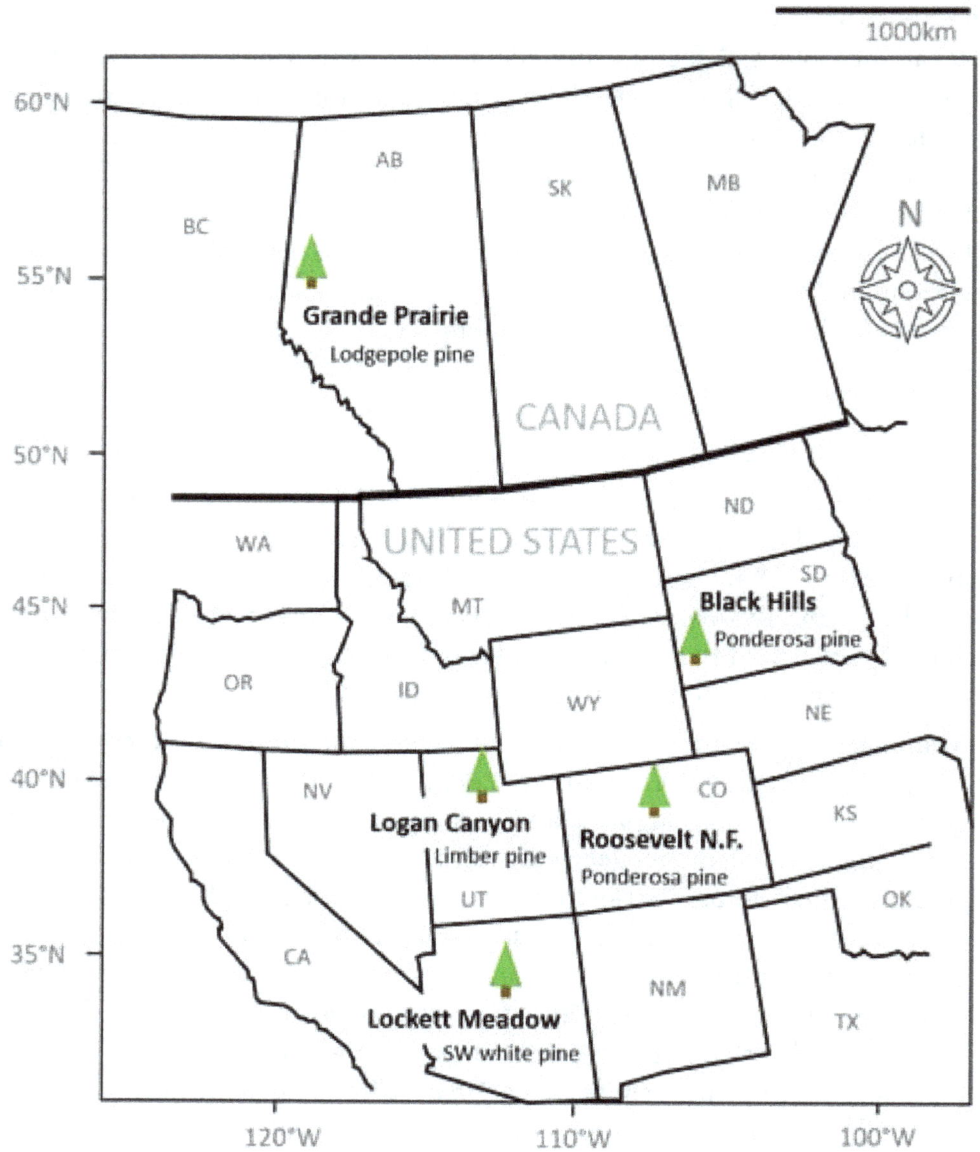

Figure 1. Map of sampling locations for the five sampled beetle populations.

2.2. Assessing Climatic Conditions

In order to evaluate the effects of sampling year and population location, we conducted a comparison between the average precipitation and temperature for the sampling year and the 30-year annual averages. These values were obtained from the PRISM Climate Group database at Oregon State University, using the field coordinates listed in Table 1 [46]. The data for Alberta, Canada was obtained from the Canadian Climate Normals database, specifically sourced from Environment and Climate Change Canada in 2019 (ref. [47]).

Table 1. Mountain Pine Beetle (MPB) population, sampling, and site information. (#) refers to the "total number of" beetles and/or mites.

Population	Sampling Year(s)	Lat (N) N to S	Descriptive Latitude	Elevation (m)	MPB Attacked Host Tree	Year	MPB Population Phase	Descriptive Precipitation	# MPB	# mites	Average Mites per Beetle	Total No. of Mite Species
Grande Prairie, Alberta, Canada (CAN)	2009	54.69	northern	~750	*Pinus contorta*	2009	Epidemic	most dry	910	801	0.88	3
Black Hills, South Dakota (USA)	2007	43.5	middle	~2200	*Pinus ponderosa*	2011	Epidemic	dry	350	1106	3.16	7
Logan Canyon, Utah (USA)	2016/2017	41.93	middle	~2300	*Pinus flexilis*	2016–2017	Endemic	wet	161	886	5.50	5
Roosevelt N.F., Colorado (USA)	2012/2013	40.4	middle	~2250	*Pinus ponderosa*	2012–2013	Epidemic to Endemic	dry	565	1460	2.58	5
Lockett Meadow, Arizona (USA)	2016/2017	35.36	southern	~2600	*Pinus strobiformis*	2016–2017	Endemic	wettest	239	595	2.49	5

Alberta, the most northerly populated area, saw the lowest minimum and maximum temperatures overall. Additionally, it had notably low average minimum temperatures for the sample year (Figure 2). South Dakota had the highest total average maximum and lowest temperatures. The climate data that was obtained revealed a scarcity of rainfall in both regions. Colorado had comparable climatic conditions as South Dakota in 2012, but witnessed a significant increase in total precipitation in 2013. Utah had comparable average maximum temperatures to Colorado, but it had lower average minimum temperatures. Arizona, located in the southernmost region, saw higher maximum and minimum temperatures than Utah. However, Arizona received less total precipitation than Utah (Figure 2).

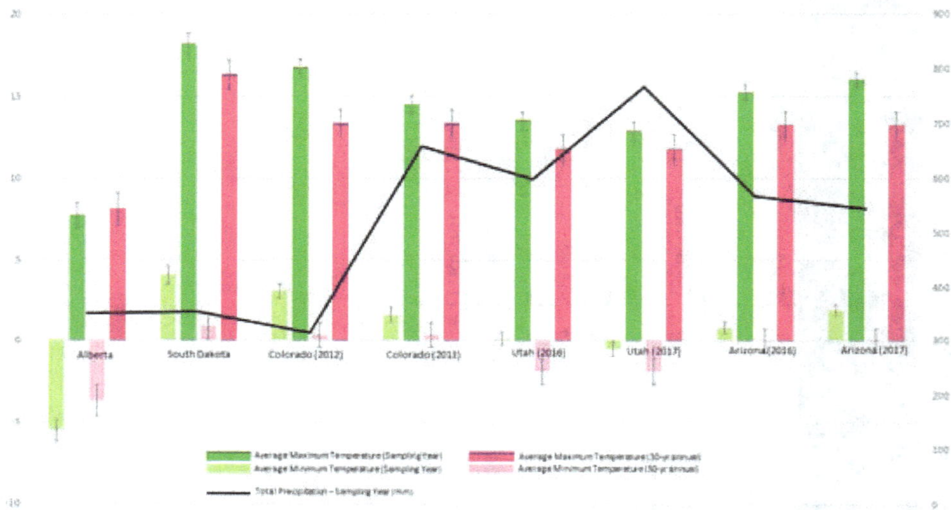

Figure 2. Basic climate information for sampling locations and years compared to 30-year annual averages. Obtained from the PRISM Climate Group database [46], Oregon and the Canadian Climate Normals database [47].

2.3. Mite Data Analysis

Analyses were carried out in R. ver. 3.6.2 (R Core Team, 2019) using raw mite abundance data. Species accumulation curves for each population and the overall total sample pool were estimated using the specaccum function in the vegan package [**48**]. To test for differences in mite loads across populations, we used generalized linear models (GLM) using the glm function in the MASS package [**49**]. To test for pairwise differences in the assemblage of species across sampling locations based on the presence and absence of species and their abundances, a negative binomial GLM with pairwise comparisons using the manyglm function in the mvabund package [**50**] was used. The use of the negative binomial distribution was determined by model selection using the Akaike information criterion (AIC) [**51**] where Poisson, zero-inflated binomial and negative binomial distribution models were compared.

The multivariate species data were fit and analyzed with the mvabund package. The principle model fitting function, manyglm fits a GLM via resampling for each sampled species using species abundances. It allows for multiple species testing and uses a likelihood ratio test (LRT) and resampled p-values to detect significance where the null-hypothesis (H_0) considers beetle population to have no effect on mite species composition. To account for correlation in testing 1000 resampling iterations were used in our analysis via 'pit.trap' resampling in testing [**50**].

Beta diversity metrics for each population were calculated with the betapart package [**52**]. This package uses multiple-site dissimilarity measures to assess the spatial patterns of beta diversity. It also accounts for compositional heterogeneity across sites, thereby accounting for multivariate structure of dissimilarity in species composition [**53**].

Beta diversity calculations were conducted using both the Sorenson and Jaccard dissimilarity indices with no differences in results observed between the two. The Sorenson dissimilarity is used in all visualizations where: β_{sor} is the overall Sorenson dissimilarity index, β_{sne} is dissimilarity explained by nestedness, and β_{sim} is dissimilarity explained by turnover. β_{sor}, β_{sne}, and β_{sim} are measured on a 0–1 scale where $\beta_{sor} = 0$ indicates all species are shared and $\beta_{sor} = 1$ indicates no species are shared. β_{sor} may also be expressed as ($\beta_{sne} + \beta_{sim}$). Degrees of nestedness were visually represented in a cluster dendrogram made within the betapart package. Finally, indicator species associated with each sampling location were assessed with the indicspecies package [**54**]. The indicator species analysis [**55**] highlights the occurrence of species that determine the community assemblage or its diversity within the given sampling area.

3. RESULTS

3.1. Mite Abundance, Taxa Found, and Species Richness

A total of 4848 mites were collected across all sampled mountain pine beetle populations (**Table 1**). Mites represented 8 genera and 12 species (**Table 2**). No species were endemic to any population, while two species (Tarsonemus endophloeus Lindquist and Proctolaelaps subcorticalis Lindquist) were found in all populations. Mites from over 100 beetles were sampled from each population with species accumulation curves beginning to approach or fully approaching asymptote in all populations (**Figure 3**). The number of mites per beetle varied significantly (GLM analysis; $p < 0.01$) in all but three comparisons across populations (**Table 3**). Mite loads were highest in middle to southern populations; and decreased in northern populations. Alberta, the northern-most population, had significantly lower mite loads than all other populations. Percentage of phoretic mites per beetle was lowest in beetle populations in Alberta, Canada. Arizona and Utah had the greatest percentage of beetles carrying mites, followed by South Dakota where half the beetles carried mites (**Figure 4**).

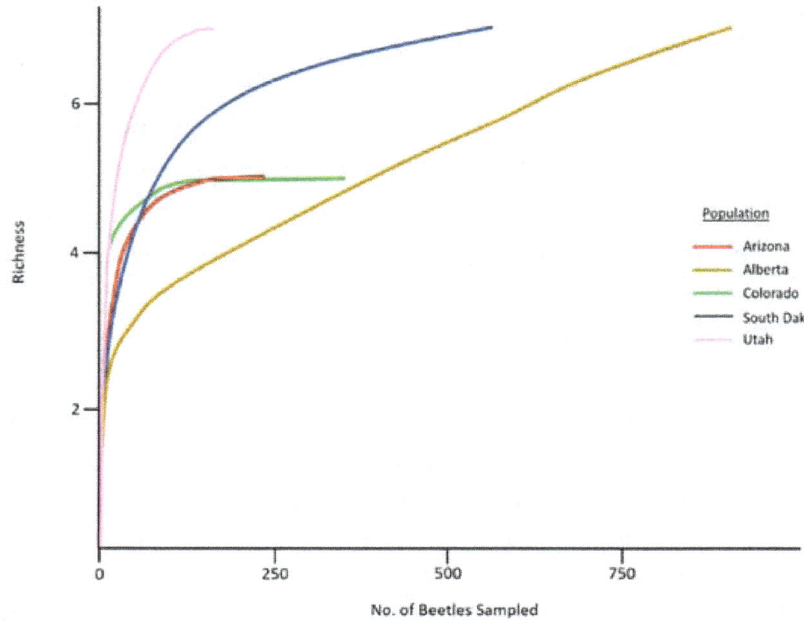

Figure 3. Mite species accumulation curves for our five sampled mountain pine beetle populations.

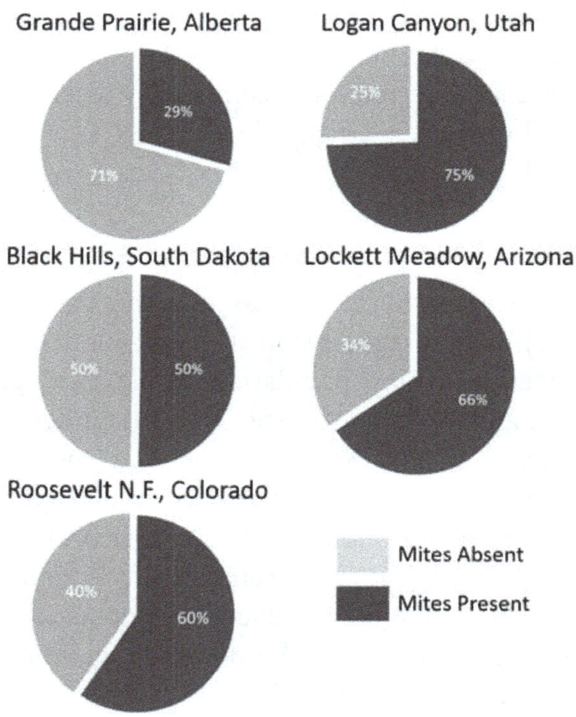

Figure 4. Percent beetles carrying mites in each sampled population.

Table 2. Average abundance per beetle (±standard error), functional and taxonomic information for mite species captured. Indicator species (p < 0.05) for each population are indicated with an asterisk (*). Species with average abundance <0.1 are considered rare species. Significant difference between abundances listed below across populations were confirmed (see generalized linear models (GLM) analysis; Table 3).

Taxa Information		Average Abundance Per Beetle (±S.E.)				
Mite Species	Functional Group	Alberta	South Dakota	Utah	Colorado	Arizona
Tarsonemus ips Lindq.	Fungivore	0.34 (0.03)	3.26 (0.14)	3.82 (0.46)	1.19 (0.15)	0.79 (0.1) *
Tarsonemus endophloeus Lindq.	Fungivore		0.32 (0.01)	0.04 (0.02)	0.27 (0.05) *	
Proctolaelaps subcorticalis Lindq	Predator	0.5 (0.07) *	1.73 (0.07) *	0.03 (0.03)	0.39 (0.07) *	0.01 (0.01)
Dendrolaelaps quadrisetus Berl.	Predator			0.22 (0.05)		0.08 (0.02)
Trichouropoda sp. (*T. utahensis* Wis. and Hirch. in Utah, Arizona; and *T. utahensis* + *T. maeandralis* Hirsch. in Colorado)	Omnivore			1.39 (0.22) *	2.3 (0.35) *	1.32 (0.16)
Nanacarus sp.	Omnivore		0.01 (0.00)			
Histiogaster sp.	Detritivore		0.43 (0.02)			0.28 (0.19) *
Histiogaster arborsignis Woodr.	Detritivore	0.04 (0.01) *			0.02 (0.01)	
Schweibea sp.	Unknown		0.18 (0.01) *			
Parawinterschmidtia sp.	Unknown		0.04 (0.00)			
Overall mites/beetle		0.9	3.2	5.5	2.6	2.49
Total Species Richness		3	7	5	5	5

Table 3. GLM test statistics for total mite abundance per beetle across five sampled populations. Asterisk (*) shows significane at $p \leq 0.05$.

Analysis of Deviance Table		
Population	t-value	p-value
Intercept (Arizona)	7.079	<0.01 *

Analysis of Deviance Table		
Alberta	-6.8	<0.01 *
South Dakota	-1.542	0.1
Colorado	3.154	<0.01 *
Utah	4.091	<0.01 *
Pairwise Comparisons		
Population Pair	*z*-value	*p*-value
Arizona-Canada	6.8	<0.01 *
Arizona-Colorado	-3.154	0.01 *
Arizona-South Dakota	1.542	0.53
Arizona-Utah	-4.091	<0.01 *
Canada-Colorado	-12.016	<.01
Canada-South Dakota	-6.685	<.01
Canada-Utah	-11.037	<0.01 *
Colorado-South Dakota	5.687	<0.01 *
Colorado-Utah	-1.574	0.51
South Dakota-Utah	-6.122	<0.01 *

A pairwise comparison of the total mite species richness per beetle showed no difference between populations (p > 0.05; **Table 4**). Specific indicator species in each population varied, with some overlap occurring across Alberta, South Dakota, and Colorado. There was no overlap in indicator species between the southern-most and northern-most populations (**Table 2**).

Table 4. Pairwise population comparisons of total mite species richness per beetle.

Population Pair	|*t*-Value|	Std. Error	*p*-Value
Arizona-Alberta	0.164	0.01	0.99
Arizona-Colorado	0.797	0.01	0.93
Arizona-South Dakota	0.032	0.012	1
Arizona-Utah	1.412	0.01	0.62
Alberta-Colorado	1.089	0.01	0.81
Alberta-South Dakota	0.157	0.012	0.99
Alberta-Utah	1.706	0.01	0.43
Colorado-South Dakota	0.961	0.01	0.87
Colorado-Utah	0.762	0.012	0.94
South Dakota-Utah	1.604	0.01	0.5

3.2. Partitioning Compositional Differences Using Measures of Beta Diversity

The mite assemblages associated with the five bark beetle populations differed geographically based on the presence and absence of species and their relative abundances (**Table 5**; LRT = 7159; p = 0.001). Tarsonemus endophloeus was present in mid-latitude populations (Utah, Colorado, and South Dakota), with highest abundances in South Dakota and Colorado. Tarsonemus endophloeus was rare in Utah and absent from the southernmost (Arizona) and northernmost (Alberta) populations in our study (**Figure 5**; **Table 2**).

Dendrolaelaps quadrisetus Berlese was found only in Arizona and Utah; however, these were captured in 2016 and not in 2017. Trichouropoda utahensis Hirschmann and Wisniewski, in particular, was unique to middle and southern mountain pine beetle populations from Utah, Arizona, and in Colorado (where T. maeandralis Hirschmann also was found). It is also indicative of Utah where it was most abundant (**Figure 5**; **Table 2**). Mites in the genus Trichouropoda are not reported in Alberta in our study location; however, they have been observed on mountain pine beetles from the southwest of the province near Banff National Park [**56**]. Some rare species, which include Histiogaster arborsignis Woodring, Nanacarus sp., and Parawinterschmidtia sp., were captured in very low numbers.

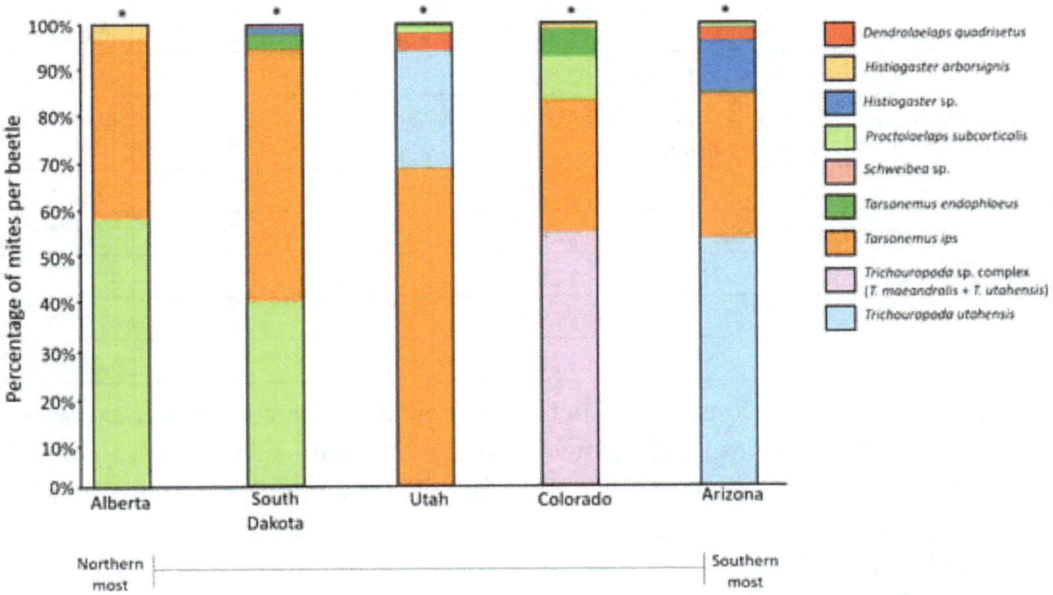

Figure 5. Mite assemblages shown as percentage of each mite species (excluding extremely rare species) sampled across five populations. Asterisk (*) indicates significantly different assemblages between populations (for full statistics, see Table 5).

Table 5. (A) Multivariate test statistic showing likelihood ratio test (LRT) values for population effect on mite composition based on the presence/absence of species and abundances where D_f = degrees of freedom; (B) Wald test statistics for population differences shown in (A); (C) LRT for differences in mite species across all populations. Asterisk (*) shows significance at p ≤ 0.05.

(A) Multivariate Test—Analysis of Deviance Table				
	D_f	D_f Difference	LRT	*p*-value
(Intercept)	2030	4	7159	0.01 *
Population	2026			
(B) Wald Test Statistics by Population				
Population	Wald Value			*p*-Value
Intercept (Arizona)	6.311			<0.01 *
Alberta	6.646			<0.01 *
South Dakota	4.937			<0.01 *
Colorado	3.922			<0.01 *
Utah	5.31			<0.01 *
(C) Univariate Comparisons of Species Abundances Across All Populations				
Mite Species	LRT			*p*-value
Tarsonemus ips Lindq.	192.849			<0.01 *
Tarsonemus endophloeus Lindq.	116.192			<0.01 *
Trichouropoda utahensis Wis. and Hirsch.	223.699			<0.01 *
Trichouropoda sp. complex (*T. maeandralis* Hirsch. + *T. utahensis*)	459.59			<0.01 *
Proctolaelaps subcorticalis Lindq.	65.354			<0.01 *
Histiogaster arborsignis Woodr.	27.042			<0.01 *
Histiogaster sp.	24.011			0
Schweibea sp.	18.026			<0.01 *
Other/Rare spp.	9.689			0.04

Differences in mite composition between populations can be further explained by measures of nestedness and turnover (species replacement). On a scale of 0–1 where 1 = complete dissimilarity, the total dissimilarity in mite composition (β_{sor}) across populations was 0.54. The majority of dissimilarity resulted from turnover (β_{sim}) which explained 0.43 of the total dissimilarity, while nestedness (β_{sne}) accounted for 0.114 of the total dissimilarity.

The highest rates of turnover occurred between the southernmost beetle populations (Arizona and Utah) and the other populations, particularly Alberta (**Figure 6**a). High rates of turnover were also seen between southern-most population (Arizona) and each of the mid-latitude populations (Colorado and

South Dakota) as well as between the northernmost population (Alberta) and the mid-latitude populations (**Figure 6**a). The mite assemblages of Arizona and Utah were completely nested, i.e., the mites found in Utah were a subset of those found in Arizona. Alberta mite assemblages were a subset of those found in South Dakota and other middle populations (**Figure 6**a,b).

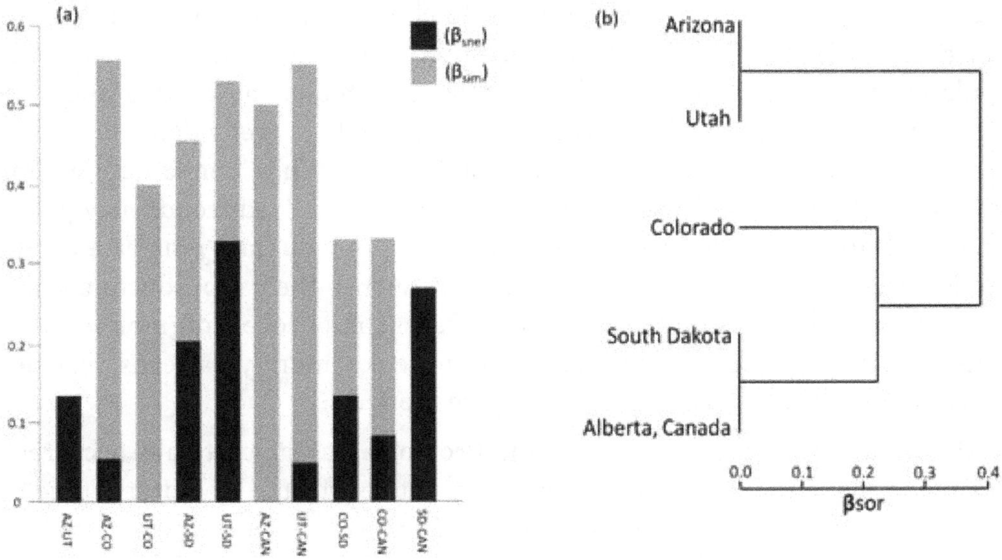

Figure 6. (a) Sorenson dissimilarity indices of Beta (β) diversity for mite communities partitioned into nestedness (β_{sne}) and turnover (β_{sim}) from southern-most (Arizona [AZ]) to northern-most (Alberta, Canada [CAN]) sampled populations; (b) cluster dendrogram showing degree of nestedness based on overall dissimilarity (β_{sor}).

4. DISCUSSION

Geographically distant mountain pine beetle populations have different symbiotic mite assemblages despite some commonalities reflected in species overlap. Sorenson's dissimilarity values for beta diversity indicate that the mite assemblage of beetle populations geographically closer to one another are more similar than populations that are geographically more distant. This was most evident between Alberta and all other populations. Turnover and nestedness patterns varied although species richness did not show any significant variability across populations.

4.1. Differences in Species Richness and Abundance

No significant differences were detected in mite species richness, presumably because richness per beetle was generally very low for all populations. Species accumulation curves approached asymptote in all populations (although not at the same rates), indicating that our sampling efforts adequately captured the overall mite richness associated with these populations. However, the disparities in species accumulation curves between populations may also be due to the possible presence of cryptic species that could not be uncovered. The number of mites per beetle (mite load) did not vary on a latitudinal gradient (as predicted), suggesting that other factors may better explain these differences. We observed that post-outbreak and 'endemic' stage populations had higher mite abundances, and observably higher total richness (**Table 2**) suggesting that mite loads may be indicative of beetle population phase. Hofstetter, et al. [**57**] show that higher mite loads dominated by Tarsonemus mites correlated with decreased beetle progeny densities because of mite introduction of antagonistic blue-stain fungi. While our sampling efforts could not test particular hypotheses, our observed mite load trends may provide insight on population specific ecological dynamics in mountain pine beetles. High mite loads can affect beetle flight [**27,33**] or influence their reproductive success within trees [**58**]. The specific effects of different mite levels on mountain pine beetles has not been tested. Our analysis of mite community patterns provides a foundational step towards understanding these effects.

The finding that Tarsonemus ips and Proctolaelaps subcorticalis occurred in all populations suggests that these mites may not be affected by latitude, pine tree host, climate, or beetle population stage. Trichourpoda utahensis was only found in Arizona and Utah; and although also found in Colorado, occurred alongside a second Trichouropoda species, T. maeandralis (which was not the case for Arizona and Utah). Trichouropoda mites carry fungi that differ from the mutualistic fungi carried by the mountain pine beetles, and its relationship with collapsing beetle population has been suggested [**16**].

4.2. Beta Diversity Patterns

As expected, populations relatively closer together shared more species between them than those geographically distant. High nestedness at these sites may indicate a low overall phoretic mite biodiversity [**59**]. Utah and Arizona (and likewise South Dakota and Alberta) showed nested patterns where mite species were likely removed rather than replaced. This could be due to the large latitudinal and subsequent climate shift between these populations resulting in loss of species with changes in local climate.

High rates of species turnover are explained by environmental differences causing some species to be replaced by others more suited to the new environment [**60**]. Although not specifically tested in this study, we hypothesize that environmental filtering, specifically temperature, serves as a likely explanation for differences in mite community assemblages across latitude. Beta diversity patterns driven by turnover are associated with temperature seasonality, which may be experienced across a latitudinal gradient [**61**]. Other complementary abiotic factors may contribute to high levels of turnover, particularly dispersal ability. Hill, et al. [**62**], showed that spatial patterns and environmental heterogeneity play a large role in explaining high levels of dissimilarity [in community assemblage] among sites in passively transported invertebrates and are better explained by increased turnover. However, dispersal limitations are often defined for arthropods by temperature [**63**], which may be the ultimate driver of the latitudinal patterns.

Differences in host tree species may also help explain mite composition patterns across bark beetle populations. Our data analysis could not accommodate testing for differences in mite composition by host tree because host tree and location are confounding factors except for South Dakota and Colorado which shared the same host tree species (**Table 1**), and beetles collected via flight traps may come from a variety of tree species. Fungal symbiota associated with the mountain pine beetle induce different secondary chemical responses in different host tree species which may contribute to the tree's attractiveness as a host environment for the beetle and its associates [**64,65**]. In our study, although Colorado and South Dakota populations shared the same host (Pinus ponderosa), the overall mite compositions associated with these populations differed significantly. However, it should be noted that sampling years varied. However, these two populations did differ in their beetle population stage—Colorado was a collapsing population from an epidemic while the South Dakota population was increasing to epidemic status. Vissa et al. [**35**] reported vastly different mite loads associated with beetles attacking two different tree species within a single region in Portugal, suggesting that dominant host tree type can also affect beetle-associated mite abundances. Here we found that the population from limber pine (Utah) had beetles with the greatest mite loads. Tree genotype has also been shown to drive different defense production, insect preference, and long-term population dynamics [**66,67**].

Temperature affects patterns in species diversity across latitudinal gradients [**68**]. Our colder population of Alberta had the lowest mite diversity while the others, similarly warmer populations, had contrasting higher mite diversities. Evans, et al. [**69**] showed that a Trichouropoda sp. associated with the southern pine beetle had the highest reproductive success in warmer temperatures. This was reflected in our results where populations with higher average minimum and maximum temperatures also had more mites per beetle. Species of Trichouropoda were also absent in Alberta but prevalent in Colorado, Utah, and Arizona, where both temperature and precipitation were higher during development years; however, this may be confounded with the population stages in these populations.

4.3. Implications for Shifts in Functional Groups of Mites

The concept of environmental filtering in symbiotic mite communities associated with bark beetle systems is not well explored. Swenson et al. [**70**] hypothesized that abiotic (i.e., environmental) filtering not only alters species assemblages in plants, but also alters the functional diversity. Thus, in climatically 'extreme' regions, not only are there fewer species [**70**], but the functional groups can be altered as well [**71**]. While we were unable to identify specific shifts in the functional groups of mites, the observed compositional differences may contain vital information on the resulting functional shifts.

High taxonomic turnover does not directly translate to a significant functional turnover [**72**]. Fish faunas with high taxonomic beta diversity were associated with low functional beta diversity, suggesting that although species assemblages varied, they maintained shared functional attributes [**72**]. Our methods were unable to tease apart the functional diversity patterns associated with symbiotic mite populations or allow us to specifically identify which mite species are being replaced and by whom; however, we did observe some uniformity in the abundant functional groups across all populations. Fungivores (particularly Tarsonemus mites) were abundant in each epidemic population, followed by a generalist predator (P. subcorticalis), suggesting that these species likely affect beetle population dynamics (either by promoting the spread of fungi or by direct predation).

5. CONCISE SUMMARY

This study is the first to investigate and categorize the variations in the beta diversity of symbiotic mite communities that are linked to the mountain pine beetle, an economically significant bark beetle in North America. The study covers a broad range of the beetle's distribution. It can potentially be used as a model to understand the diversity patterns of other symbiotic organisms associated with major insect pests in pine forest ecosystems. The mite biodiversity associated with mountain pine beetles showed geographical variation, but no mite assemblage was completely unique. These differences can be best explained by environmental variations influenced by geography. The evaluation of mite species composition patterns offers fundamental data for subsequent efforts in finding crucial species interactions that influence forest insect systems.

6. AT A GLANCE

The mountain pine beetle, scientifically known as Dendroctonus ponderosae (Coleoptera: Scolytinae), is a bark beetle species that holds significant economic value. It is found across a large geographic area, extending from the southwestern United States to northern Canada. This beetle is responsible for significant death of trees in 13 different kinds of pine. Mites, specifically Acari, are frequently found and plentiful symbiotic organisms that can impact the overall health and well-being of mountain beetles through both beneficial and detrimental interactions. We provide a distinctive evaluation of the mite companions of mountain pine beetles utilizing metrics of alpha and beta diversity. We collected phoretic mites from five beetle populations: Arizona, Colorado, South Dakota, Utah (USA), and Alberta (Canada). These populations differed in terms of the tree species that the beetles inhabited, the climate of their respective locations, and the size of the beetle populations. A total of 4848 mites were gathered, representing 8 different genera and 12 distinct species. Between fifty and seventy percent of beetles were found to be carrying mites when in flight, with the greatest number of mites observed in populations located in the middle and southern regions. The mite loads were found to decrease in populations located in the northern regions. The mite communities, in terms of both the number of species and their composition, exhibited variations along a latitudinal gradient from south to north. These variations were mostly caused by the replacement of species, resulting in changes in the overall mite assemblages. The variation in mite composition became more pronounced as the distance between groups grew. We examine climate variation, environmental filtration, and host tree disparities as potential factors influencing variations in mite composition among beetle populations, and explore the consequences for functional alterations. The outcomes of our study may serve as a framework for predicting the diversity patterns of mite symbionts that are linked to other significant insect pests in coniferous forest ecosystems.

REFERENCES

1. Hart, S.J.; Veblen, T.T.; Schneider, D.; Molotch, N.P. Summer and winter drought drive the initiation and spread of spruce beetle outbreak. Ecology 2017, 98, 2698–2707. [**Google Scholar**] [**CrossRef**] [**PubMed**]

2. Biedermann, P.H.; Müller, J.; Grégoire, J.C.; Gruppe, A.; Hagge, J.; Hammerbacher, A.; Hofstetter, R.W.; Kandasamy, D.; Kolarik, M.; Kostovcik, M.; et al. Bark beetle population dynamics in the Anthropocene: Challenges and solutions. Trends Ecol. Evol. 2019, 34, 914–924. [**Google Scholar**] [**CrossRef**] [**Green Version**]

3. De Groot, M.; Ogris, N. Short-term forecasting of bark beetle outbreaks on two economically important conifer tree species. For. Ecol. Manag. 2019, 450, e117495. [**Google Scholar**] [**CrossRef**]

4. Raffa, K.F.; Aukema, B.H.; Bentz, B.J.; Carroll, A.L.; Hicke, J.A.; Turner, M.G.; Romme, W.H. Cross-scale drivers of natural disturbances prone to anthropogenic amplification: The dynamics of bark beetle eruptions. BioScience 2008, 58, 501–517. [**Google Scholar**] [**CrossRef**] [**Green Version**]

5. Bentz, B.J.; Régnière, J.; Fettig, C.J.; Hansen, M.E.; Hayes, J.L.; Hicke, J.A.; Kelsey, R.G.; Negrón, J.F.; Seybold, S.J. Climate change and bark beetles of the western United States and Canada: Direct and indirect effects. BioScience 2010, 60, 601–613. [**Google Scholar**] [**CrossRef**]

6. Bentz, B.J.; Bracewell, R.R.; Mock, K.E.; Pfrender, M.E. Genetic architecture and phenotypic plasticity of thermally-regulated traits in an eruptive species, Dendroctonus ponderosae. Evol. Ecol. 2011, 25, 1269–1288. [**Google Scholar**] [**CrossRef**]

7. Cullingham, C.I.; Cooke, J.E.K.; Dang, S.; Davis, C.S.; Cooke, B.J.; Coltman, D.W. Mountain pine beetle host-range expansion threatens the boreal forest. Mol. Ecol. 2011, 20, 2157–2171. **[Google Scholar]** **[CrossRef]** **[PubMed]** **[Green Version]**

8. Raffa, K.F.; Powell, E.N.; Townsend, P.A. Temperature-driven range expansion of an irruptive insect heightened by weakly coevolved plan defenses. Proc. Natl. Acad. Sci. USA 2013, 110, 2193–2198. **[Google Scholar]** **[CrossRef]** **[Green Version]**

9. Wood, S.L. The bark and ambrosia beetles of North and Central America (Coleoptera: Scolytinae), a taxonomic monograph. Great Basin Nat. Mem. 1982, 6, 1–1356. **[Google Scholar]**

10. Rosenberger, D.W.; Venette, R.C.; Maddox, M.P.; Aukema, B.H. Colonization behaviors of mountain pine beetle on novel hosts: Implications for range expansion into northeastern North America. PLoS ONE 2017, 12, e0176269. **[Google Scholar]** **[CrossRef]** **[Green Version]**

11. Reid, R.W. Biology of the mountain pine beetle, Dendroctonus monticolae Hopkins, in the east Kootenay region of British Columbia, I. Life cycle, brood development, and flight periods. Can. Entomol. 1962, 94, 531–538. **[Google Scholar]** **[CrossRef]**

12. Amman, G.D.; McGregor, M.D.; Dolph, R.E. Reprinted. (Updated 2002). Mountain pine beetle. USDA Forest Service. For. Insect Dis. Leafl. 1990, 2, 1–12. **[Google Scholar]**

13. Six, D.L.; Bracewell, R. Dendroctonus. Bark Beetle; Elsevier: Amsterdam, The Netherlands, 2015; pp. 305–350. **[Google Scholar]**

14. Hofstetter, R.W.; Dinkins-Bookwalter, J.; Davis, T.S.; Klepzig, K.D. Symbiotic associations of bark beetles. Bark Beetles; Elsevier: Amsterdam, The Netherlands, 2015; pp. 209–245. **[Google Scholar]**

15. Vissa, S.; Hofstetter, R.W. The role of mites in bark and ambrosia beetle–fungal interactions. In Insect Physiology and Ecology; Intech Press: London, UK, 2017; pp. 135–156. **[Google Scholar]**

16. Mercado, J.; Ortiz-Santana, B.; Kay, S. Fungal Frequency and Mite Load Trends Interact with a Declining Mountain Pine Beetle Population. Forests 2018, 9, 484. **[Google Scholar]** **[CrossRef]** **[Green Version]**

17. Six, D.L.; Paine, T.D. Effects of Mycangial Fungi and Host Tree Species on Progeny Survival and Emergence of Dendroctonus ponderosae (Coleoptera: Scolytidae). Environ. Entomol. 1998, 27, 1393–1401. **[Google Scholar]** **[CrossRef]**

18. Mock, K.E.; Bentz, B.J.; O'Neill, E.M.; Chong, J.P.; Orwin, J.; Pfrender, M.E. Landscape-scale genetic variation in a forest outbreak species, the mountain pine beetle (Dendroctonus ponderosae). Mol. Ecol. 2007, 16, 553–568. **[Google Scholar]** **[CrossRef]** **[PubMed]**

19. Hunt, D.W.A.; Borden, J.H.; Rahe, J.E.; Whitney, H.S. Nutrient-mediated germination of Beauveria bassiana conidia on the integument of the bark beetle Dendroctonus ponderosae (Coleoptera: Scolytidae). J. Invertebr. Pathol. 1984, 44, 304–314. **[Google Scholar]** **[CrossRef]**

20. Wegensteiner, R. Pathogens in bark beetles. In Bark and Wood Boring Insects in Living Trees in Europe: A Synthesis; Springer: Dordrecht, The Netherlands, 2007; pp. 291–313. **[Google Scholar]**

21. Mercado, J.E.; Hofstetter, R.W.; Reboletti, D.M.; Negro, F. Phoretic symbionts of the mountain pine beetle (Dendroctonus ponderosae Hopkins). For. Sci. 2014, 60, 512–526. **[Google Scholar]** **[CrossRef]** **[Green Version]**

22. Klepzig, K.D.; Moser, J.C.; Lombardero, F.J.; Hofstetter, R.W.; Ayres, M.P. Symbiosis and competition: Complex interactions among beetles, fungi, and mites. Symbiosis 2001, 30, 83–96. [**Google Scholar**]

23. Cardoza, Y.J.; Moser, J.C.; Klepzig, K.D.; Raffa, K.F. Multipartite symbioses among fungi, mites, nematodes, and the spruce beetle, Dendroctonus rufipennis. Environ. Entomol. 2008, 37, 956–963. [**Google Scholar**] [**CrossRef**]

24. Knee, W.; Beaulieu, F.; Skevington, J.H.; Kelso, S.; Cognato, A.I.; Forbes, M.R. Species boundaries and host range of tortoise mites (Uropodoidea) phoretic on bark beetles (Scolytinae), using morphometric and molecular markers. PLoS ONE 2012, 7, e47243. [**Google Scholar**] [**CrossRef**]

25. Knee, W.; Beaulieu, F.; Skevington, J.H.; Kelso, S.; Forbes, M.R. Cryptic species of mites (Uropodoidea: Uroobovella spp.) associated with burying beetles (Silphidae: Nicrophorus): The collapse of a host generalist revealed by molecular and morphological analyses. Mol. Phylogenet. Evol. 2012, 65, 276–286. [**Google Scholar**] [**CrossRef**]

26. Moser, J.C. Mite predators of the southern pine beetle. Ann. Entomol. Soc. Am. 1975, 68, 1113–1116. [**Google Scholar**] [**CrossRef**]

27. Moser, J.C. Phoretic carrying capacity of flying southern pine beetles (Coleoptera: Scolytidae). Can. Entomol. 1976, 108, 807–808. [**Google Scholar**] [**CrossRef**] [**Green Version**]

28. Moser, J.C. Use of sporothecae by phoretic Tarsonemus mites to transport ascospores of coniferous bluestain fungi. Trans. Br. Mycol. Soc. 1985, 84, 750–753. [**Google Scholar**] [**CrossRef**]

29. Hofstetter, R.W.; Moser, J.; Blomquist, S. Mites associated with bark beetles and their hyperphoretic ophiostomatoid fungi. Biodivers. Ser. 2014, 12, 165–176. [**Google Scholar**]

30. Moser, J.C.; Perry, T.J.; Solheim, H. Ascospores hyperphoretic on mites associated with Ips typographus. Mycol. Res. 1989, 93, 513–517. [**Google Scholar**] [**CrossRef**]

31. Moser, J.C.; Konrad, H.; Blomquist, S.R.; Kirisits, T. Do mites phoretic on elm bark beetles contribute to the transmission of Dutch elm disease? Naturwissenschaften 2010, 97, 219–227. [**Google Scholar**] [**CrossRef**] [**Green Version**]

32. Lindquist, E.E. New species of Tarsonemus (Acarina: Tarsonemidae) associated with bark beetles. Can. Entomol. 1969, 101, 1219–1314. [**Google Scholar**] [**CrossRef**]

33. Kinn, D.; Witcosky, J. Variation in southern pine beetle attack height associated with phoretic Uropodid mites. Can. Entomol. 1978, 110, 249–251. [**Google Scholar**] [**CrossRef**]

34. Wilson, D.S.; Knollenberg, W.G. Adaptive indirect effects: The fitness of burying beetles with and without their phoretic mites. Evol. Ecol. 1987, 1, 139–159. [**Google Scholar**] [**CrossRef**]

35. Vissa, S.; Hofstetter, R.W.; Bonifácio, L.; Khaustov, A.; Knee, W.; Uhey, D.A. Phoretic mite communities associated with bark beetles in the maritime and stone pine forests of Setúbal, Portugal. Exp. Appl. Acarol. 2019, 77, 117–131. [**Google Scholar**] [**CrossRef**] [**PubMed**]

36. Lombardero, Â.J.; Klepzig, K.D.; Moser, J.C.; Ayres, M.P. Biology, demography and community interactions of Tarsonemus (Acarina: Tarsonemidae) mites phoretic on Dendroctonus frontalis (Coleoptera: Scolytidae). Agric. For. Entomol. 2001, 2, 193–202. [**Google Scholar**] [**CrossRef**] [**Green Version**]

37. Lombardero, M.J.; Ayres, M.P.; Hofstetter, R.W.; Moser, J.C.; Klepzig, K.D. Strong indirect interactions of Tarsonemus mites (Acarina: Tarsonemidae) and Dendroctonus frontalis

(Coleoptera: Scolytidae). Oikos 2003, 102, 243–252. [**Google Scholar**] [**CrossRef**] [**Green Version**]

38. Hofstetter, R.W.; Dempsey, T.D.; Klepzig, K.D.; Ayres, M.P. Temperature-dependent effects on mutualistic and phoretic associations. Community Ecol. 2007, 8, 47–56. [**Google Scholar**] [**CrossRef**]
39. Walter, D.E.; Proctor, H.C. Mites on plants. In Mites: Ecology, Evolution & Behavior; Springer: Dordrecht, The Netherlands, 2013; pp. 281–339. [**Google Scholar**]
40. Walter, D.E.; Proctor, H.C. Mites: Ecology, Evolution and Behaviour; CABI Publishing: Wallingford, UK, 1999; pp. 52–55. [**Google Scholar**]
41. Hofstetter, R.W.; Moser, J.C. The role of mites in insect-fungus associations. Ann. Rev. Entomol. 2014, 59, 537–557. [**Google Scholar**] [**CrossRef**] [**PubMed**] [**Green Version**]
42. Mori, B.A.; Proctor, H.C.; Walter, D.E.; Evenden, M.L. Phoretic mite associates of mountain pine beetle at the leading edge of an infestation in northwestern Alberta, Canada. Can. Entomol. 2011, 143, 44–55. [**Google Scholar**] [**CrossRef**]
43. Lindgren, B.S. A multiple funnel trap for scolytid beetles (Coleoptera). Can. Entomol. 1983, 115, 299–302. [**Google Scholar**] [**CrossRef**]
44. Reboletti, D.M. A Multi-Partite Mutualism: Bark Beetles, Fungi, and Mites. Ph.D. Thesis, Northern Arizona University, Flagstaff, AZ, USA, 2008. [**Google Scholar**]
45. Davis, T.S.; Hofstetter, R.W. The effects of gallery density and ratio on the fitness and fecundity of two sympatric bark beetles. Environ. Entomol. 2009, 38, 639–650. [**Google Scholar**] [**CrossRef**]
46. PRISM Climate Group, Oregon State University. Available online: **http://prism.oregonstate.edu** (accessed on 31 August 2020).
47. Canada's Changing Climate Report. Natural Resources Canada, Government of Canada, 2019. Available online: **https://www.nrcan.gc.ca/maps-tools-publications/publications/climate-change-publications/canada-changing-climate-reports/canadas-changing-climate-report/21177** (accessed on 31 August 2020).
48. Oksanen, J.; Kindt, R.; Legendre, P.; O'Hara, B.; Stevens, M.H.H.; Oksanen, M.J.; Suggests, M.A.S.S. The vegan package. Community Ecol. Package 2007, 10, 719. [**Google Scholar**]
49. Venables, W.N.; Ripley, B.D. Exploratory multivariate analysis. In Modern Applied Statistics with S; Springer: New York, NY, USA, 2002; pp. 301–330. [**Google Scholar**]
50. Wang, Y.I.; Naumann, U.; Wright, S.T.; Warton, D.I. mvabund–an R package for model-based analysis of multivariate abundance data. Methods Ecol. Evol. 2012, 3, 471–474. [**Google Scholar**] [**CrossRef**]
51. Akaike, H. Maximum likelihood identification of Gaussian autoregressive moving average models. Biometrika 1973, 60, 255–265. [**Google Scholar**] [**CrossRef**]
52. Baselga, A. Partitioning abundance-based multiple-site dissimilarity into components: Balanced variation in abundance and abundance gradients. Methods Ecol. Evol 2017, 7, 799–808. [**Google Scholar**] [**CrossRef**] [**Green Version**]
53. Baselga, A.; Orme, C.D.L. Betapart: An R package for the study of beta diversity. Methods Ecol. Evol. 2012, 3, 808–812. [**Google Scholar**] [**CrossRef**]
54. Cáceres, M.D.; Legendre, P. Associations between species and groups of sites: Indices and statistical inference. Ecology 2009, 90, 3566–3574. [**Google Scholar**] [**CrossRef**]

55. McCune, B.; Grace, J.B. Analysis of Ecological Communities; MjM Software: Gleneden Beach, OR, USA, 2002. [**Google Scholar**]

56. Peralta Vázquez, G.H. Ecology of Mite Phoresy on Mountain Pine Beetles. Ph.D. Thesis, University of Calgary, Calgary, AB, Canada, 2018. [**Google Scholar**]

57. Hofstetter, R.W.; Cronin, J.T.; Klepzig, K.D.; Moser, J.C.; Matthew, A.E.; Ayres, P. Antagonisms, mutualisms and commensalisms affect outbreak dynamics of the southern pine beetle. Oecologia 2006, 147, 679–691. [**Google Scholar**] [**CrossRef**]

58. Pfammatter, J.A.; Malas, K.M.; Raffa, K.F. Behaviours of phoretic mites (Acari) associated with Ips pini and Ips grandicollis (Coleoptera: Curculionidae) during host-tree colonization. Agric. For. Entomol. 2016, 18, 108–118. [**Google Scholar**] [**CrossRef**]

59. Wright, D.H.; Reeves, J.H. On the meaning and measurement of nestedness of species assemblages. Oecologia 1992, 92, 416–428. [**Google Scholar**] [**CrossRef**]

60. Brown, J.H.; Ernest, S.K.M.; Parody, J.M.; Haskell, J.P. Regulation of diversity: Maintenance of species richness in changing environments. Oecologia 2001, 126, 321–332. [**Google Scholar**] [**CrossRef**]

61. Archibald, S.B.; Greenwood, D.R.; Mathewes, R.W. Seasonality, montane beta diversity, and Eocene insects: Testing Janzen's dispersal hypothesis in an equable world. Palaeogeogr. Palaeoclimatol. Palaeoecol. 2013, 371, 1–8. [**Google Scholar**] [**CrossRef**]

62. Hill, M.J.; Heino, J.; Thornhill, I.; Ryves, D.B.; Wood, P.J. Effects of dispersal mode on the environmental and spatial correlates of nestedness and species turnover in pond communities. Oikos 2017, 126, 1575–1585. [**Google Scholar**] [**CrossRef**] [**Green Version**]

63. Bradshaw, W.E.; Holzapfel, C.M. Genetic response to rapid climate change. Mol. Ecol. 2008, 17, 157–166. [**Google Scholar**] [**CrossRef**] [**PubMed**]

64. Jost, R.W.; Rice, A.V.; Langor, D.W.; Boluk, Y. Monoterpene emissions from lodgepole and jack pine bark inoculated with mountain pine beetle-associated fungi. J. Wood Chem. Technol. 2008, 28, 37–46. [**Google Scholar**] [**CrossRef**]

65. Arango-Velez, A.; El Kayal, W.; Copeland, C.C.J.; Zaharia, L.I.; Lusebrink, I.; Cooke, J.E.K. Differences in defence responses of Pinus contorta and Pinus banksiana to the mountain pine beetle fungal associate Grosmannia clavigera are affected by water deficit. Plant Cell Environ. 2016, 39, 726–744. [**Google Scholar**] [**CrossRef**] [**Green Version**]

66. Holeski, L.M.; Hillstrom, M.L.; Whitham, T.G.; Lindroth, R.L. Relative importance of genetic, ontogenetic, induction, and seasonal variation in producing a multivariate defense phenotype in a foundation tree species. Oecologia 2012, 170, 695–707. [**Google Scholar**] [**CrossRef**] [**PubMed**]

67. McIntyre, P.J.; Whitham, T.G. Plant genotype affects long-term herbivore population dynamics and extinction: Conservation implications. Ecology 2003, 84, 311–322. [**Google Scholar**] [**CrossRef**]

68. Mittelbach, G.G.; Schemske, D.W.; Cornell, H.V.; Allen, A.P.; Brown, J.M.; Bush, M.B.; Harrison, S.P.; Hurlbert, A.H.; Knowlton, N.; Lessios, H.A.; et al. Evolution and the latitudinal diversity gradient: Speciation, extinction and biogeography. Ecol. Lett. 2007, 10, 315–331. [**Google Scholar**] [**CrossRef**]

69. Evans, L.M.; Hofstetter, R.W.; Ayres, M.P.; Klepzig, K.D. Temperature alters the relative abundance and population growth rates of species within the Dendroctonus frontalis (Coleoptera: Curculionidae) community. Environ. Entomol. 2011, 40, 824–834. [**Google Scholar**] [**CrossRef**]

70. Swenson, N.G.; Enquist, B.J.; Pither, J.; Kerkhoff, A.J.; Boyle, B.; Weiser, M.D.; Elser, J.J.; Fagan, W.F.; Forero-Montaña, J.; Fyllas, N.; et al. The biogeography and filtering of woody plant functional diversity in North and South America. Glob. Ecol. Biogeogr. 2012, 21, 798–808. [**Google Scholar**] [**CrossRef**]

71. Flores, O.; Seoane, J.; Hevia, V.; Azcárate, F.M. Spatial patterns of species richness and nestedness in ant assemblages along an elevational gradient in a Mediterranean mountain range. PLoS ONE 2018, 13, e0204787. [**Google Scholar**] [**CrossRef**] [**Green Version**]

72. Villéger, S.; Grenouillet, G.; Brosse, S. Decomposing functional β-diversity reveals that low functional β-diversity is driven by low functional turnover in European fish assemblages. Glob. Ecol. 2013, 22, 671–681. [**Google Scholar**] [**CrossRef**]

16 MASS-TRAPPING OF IPS SEXDENTATUS: MINIMIZING ADVERSE EFFECTS ON SAPROXYLIC BEETLES LARGER THAN THE INTENDED TARGET

1. INTRODUCTION: AN OVERVIEW

The prevalence of bark beetle outbreaks in European conifer forests is rapidly escalating due to advantageous circumstances, resulting in significant consequences [1,2,3]. Climate change has led to an increased occurrence of extreme weather events such as windstorms and severe droughts in recent years [1]. These disruptions enhance the availability of appropriate breeding materials for bark beetles, such as trees that are stressed or uprooted [1]. As a result, the number of bark beetles may increase, and during epidemic outbreaks, aggressive bark beetle species can move from trees that are stressed and weakened to healthy ones. This behavior has been observed in the six-toothed pine bark beetle, Ips sexdentatus (Böerner) (Coleoptera: Curculionidae: Scolytinae) [4].

I. sexdentatus is found in pine forests across the whole Eurasian continent. It is commonly regarded as a less harmful pest compared to the closely related I. typographus (Linnaeus), however there are instances where it can become a more significant threat. For instance, in Italy, the bark beetle has recently thrived due to the presence of the invasive scale Matsucoccus feytaudi Ducasse. This has resulted in the collapse of the maritime pine forests (Pinus pinaster Aiton). Due to the favorable climate in Southern Europe and the abundance of suitable hosts, namely stressed coastal pines, populations of I. sexdentatus have experienced a rapid growth. In these circumstances, their assaults also encompass robust trees of different pine species [7,8,9]. Coastal pine forests play a crucial role in Italy due to their provision of various goods and services, particularly in the realm of recreational tourism [10]. Consequently, safeguarding these woods is imperative to preserve their social, environmental, and economic roles.

There are only a limited number of methods that can be used to control bark beetle populations directly. These include promptly chopping down trees that are affected, using trap-trees and trap-logs, and employing pheromone-baited mass-trapping [11]. Mass-trapping is crucial in situations when timely cutting of contaminated trees is not possible, especially when phytosanitary felling cannot be done quickly [12]. Multiple studies have documented the effectiveness of mass-trapping as a method for controlling bark beetles. This method can be used on its own or in conjunction with other strategies [13,14,15,16,17]. Nevertheless, the usefulness of this method remains a subject of debate, as it is difficult to definitively demonstrate its ability to reduce tree mortality. This is especially true because other living organisms and non-living variables can also influence the outcomes.

Mass-trapping, meanwhile, can result in substantial adverse consequences. Multiple studies have demonstrated the adverse effects of pheromone baits on non-target insect species, which can also be attracted to these baits [18,19,20]. Predators are a specific danger when it comes to non-target catches [18,19]. Bark beetle predators utilize aggregation pheromones called kairomones to locate their prey, making them susceptible to attraction by pheromone traps [21,22]. Thanasimus formicarius and Temnoscheila caerulea, which belong to the Coleoptera families Cleridae and Trogossitidae respectively, are commonly found in large quantities in pheromone traps designed for I. sexdentatus. In addition, mass-trapping operations may inadvertently capture other saproxylic insects that use the same pheromones to find host plants, which are usually weakened by the targeted bark beetle species [19]. Therefore, unintended captures of non-target species can have a detrimental impact on the natural control of bark beetle populations, as well as the preservation of other saproxylic species [19].

Multiple studies are dedicated to the advancement of more precise trapping methods in order to reduce these adverse consequences [8,9,20]. Enhancing the selectivity of mass-trapping can be achieved by identifying the optimal combinations of components. While bark beetles' predators often rely on the prey's pheromones to locate them, it is worth noting that the prey and predators may have different preferences for specific pheromone components or optical isomers. This distinction can be utilized to decrease the number of predators caught. Regarding I. sexdentatus, Etxebeste et al. [8] found that the makeup of unintended catches linked to this species significantly differed depending on the specific pheromone mixes used. Opting for the appropriate trap can also lead to a decrease in the adverse consequences of mass trapping. Martin et al. [9] discovered that slot traps exhibit greater selectivity compared to multi-funnel traps, resulting in fewer captures of predators, and specifically facilitating the escape of T. formicarius. Traps can be altered to hinder the entry of beetles that are not the intended target, particularly larger predators of bark beetles. Martin et al. [9] and Ross and Daterman [28] examined the impact of incorporating a 6-mm mesh screen and opening escape windows on multiple-funnel traps and slot traps to reduce predator captures. This study documented the capture of non-target beetles during a mass-trapping initiative in a pine forest in Tuscany, Italy. We assessed the effectiveness of trap modifications that were comparable to those employed by Martín et al. [9].

2. MATERIALS AND METHODS

2.1. Study Area and Sampling Design

The present study was carried out during 2016–2019 in a Tuscan coastal pine stand, approximately 70 hectares, between Marina di Grosseto and Principina a Mare, in the Province of Grosseto (Italy) (42°42′17.22″ N, 10°59′47.59″ E). This private forest plot was mainly composed of P. pinaster, with a small amount of stone pine (Pinus pinea L.); here, since 2006, the former species has been gradually infested by M. feytaudi, which has caused serious damages and has stressed the trees. Since 2014, the huge availability of weakened pines caused a high increase in the population density of I. sexdentatus. Another aggressive bark beetle species, Tomicus destruens (Wollaston), was also present in this stand and contributed to pine mortality. In addition, the bark beetle Orthotomicus erosus Wollaston was abundant, although it was mostly found on pines already infested by I. sexdentatus or T. destruens, and/or highly stressed by abiotic factors. This area has a high touristic value due to its many campsites; therefore, to reduce the population density of I. sexdentatus and the number of attacked pines, prompt cutting of infested trees, as well as mass-trapping were carried out. On 28 April 2016, a total of 25 Super Forest bark beetle slot traps (Serbios s.r.l., Badia Polesine, Rovigo, Italy) were placed in the pine stand along the south-western borders of the forest plot, spacing each trap by about 75 m (**Figure 1**). This pattern addressed the higher I. sexdentatus infestation in the pine stand closer to the coastline—and outside our study area—where no control measures were carried out. Super Forest traps are a variant of the common Theysohn design, with fewer, longer slots, but with the same amount of slot rows (seven) (**Figure 2**A).

Traps were baited with the commercial blend of the I. sexdentatus aggregation pheromone manufactured by Serbios s.r.l., whose main compounds are ipsenol, ipsdienol, and 2-methyl-3-buten-2-ol. Catches were collected every 10 days from March (except in 2016 when traps were installed in April) to December. Bait was renewed every 30–50 days, according to the dispensers' lifespan (as indicated by the manufacturer) and to the weather conditions in the study area, since warmer climates of Tuscany deplete these dispensers considerably faster during summer months [**29**]. Thus, sticking to manufacturer instructions would lead to errors in interpreting flight patterns. In order to ensure a continuous presence of adequately baited traps in the field, a baiting rotation system based on alternating pairs was set up: group A consisted of six pairs plus the 25th trap; group B included the remaining six pairs (**Figure 1**). The two groups were baited 20 days apart, always following the same renewal schedule based on the dispensers' lifespan and on weather conditions.

Figure 1. Study area with trap pattern as applied in 2019. Part of the coastal pine stands in the Municipality of Grosseto (Tuscany, Italy) is shown; the green line delimits the study area's borders. Bait rotation alternated groups A (in red) and B (in yellow; symbols indicate modified (stars) and standard (circles) traps.

Figure 2. Standard trap without modifications (A) and detail of a modified trap (B): through one of the three escape windows the 6mm mesh screen can be seen inside.

Besides the target beetle, in the first three years of the study (2016–2018), special attention was reserved for some of the other beetles trapped: the major non-target bark beetles and the two major bark beetle predators, T. formicarius and T. caerulea. In 2019, additional non-target beetles were considered, we grouped them into two categories according to their mean length: (1) all bark beetles; and (2) all beetles on average > 8 mm, among which T. caerulea and T. formicarius were grouped. In this year, 13 traps were modified by applying a 6-mm mesh screen on top of the collection container as in Martìn et al. [**9**], and by providing three additional 60 mm × 8 mm escape windows immediately above the mesh screen (one central window on one side and two lateral windows on the other side) (**Figure** 2B). These traps were called "modified traps" and their captures were considered separately than those from the "standard traps," which were not modified. "Modified traps" were evenly distributed in the bait renewal rotation by alternating them with "standard traps." Finally, a random sub-sample of 30 specimens per each species (when possible) per each kind of trap (modified or standard) was measured, recording their body length.

2.2. Statistical Analysis

The analysis of variance (ANOVA) for repeated measures (catches in the same trap on different dates were considered repeated measures) was used to compare catches in standard traps and modified traps of (1) I. sexdentatus; (2) non-target bark beetles (all species together); (3) all bark beetles together; (4) beetles > 8 mm; and finally (5) species of this last category, with enough catches (more than 30), were analyzed separately. All the data analyzed with ANOVA were log transformed. Simpson's index was used to compare species diversity in the two kinds of traps employed in 2019. As regards the two main predators, T. caerulea and T. formicarius, the 2019 mean catches, separated according to the kind of trap, were then compared with ANOVA with those of previous years when the traps had not yet been modified but were in the same positions. Pearson correlation was applied to assess the correlation between the main predators and bark beetle catches. In addition, the percentages of these two predators were compared using the chi square test, taking into account the total bark beetles and predators caught by standard and modified traps. The size of beetles > 8 mm trapped in 2019 in modified traps were compared with the size of those from standard traps with ANOVA; the same test was separately carried out for those species with enough catches (more than 30) in each kind of trap.

3. RESULTS

3.1. Catches in I. sexdentatus Pheromone Traps

During the first three years of the study (2016–2018), 42,823 I. sexdentatus adults were caught in pheromone traps (**Table 1**). However, these were not the most numerous catches; in fact, I. sexdentatus was less than a tenth compared to the non-target beetle species. The most-trapped beetles (469,545 specimens) were non-target bark beetles (**Table 1**): O. erosus, Hylurgus ligniperda (Fabr.), and T. destruens. While only few T. destruens specimens were caught (62 in total in the three years), the other two species were trapped frequently and in great numbers. More specifically, O. erosus was the most-caught species in the first three-year period, with 452,756 individuals, which represented almost 95% of all beetles snared. Among the non-target beetles were two of the bark beetles' major predators: T. caerulea and T. formicarius; each year more than 2000 specimens were caught, with a total of 7500 in the three years. The predator/prey (I. sexdentatus) ratios for T. formicarius were 0.13, 0.12 and 0.08 in 2016, 2017, and 2018 respectively, while for T. caerulea they were 0.09, 0.05, 0.07. Finally, other beetles, larger than bark beetles, were also captured, but these were neither identified nor counted in the first three years of the study.

Table 1. Total number of beetles trapped with the Ips sexdentatus aggregation pheromone during the 2016–2018 period in the study site (Municipality of Grosseto, Tuscany, Italy).

Year	*I. sexdentatus*	Non-Target Beetles				Predator/Prey Ratios	
		Other Bark Beetles	*T. formicarius*	*T. caerulea*	Total	*T. formicarius*	*T. caerulea*
2016	11,042	231,284	1447	1013	233,744	0.13	0.09
2017	12,774	120,574	1587	621	122,782	0.12	0.05
2018	19,007	117,687	1450	1382	120,519	0.08	0.07
Total	42,823	469,545	4484	3016	477,045		

During 2019, the last year of the study, 182,641 specimens were caught across both types of traps; this tally takes into account only the species that had been counted during the first three-year period (I. sexdentatus, T. formicarius, T. caerulea, and other bark beetles). Again, the great majority were non-target bark beetles (170,219) (**Table 2**): about 98% of these trapped beetles were O. erosus, which was the most captured species, with 167,773 individuals; as regards the other bark beetles, 1855 adults of H. ligniperda and one T. destruens were recorded. Besides these main species, other minor bark beetles were recorded with a total of 590 specimens: Carphoborus pini Eichhoff (33), Coccotrypes dactyliperda Fabricius (8), Crypturgus mediterraneus Eichhoff (31), Crypturgus pusillus Gyllenhal (143), Hylurgus micklitzi Wachtl (97), Pityogenes bidentatus Herbst (251), Pityophthorus pubescens Marsham (21), and Xylocleptes biuncus Reitter (6). Again, the two most important predator species, T. caerulea (2409) and T. formicarius (925), were trapped. The predator/prey ratios calculated on standard traps were 0.13 for T. formicarius and 0.33 for T. caerulea. In 2019, however, other large non-target beetles were also taken into account, for a total of 1,662 recorded specimens. This group included specimens from five different families: Melolonthidae, with one species, Amadotrogus grassii (Mainardi) (112); Elateridae, with two species, Lacon punctatus (Herbst) (191) and Melanotus (Melanotus) crassicollis (Erichson) (342); Buprestidae, with two species, Buprestis (Buprestis) novemmaculata Linnaeus (92), and Chalcophora detrita (Klug) (5); Cerambycidae, with three species Monochamus galloprovincialis (Olivier) (8), Acanthocinus griseus (Fabricius) (838), Oxypleurus nodieri Mulsant (3), and finally Curculionidae not in the sub-family of Scolytinae, with one species, Brachyderes (Brachyderes) incanus (Linnaeus) (71).

Table 2. Number of beetles trapped with the Ips sexdentatus aggregation pheromone during 2019 in the study site (Municipality of Grosseto, Tuscany, Italy) with standard and modified slot traps.

Type of Trap	N. of Traps	*I. sexdentatus*	Non-Target Beetles				Total
			Other Bark Beetles	Larger Beetles			
				T. formicarius	*T. caerulea*	Others	
Standard	12	3982	53,622	505	1333	1392	56,852
Modified	13	5106	116,597	420	1076	270	118,362
Total	25	9088	170,219	925	2409	1662	175,214

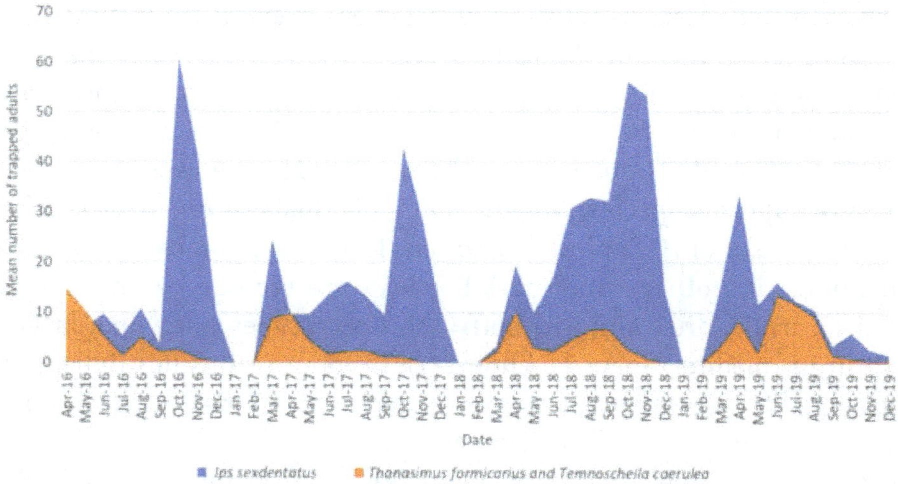

Figure 3. Mean number of trapped adults of Ips sexdentatus and its main predators (Thanasimus formicarius and Temnoscheila caerulea) in the study site (Municipality of Grosseto, Tuscany, Italy) using standard slot traps, from 2016 to 2019.

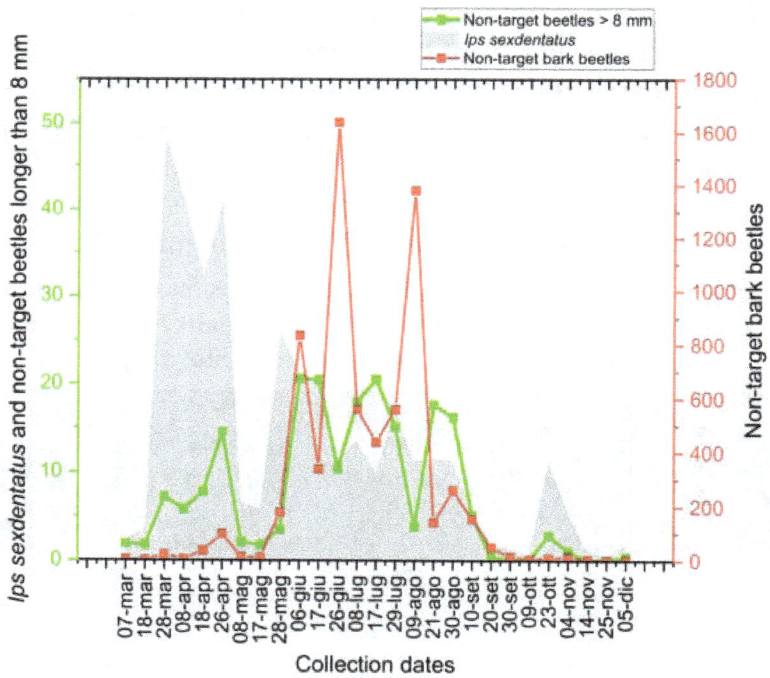

Figure 4. Mean number (standard and modified traps together) of trapped adults of Ips sexdentatus, non-target bark beetles, and non-target beetles > 8 mm in the study site (Municipality of Grosseto, Tuscany, Italy) during 2019.

Catches of target and non-target beetles were differently distributed during the year. I. sexdentatus adults in 2016, 2017, and 2018 had three peaks in activity: spring

(March/April) (in 2016 this first peak was not noted because traps were deployed later), summer (July/August), and particularly autumn (October/November). On the contrary, in 2019 the autumnal catch showed a remarkable drop (**Figure 3**). The two main I. sexdentatus predators (T. caerulea and T. formicarius), instead, were more active in spring, particularly in April, and summer, while in autumn only a very low number of adults were trapped. In fact, in the spring/summer period during the years 2016/2018, the mean predator/prey ratio was 0.43, while during the autumn it was 0.03. This activity distribution was also true for all other non-target beetles trapped. In fact, during the study period non-target bark beetles were caught particularly during the spring/summer period, with a non-target/target bark beetle ratio of 21, while during autumn these non-target catches were much lower, plummeting to a ratio of 0.51. Accordingly, beetles > 8 mm and non-target bark beetles were caught mainly during the 2019 spring/summer period, while very low catches occurred in the autumn (**Figure 4**).

3.2. Standard Traps Versus Modified Traps

The mean number of I. sexdentatus catches in standard traps was not different when compared to the average catches of I. sexdentatus in modified traps (ANOVA, df = 1, F = 0.528, p = 0.475). The mean number of non-target bark beetle catches did not differ between the two types of traps (ANOVA, df = 1, F = 2.325, p = 0.141). On the contrary, the mean number of all non-target beetles > 8 mm was lower in the modified traps (ANOVA, df = 1, F = 27.744, p < 0.001). In fact, the percentage of all non-target beetles > 8 mm trapped by the standard traps was 44.79% of the total (7,212 specimens, not including the non-target bark beetles) (**Figure 5**A), while in the modified ones this percentage dropped to 25.70% of the total (6871 specimens) (**Figure 5**B). In addition, the modified traps had a lower species diversity (Simpson's index = 0.56) compared to the standard ones (Simpson's index = 0.75).

For 10 out of the 11 species included in the "non-target beetles > 8 mm", a decrease in catches was recorded with the modified traps, although this decrease was statistically significant only for seven species: L. punctatus (ANOVA, df = 1, F = 34.181, p < 0.001), M. crassicollis (ANOVA, df = 1, F = 30.968, p < 0.001), B. novemmaculata (ANOVA, df = 1, F = 14.493, p < 0.01), A. griseus (ANOVA, df = 1, F = 18.985, p < 0.001), and B. incanus (ANOVA, df = 1, F = 6.163, p < 0.05); both C. detrita and M. galloprovincialis were caught only in standard traps.

The two main bark beetle predators were the most-caught species in this category (non-target beetles > 8 mm), though the efficacy of modified traps was less evident. Indeed, catches in 2019 in the two types of traps were not statistically different (ANOVA, df = 1, F = 3.882, p = 0.061). Furthermore, the predator/prey ratios for T. formicarius in standard and modified traps were 0.13 and 0.08, respectively, while for T. caerulea the predator/prey ratios were 0.33 and 0.21. Predator (T.

formicarius and T. caerulea) catches were positively correlated to bark beetle (I. sexdentatus and non-target bark beetles) catches (Pearson correlation, $r = 0.404$, $p < 0.01$), while no differences in all bark beetle captures emerged between standard and modified traps (ANOVA, df = 1, $F = 2.314$, $p = 0.142$). However, the percentage of predators in modified traps (1.22% of the total bark beetles and predators) was significantly lower than that in standard traps (3.10%) (Chi square test, $\chi^2 = 790.909$, $p < 0.001$).

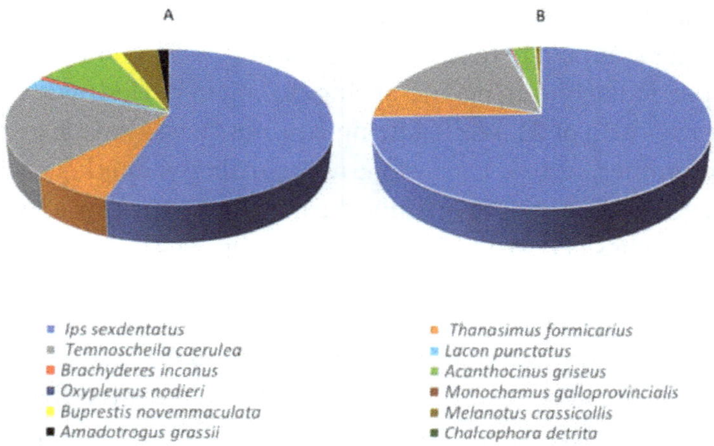

Figure 5. Percentages of I. sexdentatus and non-target beetles > 8 mm caught in standard traps (A) and modified traps (B) baited with the I. sexdentatus aggregation pheromone in the study site (Municipality of Grosseto, Tuscany, Italy) during 2019.

Figure 6. Mean catches per trap per control date of T. caerulea and T. formicarius in standard and modified traps in the study site (Municipality of Grosseto, Tuscany, Italy). Trap modifications were not present until 2019; however, trap position never changed during the four-year study. Bars indicate standard error.

In addition, considering standard traps throughout the whole study period, the mean number of catches for 2019 were significantly higher than the previous three years' (GLM, Wald Chi-Square = 46,295, $p < 0.001$). In modified traps (in 2016–2018 the same traps, located in the same sites, but not yet modified, were used) there was not a statistically significant difference (**Figure 6**).

Trap modification favored the larger species (**Figure 7**); in fact, beetles captured in standard traps had a higher mean size (14.22 mm) compared to those in the modified traps (12.14 mm) (ANOVA, df = 1, F = 15.014, $p < 0.001$). However, T. caerulea did not benefit from trap modification despite its size (**Figure 7**). Comparing the size of trapped specimens of each species in standard and modified traps, a general significant difference was observed: for each species, specimens captured by modified traps were smaller in size (ANOVA, df = 7, F = 14.972, $p < 0.05$). However, only in the case of A. griseus did this difference end up being statistically significant (Wald Chi square = 8.878, $p < 0.01$).

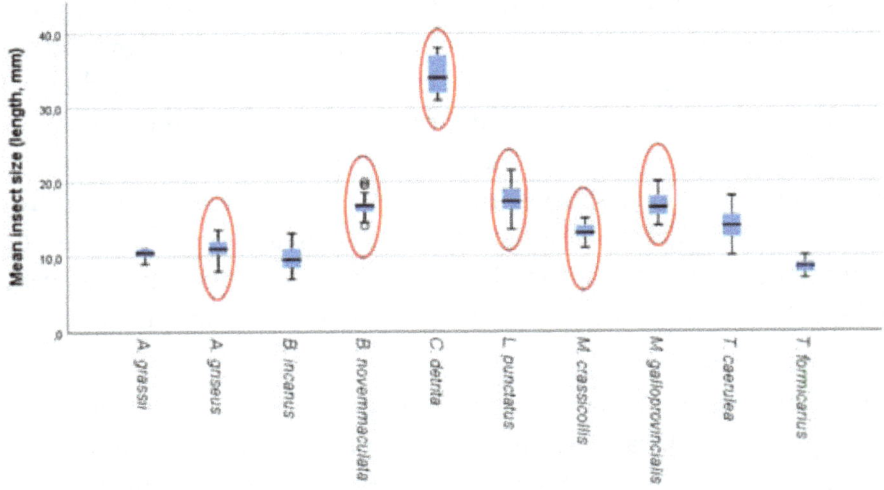

Figure 7. Body length (mm) of the trapped insect species. Red ovals represent species with a subsequent significant catch reduction in modified traps. Bars indicate standard error.

4. DISCUSSION

As already assessed in other studies [**19,20**] I. sexdentatus aggregation pheromone attracts many other saproxylic beetles, among which are several bark beetles receptive to the same pheromonal components [**30**]. In our study, the most-trapped beetles were O. erosus and H. ligniperda, both of which had been shown to be captured when the I. sexdentatus pheromone was used as bait [**8,9,20**]. Furthermore, Serez [**31**] demonstrated how commercially available pheromones for I. sexdentatus can be used to attract O. erosus. The numbers of specimens trapped in our studies were in agreement with the observations of Serez [**31**] in Turkish areas infested by O. erosus; moreover, O. erosus has been recently confirmed as an increasingly dangerous pest in Mediterranean pine forests [**32**]. However, in our study area, O. erosus seems less aggressive. In fact, despite the abundant catches and the fact that O. erosus is able to kill whole trees, in our study only small pine branches were attacked, while no pines showed large areas of the trunk or of the main branches infested by this species. Interestingly, T. destruens, the other bark beetle which causes serious damage in the study area's pine forest, was trapped only in low numbers (63 specimens total over the four years).

Other saproxylic beetles trapped in I. sexdentatus traps were bark beetles' predators, which can be kairomonally attracted by the bait intended for their prey, as stated by other studies [**8,19**]. In fact, these species can use aggregation pheromones, together with volatiles, to locate their bark beetle prey [**18**]. In our study, four predators, T. formicarius, T. caerulea, L. punctatus, and M. crassicollis were trapped. The first three species had already been collected in traps baited with pheromone blends of I. sexdentatus in other studies [**8,19,20,33**]. T. formicarius and T. caerulea are important bark beetle natural enemies that contribute to their population control by preying on their larvae under the bark [**9**]; thus, catching these predators during mass-trapping constitutes a problem for bark beetle management. Bark beetle control methods must be sustainable, impacting only (or almost only) the target species. In this study, the predator/prey ratios were considerably high compared to other studies [**34**]. This could cause a boomerang effect, since it was estimated that a reduction of predator densities could lead, within 4–20 bark-beetle generations, to a doubling of the pest population, prolonging bark beetle outbreaks [**35**].

Yet another important consideration is that other saproxylic beetles, including vulnerable species, also locate their plant hosts by exploiting bark beetle pheromones [**36**]. In our study's catches, almost all beetles > 8 mm were saproxylic. Among these, we found several vulnerable species: Melolonthidae A. grassii is considered rare; one species of Buprestidae, C. detrita, is categorized as "endangered" in the European red list of the Italian saproxylic beetles; and one, Cerambycidae O. nodieri, which is "near threatened" and is also considered a facultative predator of bark beetles [**37**]; all other species were of lesser concern [**38**]. Many of the captured species had already been caught in similar studies with traps baited with conifer monoterpenes and/or bark beetle pheromones containing ipsenol and 2-methyl-3-buten-2-ol, or ipsdienol [**8,9,19,20,36**]. Therefore, besides the problem of predator decline, mass-trapping might reduce the overall biodiversity of saproxylic insects, with even higher impacts on endangered species.

The hypothesis that low I. sexdentatus populations may negatively affect target/non-target ratios in traps was proposed by Panzavolta et al. [**20**], who, in their two-year study, observed low population densities and low percentages of the target bark beetle in traps (ranging between 2% and 35%). In the present study, higher averages of I. sexdentatus specimens were caught by pheromone traps, and more severe damage to the pine forest was observed, both indicative signs of a larger population density. In fact, considering only the study period of Panzavolta et al. [**20**] (from June to the end of I. sexdentatus flight), we trapped a mean of about 492 specimens per trap per year, while in their study only 44.3 and 3.7 mean catches were obtained. However, in the current study, higher I. sexdentatus catches corresponded to higher non-target catches. In fact, during the four years, I. sexdentatus accounted for only 8% of the total specimens collected in the pheromone traps. Therefore, the low specificity of the pheromone blend used is confirmed, regardless of I. sexdentatus population level.

This study confirmed the effectiveness of the trap modifications proposed by Martín et al. [**9**] if all saproxylic beetles > 8 mm were considered together, although modifications applied are still not enough to negate non-target catches of some species taken individually. More specifically, results were not satisfactory for all natural enemies: only for L. punctatus and M. crassicollis was the catch reduction in modified traps significant in comparison with standard traps, while the two main predators T. formicarius and T. caerulea were caught in high numbers in modified traps as well, in agreement with Martin et al. [**9**]. Traps in our study were also provided with escape windows, not present in the slot traps modified by Martin et al. [**9**], but this did not lead to fewer catches of this non-target species when compared with their study, as we recorded a ratio of one trapped T. caerulea for every five I. sexdentatus. However, we found that catches of T. formicarius and T. caerulea were positively correlated with those of bark beetles (all trapped species). Probably, as suggested by Bakke and Kvamme [**39**], predators may be also attracted by already caught bark beetles inside the trap, which may release pheromones with their frass. In fact, some bark beetle species are able to produce pheromones after emergence without feeding, as demonstrated for I. paraconfusus Lanier and I. pini (Say) [**40**]. In our study, traps with high bark beetle catches resulted uniformly distributed between modified and standard traps (differences in mean catches were not statistically significant). However, if we consider predators in proportion to bark beetles, a statistically significant reduction of predator catches emerged in modified traps, showing an encouraging outcome for trap modifications.

Another important strategy to reduce the impact of mass-trapping on non-target beetle species is to adjust the period of traps set in the field, exploiting the different seasonal abundance of target and non-target beetles. Overall, we observed the lowest non-target catches during autumn, both as regards beetles > 8 mm long and non-target bark beetles (**Figure 4**). This is in agreement with other studies that confirmed the lowest flight activity of bark beetles' predators during autumn

[**20,27**], or at least their higher activity during spring [**18,25**]. The highest catches of I. sexdentatus in our study area occurred, on the contrary, in autumn, at least in the 2016–2018 period. This was not confirmed in 2019; however, this was likely due to the phytosanitary felling of numerous attacked pines in the study area which was carried out at the end of 2018. As a result, I. sexdentatus population density was reduced, and the study area traits were also modified significantly.

5. CONCISE SUMMARY

The findings of our study highlight the limited specificity of the I. sexdentatus aggregation pheromone, as well as its detrimental impacts when employed in mass-trapping initiatives. Modifications to slot traps can help reduce the adverse impact of mass-trapping on non-target saproxylic beetles, especially those larger than 8 mm. However, the capture of key predators of bark beetles, namely T. formicarius and T. caerulea, continues to be excessively high. This negative effect may worsen in subsequent generations, resulting in a twofold increase in bark beetle populations and prolonging outbreaks. Hence, it would be highly beneficial to take advantage of the varying seasonal prevalence of desired and undesired beetles, especially in temperate regions where the desired species has its highest population peak during autumn, a time when other saproxylic species are often less common. To minimize the adverse impacts of managing I. sexdentatus in the forest, particularly in reducing unintended catches, we suggest using slot traps that have been altered with a 6-mm mesh screen and three extra escape windows. Additionally, it is advisable to focus mass-trapping efforts during the autumn season.

6. AT A GLANCE

We conducted a study to examine the adverse consequences of employing mass trapping techniques for bark beetles, with a particular focus on unintended captures of non-target organisms that are natural predators of the target species. Slot traps were subjected to testing after being upgraded with mesh screens and escape windows in order to enhance their selectivity. Context and Goals: Thanasimus formicarius and Temnoscheila caerulea, which are natural enemies of bark beetles, are often caught in large quantities in pheromone traps designed for Ips sexdentatus. This can result in significantly larger populations of pests in the following 4-20 generations of beetles, along with other saproxylic insects. Between 2016 and 2019, while conducting mass-trapping of I. sexdentatus in a pine forest in Tuscany, Italy, the number of unintended catches was recorded. An assessment was conducted on trap improvements to reduce unintended captures, particularly those involving the predators of bark beetles. Experimental procedures and techniques: Exactly 25 bark-beetle slot traps were evenly distributed at intervals of approximately 75 meters in a pine forest that was infested by I. sexdentatus. The traps were lured with the aggregation pheromone of I. sexdentatus, which mostly consists of ipsenol, ipsdienol, and 2-methyl-3-buten-2-ol. Specimens were gathered at 10-day intervals from March through December. In 2019, a total of 13 traps were altered by adding a 6-mm mesh screen on the collection container and installing three escape windows measuring 60 mm × 8 mm just above the screen. The "modified traps" and their captures were evaluated independently from the 12 remaining "standard traps." All species of bark beetles were documented, along with all kinds of beetles that were larger than 8 mm. In sum, target catches accounted for less than 10% of the total beetle catches. The bark beetle Orthotomicus erosus Wollaston was the species that was collected the most. The alteration of traps enabled the larger species to escape, leading to a decrease in the average size of the captured animals. Despite the continued occurrence of non-target catches among predators, the redesigned traps resulted in a statistically significant decrease in the number of main predators (T. formicarius and T. caerulea) caught in relation to bark beetles. This outcome is encouraging. Conclusion: Modifying traps can help alleviate the issue of unintentional captures during mass capturing, particularly by lowering captures of beetle species that are larger than the intended target. It is crucial to only schedule mass trapping during seasons when the target adults are more active than the primary predator adults. This will help to reduce their catches and minimize the detrimental impact on pest control and biodiversity.

REFERENCES

1. Hroššo, B.; Mezei, P.; Potterf, M.; Majdák, A.; Blaženec, M.; Korolyova, N.; Jakuš, R. Drivers of Spruce Bark Beetle (Ips typographus) infestations on downed trees after severe windthrow. Forests 2020, 11, 1290. [**Google Scholar**] [**CrossRef**]
2. Nikolov, C.; Konôpka, B.; Kajba, M.; Galko, J.; Kunca, A.; Janský, L. Post-disaster forest management and bark beetle outbreak in Tatra National Park, Slovakia. Mt. Res. Dev. 2014, 34, 326–335. [**Google Scholar**] [**CrossRef**]
3. Hlásny, T.; Mátyás, C.; Seidl, R.; Kulla, L.; Merganičová, K.; Trombik, J.; Dobor, L.; Barcza, Z.; Konôpka, B. Climate change increases the drought risk in Central European forests: What are the options for adaptation? For. J. 2014, 60, 5–18. [**Google Scholar**] [**CrossRef**] [**Green Version**]
4. Rossi, J.P.; Samalens, J.C.; Guyon, D.; van Halder, I.; Jactel, H.; Menassieu, P.; Piou, D. Multiscale spatial variation of the bark beetle Ips sexdentatus damage in a pine plantation forest (Landes de Gascogne, Southwestern France). For. Ecol. Manag. 2009, 257, 1551–1557. [**Google Scholar**] [**CrossRef**]
5. Romero, S.L.; Ochoa, P.R.; Bilbao, J.C.I.; Lafuente, A.G. Los Escolítidos de las Coníferas del País Vasco: Guía Práctica Para su Identificación y Control, 1st ed.; Servicio Central de Publicaciones del Gobierno Vasco: Vitoria-Gasteiz, Spain, 2007; pp. 136–139. ISBN 9788445726501. [**Google Scholar**]
6. Eppo Global Databae. Available online: **https://gd.eppo.int/taxon/IPSXSE/documents** (accessed on 10 December 2020).

7. Hlásny, T.; Krokene, P.; Liebhold, A.; Montagné-Huck, C.; Müller, J.; Qin, H.; Raffa, K.; Schelhaas, M.-J.; Seidl, R.; Svoboda, M.; et al. Living with Bark Beetles: Impacts, Outlook and Management Options; From Science to Policy 8; European Forest Institute: Joensuu, Finland, 2019; ISBN 9789525980752. [**Google Scholar**]

8. Etxebeste, I.; Lencina, J.L.; Pajares, J. Saproxylic community, guild and species responses to varying pheromone components of a pine bark beetle. Bull. Entomol. Res. 2013, 103, 497–510. [**Google Scholar**] [**CrossRef**] [**PubMed**]

9. Martín, A.; Etxebeste, I.; Pérez, G.; Álvarez, G.; Sánchez, E.; Pajares, J. Modified pheromone traps help reduce bycatch of bark-beetle natural enemies. Agric. For. Entomol. 2013, 15, 86–97. [**Google Scholar**] [**CrossRef**]

10. Del Perugia, B.; Travaglini, D.; Bottalico, F.; Nocentini, S.; Rossi, P.; Salbitano, F.; Sanesi, G. Are Italian stone pine forests (Pinus pinea L.) an endangered coastal landscape? A case study in Tuscany. L'Italia For. E Mont. 2017, 72, 103–121. [**Google Scholar**] [**CrossRef**] [**Green Version**]

11. Fettig, C.J.; Hilszczański, J. Management strategies for bark beetles in conifer forests. In Bark Beetles: Biology and Ecology of Native and Invasive Species, 1st ed.; Vega, F.E., Hofstetter, R.W., Eds.; Academic Press: London, UK, 2015; pp. 555–584. [**Google Scholar**]

12. Gregoire, J.C.; Raty, L.; Drumont, A.; de Windt, N. Pheromone mass trapping: Does it protect windfalls from attack by Ips typographus L. (Coleoptera: Scolytidae)? In Proceedings of the "Integrating cultural tactics into the management of bark beetle and reforestation pests", Vallombrosa, Italy, 1–3 September 1996; Gen. Tech. Rep. NE-236. U.S. Department of Agriculture, Forest Service, Northeastern Forest Experiment Station: Radnor, PA, USA, 1997; pp. 1–8. [**Google Scholar**] [**CrossRef**]

13. Vité, J.P. The European struggle to control Ips typographus-past, present and future. Ecography 1989, 12, 520–525. [**Google Scholar**] [**CrossRef**]

14. Niemeyer, H. Monitoring Ips typographus and Pityogenes chalcographus (Col., Scolytidae) in Lower Saxony and Schleswig-Holstein. J. Appl. Entomol. 1992, 114, 98–102. [**Google Scholar**] [**CrossRef**]

15. Jakuš, R. A method for the protection of spruce stands against Ips typographus by the use of barriers of pheromone traps in north-eastern Slovakia. Anz. Schadlingskd. Pfl. 1998, 71, 152–158. [**Google Scholar**] [**CrossRef**]

16. Schlyter, F.; Zhang, Q.H.; Liu, G.T.; Ji, L.Z. A successful case of pheromone mass trapping of the bark beetle Ips duplicatus in a forest island, analysed by 20-year time-series data. Integr. Pest Manag. Rev. 2001, 6, 185–196. [**Google Scholar**] [**CrossRef**]

17. Faccoli, M.; Stergulc, F. Damage reduction and performance of mass trapping devices for forest protection against the spruce bark beetle, Ips typographus (Coleoptera Curculionidae Scolytinae). Ann. For. Sci. 2008, 65, 309. [**Google Scholar**] [**CrossRef**] [**Green Version**]

18. Aukema, B.H.; Dahlsten, D.L.; Raffa, K.F. Exploiting behavioral disparities among predators and prey to selectively remove pests: Maximizing the ratio of bark beetles to predators removed during semiochemieally based trap-out. Environ. Entomol. 2000, 29, 651–660. [**Google Scholar**] [**CrossRef**]

19. Etxebeste, I.; Álvarez, G.; Pérez, G.; Pajares, J.A. Field response of the six-toothed pine bark beetle, Ips sexdentatus (Col.: Curculionidae, Scolytinae), to pheromonal blend candidates. J. Appl. Entomol. 2012, 136, 431–444. **[Google Scholar]** **[CrossRef]**

20. Panzavolta, T.; Bracalini, M.; Bonuomo, L.; Croci, F.; Tiberi, R. Field response of non-target beetles to Ips sexdentatus aggregation pheromone and pine volatiles. J. Appl. Entomol. 2014, 138. **[Google Scholar]** **[CrossRef]**

21. Raffa, K.F. Mixed messages across multiple trophic levels: The ecology of bark beetle chemical communication systems. Chemoecology 2001, 11, 49–65. **[Google Scholar]** **[CrossRef]**

22. Wood, D.L. The role of pheromones, kairomones, and allomones in the host selection and colonization behavior of bark beetles. Annu. Rev. Entomol. 1982, 27, 411–446. **[Google Scholar]** **[CrossRef]**

23. Ozcan, G.E.; Eroglu, M.; Akinci, H.A. Use of pheromone-baited traps for monitoring Ips sexdentatus (Boerner) (Coleoptera: Curculionidae) in oriental spruce stands. African J. Biotechnol. 2011, 10, 16351–16360. **[Google Scholar]** **[CrossRef]**

24. Pajares, J.A.; Ibeas, F.; Díez, J.J.; Gallego, D. Attractive reponses by Monochamus galloprovincialis (Col., Cerambycidae) to host and bark bettle semiochemicals. J. Appl. Entomol. 2004, 128, 633–638. **[Google Scholar]** **[CrossRef]**

25. Aukema, B.H.; Dahlsten, D.L.; Raffa, K.F. Improved population monitoring of bark beetles and predators by incorporating disparate behavioral responses to semiochemicals. Environ. Entomol. 2000, 29, 618–629. **[Google Scholar]** **[CrossRef]**

26. Zhou, J.; Ross, D.W.; Niwa, C.G. Kairomonal response of Thanasimus undatulus, Enoclerus sphegeus (Coleoptera: Cleridae), and Temnochila chlorodia (Coleoptera: Trogositidae) to bark beetle semiochemicals in Eastern Oregon. Environ. Entomol. 2001, 30, 993–998. **[Google Scholar]** **[CrossRef]**

27. Williams, K.K.; McMillin, J.D.; Degomez, T.E. Relative and seasonal abundance of three bark beetle predators (Coleoptera: Trogositidae, Cleridae) Across an elevation gradient in ponderosa pine forests of North Central Arizona. West. North Am. Nat. 2009, 69, 351–363. **[Google Scholar]** **[CrossRef]** **[Green Version]**

28. Ross, D.W.; Daterman, G.E. Pheromone-Baited Traps for Dendroctonus pseudotsugae (Coleoptera: Scolytidae): Influence of selected release rates and trap designs. J. Econ. Entomol. 1998, 91, 500–506. **[Google Scholar]** **[CrossRef]**

29. Bracalini, M.; Croci, F.; Tiberi, R.; Panzavolta, T. Studying Ips sexdentatus (Börner) outbreaks in Italian coastal pine stands: Technical issues behind trapping in warmer climates. In Proceedings of the XI European Congress of Entomology, Naples, Italy, 2–6 July 2018; p. 171. **[Google Scholar]**

30. Seybold, S.J.; Huber, D.P.W.; Lee, J.C.; Graves, A.D.; Bohlmann, J. Pine monoterpenes and pine bark beetles: A marriage of convenience for defense and chemical communication. Phytochem. Rev. 2006, 5, 143–178. **[Google Scholar]** **[CrossRef]**

31. Serez, M. Verwendung des Aggregationspheromon-Präparats "Ipslure" gegen den mediterranen Kiefernborkenkäfer, Ips (Orthotomicus) erosus (Woll.) (Col., Scolytidae). Schadlingskd. Pfl. 1987, 60, 94–95. **[Google Scholar]** **[CrossRef]**

32. Pernek, M.; Lacković, N.; Lukić, I.; Zorić, N.; Matošević, D. Outbreak of Orthotomicus erosus (Coleoptera, Curculionidae) on Aleppo pine in the Mediterranean region in Croatia. South East Eur. For. 2019, 10, 19–27. [**Google Scholar**] [**CrossRef**] [**Green Version**]

33. Larrañaga, I.E.; Hernández, A.B.M.; Escolar, G.P.; Fernández, M.F.; Casero, J.J.D.; Alonso, J.A.P. Evaluación de compuestos semioquímicos para su empleo en estrategias de aumento de enemigos naturales de Ips sexdentatus (Col.: Scolytidae). Cuad. Soc. Esp. Cienc. Forestales 2008, 32, 27–32. [**Google Scholar**]

34. Warzeè, N.; Gilbert, M.; Grégoire, J.-C. Predator/prey ratios: A measure of bark-beetle population status influenced by stand composition in different French stands after the 1999 storms. Ann. For. Sci. 2006, 63, 301–308. [**Google Scholar**] [**CrossRef**] [**Green Version**]

35. Krista, L.; Ryall, L.F. Habitat loss decreases predator-prey ratios in a pine-bark beetle system. Oikos 2005, 110, 265–270. [**Google Scholar**]

36. Allison, J.D.; Borden, J.H.; Seybold, S.J. A review of the chemical ecology of the Cerambycidae (Coleoptera). Chemoecology 2004, 14, 123–150. [**Google Scholar**] [**CrossRef**]

37. Gallego, D.; Galián, J.; Diez, J.J.; Pajares, J.A. Kairomonal responses of Tomicus destruens (Col., Scolytidae) to host volatiles α-pinene and ethanol. J. Appl. Entomol. 2008, 132, 654–662. [**Google Scholar**] [**CrossRef**]

38. Audisio, P.; Baviera, C.; Carpaneto, G.M.; Biscaccianti, A.B.; Battistoni, A.; Teofili, C.; Rondini, C. Lista Rossa IUCN dei Coleotteri Saproxilici Italiani, 1st ed.; Comitato Italiano IUCN e Ministero dell'Ambiente e della Tutela del Territorio e del Mare: Rome, Italy, 2014. [**Google Scholar**]

39. Bakke, A.; Kvamme, T. Kairomone response in Thanasimus predators to pheromone components of Ips typographus. J. Chem. Ecol. 1981, 7, 305–312. [**Google Scholar**] [**CrossRef**] [**PubMed**]

40. Seybold, S.J.; Quilici, D.R.; Tillman, J.A.; Vanderwelt, D.; Wood, D.L.; Blomquist, G.J. De novo biosynthesis of the aggregation pheromone components ipsenol and ipsdienol by the pine bark beetles Ips paraconfusus Lanier and Ips pini (Say) (Coleoptera: Scolytidae). Proc. Natl. Acad. Sci. USA 1995, 92, 8393–8397. [**Google Scholar**] [**CrossRef**] [**PubMed**] [**Green Version**]

17 AN EVALUATION OF THE IMPACT OF FOREST MANAGEMENT ON ARTHROPODS THAT INHABIT TERRESTRIAL LEAF LITTER

1. INTRODUCTION: AN OVERVIEW

Arthropods that live in leaf litter (as shown in Table 1) are important and abundant organisms in ecosystems, contributing significantly to both the total mass and variety of species [1,2]. Forested ecosystems rely heavily on them [3,4,5] as they offer a diverse range of ecosystem services. Specifically, arthropods that live in litter fulfill the roles of predators [6,7,8,9,10], decomposers of organic matter [11,12], and facilitators of nutrient mineralization in the soil [13]. The feeding activity of these arthropods promotes the proliferation of microorganisms [14] and contributes to the accumulation of fecal matter in the substrate [15]. Certain arthropods, such as termites and ants, that live in leaf litter are recognized as "ecosystem engineers" because of their significant impact on the ecosystem [16,17,18].

Table 1. Arthropods commonly encountered in forest leaf litter.

Class	Order	Family
Arachnida	Acari	Ixodidae
Crustacea	Araneae	Oribatida
Chilopoda	Opiliones	Gnaphosidae
Diplopoda	Pseudoscorpionida	Lycosidae
Hexapoda	Amphipoda	Salticidae
	Decapoda	Theridiidae
	Isopoda	Cantharidae
	Coleoptera	Carabidae
	Collembola	Chrysomelidae

Class	Order	Family
	Dermaptera	Curculionidae
	Diplura	Elateridae
	Diptera	Histerida
	Orthoptera	Psephalidae
	Hemiptera	Ptilodatilidae
	Hymenoptera	Scydmaenidae
	Lepidoptera	Staphylinidae
	Thysanoptera	Tenebrionidae
		Gryllidae
		Cercopidae
		Cicadellidae
		Pentatomidae
		Formicidae
		Microhymenoptera
		Geomteridae
		Noctuidae

Leaf litter arthropods, because of their wide range of evolutionary and functional characteristics, can exert substantial effects on other levels of the food chain [19]. Leaf litter composition has been demonstrated to modify the alkaloid profiles of ants that inhabit the litter. This, in turn, affects the chemical defenses of poison frogs who consume these ants [20]. Leaf litter-dwelling arthropods have also been utilized as indicators of ecosystem and soil health [5,21,22,23,24]. Certain terrestrial taxa, such as beetles (Coleoptera) and spiders (Araneae), exhibit high sensitivity to alterations in their habitat [25,26]. Forest management practices have a greater impact on these groups compared to others, both directly and indirectly.

The impact of forest management on arthropod groups varies depending on their area of origin. The majority of research on this topic has focused on tropical systems [27,28,29,30]. Nevertheless, there are other studies conducted in Europe [23] and North America [31,32] that provide insights into the distinct effects of forest management in temperate forests. Arthropods in tropical systems typically decrease in response to forest harvests and changes in land use, such as urbanization. On the other hand, temperate systems experience an increase in arthropod diversity under comparable circumstances [33]. However, these impacts vary depending on the taxa. Acari, particularly oribatid mites, exhibit a favorable response to a moderate level of urbanization, but Collembola and Diplopoda demonstrate an unfavorable response to urbanization at all levels. Previous studies have demonstrated variations in the diversity and abundance of Collembola, depending on the kind of leaf litter, irrespective of management or land-use practices [36]. Therefore, responses may also vary according to the plant species. The presence of a wide range of species in locations such as urban forests does not automatically result in a corresponding high level of variety in the surrounding areas. Artificial obstacles such as roads and sidewalks significantly hinder the movement of arthropods in these environments, resulting in uneven and unpredictable distribution of species variety [37].

Forest management style and regionality have a significant influence on the degree and direction of impacts on arthropods. However, there are some overall patterns that can be seen. Plantation forestry, despite certain misconceptions, does indeed promote the variety of arthropods found in leaf litter [38,39]. Furthermore, arthropods have shown resilience to both site preparation [40] and harvesting procedures [41]. Nevertheless, these interactions vary depending on the taxonomic group, with mobile species experiencing differing effects compared to sessile species [42]. Winter-active spiders in plantation woods exhibited variations in community composition, which were influenced by the type of plantation species and whether it was indigenous to the area [43]. The primary aims of this research are to examine the impacts of various forest management practices on arthropods found in leaf litter, and to identify areas of limited understanding concerning arthropods that inhabit forested environments.

In order to evaluate the existing literature about forest management practices and their impact on arthropods that reside in leaf litter, I conducted a search using Google Scholar and ISI Web of Knowledge. I utilized the keywords "leaf litter," "arthropods," and "forest management," both individually and in various combinations. Articles were deemed suitable if they included quantitative data on leaf litter arthropods and specified the type of forest management employed, such as clear-cutting or slash management. This review did not include studies that examined the effects of arthropods on forest productivity or articles that assessed soil arthropods.

2. EFFECTS OF FOREST HARVESTING

Forest management is frequently done in such a way as to mimic natural disturbances like fire, insect outbreaks, or hurricanes [44]. Natural disturbances may occur frequently and create partial-stand clearings, resulting in a mosaic of stand types and ages. Natural disturbances may also occur as whole-stand replacing events [45]. Forest management affects microclimates of the forest floor through altering light availability, wind, temperature, and moisture, thereby indirectly affecting the arthropods that inhabit the leaf litter [46].

Harvest methods which mimic large disturbances (e.g., clear-cuts, gap cuts) create significant amounts of edge habitats which tend to have more extreme microclimates [47] and support a lower abundance of leaf litter arthropods [48]. Clear cuts also increase the temperature extremes of the forest floor and remove organic matter inputs, both of which negatively impact leaf litter arthropods [49]. Compared to unforested habitats, however, forests may have lower diversity [50] and more clustered phylogenetic relationships of leaf litter communities [32]. Single-tree selection also has significant impacts on some leaf litter dwelling arthropods and has been found to shift carabid communities towards those more associated with grassland habitats, even without substantial tree removal [51].

In addition, soil compaction from logging equipment results in a reduction in arthropod diversity [52]. In intensively managed forests like oil palm plantations, abundance and diversity of leaf litter arthropods was significantly lower compared to untouched forests [53]. Forest harvesting may also have positive effects on leaf litter arthropod communities, however. Ground-dwelling arthropods are more abundant near course woody debris that is left over from harvest operations [31,54]. This increase in course woody debris may explain observed increases in functional diversity of ants, specifically [55].

While some information is known on the impacts of forest management on leaf litter arthropods, little is known on how specific operations (e.g., single-tree selection, gap harvests) affects individual groups or their interactions. There is also little known about how arthropod communities respond to forest harvests over time or how their responses compare to that of natural disturbances.

3. EFFECTS OF FOREST RESTORATION

Rainforest restoration presents unique forest management challenges with a variety of techniques available [56,57], and some ground-dwelling groups, like ants, have been developed as biological indicators of the progress of restoration efforts [29,30]. While restoration methods undoubtedly affect leaf litter arthropods, little has been done to examine the effects of these efforts on said communities.

In general, decomposers are less abundant in anthropogenic sites that have been cleared for pasture or other agricultural use [58,59]. However, some authors have found no differences among certain communities, such as ants, among primary forests and forests disturbed for agricultural uses [60]. Also, increased amounts of detritus leads to higher abundance of all arthropod feeding guilds in leaf litter communities, indicating that there are bottom-up limitations in rainforests [19]. These limitations have implications for methods of restoration in degraded rainforests. For example, reforestation efforts that lead to greater structural diversity of plant life likely lead to greater diversity of leaf litter as well. This diversity creates a wide range of microclimates which are beneficial to many ground-dwelling beetles [61]. Compared to other restoration methods, island plantations most closely resembled untouched forests, however, leaf litter arthropods recovered eight years post-harvest even with no restoration [27]. In contrast, methods such as brushwood transposition, wherein woody material is transported to the damaged area to create a heterogenous environment, had low leaf litter arthropod diversity [62]. In North America, second-growth forests support highly diverse populations of leaf-litter inhabiting beetles [63,64] indicating considerable resilience to forest harvesting and conservation efforts.

It is of paramount importance to keep in mind that not all arthropod species are affected in the same ways by these restoration efforts. For example, common species appear to be unaffected while rarer species are more effected by both deforestation and reforestation [65]. Considering the amazing biodiversity of arthropods in rainforests, the effects on rare species are highly concerning. There are significant gaps in knowledge regarding what those rare species are, how they interact with more common species, and how these transition periods of reforestation affect them.

4. INVASIVE SPECIES CONTROL

Control of forest pests often involves intense management including sanitation harvests and chemical control. In general, control of invasive species negatively impacts leaf litter arthropod communities, although in different ways. As previously mentioned, sanitation harvests, which are often clear cuts, negatively impact leaf litter dwelling arthropod communities through an increase in forest floor temperatures and a reduction in organic matter inputs [49]. Chemical control also impacts arthropods on the forest floor. For example, diflubenzuron, a commonly used chemical control of the invasive gypsy moth, caused reductions in mites, thrips, centipedes and spiders [3,66].

A lack of invasive species control also impacts ground dwelling arthropods in the leaf litter. For example, increased ash wood on the ground from emerald ash borer (EAB) infestation causes increases in leaf litter arthropods [67]. Tree-killing invasive insects, like EAB, may also result in changes to leaf litter-dwelling arthropod communities by temporarily increasing the litterfall input, which is then followed by significant light gaps in the canopy [68]. Invasive plants may represent an initial pulse of novel resources for leaf litter arthropods, thereby showing some benefit early on. However, following decomposition, there is an overall negative effect of invasive plants on leaf litter arthropods [69].

There are several significant gaps in knowledge with regards to invasive species control and leaf litter arthropods. The current literature mainly covers the effects of chemicals or impacts of the invasive species itself on leaf litter arthropod groups. However, there is little known about how leaf litter affects the resiliency of forests to invasive pests or how invasive plants affect these communities.

5. EFFECTS OF FIRE

Compared to logging operations, fire has a much bigger impact on leaf litter arthropods [70]. Fire has been used for thousands of years by indigenous peoples around the world in agricultural, forest, and wildlife management [71,72,73,74]. Decades of fire suppression has led to an increase in fuel load and significant changes to the landscape, such as loss of dominant fire-dependent species, and there have been recent efforts to get fire back on the ground and reincorporated as an essential part of forest management [75]. Fire has complex effects on leaf litter arthropods through the creation of more structurally diverse habitats [76,77] and through changes to the forest floor microclimate [78]. Fire increases temperature and changes soil pH, as well as fluctuations in soil moisture [79]. The exposure to extreme temperatures, increased light, and reduced moisture frequently result in habitat loss [80], especially for those fire-intolerant species that are prone to inhabiting mid- and late-successional forests [81,82]. Changes to plant communities through fire can also modify soil moisture, invariably affecting ground-dwelling arthropods [83].

Fire, like harvesting operations and invasive species control, impacts different taxa in different ways [84,85,86]. For example, species richness of beetles increased following fire in the Sierra Nevada mountain range, but the intensity and directionality of impacts were taxon specific with no general patterns [87]. Some general patterns have emerged, however, from the literature. There is typically a significant, immediate reduction in arthropod abundance following fire [28,88,89]. These communities, however, appear to be resilient to fire [86,90,91,92,93,94,95] and abundance and diversity of leaf litter arthropods actually increases following burning [96,97].

Leaf litter arthropods also have complicated relationships with fungi [98] and fire has been shown to reduce total fungal biomass in forests [99]. Arthropod feeding on senescent hyphal structures of fungi stimulates fungal growth [97], and the presence of leaf litter decomposing arthropods is significantly related to mycorrhizal growth in some tree species like red maple [14].

While it is known that fire has significant effects on leaf litter dwelling arthropods, there is a significant gap in knowledge as to how various arthropod characteristics affect their interactions with fire. How different environmental and landscape characteristics affect leaf litter following a burn and what this means for associated arthropod communities remains unexplored.

6. CONCISE SUMMARY

Although we have identified several overarching patterns regarding the impact of forest management on arthropods found in leaf litter, there is a significant amount of information yet to be discovered. Forest management operations have a definite impact on arthropods, although these creatures seem to be able to withstand and recover from the effects of human activities. Nevertheless, the impact of particular management strategies on specific species remains uncertain, which is crucial for assessing the overall health of the forest. Further research is required to comprehend the impact of different forest management practices on specific groups such as beetles, spiders, ants, and other bioindicators.

7. AT A GLANCE

Arthropods that live in leaf litter are highly common and diverse in forest ecosystems. They play a crucial role in several ecosystem functions such as decomposition, mineralization, pollination, and predation. They play a crucial role in maintaining the health and management of forests and, consequently, are influenced by our forest management practices. There is currently no existing evaluation of the literature on the effects of various forest management strategies, such as clear cutting, burning, and chemical treatment of invasive species, on ground dwelling arthropods. The aim of this paper is to assess the impact of various forest management techniques on arthropods in leaf litter ecosystems.

REFERENCES

1. Anderson, J.M. Succession, Diversity and Trophic Relationships of some Soil Animals in Decomposing Leaf Litter. *J. Anim. Ecol.* 1975, *44*, 475. [Google Scholar] [CrossRef]
2. Stork, N.E.; Blackburn, T.M. Abundance, Body Size and Biomass of Arthropods in Tropical Forest. *Oikos* 1993, *67*, 483. [Google Scholar] [CrossRef]
3. Perry, W.B.; Christiansen, T.A.; Perry, S.A. Response of Soil and Leaf Litter Microarthropods to Forest Application of Diflubenzuron. *Ecotoxicology* 1997, *6*, 87–99. [Google Scholar] [CrossRef]
4. Niemelä, J.; Haila, Y.; Punttila, P. The Importance of Small-scale Heterogeneity in Boreal Forests: Variation in Diversity in Forest-floor Invertebrates Across the Succession Gradient. *Ecography* 1996, *19*, 352–368. [Google Scholar] [CrossRef]
5. Kremen, C.; Colwell, R.K.; Erwin, T.L.; Murphy, D.D.; Noss, R.A.; Sanjayan, M.A. Terrestrial Arthropod Assemblages: Their use in Conservation Planning. *Conserv. Biol.* 1993, 796–808. [Google Scholar] [CrossRef] [Green Version]
6. Cameron, E.A.; Reeves, R.M. Carabidae (*Coleoptera*) Associated with Gypsy Moth, *Lymantria Dispar* (L.) (Lepidoptera: *Lymantriidae*), Populations Subjected to Bacillus Thuringiensis Berliner Treatments in Pennsylvania. *Can. Entomol.* 1990, *122*, 123–129. [Google Scholar] [CrossRef]
7. DuDevoir, D.S.; Reeves, R.M. Feeding Activity of Carabid Beetles and Spiders on Gypsy Moth Larvae (Lepidoptera: *Lymantriidae*) at High-Density Prey Populations. *J. Entomol. Sci.* 1990, *25*, 341–356. [Google Scholar] [CrossRef]
8. Jennings, D.T.; Dimond, J.B.; Watt, B.A. Population Densities of Spiders (Araneae) and Spruce Budworms (Lepidoptera, *Tortricidae*) on Foliage of Balsam Fir and Red Spruce in East-Central Maine. *J. Arachnol. J. Èntomol. Sci.* 1990, *25*, 341–356. [Google Scholar]

9. Mason, R.R.; Jennings, D.T.; Paul, H.G.; Wickman, B.E. Patterns of Spider (Araneae) Abundance during an Outbreak of Western Spruce Budworm (Lepidoptera: *Tortricidae*). *Environ. Entomol.* 1997, *26*, 507–518. [Google Scholar] [CrossRef]
10. Raymond, B.; Vanbergen, A.; Watt, A.; Hartley, S.E.; Cory, J.S.; Hails, R.S. Escape from Pupal Predation as a Potential Cause of Outbreaks of the Winter Moth, Operophtera Brumata. *Oikos* 2002, *98*, 219–228. [Google Scholar] [CrossRef]
11. Reichle, D.E. The Role of Soil Invertebrates in Nutrient Cycling. *Ecol. Bull.* 1977, *25*, 145–156. [Google Scholar]
12. Petal, J. The role of ants in ecosystems. In *Production Ecology of Ants and Termites*; Brian, M.V., Ed.; Cambridge University Press: London, UK, 1978; pp. 293–325. [Google Scholar]
13. Petersen, H.; Luxton, M. A Comparative Analysis of Soil Fauna Populations and their Role in Decomposition Processes. *Oikos* 1982, *39*, 288. [Google Scholar] [CrossRef]
14. Klironomos, J.N.; Kendrick, W.B. Stimulative Effects of Arthropods on Endomycorrhizas of Sugar Maple in the Presence of Decaying Litter. *Funct. Ecol.* 1995, *9*, 528. [Google Scholar] [CrossRef]
15. Webb, D.P. Regulation of deciduous forest litter decomposition by soil arthropod feces. In *The Role of Arthropods in Forest Ecosystems*; Springer: Berlin/Heidelberg, Germany, 1977; pp. 57–69. [Google Scholar]
16. Pardeshi, M.; Prusty, B.K. Termites as Ecosystem Engineers and Potentials for Soil Restoration. *Curr. Sci.* 2010, *99*, 11. [Google Scholar]
17. Jouquet, P.; Dauber, J.; Lagerlöf, J.; Lavelle, P.; Lepage, M. Soil Invertebrates as Ecosystem Engineers: Intended and Accidental Effects on Soil and Feedback Loops. *Appl. Soil Ecol.* 2006, *32*, 153–164. [Google Scholar] [CrossRef]
18. Ulyshen, M.D. Wood Decomposition as Influenced by Invertebrates. *Biol. Rev.* 2016, *91*, 70–85. [Google Scholar] [CrossRef]
19. Chen, B.; Wise, D.H. Bottom-up Limitation of Predaceous Arthropods in a Detritus-based Terrestrial Food Web. *Ecology* 1999, *80*, 761–772. [Google Scholar] [CrossRef]
20. Moskowitz, N.A.; Dorritie, B.; Fay, T.; Nieves, O.C.; Vidoudez, C.; Cambridge Rindge and Latin 2017 Biology Class; Masconomet 2017 Biotechnology Class; Fischer, E.K.; Trauger, S.A.; Coloma, L.A.; et al. Land use Impacts Poison Frog Chemical Defenses through Changes in Leaf Litter Ant Communities. *Neotrop. Biodivers.* 2020, *6*, lxxv–xvii. [Google Scholar] [CrossRef] [Green Version]
21. Langor, D.W.; Spence, J.R. Arthropods as Ecological Indicators of Sustainability in Canadian Forests. *For. Chron.* 2006, *82*, 344–350. [Google Scholar] [CrossRef]
22. McGeoch, M.A.; Chown, S.L. Scaling Up the Value of Bioindicators. *Trends Ecol. Evol.* 1998, *13*, 46–47. [Google Scholar] [CrossRef]
23. Rainio, J.; Niemelä, J. Ground Beetles (Coleoptera: *Carabidae*) as Bioindicators. *Biodivers. Conserv.* 2003, *12*, 487–506. [Google Scholar] [CrossRef]
24. Andersen, A.N.; Majer, J.D. Ants show the Way Down Under: Invertebrates as Bioindicators in Land Management. *Front. Ecol. Environ.* 2004, *2*, 291–298. [Google Scholar] [CrossRef]
25. Buddle, C.M.; Langor, D.W.; Pohl, G.R.; Spence, J.R. Arthropod Responses to Harvesting and Wildfire: Implications for Emulation of Natural Disturbance in Forest Management. *Biol. Conserv.* 2006, *128*, 346–357. [Google Scholar] [CrossRef]

26. Greenberg, C.H.; McGrane, A. A Comparison of Relative Abundance and Biomass of Ground-Dwelling Arthropods Under Different Forest Management Practices. *For. Ecol. Manag.* 1996, *89*, 31–41. [Google Scholar] [CrossRef]

27. Cole, R.J.; Holl, K.D.; Zahawi, R.A.; Wickey, P.; Townsend, A.R. Leaf Litter Arthropod Responses to Tropical Forest Restoration. *Ecol. Evol.* 2016, *6*, 5158–5168. [Google Scholar] [CrossRef] [Green Version]

28. Vasconcelos, H.L.; Pacheco, R.; Silva, R.C.; Vasconcelos, P.B.; Lopes, C.T.; Costa, A.N.; Bruna, E.M. Dynamics of the Leaf-Litter Arthropod Fauna Following Fire in a Neotropical Woodland Savanna. *PLoS ONE* 2009, *4*, e7762. [Google Scholar] [CrossRef]

29. Nakamura, A.; Proctor, H.; Catterall, C.P. Using Soil and Litter Arthropods to Assess the State of Rainforest Restoration. *Ecol. Manag. Restor.* 2003, *4*, S20–S28. [Google Scholar] [CrossRef]

30. Nakamura, A.; Catterall, C.P.; House, A.P.; Kitching, R.L.; Burwell, C.J. The use of Ants and Other Soil and Litter Arthropods as Bio-Indicators of the Impacts of Rainforest Clearing and Subsequent Land Use. *J. Insect Conserv.* 2007, *11*, 177–186. [Google Scholar] [CrossRef]

31. Ulyshen, M.D.; Hanula, J.L. Litter-Dwelling Arthropod Abundance Peaks Near Coarse Woody Debris in Loblolly Pine Forests of the Southeastern United States. *Fla. Entomol.* 2009, *92*, 163–164. [Google Scholar] [CrossRef]

32. Smith, M.A.; Boyd, A.; Chan, A.; Clout, S.; des Brisay, P.; Dolson, S.; Eagalle, T.; Espinola, S.; Fairweather, A.; Frank, S. Investigating the Effect of Forestry on Leaf-Litter Arthropods (Algonquin Park, Ontario, Canada). *PLoS ONE* 2017, *12*, e0178568. [Google Scholar] [CrossRef]

33. Lange, M.; Weisser, W.W.; Gossner, M.M.; Kowalski, E.; Türke, M.; Joner, F.; Fonseca, C.R. The Impact of Forest Management on Litter-Dwelling Invertebrates: A Subtropical-temperate Contrast. *Biodivers. Conserv.* 2011, *20*, 2133–2147. [Google Scholar] [CrossRef]

34. Meyer, S.; Rusterholz, H.; Salamon, J.; Baur, B. Leaf Litter Decomposition and Litter Fauna in Urban Forests: Effect of the Degree of Urbanisation and Forest Size. *Pedobiologia* 2020, *78*, 150609. [Google Scholar] [CrossRef]

35. Tóth, Z.; Hornung, E. Taxonomic and Functional Response of Millipedes (Diplopoda) to Urban Soil Disturbance in a Metropolitan Area. *Insects* 2020, *11*, 25. [Google Scholar] [CrossRef] [PubMed] [Green Version]

36. Nascimento, E.; Reis, F.; Chichorro, F.; Canhoto, C.; Gonçalves, A.L.; Simões, S.; Sousa, J.P.; da Silva, P.M. Effects of Management on Plant Litter Traits and Consequences for Litter Mass Loss and Collembola Functional Diversity in a Mediterranean Agro-Forest System. *Pedobiologia* 2019, *75*, 38–51. [Google Scholar] [CrossRef]

37. Peng, M.; Hung, Y.; Liu, K.; Neoh, K. Landscape Configuration and Habitat Complexity Shape Arthropod Assemblage in Urban Parks. *Sci. Rep.* 2020, *10*, 16043. [Google Scholar] [CrossRef]

38. Oxbrough, A.; French, V.; Irwin, S.; Kelly, T.C.; Smiddy, P.; O'Halloran, J. Can Mixed Species Stands Enhance Arthropod Diversity in Plantation Forests? *For. Ecol. Manag.* 2012, *270*, 11–18. [Google Scholar] [CrossRef] [Green Version]

39. Kattan, G.H.; Correa, D.; Escobar, F.; Medina, C. Leaf-litter Arthropods in Restored Forests in the Colombian Andes: A Comparison between Secondary Forest and Tree Plantations. *Restor. Ecol.* 2006, *14*, 95–102. [Google Scholar] [CrossRef]

40. Bird, S.B.; Coulson, R.N.; Fisher, R.F. Changes in Soil and Litter Arthropod Abundance Following Tree Harvesting and Site Preparation in a Loblolly Pine (*Pinus Taeda* L.) Plantation. *For. Ecol. Manag.* 2004, *202*, 195–208. [Google Scholar] [CrossRef]

41. Bird, S.; Coulson, R.N.; Crossley, D.A., Jr. Impacts of Silvicultural Practices on Soil and Litter Arthropod Diversity in a Texas Pine Plantation. *For. Ecol. Manag.* 2000, *131*, 65–80. [Google Scholar] [CrossRef]

42. Cicort-Lucaciu, A.; Cupsa, D.; Sucea, F.; Ferenti, S.; Covaciu-Marcov, S. Litter-Dwelling Invertebrates in Natural and Plantation Forests in the Southern Carpathians, Romania. *Balt. For.* 2020, *26*, 1. [Google Scholar] [CrossRef]

43. Ingle, K.; Kaur, H.; Gallé-Szpisjak, N.; Bürgés, J.; Szabó, Á.; Gallé, R. Winter-Active Spider Fauna is Affected by Plantation Forest Type. *Environ. Entomol.* 2020, *49*, 601–606. [Google Scholar] [CrossRef] [PubMed]

44. Niemelä, J. Management in Relation to Disturbance in the Boreal Forest. *For. Ecol. Manag.* 1999, *115*, 127–134. [Google Scholar] [CrossRef] [Green Version]

45. Seymour, R.S.; White, A.S.; Philip, G.D. Natural Disturbance Regimes in Northeastern North America—evaluating Silvicultural Systems using Natural Scales and Frequencies. *For. Ecol. Manag.* 2002, *155*, 357–367. [Google Scholar] [CrossRef]

46. Herrera-Alvarez, X.; Blanco, J.A.; Imbert, J.B.; Alvarez, W.; Rivadeneira-Barba, G. Coarse Woody Debris' Invertebrate Community is Affected Directly by Canopy Type and Indirectly by Thinning in Mixed Scots Pine—European Beech Forests. *Forests* 2020, *11*, 975. [Google Scholar] [CrossRef]

47. Chen, J.; Franklin, J.F.; Spies, T.A. Contrasting Microclimates among Clearcut, Edge, and Interior of Old-Growth Douglas-Fir Forest. *Agric. For. Meteorol.* 1993, *63*, 219–237. [Google Scholar] [CrossRef]

48. Van Wilgenburg, S.L.; Mazerolle, D.F.; Hobson, K.A. Patterns of Arthropod Abundance, Vegetation, and Microclimate at Boreal Forest Edge and Interior in Two Landscapes: Implications for Forest Birds. *Ecoscience* 2001, *8*, 454–461. [Google Scholar] [CrossRef]

49. Seastedt, T.R.; Crossley, D.A., III. Microarthropod Response Following Cable Logging and Clear-cutting in the Southern Appalachians. *Ecology* 1981, *62*, 126–135. [Google Scholar] [CrossRef]

50. Burghouts, T.; Ernsting, G.; Korthals, G.; De Vries, T. Litterfall, Leaf Litter Decomposition and Litter Invertebrates in Primary and Selectively Logged Dipterocarp Forest in Sabah, Malaysia. Philosophical Transactions of the Royal Society of London. *Ser. B Biol. Sci.* 1992, *335*, 407–416. [Google Scholar]

51. Kriska, D.J.; Lee, H.J.; Krebs, R.A. Effect of Oak Barren Restoration on Carabidae (*Coleoptera*) within a Kame-Kettle Bog System. *Ecol. Restor.* 2020, *38*, 24–31. [Google Scholar] [CrossRef]

52. Simard, J.R.; Fryxell, J.M. Effects of Selective Logging on Terrestrial Small Mammals and Arthropods. *Can. J. Zool.* 2003, *81*, 1318–1326. [Google Scholar] [CrossRef]

53. Turner, E.C.; Foster, W.A. The Impact of Forest Conversion to Oil Palm on Arthropod Abundance and Biomass in Sabah, Malaysia. *J. Trop. Ecol.* 2009, *25*, 23–30. [Google Scholar] [CrossRef] [Green Version]

54. Castro, A.; Wise, D.H. Influence of Fallen Coarse Woody Debris on the Diversity and Community Structure of Forest-Floor Spiders (Arachnida: *Araneae*). *For. Ecol. Manag.* 2010, *260*, 2088–2101. [Google Scholar] [CrossRef]

55. Grevé, M.E.; Hager, J.; Weisser, W.W.; Schall, P.; Gossner, M.M.; Feldhaar, H. Effect of Forest Management on Temperate Ant Communities. *Ecosphere* 2018, *9*, e02303. [Google Scholar] [CrossRef]

56. Preece, N.D.; Van Oosterzee, P.; Lawes, M.J. Planting Methods Matter for Cost-effective Rainforest Restoration. *Ecol. Manag. Restor.* 2013, *14*, 63–66. [Google Scholar] [CrossRef]

57. Bechara, F.C.; Dickens, S.J.; Farrer, E.C.; Larios, L.; Spotswood, E.N.; Mariotte, P.; Suding, K.N. Neotropical Rainforest Restoration: Comparing Passive, Plantation and Nucleation Approaches. *Biodivers. Conserv.* 2016, *25*, 2021–2034. [Google Scholar] [CrossRef] [Green Version]

58. Stone, M.J.; Shoo, L.; Stork, N.E.; Sheldon, F.; Catterall, C.P. Recovery of Decomposition Rates and Decomposer Invertebrates during Rain Forest Restoration on Disused Pasture. *Biotropica* 2020, *52*, 230–241. [Google Scholar] [CrossRef]

59. Jiménez-Carmona, E.; Herrera-Rangel, J.; Renjifo, L.M.; Armbrecht, I. Restoration of Riparian Forest Corridors: Eight Years Monitoring the Diversity of Soil Ants in an Andean Rural Landscape. *Insect Conserv. Divers.* 2020, *13*, 384–392. [Google Scholar] [CrossRef]

60. Belshaw, R.; Bolton, B. The Effect of Forest Disturbance on the Leaf Litter Ant Fauna in Ghana. *Biodivers. Conserv.* 1993, *2*, 656–666. [Google Scholar] [CrossRef]

61. Ottermanns, R.; Hopp, P.W.; Guschal, M.; dos Santos, G.P.; Meyer, S.; Roß-Nickoll, M. Causal Relationship between Leaf Litter Beetle Communities and Regeneration Patterns of Vegetation in the Atlantic Rainforest of Southern Brazil (Mata Atlântica). *Ecol. Complex.* 2011, *8*, 299–309. [Google Scholar] [CrossRef]

62. Vergílio, P.C.B.; Knoll, F.R.N.; Barreto, D.M.; Dinardi, N.M.; Ueda, M.Y.; Cavassan, O. Effect of Brushwood Transposition on the Leaf Litter Arthropod Fauna in a Cerrado Area. *Revista Brasileira Ciência Solo* 2013, *37*, 1158–1163. [Google Scholar] [CrossRef] [Green Version]

63. Caterino, M.S.; Langton-Myers, S.S. Long-Term Population Persistence of Flightless Weevils (Eurhoptus Pyriformis) Across Old- and Second-Growth Forests Patches in Southern Appalachia. *BMC Evol. Biol.* 2018, *18*, 1–16. [Google Scholar] [CrossRef] [PubMed]

64. Caterino, M.S.; Vásquez-Vélez, L.M.; Myers, S.S. On the Conservation Value of Second-Growth Forests for Leaf-Litter Inhabiting Beetles. *Insect Syst. Divers.* 2017, *1*, 20–28. [Google Scholar] [CrossRef]

65. Meloni, F.; Varanda, E.M. Litter and Soil Arthropod Colonization in Reforested Semi-deciduous Seasonal Atlantic Forests. *Restor. Ecol.* 2015, *23*, 690–697. [Google Scholar] [CrossRef]

66. Rieske, L.K.; Buss, L.J. Effects of Gypsy Moth Suppression Tactics on Litter-and Ground-Dwelling Arthropods in the Central Hardwood Forests of the Cumberland Plateau. *For. Ecol. Manag.* 2001, *149*, 181–195. [Google Scholar] [CrossRef]

67. Ulyshen, M.D. Arthropod Vertical Stratification in Temperate Deciduous Forests: Implications for Conservation-Oriented Management. *For. Ecol. Manag.* 2011, *261*, 1479–1489. [Google Scholar] [CrossRef]

68. Kreutzweiser, D.; Dutkiewicz, D.; Capell, S.; Sibley, P.; Scarr, T. Changes in Streamside Riparian Forest Canopy and Leaf Litter Nutrient Flux to Soils during an Emerald Ash Borer Infestation in an Agricultural Landscape. *Biol. Invasions* 2020, *22*, 1865–1878. [Google Scholar] [CrossRef]

69. Woodworth, G.R.; Ward, J.N.; Carr, D.E. Exotic Tree and Shrub Invasions Alter Leaf-Litter Microflora and Arthropod Communities. *Oecologia* 2020, *193*, 177–187. [Google Scholar] [CrossRef]

70. Oliver, I.; Mac Nally, R.; York, A. Identifying Performance Indicators of the Effects of Forest Management on Ground-Active Arthropod Biodiversity using Hierarchical Partitioning and Partial Canonical Correspondence Analysis. *For. Ecol. Manag.* 2000, *139*, 21–40. [Google Scholar] [CrossRef]

71. Fule, P.Z.; Ramos-Gómez, M.; Cortés-Montaño, C.; Miller, A.M. Fire Regime in a Mexican Forest Under Indigenous Resource Management. *Ecol. Appl.* 2011, *21*, 764–775. [Google Scholar] [CrossRef] [Green Version]

72. Whitehead, P.J.; Bowman, D.M.; Preece, N.; Fraser, F.; Cooke, P. Customary use of Fire by Indigenous Peoples in Northern Australia: Its Contemporary Role in Savanna Management. *Int. J. Wildland Fire* 2003, *12*, 415–425. [Google Scholar] [CrossRef]

73. Shaffer, L.J. Indigenous Fire use to Manage Savanna Landscapes in Southern Mozambique. *Fire Ecol.* 2010, *6*, 43–59. [Google Scholar] [CrossRef]

74. Kimmerer, R.W.; Lake, F.K. The Role of Indigenous Burning in Land Management. *J. For.* 2001, *99*, 36–41. [Google Scholar]

75. Scheller, R.M.; Mladenoff, D.J.; Crow, T.R.; Sickley, T.A. Simulating the Effects of Fire Reintroduction Versus Continued Fire Absence on Forest Composition and Landscape Structure in the Boundary Waters Canoe Area, Northern Minnesota, USA. *Ecosystems* 2005, *8*, 396–411. [Google Scholar] [CrossRef]

76. Buddle, C.M.; Spence, J.R.; Langor, D.W. Succession of Boreal Forest Spider Assemblages Following Wildfire and Harvesting. *Ecography* 2000, *23*, 424–436. [Google Scholar] [CrossRef]

77. Gandhi, K.J.; Spence, J.R.; Langor, D.W.; Morgantini, L.E. Fire Residuals as Habitat Reserves for Epigaeic Beetles (Coleoptera: *Carabidae* and *Staphylinidae*). *Biol. Conserv.* 2001, *102*, 131–141. [Google Scholar] [CrossRef]

78. Glenn-Lewin, D.C.; Peet, R.K.; Veblen, T.T. *Plant Succession: Theory and Prediction*; Springer Science & Business Media: Berlin/Heidelberg, Germany, 1992. [Google Scholar]

79. Haimi, J.; Fritze, H.; Moilanen, P. Responses of Soil Decomposer Animals to Wood-Ash Fertilisation and Burning in a Coniferous Forest Stand. *For. Ecol. Manag.* 2000, *129*, 53–61. [Google Scholar] [CrossRef]

80. Buffington, J.D. Soil Arthropod Populations of the New Jersey Pine Barrens as Affected by Fire. *Ann. Entomol. Soc. Am.* 1967, *60*, 530–535. [Google Scholar] [CrossRef]

81. York, A. Long-term Effects of Frequent Low-intensity Burning on Ant Communities in Coastal Blackbutt Forests of Southeastern Australia. *Austral Ecol.* 2000, *25*, 83–98. [Google Scholar] [CrossRef]

82. Yanovsky, V.M.; Kiselev, V.V. Response of the endemic insect fauna to fire damage in forest ecosystems. In *Fire in Ecosystems of Boreal Eurasia*; Goldhammer, J.G., Furyaev, V.V., Eds.; Kluwer Academic Publishers: Amsterdam, The Netherlands, 1996; pp. 409–413. [Google Scholar]

83. Mitchell, R.G. Effects of prescribed fire on insect pests. In *Natural and Prescribed Fire in Pacific Northwest Forests*; Walstad, J.D., Radosevich, S.R., Sandberg, D.V., Eds.; Oregon State University Press: Corvallis, OR, USA, 1990; pp. 111–116. [Google Scholar]

84. Verble-Pearson, R.M. Effects of Fire Intensity on Litter Arthropod Communities in Ozark Oak Forests, Arkansas, USA. *Am. Midl. Nat.* 2014, *172*, 14–24. [Google Scholar] [CrossRef]

85. Swengel, A.B. A Literature Review of Insect Responses to Fire, Compared to Other Conservation Managements of Open Habitat. *Biodivers. Conserv.* 2001, *10*, 1141–1169. [Google Scholar] [CrossRef]

86. Moretti, M.; Obrist, M.K.; Duelli, P. Arthropod Biodiversity After Forest Fires: Winners and Losers in the Winter Fire Regime of the Southern Alps. *Ecography* 2004, *27*, 173–186. [Google Scholar] [CrossRef]

87. Apigian, K.O.; Dahlsten, D.L.; Stephens, S.L. Fire and Fire Surrogate Treatment Effects on Leaf Litter Arthropods in a Western Sierra Nevada Mixed-Conifer Forest. *For. Ecol. Manag.* 2006, *221*, 110–122. [Google Scholar] [CrossRef]

88. Paquin, P.; Coderre, D. Deforestation and Fire Impact on Edaphic Insect Larvae and Other Macroarthropods. *Environ. Entomol.* 1997, *26*, 21–30. [Google Scholar] [CrossRef]

89. Siemann, E.; Haarstad, J.; Tilman, D. Short-Term and Long-Term Effects of Burning on Oak Savanna Arthropods. *Am. Midl. Nat.* 1997, *137*, 349. [Google Scholar] [CrossRef]

90. Niwa, C.G.; Peck, R.W. Influence of Prescribed Fire on Carabid Beetle (*Carabidae*) and Spider (*Araneae*) Assemblages in Forest Litter in Southwestern Oregon. *Environ. Entomol.* 2002, *31*, 785–796. [Google Scholar] [CrossRef]

91. Abbott, I. Changes in the Abundance and Activity of Certain Soil and Litter Fauna in the Jarrah Forest of Western Australia After a Moderate Intensity Fire. *Soil Res.* 1984, *22*, 463–469. [Google Scholar] [CrossRef]

92. Moretti, M.; Duelli, P.; Obrist, M.K. Biodiversity and Resilience of Arthropod Communities After Fire Disturbance in Temperate Forests. *Oecologia* 2006, *149*, 312–327. [Google Scholar] [CrossRef] [Green Version]

93. Holliday, N.J. The Carabid Fauna (Coleoptera: *Carabidae*) during Postfire Regeneration of Boreal Forest: Properties and Dynamics of Species Assemblages. *Can. J. Zool.* 1992, *70*, 440–452. [Google Scholar] [CrossRef]

94. Collett, N. Short and Long-Term Effects of Prescribed Fires in Autumn and Spring on Surface-Active Arthropods in Dry Sclerophyll Eucalypt Forests of Victoria. *For. Ecol. Manag.* 2003, *182*, 117–138. [Google Scholar] [CrossRef]

95. Baker, S.C.; Richardson, A.M.; Seeman, O.D.; Barmuta, L.A. Does Clearfell, Burn and Sow Silviculture Mimic the Effect of Wildfire? A Field Study and Review using Litter Beetles. *For. Ecol. Manag.* 2004, *199*, 433–448. [Google Scholar] [CrossRef]

96. Beaudry, S.; Duchesne, L.C.; Côté, B. Short-Term Effects of Three Forestry Practices on Carabid Assemblages in a Jack Pine Forest. *Can. J. For. Res.* 1997, *27*, 2065–2071. [Google Scholar] [CrossRef]

97. Coleman, T.W.; Rieske, L.K. Arthropod Response to Prescription Burning at the Soil-litter Interface in Oak-pine Forests. *For. Ecol. Manag.* 2006, *233*, 52–60. [Google Scholar] [CrossRef]

98. Klironomos, J.N.; Kendrick, B. Relationships among Microarthropods, Fungi, and their Environment. *Plant Soil* 1995, *170*, 183–197. [Google Scholar] [CrossRef]

99. Fritze, H.; Smolander, A.; Levula, T.; Kitunen, V.; Mälkönen, E. Wood-Ash Fertilization and Fire Treatments in a Scots Pine Forest Stand: Effects on the Organic Layer, Microbial Biomass, and Microbial Activity. *Biol. Fertil. Soils* 1994, *17*, 57–63. [Google Scholar] [CrossRef] [Green Version]

18 MONITORING AND REGULATION OF THE PINE PROCESSIONARY MOTH IN THE NORTH-WESTERN ITALIAN ALPS

1. INTRODUCTION: AN OVERVIEW

The pine processionary moth, scientifically known as Thaumetopoea pityocampa, is a species of moth belonging to the family Notodontidae. It is commonly referred to as PPM and is considered one of the most significant pests that cause defoliation in pine trees, with significant economic implications. The species is highly voracious and serves as the primary herbivorous insect pest in Mediterranean forests. It is found in the Mediterranean Basin in southern Europe, the Balkan region, and North Africa [1,2]. The distribution of the moth is limited by its need for winter sunshine and its vulnerability to extreme temperatures, both in winter and summer.

Additionally, there can be significant variations in the duration of the adult flying period depending on the specific geographic location. Over the past three decades, the PPM has experienced a change in its altitudinal range of approximately 110-230 meters in the Alps [3]. Recorded observations have shown alterations in the pest's distribution, with its geographic extent being significantly impacted by winter air temperature and annual solar radiation [4]. The PPM caterpillar is a herbivore and has a strong impact on defoliation, leading to a notable deterioration in the health of pine trees [5]. The moth is commonly found in the pine forests of the Mediterranean region, although it may also thrive in urban environments.

The larvae consume a diverse array of Pinus species, particularly Pinus nigra Arnold and P. sylvestris L. However, other conifers, including cedars and even Douglas fir, may also face potential harm. The primary outcomes of infestations in pine plantations include a significant decrease in photosynthetic area, annual radial growth, biomass, and substantial economic losses [6,7,8,9,10]. Massive infestations can cause significant damage to young plantations, making them vulnerable to fungus and secondary insect pests, such as scolytids and weevils.

In addition, larvae starting from the 3rd instar stage have stinging hairs that can cause significant societal consequences. There have been documented cases of health issues in humans and domestic or farm animals in urban and suburban settings related with these hairs [12,13,14].

The PPM demonstrates recurrent episodes, occurring approximately every 6 years, in multiple southern European countries. The cycle, however, is irregular and can range from 3 to 10 years [15]. Conventional techniques for managing the PPM involve physically removing winter nests, using ballistic methods to shoot nests from the ground, injecting insecticides into the tree trunk, using bioinsecticides such as Bacillus thuringiensis Berliner var. kurstaki, and employing trunk barriers or traps. Comprehending the influence of the PPM and accurately determining the timing of adult emergence is crucial for formulating suitable control techniques. In order to achieve this objective, Guerrero (1981) [19] discovered and isolated the sexual female pheromone, (Z)-13-hexadecen-11-ynyl acetate. This pheromone is then utilized in traps that are baited with pheromones to monitor the flying activity of males [20] or to trap them in large numbers [4].

This report presents the findings of a 6-year study conducted from 2014 to 2019 in the Aosta Valley region of northwest Italy. The study aimed to examine the population density of the PPM (pine processionary moth) on P. sylvestris (Scots pine) using traps baited with pheromones. Due of the limited information available in the literature, the objective was to evaluate the seasonal flight patterns of the PPM by documenting the presence of male adult specimens. Additionally, the infestation index was determined based on the extent of defoliation. In addition, field studies assessed the efficacy of control techniques by comparing regions that were treated with those that were not treated. The control measures employed included the utilization of commercially available B. thuringiensis var. kurstaki for microbial treatments, as well as the implementation of mating disruption techniques.

2. MATERIALS AND METHODS

2.1. Study Area

Aosta Valley is a mountainous autonomous region located in the north-western part of Italy. It has a total surface area of about 3262 km^2, 1080 km^2 of which is covered by forests. The territory is mainly mountainous, and the average altitude is 2100 m, ranging from about 295 m to 4810 m a.s.l. The main valley axis is prevalently west-east oriented, presenting a semi-continental climate.

The tree vegetation consists of Larix decidua Miller, Picea abies L. Karst, mixed stands of Scots pine (P. sylvestris), Mountain pine (P. uncinata Ram.), and Austrian black pine (P. nigra), and Quercus pubescens Will., with the sporadic presence of Populus tremula L. and Betula pendula Roth.

2.2. Seasonal Flight Activity

Over a 5-year period (2015–2019), the PPM population was monitored by highly skilled and trained forestry technicians from the "Corpo Forestale della Valle d'Aosta" (CFVDA). Surveys were carried out in four (2015), and 42 municipalities (2016–2019) of the Aosta Valley.

Monitoring was performed on the PPM's host, P. sylvestris, and commercially available funnel traps were used at the sampling sites. All trap devices (Super Green, Serbios s.r.l., Badia Polesine (RO), Italy) were baited with dispensers containing the sex pheromone component (Z)-13-hexadecen-11-ynyl acetate (loading rate: 1 mg per dispenser), and placed randomly on trunks and branches of P. sylvestris stands at an average height of 2 m. Six traps were placed per hectare. All traps were installed during early June and inspected for adult males at weekly intervals until mid-September to detect the flight period of the target species in the study area. Captured males were counted and removed from the traps. Devices were rotated clockwise to minimize the influence of the individual trapping location. The lure in each trap was replaced every six weeks. The first pheromone trap-check date was the date on which the first captures were recorded.

2.3. Infestation Index

The infestation index was recorded in 42 municipalities over a 6-year period (2014–2019) during the winter season (December–February). A 1-ha plot was selected in each municipality. All the plots were characterized by the presence of pure P. sylvestris stands, with a density of approximately 300 plants/ha, and an average tree age of about 60 years. Each tree was visually inspected by eye or with binoculars when necessary. The infestation index was recorded in all pine stands by counting the number of nests and the degree of defoliation, according to six classes: 0, zero, no nest, no defoliation; 1, very low ≤ 0.1 nest per plant, < 10% defoliation; 2, low = scattered nests, 0.2–0.5 nests per plant, 10–25% defoliation; 3, medium = 0.6–2 nests per plant, 26–50% defoliation; 4, high = 3–5 nests per plant, 51–75% defoliation; 5, very high ≥ 5 nests per plant, massive defoliation (>75%).

2.4. Application of Control Strategies

The infestation index was used to evaluate tree damage and determine threshold values above which control strategies were applied to control PPM infestations. Treatments with Btk were applied in those plots with an infestation index ≥ 3. Conversely, mating disruption was applied at sites characterized by a low PPM population density (infestation index < 2).

Microbial applications were performed at 20 different sites (5 sites per year) over a 4-year period (2016–2019). Each of these sites was at least 1 km away from monitoring traps to avoid compromising male catches, with a preference for areas with pure P. sylvestris stands and avoiding areas where the pines were mixed with other species, such as L. decidua, other Pinus species, and broadleaved trees. At each site, a plot of 100 pine saplings (about 0.30 ha) was sprayed with a commercially available bioinsecticide product based on B. thuringiensis var. kurstaki (Btk) (Kristal 32 WG, HD-1 strain, Serbios s.r.l., Badia Polesine (RO), Italy) at the rate of 32,000 BIU ha^{-1}, in the presence of L1 and L2 larvae. Bioinsecticide was sprayed from the ground, up to a height of about 20 m, between mid-August and mid-September, at about 20 °C (in the twilight hours) and in partly cloudy weather conditions, using atomizer equipment supplied by CFVDA and the phytosanitary service of the Aosta Valley.

A control plot, with an area and number of trees comparable to the treated plots, was chosen in the surroundings of each treated area, but far enough away to avoid contamination by the bacterium (at least 1 km). Treatments were not performed if wind speed exceeded 7 km h^{-1} and were repeated in the case of heavy rain. The effectiveness of the microbial treatment was assessed 10 days after treatment evaluating larval mortality. To calculate the percentage of larval mortality, the number of dead larvae was divided by the total number of dead and live larvae recorded on all the saplings for each site.

Four municipalities were chosen for mating disruption over a 2-year period (Saint Nicolas and Sarre in 2016; Arvier and Villeneuve in 2017). In total, 23 and 19 plots were chosen in 2016 and 2017, respectively (11 in Saint Nicolas, 12 in Sarre, 10 in Arvier, and 9 in Villeneuve). At each site, a 1-ha plot of about 300 pine saplings was chosen. Pheromone dispensers (only lures without the traps) containing the specific sex pheromone component (loading rate: 50 mg per dispenser) were placed on branches of P. sylvestris stands at an average height of 2 m within tree canopies immediately before adult emergence. Ten and 12 dispensers/ha were used in 2016 and 2017, respectively. For each plot, a control plot with an area and number of trees comparable to the treated plot was chosen in the surrounding area, but far enough away to avoid the effects of pheromone dispersion (at least 2 km).

The effectiveness of the technique was assessed using pheromone traps. The same type of traps as those used for monitoring was placed to collect male adults in the treated and control plots (6 traps/ha; loading rate: 1 mg per dispenser). The traps were placed in a regular pattern throughout each plot to ensure even coverage. All traps were installed immediately before adult emergence and inspected for adult males at weekly intervals until mid-September. Captured males were counted and removed from the traps. The effectiveness of the treatment was assessed during December by recording the mean number (±SEM) of nests/tree in all the treated and control plots.

2.5. Statistical Analysis

All data were first tested for homogeneity of variance (Levene's test), and transformations (log transformation and arcsine square rooted) were employed to stabilize variances and normalize the data. A paired t-test was used to assess differences in larval mortality comparing treated and control plots. Furthermore, a paired t-test was used to assess differences in male catches comparing plots where mating disruption was applied and control plots. Statistical significance was set at $p < 0.05$. All analyses were performed using SPSS version 22.0 (SPSS, Chicago, IL, USA).

3. RESULTS

A network of permanent plots intended for monitoring of the PPM was set up during the survey period. In total, 277 georeferenced traps were placed in the same position in each of the five years, except for 2015, when the number of traps was 135 (**Figure 1**).

Figure 1. Map of the surveyed area in the Aosta Valley (north-western Italian Alps), showing the georeferenced pheromone baited traps placed in 2015–2019).

The total number of PPM adult males captured with funnel traps was 86,790, 150,652, 157,300, 152,118, and 61,612, in 2015, 2016, 2017, 2018 and 2019, respectively. The mean catches per week per trap ranged from 1.91 in Brusson in 2017 to 138.05 in Morgex in 2016. All data referring to the mean number of adult male PPM captured per week per trap and the total catch per week at all the surveyed sites.

In all years, the seasonal flight activity started at the beginning of June with the peak of adults on 3rd July (2015, 2016), 10th July (2017), 26th June (2018), and 17th July (2019). Males were found in traps until mid-September (**Figure 2**).

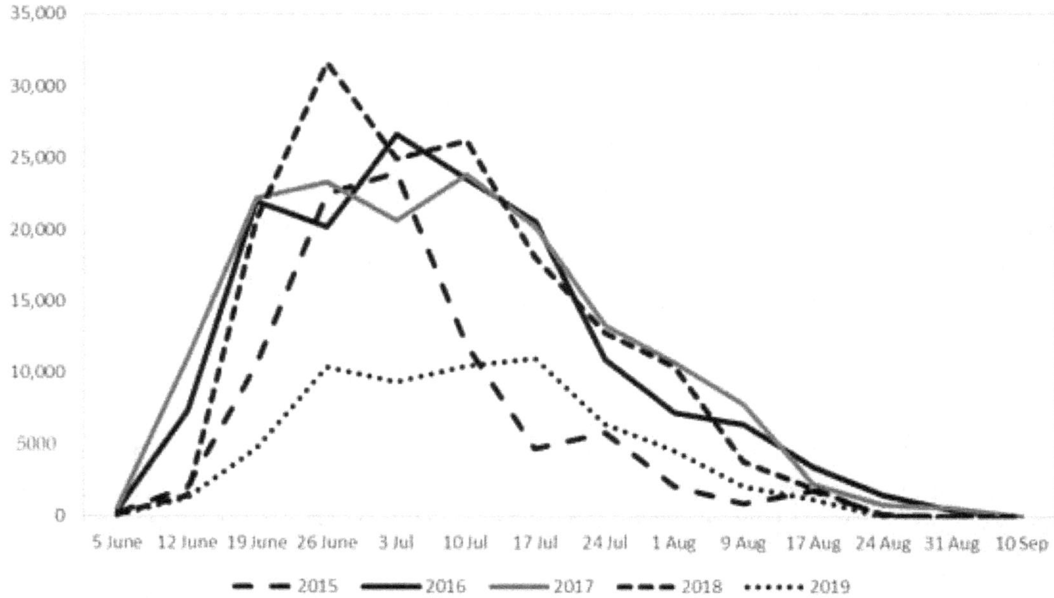

Figure 2. Total number of adults male of Thaumetopoea pityocampa collected with the pheromone-baited traps in the five-year period. (Data collected always refers to 277 traps, except for 2015 when the traps were 135).

In total, 9618 ha were visually monitored and assigned an infestation index. The percentage of woodland stands strongly defoliated by PPM (infestation index ≥ 3) increased progressively between 2015 and 2016, increasing from 8% to 19% of the total area monitored. In spring 2016, "very high" defoliation (76–100% loss of leaves) affected an area of 45 hectares, corresponding to 1% of the infested pine forests, while defoliation was "high" (with leaf loss between 51 and 75%) in 10% of the pine forests. The infestation index subsequently decreased to 16% and 13% in 2017 and 2018, and then suffered an abrupt decline to 4% in 2019 (**Table 1**). The most serious damage occurred in the municipalities of Morgex, Sarre, and Saint-Pierre.

In total, 367, 333, 208, and 253 ha were treated in 2016–2019, respectively. Btk treatments involved about 312 km of roads, covering a total area of about 1116 hectares of pine forests.

The mortality rate in the plots treated with B. thuringiensis was significantly higher than that recorded in the control plots. Where the bacterium was applied, the rate of larval mortality ranged from 79.47% to 98.43%, with an average of 90.47%. Conversely, in the control plots, the larval mortality ranged from 0.34% to 2.18%, with an average of 1.56% (**Figure 3**).

Table 1. Infestation index recorded, and the total area (ha) treated with Bacillus thuringiensis var. kurstaki, and mating disruption in 2014–2019.

		2014		2015		2016		2017		2018		2019	
		No. Municipality	ha	No. Municipality	ha	No. Municipality	ha	No. Municipality	ha	No. Municipality	ha	No. Municipality	ha
		42		42		42		42		42		42	
Infestation index	0		7598		7310		5482		4809		4713		4520
	1		1058		1154		1539		2597		2212		4040
	2		481		385		769		673		866		673
	3		385		481		866		673		614		289
	4		96		192		673		577		439		96
	5		0		96		289		289		197		0
Total treated (ha)			-		-		367		333		208		253
Bacillus thuringiensis var. kustaki		-	-	-	-	14	344	9	314	12	208		250
Mating disruption		-	-	-	-	2	23	2	19	-	-		-

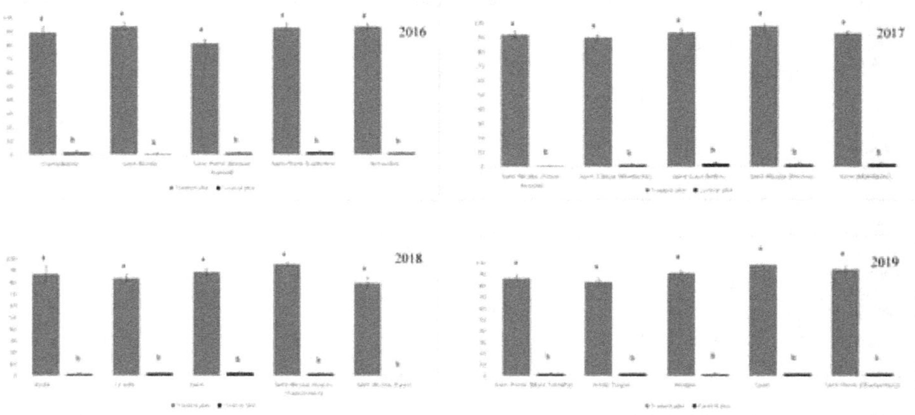

Figure 3. Percentage mortality of PPM larvae (±SEM) recorded inside the nests collected in treated and control plots after applications of Bacillus thuringiensis var. kurstaki in 2016–2019. Bars with different letters are significantly different, according to the paired t-test ($p < 0.05$).

Mating disruption was performed in 23 and 19 ha in 2016 and 2017, respectively. The mean number (±SEM) of PPM male adults collected per trap and per week is given in **Figure 4**. The mean number of PPM males was significantly lower in the plots where MD was performed, ranging from 8.36 ± 2.37 to 13.47 ± 4.68. Conversely, the mean number of males captured in the control plots was, on average, 119.16 ± 12.68.

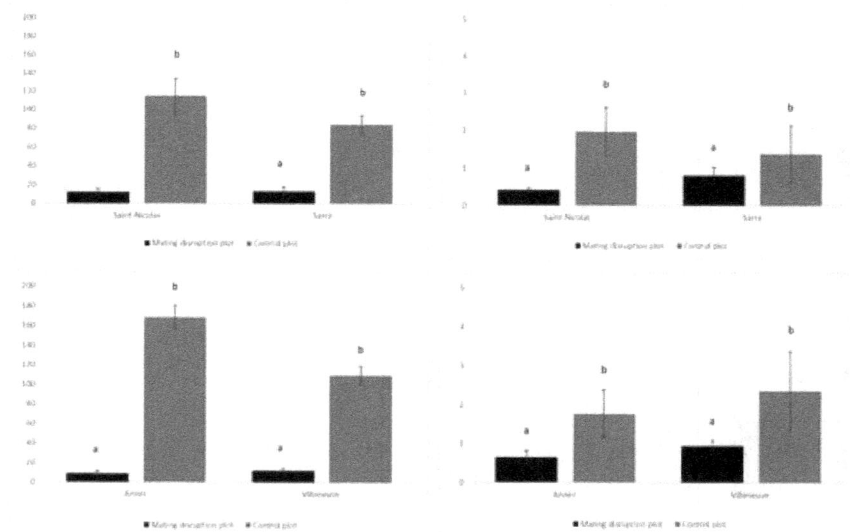

Figure 4. Mean number (±SEM) of male adults/per trap/per week of Thaumetopoea pityocampa collected with pheromone-baited traps, and mean number (±SEM) of nests/tree recorded in plots where mating disruption (MD) was applied compared to control plots in 2016 (top) and 2017 (bottom). Bars with different letters are significantly different, according to the paired t-test ($p < 0.05$).

The total number of nests recorded per tree was significantly lower in MD plots, ranging from 0.41 ± 0.05 to 0.94 ± 0.14. In the control plots, there was an average of 4.37 ± 0.76 nests per tree.

4. DISCUSSION

Although T. pityocampa is a key pest of pine trees in the Alps, there is little information about its phenology and seasonal occurrence in the north-western Italian Alps. In the Aosta Valley, PPM infestation has been consistently investigated in forest monitoring programmes from the 1960s onwards, but the defoliator moth has only been intensively studied since the 2000s. The Aosta Valley pine forests mainly consist of black pine and Scots pine; defoliation damage ranging from mild to very severe, due to PPM feeding activity, has been observed in about 50% of this area over the years (IR and LD). In 2016, the Department of Agriculture and Natural Resources, therefore, activated a plan to control PPM outbreaks.

Generally, mixed forests are less sensitive to pests and disease than pure stands. Although olfactory cues are commonly used to locate host trees, visual cues may affect the degree of infestation as well. Even in the case of PPM, mixed pine stands have been found to suffer less damage than pure pine stands **[21]**, as non-host trees may act as physical barriers **[22,23]**. However, over the last century, several species of the genus Pinus have been widely used for afforestation **[24]**, and this has clearly facilitated the spread of the PPM in the surveyed area. P. nigra is considered the primary native host of PPM, and the planting of this species has almost certainly promoted PPM population growth and outbreaks. Furthermore, PPM is reported to be sensitive to the severity and duration of the cold winter period, since it overwinters at the larval stage. The lethal temperature threshold is between approximately −10 °C and −16 °C **[3]**, and the mean temperature increase over the last few decades has, thus, favored a higher survival rate from year to year, as already reported by Robinet **[25]**. In Aosta Valley, Mont Avic was used as a study site to monitor the expansion of PPM in the Italian Alps, where both an elevation gain (113.7 ± 23.0 m) and distance gain (232.7 ± 22.4 m) were observed in 2003 **[26]**.

In our survey area, the network of permanent plots established in 2015 revealed that the PPM population progressively increased to reach its highest values in 2017, and then decreased rapidly afterward. The seasonal flight activity started at the beginning of June and continued until mid-September. The trapping of male adults in pheromone-baited traps peaked between the end of June and the beginning of July, with a similar trend in each year that was consistent with the literature **[27,28]**. Reliable population density data from monitoring and assessment surveys are required for effective suppression of PPM infestations. Although the absolute number of catches per trap may not be entirely representative of the population, being strongly influenced by sex ratio and trap frequency and position [**24**], monitoring traps may be useful to identify low- or high-density population areas where physical removal of the nests, insecticides, microbial treatments and/or mating disruption can be applied. Since several control methods are available, a practical evaluation of their effectiveness is essential for decision making in current PPM control protocols.

The Aosta Valley is a renowned tourist area, and in the past decades, there has been an increasing need for effective pest management programs, with traditional control methods (e.g., intensive pruning or chemical pesticides) strongly discouraged. The most common environmentally friendly control strategies (e.g., mechanical, biological) have now been adopted in the forestry environment.

Microbial insecticides offer a great deal of promise for pest management, being effective in controlling several defoliating forestry insect pests (e.g., spruce budworm, gypsy moth), without posing the serious environmental hazards associated with conventional pesticides [**29,30**]. Several microbial antagonists (e.g., bacteria, fungi, and viruses) have been reported for the control of PPM infestations [**16**], but the only large database based on scientific investigations concerns the use of Bacillus-based preparations. Applications of commercially available bioinsecticide products based on Btk have increased in the last few decades, targeting PPM [**16,31,32**], and other forest-defoliating lepidopterans [**33,34,35**]. At the surveyed sites, different control strategies were applied with the aim of containing and mitigating PPM outbreaks. Btk applications and mating disruption showed promising results in terms of effectiveness in controlling the infestations. Where Btk treatment was performed, the PPM mortality was very high. The average larval mortality was 90.47% in the treated plots versus 1.56% in the control plots. Similar values, with PPM larval mortality rates up to 90%, have previously been reported for Btk ground application in EU and Mediterranean countries [**16,32**]. In the literature, the susceptibility of the larvae to Btk was found to be directly proportional to the application dose, and some authors reported that a second Btk spray within 10–15 days of the first application might be useful [**16**]. In the current study, only one application was performed targeting L1–L2 larvae, using the dose of 32,000 BIU ha^{-1}, and the average larval mortality rate proved very high at all the surveyed sites and in all years, making a second application unnecessary. Similar to Btk applications, the use of pheromones to disrupt the mating of PPM was effective at all the surveyed sites, suggesting that pheromone lures may effectively decrease male captures. The number of males in pheromone-baited traps was significantly lower in all MD plots (on average, 11.47 males/ captured/week in all sites) in comparison with the control (on average, 119.16 at all sites) during 2016–2017. The number of nests per tree also confirmed the effectiveness of this technique, as also demonstrated by Trematerra [**36**]. The values recorded were significantly lower in all MD plots (on average, 0.7 nests/tree) compared to the control plots (on average, 1.87 nests/tree), highlighting the success of this technique. Although some constraints regarding the size of the plot have been reported for agricultural pests because of the risk of adult immigration [**37**], our investigations were performed in 1-ha plots, and the satisfactory results are consistent with those reported elsewhere [**36**].

PPM has a typical cyclical pattern, where intermittent outbreaks are interspersed with periods of lower population densities. In the literature, peak defoliations are reported to occur every 6–11 years on average [**38,39,40**], while in the Italian Alps, no regular cycles have been detected [**41**] (although the different statistical methods applied may have influenced the final output). A negative gradation phase characterized by a decline in the PPM population commonly occurs when there is a significant reduction in food availability. In the current study, the PPM population experienced a clear decline in 2019, but given that pine stands were not severely defoliated, it is more likely that this was due to the effectiveness of the set of control strategies applied rather than to the PPM's cyclical pattern. The pest is a strong public-health concern, due to the presence of urticating hairs that are present from the L3 larvae onwards, and which may affect humans, pets, and livestock in urban areas and forestry [**13**]. Btk applications and mating disruption have turned out to be the most appropriate control strategies for forest managers targeting the PPM. The adoption of such control strategies should be strongly encouraged by local authorities and administration policies, especially in areas frequented by people and in pine stands used for biomass production. Repeated annual applications of microbials and/or mating disruption could dramatically reduce population densities to levels below the economic injury level, with populations subsequently maintained at low levels with little management effort [**42**].

Currently, there is ongoing research in Europe [40] and certain regions of Italy [42,43] to examine the role of natural enemies. So far, researchers have identified at least 18 parasitoid and 15 predator species that have been observed preying on various stages of the PPM. Currently, there has been no research conducted on the introduction of natural predators in the Aosta Valley. Consequently, there is limited knowledge regarding their existence and the possible effects they may have along latitudinal or longitudinal gradients. There are still unanswered problems that need to be addressed, and future studies should concentrate on investigating the presence and adjustment of invertebrate natural predators in relation to the population dynamics of PPM. Further research is required to better understand how natural enemies react to PPM outbreaks and their role in regulating the population. This specific research aims to enhance our understanding of the potential contribution and beneficial role of generalist parasitoids in controlling the defoliator moth in the surveyed area. Additional research should be conducted on climatic circumstances. The diel activity and seasonal phenology patterns of both the PPM and oak processionary moth have been linked to global climate change [44]. Moth emergence may be significantly influenced by meteorological circumstances, as previously seen in the related species T. wilkinsoni Tams [45]. The additional investigations, along with the existing data on the seasonal occurrence and effectiveness of control strategies, will contribute to a comprehensive global understanding of PPM in the Aosta Valley. This will enable the adoption of long-term management strategies that are both effective and sustainable.

5. AT A GLANCE

The pine processionary moth (PPM), scientifically known as Thaumetopoea pityocampa, is a prominent defoliator of coniferous trees in Southern Europe and North Africa. It belongs to the Lepidoptera order and the Notodontidae family. This species exhibits oligophagy, feeding exclusively on pines and cedars found in Mediterranean nations. This study, conducted from 2014 to 2019 in Aosta Valley, located in northwest Italy, focused on Pinus sylvestris L. The main objectives were: (i) To monitor the population of PPM (Pine Processionary Moth) using pheromone-baited funnel traps; (ii) To evaluate the infestation index, which ranged from 0 to 5 classes, based on the extent of defoliation; and (iii) To implement control strategies, specifically Bacillus thuringiensis var. kurstaki (Btk) and mating disruption (MD). A total of 9618 hectares were visually surveyed and assigned an infestation index. The proportion of woodland stands that experienced severe defoliation, with an infestation index of 3 or higher, showed a steady increase from 2015 to 2016, affecting 8% to 19% of the total monitored area. This proportion then decreased to 16% in 2017 and further to 13% in 2018. However, there was a sudden decline to only 4% in 2019. Both Btk treatments and mating disruption effectively decreased the infestation. The application of Btk resulted in larval mortality rates ranging from 79.47% to 98.43%. In contrast, the control plots had an average larval mortality rate of 1.56%. The average number of male PPM collected in traps was dramatically reduced in the plots where MD was conducted, ranging from 8.36 ± 2.37 to 13.47 ± 4.68. The average number of males collected in the control plots was 119.16 ± 12.68. In the MD plots, the total number of nests seen per tree was much lower, with values ranging from 0.41 ± 0.05 to 0.94 ± 0.14. The control plots had an average of 4.37 ± 0.76 nests per tree. Previous research has shown that both Btk and mating disruption are successful methods for managing infestations of PPM, a type of insect that damages forests by defoliating trees. Therefore, it is highly recommended that forest managers actively promote and implement the use of microbial pesticides.

REFERENCES

1. Chenchouni, H.; Zanati, K.; Rezougui, A.; Briki, A.; Arar, A. Population monitoring of Pine Processionary Moth (Thaumetopoea pityocampa) by pheromone trapping at the southern limit of distribution of Pinus halepensis in Eastern Algeria. For. Sci. Technol. 2010, 6, 67–79. [**Google Scholar**] [**CrossRef**]
2. Devkota, B.; Schmidt, G.H. Larval development of Thaumetopoea pityocampa (Den. & Schiff.) (Lep. Thaumetopoeidae) from Greece as influenced by different host plants under laboratory conditions. J. Appl. Èntomol. 1990, 109, 321–330. [**Google Scholar**] [**CrossRef**]
3. Battisti, A.; Stastny, M.; Netherer, S.; Robinet, C.; Schopf, A.; Roques, A.; Larsson, S. Expansion of geographic range in the pine processionary moth caused by increased winter temperatures. Ecol. Appl. 2005, 15, 2084–2096. [**Google Scholar**] [**CrossRef**]
4. Buffo, E.; Battisti, A.; Stastny, M.; Larsson, S. Temperature as a predictor of survival of the pine processionary moth in the Italian Alps. Agric. For. Èntomol. 2007, 9, 65–72. [**Google Scholar**] [**CrossRef**]
5. Arnaldo, P.S.; Torres, L.M. Effect of different hosts on Thaumetopoea pityocampa populations in northeast Portugal. Phytoparasitica 2006, 34, 523–530. [**Google Scholar**] [**CrossRef**]
6. Arnaldo, P.; Chacim, S.; Lopes, D.; Arnaldo, P.S. Effects of defoliation by the pine processionary moth Thaumetopoea pityocampa on biomass growth of young stands of Pinus pinaster in northern Portugal. iFor. BiogeoSci. For. 2010, 3, 159–162. [**Google Scholar**] [**CrossRef**]
7. Buxton, R.D. Forest management and the Pine Processionary Moth. Outlook Agric. 1983, 12, 34–39. [**Google Scholar**] [**CrossRef**]
8. Çatal, Y. The effects of pine processionary moth (PPM) defoliation degree on radial growth of brutian pine (Pinus brutia). Afr. J. Agric. Res. 2011, 6, 4931–4936. [**Google Scholar**] [**CrossRef**]
9. Gatto, P.; Zocca, A.; Battisti, A.; Barrento, M.J.; Branco, M.; Paiva, M.R. Economic assessment of managing processionary moth in pine forests: A case-study in Portugal. J. Environ. Manag. 2009, 90, 683–691. [**Google Scholar**] [**CrossRef**] [**Green Version**]
10. Kanat, M.; Alma, M.H.; Sivrikaya, F. Effect of defoliation by Thaumetopoea pityocampa (Den. & Schiff.) (Lepidoptera: Thaumetopoeidae) on annual diameter increment of Pinus brutia Ten. in Turkey. Ann. For. Sci. 2005, 62, 91–94. [**Google Scholar**] [**CrossRef**] [**Green Version**]
11. Morr, I. Diagnostic protocols for regulated pests. Protocoles de diagnostic pour les organismes réglementés. EPPO Bull. 2004, 34, 271–279. [**Google Scholar**]
12. Colacci, M.; Kavallieratos, N.G.; Athanassiou, C.G.; Boukouvala, M.C.; Rumbos, C.I.; Kontodimas, D.C.; Pardo, D.; Sancho, J.; Benavent-Fernández, E.; Gálvez-Settier, S.; et al. Management of the Pine Processionary Moth, Thaumetopoea pityocampa (Lepidoptera: Thaumetopoeidae), in urban and suburban areas: Trials with trunk barrier and adhesive barrier trap devices. J. Econ. Èntomol. 2017, 111, 227–238. [**Google Scholar**] [**CrossRef**] [**PubMed**]
13. Rossi, J.-P.; Imbault, V.; Lamant, T.; Rousselet, J. A citywide survey of the pine processionary moth Thaumetopoea pityocampa spatial distribution in Orléans (France). Urban For. Urban Green. 2016, 20, 71–80. [**Google Scholar**] [**CrossRef**]

14. Vega, J.M.; García-Ortiz, J.C.; Palla, P.S.; Sanchís, M.E.; Vega, J.; González-Muñoz, M.; Moneo, I.; Battisti, A.; Roques, A. Prevalence of cutaneous reactions to the pine processionary moth (Thaumetopoea pityocampa) in an adult population. Contact Dermat. 2011, 64, 220–228. [**Google Scholar**] [**CrossRef**]

15. Cayuela, L.; Hódar, J.A.; Zamora, R. Is insecticide spraying a viable and cost-efficient management practice to control pine processionary moth in Mediterranean woodlands? For. Ecol. Manag. 2011, 261, 1732–1737. [**Google Scholar**] [**CrossRef**]

16. Battisti, A.; Longo, S.; Tiberi, R.; Triggiani, O. Results and perspectives in the use of Bacillus thuringiensis Berl. var. kurstaki and other pathogens against Thaumetopoea pityocampa (Den. et Schiff.) in Italy (Lep., Thaumetopoeidae). Anz. Schadl. 1998, 71, 72–76. [**Google Scholar**]

17. Shevelev, A.B.; Battisti, A.; Volynskaya, A.M.; Inovikova, S.; Kostina, L.A.; Zalunin, I. Susceptibility of the pine processionary caterpillar Thaumetopoea pityocampa (Lepidoptera: Thaumetopoeidae) toward delta-endotoxins of Bacillus thuringiensis under laboratory conditions. Ann. Appl. Biol. 2001, 138, 255–261. [**Google Scholar**] [**CrossRef**]

18. Parlak, S.; Özçankaya, I.M.; Batur, M.; Akkaş, M.E.; Boza, Z.; Toprak, Ö. Efficiency of funnel traps in controlling pine processionary moth. J. Plant Dis. Prot. 2018, 125, 539–548. [**Google Scholar**] [**CrossRef**]

19. Guerrero, A.; Camps, F.; Coll, J.; Riba, M.; Einhorn, J.; Descoins, C.; Lallemand, J. Identification of a potential sex pheromone of the processionary moth, Thaumetopea pityocampa (Lepidoptera, Notodontidae). Tetrahedron Lett. 1981, 22, 2013–2016. [**Google Scholar**] [**CrossRef**]

20. Bonsignore, C.P.; Manti, F. Influence of habitat and climate on the capture of male pine processionary moths. Bull. Insectology 2013, 66, 27–34. [**Google Scholar**]

21. Geri, C. Repartition et evolution des populations de la processionnaire du pin, Thaumetopoea pityocampa Schiff (Lep, Thaumetopoeidae) dans les montagnes. Acta Oecologica Oecologia 1983, 5, 3–22. [**Google Scholar**]

22. Jactel, H.; Birgersson, G.; Andersson, S.; Schlyter, F. Non-host volatiles mediate associational resistance to the pine processionary moth. Oecologia 2011, 166, 703–711. [**Google Scholar**] [**CrossRef**] [**PubMed**]

23. Demolin, G. Comportement des adultes de Thaumetopoea pityocampa Schiff. Dispersion spatiale, importance écologique. Ann. Sci. For. 1969, 26, 81–102. [**Google Scholar**] [**CrossRef**] [**Green Version**]

24. Masutti, L.; Battisti, A. Thaumetopoea pityocampa (Den. & Schiff.) in Italy Bionomics and perspectives of integrated control. J. Appl. Èntomol. 1990, 110, 229–234. [**Google Scholar**] [**CrossRef**]

25. Robinet, C.; Rousselet, J.; Pineau, P.; Miard, F.; Roques, A. Are heat waves susceptible to mitigate the expansion of a species progressing with global warming? Ecol. Evol. 2013, 3, 2947–2957. [**Google Scholar**] [**CrossRef**]

26. Battisti, A.; Stastny, M.; Buffo, E.; Larsson, S. A rapid altitudinal range expansion in the pine processionary moth produced by the 2003 climatic anomaly. Glob. Chang. Biol. 2006, 12, 662–671. [**Google Scholar**] [**CrossRef**]

27. Mirchev, P.; Georgiev, G.; Geshev, G. Dispersal of male butterflies of pine processionary moth (Thaumetopoea pityocampa). Silva Balc. 2013, 14, 102–108. [**Google Scholar**]

28. Peter, B.; Manti, F.; Castiglione, E.; Battisti, A. Pupal traits and adult emergence in the pine processionary moth Thaumetopoea pityocampa (Lepidoptera: Notodontidae) are affected by pupal density. Eur. J. Èntomol. 2019, 116, 320–329. **[Google Scholar]** **[CrossRef]**

29. Van Frankenhuyzen, K. Development and current status of Bacillus thuringiensis for control of defoliating forest insects. For. Chron. 1990, 66, 498–507. **[Google Scholar]** **[CrossRef]** **[Green Version]**

30. Cunningham, J.C.; van Frankenhuyzen, K. Microbial insecticides in forestry. For. Chron. 1991, 67, 473–480. **[Google Scholar]** **[CrossRef]** **[Green Version]**

31. Avtzis, N.D. The use of Bacillus thuringiensis against Thaumetopoea pityocampa Schiff. (Lepidoptera: Thaumetopoeidae) in Greece. In Proceedings of the Population Dynamics, Impacts and Integrated Management of Forest Defoliating Insects, USDA Forest Service, General Technical Report NE-247, Banská Štiavnica, Slovakia, 18–23 August 1998; pp. 311–316. **[Google Scholar]**

32. Cebeci, H.; Öymen, R.; Acer, S. Control of pine processionary moth, Thaumetopoea pityocampa with Bacillus thuringiensis in Antalya, Turkey. J. Environ. Biol. 2010, 31, 357–361. **[Google Scholar]**

33. Dubois, N.R.; Reardon, R.C.; Kolodny-Hirsch, D.M. Field efficacy of the NRD-12 Strain of Bacillus thuringiensis against gypsy moth (Lepidoptera: Lymantriidae). J. Econ. Èntomol. 1988, 81, 1672–1677. **[Google Scholar]** **[CrossRef]**

34. Roversi, P.F. Aerial Spraying of Bacillus thuringiensis var. kurstaki for the control of Thaumetopoea processionea in Turkey oak woods. Phytoparasitica 2008, 36, 175–186. **[Google Scholar]** **[CrossRef]**

35. Hajek, A.; Van Frankenhuyzen, K. Use of Entomopathogens against forest pests. In Microbial Control of Insect and Mite Pests; Elsevier BV: Amsterdam, The Netherlands, 2017; Volume 21, pp. 313–330. **[Google Scholar]**

36. Trematerra, P.; Colacci, M.; Athanassiou, C.G.; Kavallieratos, N.G.; Rumbos, C.; Boukouvala, M.C.; Nikolaidou, A.J.; Kontodimas, D.C.; Benavent-Fernández, E.; Gálvez-Settier, S. Evaluation of Mating disruption for the control of Thaumetopoea pityocampa (Lepidoptera: Thaumetopoeidae) in suburban recreational areas in Italy and Greece. J. Econ. Èntomol. 2019, 112, 2229–2235. **[Google Scholar]** **[CrossRef]**

37. Carde, R.T.; Minks, A.K. Control of moth pests by mating disruption: Successes and constraints. Annu. Rev. Èntomol. 1995, 40, 559–585. **[Google Scholar]** **[CrossRef]**

38. Robinet, C.; Baier, P.; Pennerstorfer, J.; Schopf, A.; Roques, A. Modelling the effects of climate change on the potential feeding activity of Thaumetopoea pityocampa (Den. & Schiff.) (Lep. Notodontidae) in France. Glob. Ecol. Biogeogr. 2007, 16, 460–471. **[Google Scholar]** **[CrossRef]**

39. Geri, C.; Miller, C.; Xeuxet, D. Mesure des populations de processionnaire du pin (Thaumetopoea pityocampa Schiff—Lepidoptere Thaumetopoeidae) au Mont-Ventoux. Ann. Sci. For. 1985, 42, 143–184. **[Google Scholar]** **[CrossRef]**

40. Li, S.; Daudin, J.-J.; Piou, D.; Robinet, C.; Jactel, H. Periodicity and synchrony of pine processionary moth outbreaks in France. For. Ecol. Manag. 2015, 354, 309–317. **[Google Scholar]** **[CrossRef]**

41. Tamburini, G.; Marini, L.; Hellrigl, K.G.; Salvadori, C.; Battisti, A. Effects of climate and density-dependent factors on population dynamics of the pine processionary moth in the Southern Alps. Clim. Chang. 2013, 121, 701–712. **[Google Scholar]** **[CrossRef]**

42. Bonsignore, C.; Manti, F.; Castiglione, E. Interactions between pupae of the pine processionary moth (Thaumetopoea pityocampa) and parasitoids in a Pinus forest. Bull. Èntomol. Res. 2015, 105, 621–628. **[Google Scholar]** **[CrossRef]**
43. Tarasco, E.; Triggiani, O.; Zamoum, M.; Oreste, M. Natural enemies emerged from Thaumetopoea pityocampa (Denis & Sciffermüller) (Lepidoptera Notodontidae) pupae in Southern Italy. Redia 2015, 98, 103–108. **[Google Scholar]**
44. De Boer, J.G.; Harvey, J.A. Range-Expansion in processionary moths and biological control. Insects 2020, 11, 267. **[Google Scholar]** **[CrossRef]** **[PubMed]**
45. Houri, A.; Doughan, D. Behaviour Patterns of the Pine Processionary Moth (Thaumetopoea wilkinsoni Tams; Lepidoptera: Thaumetopoeidae). Am. J. Agric. Biol. Sci. 2006, 1, 1–5. **[Google Scholar]** **[CrossRef]** **[Green Version]**

19 CONTROLLING GYPSY MOTH INFESTATION IN CORK OAK FOREST WITH LDMNPV BACULOVIRUS

1. INTRODUCTION: AN OVERVIEW

Lymantria dispar (L.) (Lepidoptera: Erebidae), also referred to as the gypsy moth, is a species of moth that only produces one generation per year. The larvae of this species, which emerge from eggs that have survived the winter, inflict considerable harm to the leaves of cork oak trees. The recurring outbreaks, which result in extensive defoliation of forests, are caused by the combination of their feeding behavior and their high biotic potential [1]. To mitigate these harmful consequences, it is imperative to install suitable biocontainment strategies. Therefore, the use of bioinsecticides has been demonstrated to effectively combat the actions of this pest, while also minimizing any negative effects on the environment [2].

Available formulations utilizing the entomopathogenic bacterium Bacillus thuringiensis effectively utilize the precise mode of action of bacterial toxins that specifically target moth larvae [3]. Conversely, there have been occasional reports of potential harm to non-target lepidoptera living in the forest ecosystem due to side-effects. Baculoviruses are a distinct category of entomopathogens that have co-evolved with their host. They are highly specialized microorganisms that can cause lethal infections in larvae when they consume viral particles. The bioinsecticidal activity is linked to the presence of crystalline occlusion bodies. These entities, once consumed by vulnerable insects, release occlusion-derived viruses (ODVs) that target the cells lining the midgut of the host. The progress of the infection within the host's body depends on the generation of a different kind of viral particles called budded viruses (BVs) [6].

Lymantria dispar multicapsid nucleopolyhedrovirus (LdMNPV) is exclusively linked to the gypsy moth and has co-evolved with this particular species [7]. This biocontrol agent acts as a natural regulator of the defoliator population due to periodic virus outbreaks, particularly in situations of high population density [8]. Therefore, the baculovirus, which is cultivated in the laboratory using live larval material and then released in the field, can be utilized as a powerful instrument to effectively influence the population dynamics and artificially control gypsy moth epidemics [9]. Although this technique seems encouraging, the application of LdMNPV is limited to particular situations where commercial items are accessible, such as in Canada and the USA. Experimental work is necessary to introduce the use of baculovirus in various environmental conditions due to the expensive production of viral material and the limited availability of products in certain global regions, such as Europe, where local L. dispar ecotypes pose a continuous threat to various tree species in forests.

The objective of this study was to assess the effectiveness of a LdMNPV formulation, which has been used in other parts of the world, against European L. dispar in cork oak woods that are impacted by this pest. The investigation was carried out in two stages over multiple years, utilizing small-scale ground-based viral applications in one stage and larger-scale aerial treatments in another stage.

2. MATERIALS AND METHODS

2.1. Tested Formulations

A suspension concentrate formulation of LdMNPV, commercially authorized in Canada, was provided by Andermatt Biocontrol AG (Switzerland) for experimental applications from the ground or by helicopter. The concentration of the active substance was 2.8×10^{10} OB/L. Foray 76B (Sumitomo Chemical Agro Europe S.A.S), containing 20 billion international units (BIU)/L of Bacillus thuringiensis kurstaki (Btk) strain ABTS-351, was used as a reference product.

2.2. LdMNPV Applications from the Ground

Two different trials were conducted to evaluate the effects of the baculovirus: (1) time–response and (2) dose–response. Treatments and application details are summarized in **Table 1** and **Table 2**, respectively. Time–response and dose–response trials were conducted in different experimental fields in the same year.

Table 1. Treatments in the time–response trial.

Treatment	Description	Application Date		Application Rate
		2018	2019	
Untreated Check	Not treated	-	-	-
LdMNPV Early	Earlier application	9 May	4 May	2 L/ha
LdMNPV Later	Later application	16 May	11 May	2 L/ha
Foray 76B	Reference product	16 May	11 May	2 L/ha

Table 2. Treatments in the dose–response trial.

Treatment [a]	Description	Application Rate
Untreated Check	Not treated	-
LdMNPV Low	1/3 standard rate	0.66 L/ha
LdMNPV Standard	Standard rate	2 L/ha
LdMNPV High	3× standard rate	6 L/ha
Foray 76B	Reference product	2 L/ha
[a] All applications were made on one date (9 May 2018, and 11 May 2019).		

The trials were conducted in 2018 and 2019 in forests in north-western Sardinia (Italy) in compliance with Good Experimental Practice (GEP) guidelines established by the European and Mediterranean Plant Protection Organization (EPPO PP 1/210(1), Efficacy evaluation of insecticides—Defoliators of forest trees). The completely randomized experimental design involved four plots (100 m^2) per each treatment. Gypsy moth larval density was recorded before treatments and during the following three weeks (7, 14, and 21 days after LdMNPV application). Assessments were based on counting the number of larvae in eight 30 cm long branches randomly sampled from each plant. Defoliation levels in plots were also evaluated after treatments.

In the time–response trial, early application was conducted one week earlier (9 May 2018, and 4 May 2019), targeting eggs and just-hatched first instar larvae, while standard applications (16 May 2018, and 11 May 2019) targeted first and second instar. In the dose–response trial, applications were made on one date (9 May 2018, and 11 May 2019). Baculovirus applications were carried out with a motorized atomizer (M3 series, Cifarelli SpA, Italy), with a volume of 10 L per plot.

2.3. Aerial Applications

Aerial applications were carried out on 11 May 2019 on a forest area located in the Centre of Sardinia (Abbasanta, Italy). Treatments were performed in ultra-low volumes (ULVs), employing a helicopter (LAMA SA 315/B) equipped with 4 electronic Micronair rotary atomizers (model AU) treating a 20 m wide lane. Treatments were performed early in the morning so that environmental conditions ranged within sub-optimal limits. During product application, a global positioning system (GPS) was employed to trace and record the helicopter route, ensuring accurate and homogeneous distribution. Untreated check plots were compared with plots (around 100 ha each) treated with LdMNPV or Btk (Foray 76B). Direct assessments were based on counting the number of larvae on four 30 cm long branches per each of ten plants randomly sampled in each experimental plot. In addition, samples of larvae (n = 100) were collected from each plot and maintained in the laboratory on foliage collected from the same plots after treatment, in order to compare insect survival. The experiment involved three replicates.

2.4. Data Elaboration and Statistical Analysis

Overtime differences in average larval density among treatments in application experiments from the ground were tested using repeated measures ANOVA (PROC MIXED), and means were separated by LSMEANS comparison (adjust = Tukey), using SAS software (version 9.1) [**10**] with the significance level set at $\alpha = 0.05$. Analysis of variance (ANOVA) followed by least significant difference (LSD) test ($p < 0.05$) was used to compare efficacy data on a specific date and defoliation levels among treatments.

For different datasets in this study, in order to verify assumptions of normality and heteroscedasticity, the Shapiro–Wilk [**11**] and Levene's tests [**12**] were performed, respectively. If necessary, data were transformed as the arcsine of the square root of the percentage.

Field treatment efficacy was evaluated in terms of larval density reduction, where percent reduction in treatment x after t days (ΔDxt) was calculated as:

$$\Delta D_{xt} = \frac{D_{x0} - D_{xt}}{D_{x0}} \times 100 \tag{1}$$

where $Dx0$ is the initial larval density in treatment x at sampling time 0 (i.e., before application), and Dxt is the larval density t days after applications in treatment x. Efficacy differences between treatments were tested separately for each sampling date (i.e., 7, 14 and 21 days after applications) using one-way ANOVA. Tukey's test at a significance level of 0.05 was used for means separation if necessary.

Aerial application trial data obtained from laboratory observations on field-collected larvae were analyzed by a mixed effects Cox proportional hazard model using survival [**13**], and coxme [**14**] packages in R software [**15**]. In each model, treatments were considered as fixed factors and the larval cage (i.e., replicate) as a random effect factor. Further post hoc analysis was performed using the multcomp package in R [**16**], applying a Bonferroni correction for multiple testing. Moreover, larval density reduction (%) was corrected for natural mortality to take into account the effect of natural population decreases (Cxt) using the Schneider-Orelli formula [**17**]:

$$C_{xt} = \frac{\Delta D_{xt} - \overline{D}_{Ct}}{100 - \overline{D}_{Ct}} \times 100 \tag{2}$$

where ΔDxt is the larval density reduction (%) in treatment x after t days, and $\overline{D}_{Ct}$ is the average larval density reduction (%) in untreated control t days after applications. After this correction, transformed data were used to evaluate the merely effect of Btk or LdMNPV against gypsy moth larvae as assessed in the laboratory. Student's t-test at the 0.05 level of significance was used to test for differences between different treatments 7, 14, and 21 days after application.

3. RESULTS

3.1. LdMNPV Applications from the Ground

Comparing the different plots involved in trials, a homogeneous larval density was observed before insecticidal applications in both years for time-response (2018: $F_{3,15} = 0.31$; $p = 0.82$; 2019: $F_{3,15} = 1.33$; $p = 0.31$) and dose-response (2018: $F_{4,15} = 0.26$; $p = 0.90$; 2019: $F_{4,15} = 0.16$; $p = 0.95$) trials.

In the time-response trials conducted in 2018, no significant changes in larval density were observed one week after the application of LdMNPV in the "LdMNPV early" experimental thesis. On the other hand, a significant larval density reduction was found during the following two weeks in the same plots, in comparison with the untreated check ($F_{9,63} = 15.07$; $p < 0.01$). No significant changes in larval density were instead associated with the "LdMNPV later" thesis (**Figure 1**). A significant dose-dependent effect was observed in the trial conducted in 2018 ($F_{12,79} = 19.13$; $p < 0.01$). A higher larval density reduction was associated with a higher LdMNPV dose, and this decrease became more significant as time advanced (**Figure 2**).

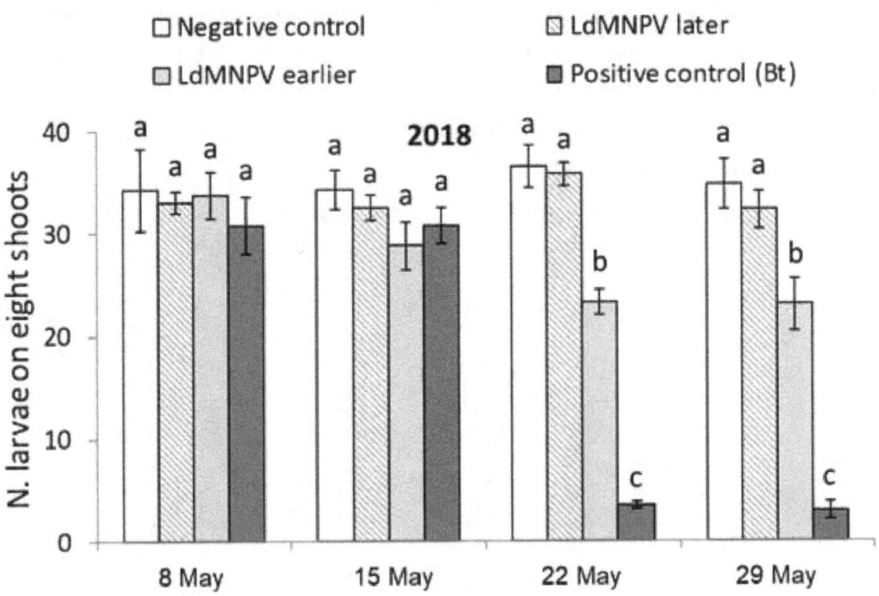

Figure 1. Larval density (mean ± SE) assessed by sampling 8 shoots/plot in the time-response trial with LdMNPV applications from the ground in 2018. Different letters (a, b, c) above bars indicate significant differences among means within each sampling date (ANOVA Mixed Proc., LSMEANS, p < 0.05).

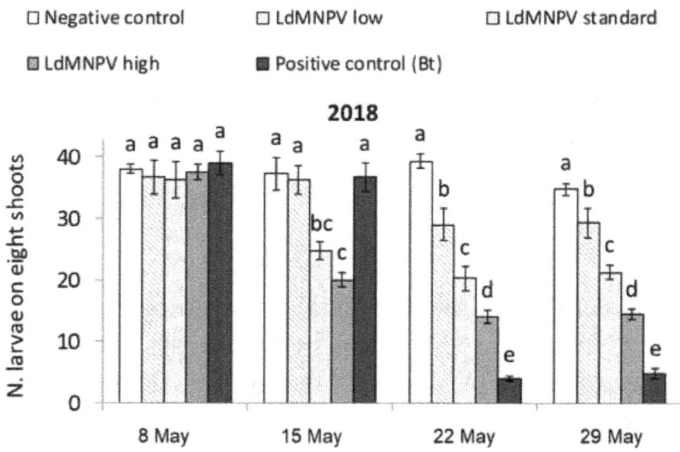

Figure 2. Larval density (mean ± SE) assessed by sampling 8 shoots/plot in the dose–response trial with LdMNPV applications from the ground in 2018. Different letters (a, b, c, d, e) above bars indicate significant differences among means within each sampling date (ANOVA Mixed Proc., LSMEANS, p < 0.05).

These results were comparable to the output of trials conducted in 2019. In the case of the time–response trial, a significant larval density decrease was achieved by both early and later applications of the baculovirus ($F_{9,63}$ = 16.62; p < 0.01), with a higher and faster effect of the earlier treatment in comparison to the untreated check (**Figure 3**). A good efficacy was also observed in the dose–response trial, in which the LdMNPV doses assayed showed a significant biocontrol action on gypsy moth larvae with a dose-dependent effect, in comparison with the untreated control ($F_{12,79}$ = 9.43; p < 0.01). A greater protection of trees was associated with the highest doses applied (**Figure 4**).

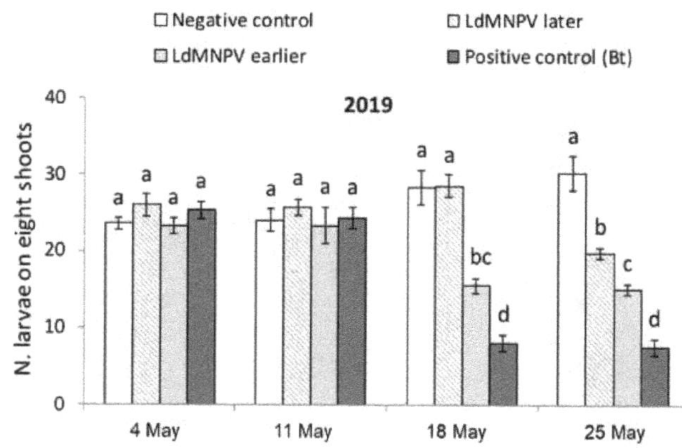

Figure 3. Larval density (mean ± SE) assessed by sampling 8 shoots/plot in the time–response trial with LdMNPV applications from the ground in 2019. Different letters (a, b, c, d) above bars indicate significant differences among means within each sampling date (ANOVA Mixed Proc., LSMEANS, p < 0.05).

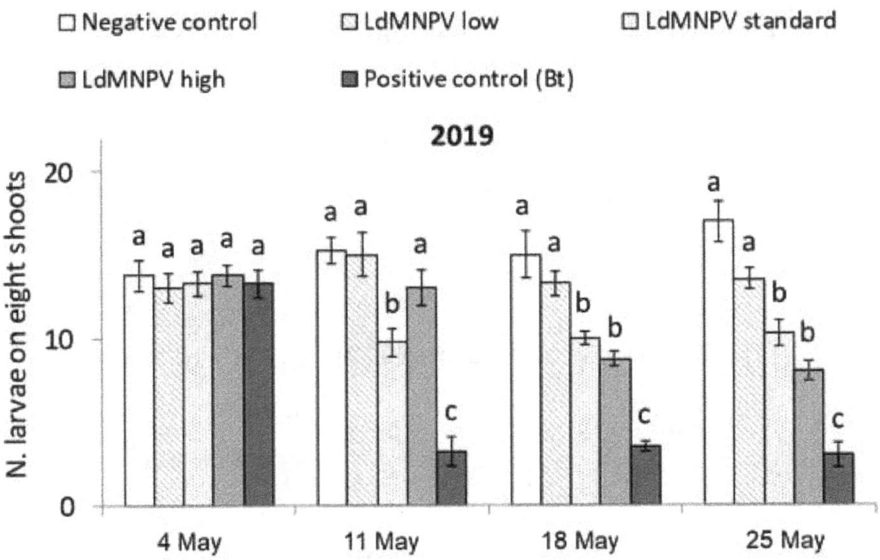

Figure 4. Larval density (mean ± SE) assessed by sampling 8 shoots/plot in the dose-response trial with LdMNPV applications from the ground in 2019. Different letters (a, b, c) above bars indicate significant differences among means within each sampling date (ANOVA Mixed Proc., LSMEANS, $p < 0.05$).

In general, a higher percentage of defoliation was found in the untreated check, while a significant protection was associated with higher LdMNPV doses (2018: $F_{4,19} = 24.50$; $p < 0.001$; 2019: $F_{4,19} = 15.43$; $p < 0.01$) and earlier treatments (2018: $F_{3,15} = 33.08$; $p < 0.01$; 2019: $F_{3,15} = 47.61$; $p < 0.01$) **(Figure 5)**.

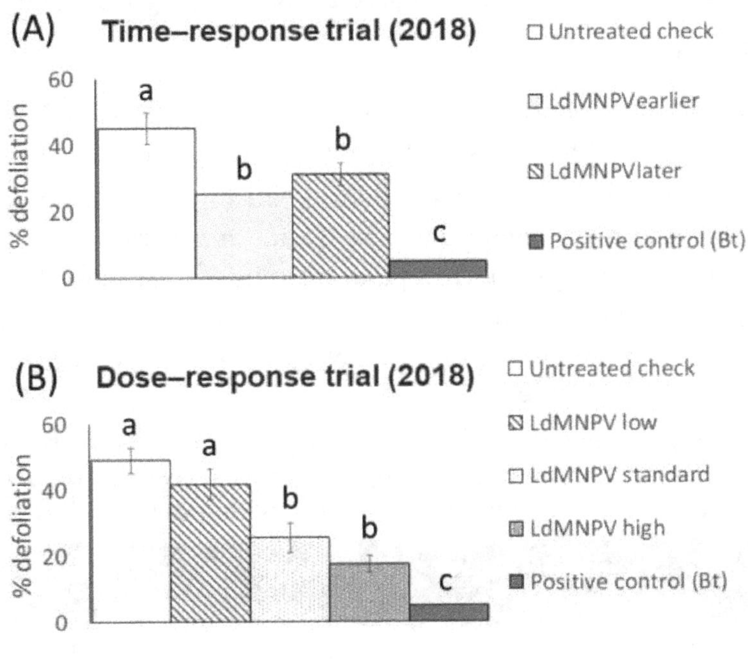

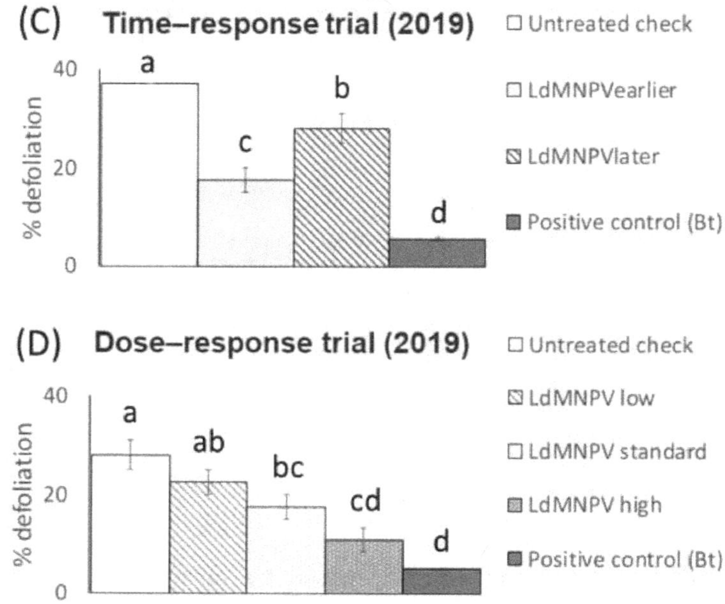

Figure 5. Defoliation percentage (mean ± SE) in different plots treated with LdMNPV from the ground in 2018 (A,B) and 2019 (C,D). Different letters (a, b, c, d) above bars indicate significant differences among means (ANOVA, Tukey test, $p < 0.05$).

In all trials, the decrease in larval density and the protection against defoliation in plots treated with the Btk reference product was the best and associated with greater and faster action (**Figure 1**, **Figure 2**, **Figure 3**, **Figure 4** and **Figure 5**).

3.2. Aerial Applications

LdMNPV formulation applied at a dose of 2 L/ha appeared to be well and homogeneously distributed in the treated plots.

In 2019, a general drop in larval density during the season was observed in the experimental area involved in the aerial application study, outlining a retrogradation phase of gypsy moth population in this forest ecosystem in Sardinia. Accordingly, such a reduction was observed in all plots, with no differences among treatments 7 ($F_{2,8} = 3.52$, $p = 0.13$), 14 ($F_{2,8} = 0.95$, $p = 0.46$), and 21 ($F_{2,8} = 1.85$, $p = 0.27$) days after applications (**Table 3**).

Table 3. Percentage (mean ± SE) of larval density reduction in the field at different time intervals after bioinsecticidal application, in respect to pre-treatment. Percentage data are corrected using the Schneider-Orelli formula.

Days [a]	Treatment			F	*p*
	Foray 76B	LdMNPV	Untreated Check		
7	64.40 ± 8.82 [b]	31.71 ± 4.00	39.66 ± 17.02	3.52	0.13

Days [a]	Treatment			F	p
	Foray 76B	LdMNPV	Untreated Check		
14	70.28 ± 8.94	34.34 ± 34.33	46.87 ± 18.57	0.95	0.46
21	75.30 ± 9.11	45.59 ± 5.83	50.58 ± 20.39	1.85	0.27

[a] Days after application. [b] No significant differences among means were observed (ANOVA, $p > 0.05$).

On the other hand, significant differences in survival rate were observed in the laboratory on the field-collected larvae from different plots ($\chi^2 = 486.79$, $p < 0.01$), with a significant reduction associated with larvae from plots treated with either LdMNPV or Btk (**Figure 6**). In more detail, the survival rate achieved at the end of the observation period was higher for LdMNPV (12%) than Btk-treated larvae (0.7%) ($z = -15.73$, $p < 0.01$). The highest survival rate (52%) was instead associated with larvae from untreated plots ($z = -9.08$, $p < 0.01$). The reduction in surviving larvae attributable exclusively to Btk and LdMNPV was significantly different between these formulations, either 7 ($t = 7.16$, $p < 0.01$) and 14 ($t = 13.44$, $p < 0.01$) days after applications. Instead, no statistical differences in corrected larval reduction were found between Btk and LdMNPV, 21 days after application ($t = 2.75$, $p = 0.10$) (**Table 4**).

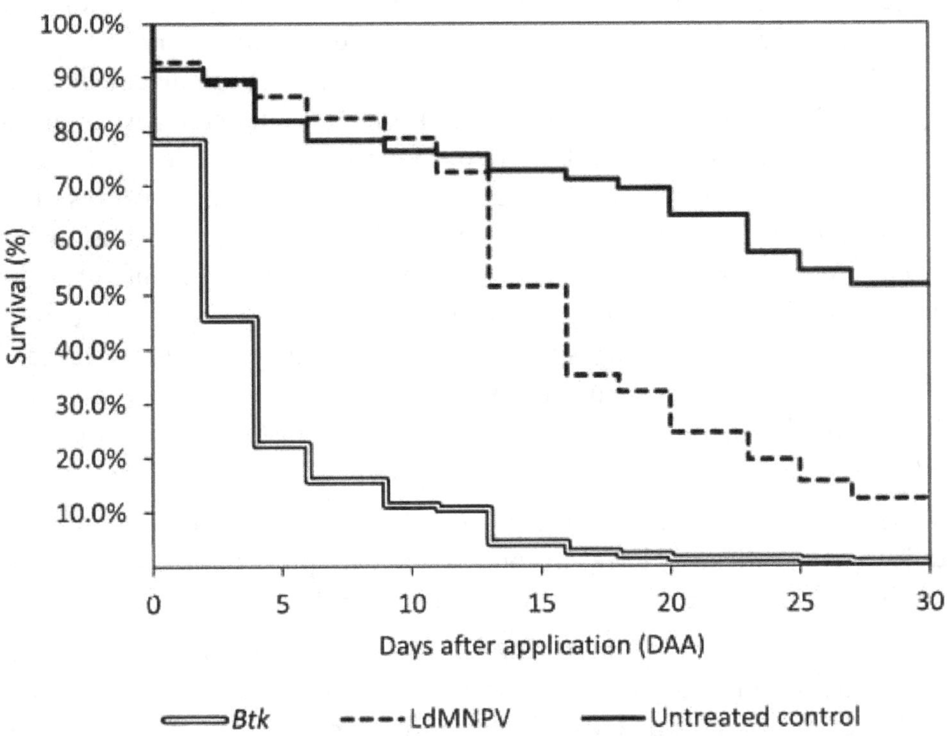

Figure 6. Survival rate of field-collected Lymantria dispar larvae from plots treated with Foray76B, LdMNPV formulation, or untreated (control).

Table 4. Reduction percentage (mean ± SE) of surviving larvae in the laboratory attributable exclusively to treatments at different time intervals from bioinsecticidal application.

Days [a]	Treatment [b]		t	*p*
	Foray 76B	LdMNPV		
7	75.44 ± 4.55 [a]	10.05 ± 2.67 [b]	7.16	0.004
14	86.74 ± 1.80 [a]	21.94 ± 2.12 [b]	13.44	<0.001
21	96.33 ± 0.41 [a]	70.19 ± 5.47 [a]	2.75	0.010

[a] Days after application. [b] Different letters in each line indicate significantly different means (Student's *t*-test, $p < 0.05$).

4. DISCUSSION

Baculoviruses represent natural and selective bioinsecticides and have successfully been used against several Lepidopteran pests worldwide. However, their use is limited to niche contexts, due to their narrow host range, a delayed insecticidal action in respect to synthetic chemicals, and economical issues related to industrial production technologies still necessarily relying on the use of living insects as substrates for virus replication [**18**].

Lymantria dispar multicapsid nucleopolyhedrovirus (LdMNPV) formulation used in this study showed good efficacy against gypsy moth larval populations in Sardinian forest areas, where this pest is the cause of important defoliations during its periodic outbreaks [**19**]. In the experiments conducted with applications from the ground, the lethal effects were dose- and time-dependent, with a higher efficacy achieved with higher doses and earlier treatments. These results align with a pathogenic process that begins with the ingestion of occlusion bodies (OBs) releasing occlusion-derived viruses (ODVs) that act in the midgut, infecting epithelial cells [**20**]. Accordingly, a stronger and faster effect is expected as a consequence of the earlier ingestion of a higher number of viral particles [**21**]. It follows that in order to ensure baculovirus' short-time effectiveness, an early application in the season, possibly against the first instar larvae, is of primary importance.

While a good baculovirus efficacy was achieved in these experiments, larval mortality was significantly lower in comparison with plots treated with Btk, which was confirmed to be a powerful bioinsecticide against gypsy moth [**22,23**].

Higher scale experiments involving larger areas and aerial applications of the bioinsecticidal products employing standard doses (2 L/ha) confirmed a reduced survival rate of baculovirus-treated larvae, in respect to the untreated control. Additionally, in this case, Btk treatments generated a higher lethal-effect. This greater knock-down power relates to the mechanism of action of solubilized and activated bacterial crystal toxins (Cry proteins) interacting with and disrupting midgut epithelial cells, which leads to insect paralysis and death [**24**]. This direct toxicity caused by Btk is in antithesis with a slower action of the baculovirus depending on an infectious process involving replication of the virus and its spread within the insect body through the tracheal system [**20**]. Everything considered, a milder action of the virus compared to Btk clearly emerged in field trials. Despite such differences, larval population density in 2019 was affected by a natural reduction associated with all treated and untreated plots and related to gypsy moth population retrogradation in Sardinian forest. Accordingly, a more evident efficacy of baculovirus applications in large areas is expected during population progradation, when the baculovirus can express its full potential as a natural regulator of moth population dynamics [**25**].

Thus, a higher host density triggers horizontal transmission processes, determining a greater number of infected individuals [**26**]. While these microparasites can naturally regulate periodic cycles of host abundance, their artificial introduction in the forest ecosystem by early applications in the season would produce a similar effect, under appropriate density dynamic conditions. Such density-dependent containment ability has also been demonstrated in laboratory experiments, in which different degrees of resistance to the baculovirus were associated with diverse larval densities [**27**].

Besides an action normally contained during the season of application, the virus introduced into the forest environment is expected to produce an additional impact on the following generations as a result of sub-lethal effects and vertical transmission [**28**]. This expectation supports the use of baculovirus against gypsy moth even if the efficacy in the application season is limited. Following an integrated approach to forest management, baculovirus with a slow action, but a detectable midterm impact on subsequent generations, could be combined with applications of Btk that generate a more immediate knockdown effect. However, such an emerging hypothesis needs specific multiyear studies to be appropriately documented.

On the other hand, however, it is important that the application of these microbiological control agents is calibrated on the basis of the actual conditions of population dynamics at a given time, in order to produce the desired pest containment effects and make these low-environmental impact interventions even more economically viable. Thus, gypsy moth baculoviruses are good candidates to be introduced in gypsy moth multi-year management programs aiming at interfering with their natural population dynamics.

5. CONCISE SUMMARY

According to the results from small-scale studies, L. dispar shown a notable vulnerability to the LdMNPV formulation when administered in greater doses and targeting younger larvae. The highest dosage demonstrated a satisfactory level of effectiveness in safeguarding the crop, although to a lesser extent compared to the Btk reference product. The effectiveness of the treatment was not verified in larger-scale trials carried out through aerial applications, mostly because of population dynamics influenced by a natural regression phase. Nevertheless, a notable rise in mortality of larvae found in plots treated with the baculovirus was observed. Due to the increased vulnerability of younger larvae, it is advisable to apply treatments early.

Overall, the utilization of LdMNPV in forest conservation initiatives targeting the gypsy moth warrants additional evaluation in varying infestation scenarios. The effectiveness of its ability to control population dynamics during epidemics is anticipated to be optimized under progradation [29]. This ecological phenomenon could be utilized in a long-term comprehensive approach that involves the simultaneous application of Btk to control infestations and the baculovirus to regulate population dynamics.

6. AT A GLANCE

Utilizing a species-specific baculovirus is a highly promising approach to effectively manage the gypsy moth and maintain control over its population dynamics within forest ecosystems. The preservation of cork oak trees from the Lepidopteran defoliator Lymantria dispar necessitates a well-designed forest ecosystem management plan that includes the use of environmentally friendly microbial products when there is a surge in the population of these insects. The LdMNPV, a virus that specifically targets gypsy moth populations, was studied over several years to determine its effectiveness in controlling these populations. The study involved applying the virus either from the ground or by aerial treatment.

Trials to assess the effectiveness of treatments against L. dispar populations were carried out in 2018 and 2019 in Sardinia, using a randomized block design. Annually, two experiments were performed, utilizing a commercially available baculovirus formulation dispersed using an atomizer from the ground. The trials aimed to evaluate the impacts of varying dosages and application timing. In 2019, a trial was done to evaluate the effectiveness of LdMNPV when distributed through aerial application at a bigger field scale using ultra-low quantities of 2 L/ha. Both years showed a notable rise in larval death rates in plots treated with larger doses of viral occlusion bodies (OB) and with an earlier application that targeted younger larvae, when compared to untreated controls. In 2019, we noticed that the target pest went through a retrogradation phase. We did not find any significant differences in the number of larvae between the areas treated from a helicopter and the control areas. However, we did observe a noticeable decrease in the vitality of the larvae from the treated plots in the few weeks after the application. The findings of this study suggest that incorporating LdMNPV into forest protection initiatives as part of multi-year integrated strategies may be a valuable approach to controlling gypsy moth populations.

REFERENCES

1. Lentini, A.; Mannu, R.; Cocco, A.; Ruiu, P.A.; Carboneschi, A.; Luciano, P. Long-term monitoring and microbiological control programs against lepidopteran defoliators in Sardinian cork oak forests (Italy). Ann. Silvic. Res. 2020, 45, 21–30. [**Google Scholar**]
2. Martin, J.C.; Bonneau, X. Bacillus thuringiensis 30 ans de lutte contre les chenilles defoliatrices en foret. Phytoma Def. Veg. 2006, 590, 4–7. [**Google Scholar**]
3. Crickmore, N. Beyond the spore—Past and future developments of Bacillus thuringiensis as a biopesticide. J. Appl. Microbiol. 2006, 101, 616–619. [**Google Scholar**] [**CrossRef**]
4. Scriber, J.M. Non-target impacts of forest defoliator management options: Decision for no spraying may have worse impacts on non-target Lepidoptera than Bacillus thuringiensis insecticides. J. Insect Conserv. 2004, 8, 243–263. [**Google Scholar**] [**CrossRef**]
5. Cory, J.S.; Myers, J.H. The ecology and evolution of insect baculoviruses. Annu. Rev. Ecol. Evol. Syst. 2003, 34, 239–272. [**Google Scholar**] [**CrossRef**] [**Green Version**]
6. Clem, R.J.; Passarelli, A.L. Baculoviruses: Sophisticated pathogens of insects. PLoS Pathog. 2013, 9, e1003729. [**Google Scholar**] [**CrossRef**]
7. Barber, K.N.; Kaupp, W.J.; Holmes, S.B. Specificity testing of the nuclear polyhedrosis virus of the gypsy moth, Lymantria dispar (L.) (Lepidoptera: Lymantriidae). Can. Entomol. 1993, 125, 1055–1066. [**Google Scholar**] [**CrossRef**]
8. Elkinton, J.S.; Liebhold, A.M. Population dynamics of gypsy moth in North America. Annu. Rev. Entomol. 1990, 35, 571–596. [**Google Scholar**] [**CrossRef**]
9. Liebhold, A.; Elkinton, J.; Williams, D.; Muzika, R.M. What causes outbreaks of the gypsy moth in North America? Popul. Ecol. 2000, 42, 257–266. [**Google Scholar**] [**CrossRef**]
10. SAS Institute. SAS/STAT 9.1: User's Guide; SAS Institute Inc.: Cary, NC, USA, 2004. [**Google Scholar**]
11. Shapiro, S.S.; Wilk, M.B. An analysis of variance test for normality (complete samples). Biometrika 1965, 52, 591–611. [**Google Scholar**] [**CrossRef**]
12. Levene, H. Robust tests of equality of variances. In Contributions to Probability and Statistics: Essays in Honor of Harold Hotelling; Olkin, I., Ed.; Stanford University Press: Palo Alto, CA, USA, 1960; pp. 278–292. [**Google Scholar**]
13. Therneau, T.M.; Grambsch, P.M. The cox model. In Modeling Survival Data: Extending the Cox Model; Springer: New York, NY, USA, 2000; pp. 39–77. [**Google Scholar**]
14. Therneau, T.M. Coxme: Mixed Effects Cox Models. R Package Version 2.2-5. 2015. Available online: **http://CRAN.R-project.org/package=coxme** (accessed on 21 February 2020).

15. R Development Core Team. R: A Language and Environment for Statistical Computing; R Foundation for Statistical Computing: Vienna, Austria, 2016. [**Google Scholar**]
16. Hothorn, T.; Bretz, F.; Westfall, P. Simultaneous inference in general parametric models. Biom. J. J. Math. Methods Biosci. 2008, 50, 346–363. [**Google Scholar**] [**CrossRef**] [**Green Version**]
17. Puntener, W. Manual for Field Trials in Plant Protection; Ciba-Geigy Limited: Basle, Switzerland, 1981; p. 205. [**Google Scholar**]
18. Moscardi, F.; de Souza, M.L.; de Castro, M.E.B.; Moscardi, M.L.; Szewczyk, B. Baculovirus pesticides: Present state and future perspectives. In Microbes and Microbial Technology; Springer: New York, NY, USA, 2011; pp. 415–445. [**Google Scholar**]
19. Ruiu, L.; Mannu, R.; Falchi, G.; Braggio, A.; Luciano, P. Evaluation of different Bacillus thuringiensis sv kurstaki formulations against Lymantria dispar and Malacosoma neustria larvae infesting Quercus suber trees. Redia 2013, 96, 27–31. [**Google Scholar**]
20. McNeil, J.; Cox-Foster, D.; Gardner, M.; Slavicek, J.; Thiem, S.; Hoover, K. Pathogenesis of Lymantria dispar multiple nucleopolyhedrovirus in L. dispar and mechanisms of developmental resistance. J. Gen. Virol. 2010, 91, 1590–1600. [**Google Scholar**] [**CrossRef**]
21. Scholefield, J.A.; Shikano, I.; Lowenberger, C.A.; Cory, J.S. The impact of baculovirus challenge on immunity: The effect of dose and time after infection. J. Invertebr. Pathol. 2019, 167, 107232. [**Google Scholar**] [**CrossRef**]
22. Smitley, D.R.; Davis, T.W. Aerial application of Bacillus thuringiensis for suppression of Gypsy Moth (Lepidoptera: Lymantriidae) in Populus-Quercus Forests. J. Econ. Entomol. 1993, 86, 1178–1184. [**Google Scholar**] [**CrossRef**]
23. Mannu, R.; Cocco, A.; Luciano, P.; Lentini, A. Influence of Bacillus thuringiensis application timing on population dynamics of gypsy moth in Mediterranean cork oak forests. Pest Manag. Sci. 2020, 76, 1103–1111. [**Google Scholar**] [**CrossRef**]
24. Jurat-Fuentes, J.L.; Crickmore, N. Specificity determinants for Cry insecticidal proteins: Insights from their mode of action. J. Invertebr. Pathol. 2017, 142, 5–10. [**Google Scholar**] [**CrossRef**] [**PubMed**] [**Green Version**]
25. Vasiljevic, L.; Injac, M. Influence des maladies virales sur le dynamisme des populations de Lymantria dispar L. et Hyphantria cunea Drury en Yougoslavie. Bull. Soc. Entomol. Fr. 1984, 89, 800–807. [**Google Scholar**]
26. Anderson, R.M.; May, R.M. The population dynamics of microparasites and their invertebrate hosts. Philos. Trans. R. Soc. Lond. B Biol. Sci. 1981, 291, 451–524. [**Google Scholar**]
27. Reilly, J.R.; Hajek, A.E. Density-dependent resistance of the gypsy moth Lymantria dispar to its nucleopolyhedrovirus, and the consequences for population dynamics. Oecologia 2008, 154, 691–701. [**Google Scholar**] [**CrossRef**]
28. Akhanaev, Y.B.; Belousova, I.A.; Lebedeva, D.A.; Pavlushin, S.V.; Martemyanov, V.V. A Comparison of the Vertical Transmission of High- and Low-Virulence Nucleopolyhedrovirus Strains in Lymantria dispar L. Insects 2020, 11, 455. [**Google Scholar**] [**CrossRef**] [**PubMed**]
29. Myers, J.H.; Cory, J.S. Ecology and evolution of pathogens in natural populations of Lepidoptera. Evol. Appl. 2016, 9, 231–247. [**Google Scholar**] [**CrossRef**] [**Green Version**]

20 PHYLOGENETIC RELATIONSHIPS BETWEEN PHLEBIOPSIS GIGANTEA AND CERTAIN SPECIES OF BASIDIOMYCOTA WERE DETERMINED BY ANALYZING A PORTION OF THE DNA SEQUENCE OF THE ELONGATION FACTOR 1-ALPHA GENE

1. INTRODUCTION: AN OVERVIEW

Phlebiopsis gigantea (Fr.) Jülich is commonly employed as a biological agent to combat the fungus Heterobasidion annosum (Fr.) Bref., which is responsible for conifer root rot infections [1,2,3,4,5,6,7,8,9,10,11,12]. This species is referred to by various synonymous names in the literature, such as Corticium giganteum (Fr.) Fr., Peniophora gigantea (Fr.) Massee, Peniophora gigantea f. pruinosa Pilát, Phanerochaete gigantea (Fr.) S.S. Rattan, and Phlebia gigantea (Fr.) Donk. Following the taxonomic review conducted by Jülich (1978) in Parmasto and Hallenberg [13], the fungus was ultimately classified under the newly established genus Phlebiopsis, a classification that is widely accepted today. The fungus is a common saprotroph that frequently inhabits decaying sections of coniferous forests. It is primarily found in the Northern Hemisphere but has a global distribution. Notably, a few strains have been discovered in South Africa, New Zealand, and Canada [9,14].

Based on physical traits, the taxonomic position of P. gigantea is widely acknowledged according to the Index Fungorum database (www.indexfungorum.

org). The genus Phlebiopsis includes at least 23 species. Various methods have been utilized to distinguish P. gigantea, with the conventional approach relying on the examination of the spores' morphological characteristics. While conventional methods are relatively simple and quick to use [15], molecular techniques provide confirmation of fungal identification and are highly valuable for species identification [16,17].

The most often used molecular markers in fungal taxonomy are ITS 1/2, β-tubulin, histone H3, and elongation factor α (also known as EF1α). Their application enables the identification of fungal genotypes at a species level [18,19,20]. DNA sequence data comparisons are becoming more commonly employed to get information about the phylogenetic relationships among P. gigantea isolates [7,10,11,21,22]. Several studies have employed DNA sequence data from the EF1α gene to conduct phylogenetic analyses, encompassing a diverse array of fungal taxa [23,24].

The aims of this study were to acquire DNA partial sequences for the EF1α gene in P. gigantea strains and to compare them with other Basidiomycota species available in the NCBI database. This gene is a widely present protein involved in the process of translation that is substantially conserved across different species. It has been proposed to possess favorable characteristics for the purpose of phylogenetic inference [25]. It has been effective in phylogenetic investigations as a marker for Eukaryotes, Acomycetes, and Basidiomycota [23,24,25]. The study examined the partial sequences of intron and exon sections of the EF1α gene isolates of P. gigantea from Poland, Finland, Sweden, and Great Britain. It also studied the partial genome sequence (gi:752829739) deposited in GenBank by Hori et al. [26]. So far, the EF1α gene has not been examined in research investigating the differentiation of selected P. gigantea isolates based on their known linear growth and wood decay activity [3,4,11,27].

Our study identifies a new gene region that can be used to assess taxonomic categories and phylogenetic relationships. This gene region complements the previously identified gene sections, such as ITS [2,3]. Multiple variables can influence the biological activity of the fungus. Grossbard [28] stated that the existence of some fungi in soil can alter the biological characteristics of coexisting taxa. Schardl and Craven [29] reported that the temporal variation in enzyme and degradation activity of fungal isolates may indicate potential hazards in altering or losing their molecular and biochemical properties. The cause is demonstrated by the potential hybridization resulting from the breeding of individuals that are plainly homozygous. Żółciak et al. [3] and Sierota et al. [27] proposed alterations in the activity of several P. gigantea isolates throughout time, as well as in relation to their origin and wood density. In order to enhance the efficiency of competitive fungi employed in biopreparations for biological control of diseases (such as P.

gigantea), it is necessary to regularly replace strains with more effective ones [4]. Evaluating the usefulness of the EF1α region can provide helpful insights and aid in selecting the most effective P. gigantea isolates as competitors against Heterobasidion spp.

2. MATERIALS AND METHODS

2.1. Cultivation of Isolates

Twelve previously identified and tested isolates of P. gigantea [11] were used in the experiment: six from Poland (not registered as biocontrol agent) and one from Finland, one from Sweden and four from Great Britain (registered as biocontrol agent) (**Table 1**). The number of Polish isolates was limited to six due to difficulties in obtaining homogeneous single-spore cultures, while the Finnish and British isolates were accepted as previously tested and approved. Isolates were grown on potato dextrose agar (PDA) medium (Difco™, Sparks, MD, USA) in Petri dishes for ten days at 20 °C according to Kwaśna et al. [**30**].

Table 1. List of P. gigantea isolates used in the study.

Polish Isolate Codes	Location Coordinates/Forest District/Country	Host	Substrate	Collector	Collection (Institution Name, Isolate Code, Country)
PL Pg_1	x:52.09859 y: 20.85479 Chojnów FD, Poland	*Pinus sylvestris* L.	Stump wood	A. Żółciak	Forest Research Institute, 01.06.08.1.5; 02.10.23.1.2; 03.11.04.1.1; 03.11.13.1.3; 99.09.10.1.1; 04.11.19.1.2 Poland
PL Pg_2					
PL Pg_6					
PL Pg_9					
PL Pg_{11} (Accession number: KU886024)	x:51.22000 y: 20.33157 Barycz FD, Poland		Fruitbody on stump	Z. Sierota	
PL Pg_{12}	x:53.42508 y: 20.59593 Nidzica FD, Poland		Stump wood	A. Żółciak	
SE Pg_8	Råberg near Uppsala, Sweden	*Picea abies*(L.) Karst.		K. Korhonen	Lallemand Plant Care - Verdera Oy, VRA 1984 *, Finland
FI Pg_{10}	Loppi, Finland		Log wood		Lallemand Plant Care - Verdera Oy, VRA 1835 *, Finland

Polish Isolate Codes	Location Coordinates/Forest District/Country	Host	Substrate	Collector	Collection (Institution Name, Isolate Code, Country)
GB Pg_{14}	Mull, Great Britain	*Pinus contorta* Dougl. ex Loud.	No data	No data	Forestry Commission, FOC PG 410.3 *, Great Britain
GB Pg_{15}	Roslin, GB	No data			Forestry Commission, FOC PG SP log 5 *, Great Britain
GB Pg_{16} (Accession number: KU886025)	NRS, GB	*Pinus sylvestris* L.			Forestry Commission, FOC PG B 20/5 *, Great Britain
GB Pg_{17}	Buchan, GB				Forestry Commission, FOC PG BU 3 *, Great Britain
* isolates registration in European Union.					

2.2. DNA Extraction

Total fungal DNA of P. gigantea was extracted from mycelium grown on PDA by using DNeasy Plant Mini Kit (Qiagen, Hilden, Germany), according to the protocol. Quality of the DNA was checked with an Infinite 200 PRO multimode plate reader (Tecan, Group Ltd., Männedorf, Switzerland).

2.3. Primers and PCR Conditions

The PCR reactions were done in 25 μL volumes using a Veriti Thermal Cycler (Applied Biosystems, Foster City, CA, USA). Each reaction contained 1 unit (0.25 μL) of Taq DNA polymerase (recombinant) (Thermo Scientific, Life Technologies Inc., Carlsbad, CA, USA), 2.5 μL of 10 × Taq buffer, 2 mM of each dNTP (0.5 μL), 1.5 mM of $MgCl_2$ (1.5 μL), 12.5 pmol of forward/reverse primers (0.125 μL), and 20 ng (1 μL) of DNA.

The primers and the touchdown PCR reaction conditions were used according to modified procedure of Rehner [**18**]. Amplicons of the partial region of the EF1& gene were generated using two overlapping primer combinations, 526F (5′ GTC GTY GTY ATY GGH CAY GT 3′) × 1567R (5′ AC HGT RCC RAT ACC ACC RAT CTT 3′) and EFdF (5′ AAG GAY GGN CAR ACY CGN GAR CAY GC 3′) × 2218R (5′ AT GAC ACC RAC RGC RAC RGT YTG 3′). For 526F × 1567R primers the touchdown PCR was as follows: 5 min at 94 °C (an initial denaturation), 30 s at 94 °C (denaturation) and 45 s at 60 °C (annealing) in the first cycle, successively reducing the Tm by 1 °C per cycle over the next 9 cycles to a final Tm 50 °C, which was used in the remaining 36 cycles. An extension step per cycle was 1 min 30 s at 72 °C. The final extension step was 1 cycle of 7 min at 72 °C.

PCR parameters for EFdF × 2218R primers were as follows: 5 min at 94 °C, 45 cycles of (30 s at 94 °C, 30 s at 63 °C, 1 min 30 s at 72 °C) and 7 min at 72 °C. Amplicons were run in 1.5% (w/v) ethidium bromide-stained agarose gels (Sigma-Aldrich, St. Louis, MO, USA) and the bands were visualized under UV illumination. PCR products were sequenced using Sanger's sequencing method at the Polish Academy of Sciences Institute of Biochemistry and Biophysics (Warsaw, Poland).

2.4. DNA Sequence Analysis

Sequencing results were analyzed using the BLAST algorithm on The National Center for Biotechnology Information (NCBI, **https://blast.ncbi.nlm.nih.gov/**) database and MEGA version 5 [**31**]. The DNA sequences were aligned using Clustal W version 2.0. [**32**]. Genetic variability was estimated for the sequences of P. gigantea and realigned separately for intron and exon regions using FGENESH 2.6 [**33,34**].

The obtained sequences were analyzed using Tajima's Neutrality Test. The number of sequences (m), number of segregating sites (S), mean number of segregating sites (Ps), nucleotide diversity (π) and estimates of Theta (Θ) per site as well as the Tajima test statistic (D) were estimated using MEGA5 program [**35,36**].

EF1α partial gene sequences of P. gigantea and selected Basidiomycota species from the NCBI database were analyzed using the maximum likelihood approach for phylogeny reconstruction and tested by bootstrapping with 1,000 replicates. Missing and ambiguous characters were excluded from the analysis. Phylogenetic trees were generated based on maximum likelihood method [**37**]. Two sequences of P. gigantea were deposited in GenBank NCBI (Accession numbers: KU886024 and KU886025, **www.ncbi.nlm.nih.gov.genbank/**).

3. RESULTS

EF1α Partial Gene of P. gigantea Analysis

The region of the EF1α gene was successfully amplified for all P. gigantea isolates described in this study. All amplifications yielded a single band approximately 1,500 bp long. The final analyzed length partial sequence of the EF1α gene was 1,411 bp, the mRNA length was 1,403 bp. The structure of EF1α partial gene from the P. gigantea (Accession numbers: KU886024 and KU886025) is shown in **Figure 1**.

Figure 1. The structure of EF1α partial gene from the Phlebiopsis gigantea. The positions for exons are presented respectively: exon 1 (1–75), exon 2 (134–228), exon 3 (284–496), exon 4 (550–686), exon 5 (739–1056), exon 6 (1113–1403). (mRNA) 1–6 exon (s): 1–1403. The numbers indicate the positions of the bases identified using FGENESH software.

Genetic diversity was calculated for all obtained sequences including exons and introns of 12 partial sequences of the EF1α gene and partial genome sequence (gi:752829739) of P. gigantea (**Figure 1**) and separately for exons and introns (**Table 2**) The number of segregating sites (S) and mean number of segregating sites (Ps) were 32 and 0.022679 respectively. The value of the Θ, expressing the total variability was 0.007308 for analyzed 13 partial sequences mentioned above. The nucleotide diversity (π) was 0.005761. The value of Tajima's D was below zero (-0.931820).

Table 2. Genetic polymorphism and neutrality tests of EF1α partial gene sequences of 12 Polish isolates and partial genome sequence (gi:752829739) of P. gigantea, including introns and exons.

Region	Sequence	*bp*	*m*	*s*	P_s	Θ	*π*	*D*
Intron	I_1	58	13	1	0.017241	0.005556	0.004863	−0.274290
	I_2	55	13	1	0.018182	0.005859	0.005128	−0.274290
	I_3	53	13	3	0.056604	0.018240	0.013546	−0.813698
	I_4	52	13	3	0.057692	0.018591	0.013807	−0.813698
	I_5	56	13	6	0.107143	0.034526	0.025641	−0.950320
Exon	Partial E_1	75	13	0	0.000000	0.00000E+000	0.000000	*n/c*
	E_2	95	13	1	0.010526	0.003392	0.002969	−0.274290
	E_3	213	13	4	0.018779	0.006052	0.004093	-1.099317
	E_4	137	13	3	0.021898	0.007057	0.006176	-0.394391
	E_5	318	13	3	0.009434	0.003040	0.002661	-0.394391
	E_6	299	13	7	0.023411	0.007544	0.006174	-0.688629

Abbreviations: *bp*, base pairs; *m*, number of sequences; *s*, number of segregating sites; P_s = *s/m*, mean number of segregating sites; $\Theta = P_s/a_1$, estimates of Theta per site; *π*, nucleotide diversity; *D*, Tajima's *D* Tajima test statistic; *n/c*, no changes.

In the I_1 and I_2 intron region the number of segregating sites was one. The remaining introns I_3, I_4 and I_5 showed three and six segregating sites, respectively. The value of the Θ, expressing the total variability ranged from 0.005556 to 0.034526 for all intron part sequences. The nucleotide diversity (π) ranged from 0.004863 to 0.025641. The analysis of exons of the coding regions showed no segregating sites for partial E_1, however the number of segregating sites ranged from one for E_2 to seven for E_6. The value of the Θ was zero for partial E_1 and the highest 0.007544 for E_6. Among exons the nucleotide diversity (π) ranged from zero to 0.006176. The value of Tajima's D was below zero for all intron and E_2–E_6 exon regions (**Table 2**).

Bootstrap values supported the separation of species into distinct clades (**Figure 2**). Among partial 12 sequences of P. gigantea the closest one to the partial genome sequence of P. gigantea (gi: 752829739) was GB Pg 16 (KU886025) from Great Britain. The couple of single mutations differentiated GB Pg 16 sequence from the rest of identical isolates of this fungus.

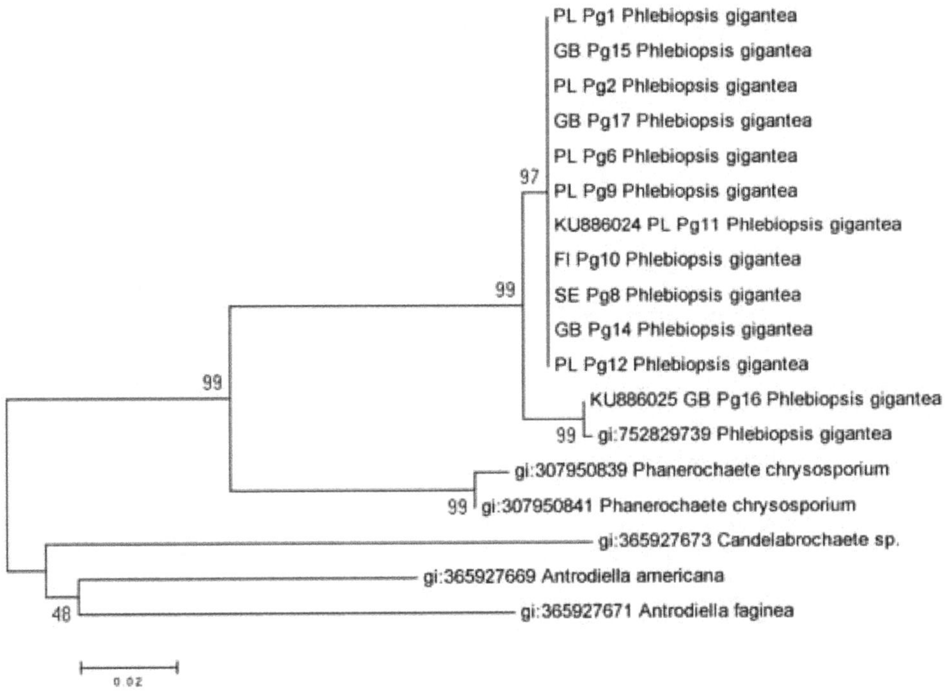

Figure 2. Phylogenetic tree for Phlebiopsis gigantea isolates and other species from the Phanerochaetaceae, based on EF1α gene partial sequences. Branch length values were shown, the tree was reconstructed using the maximum likelihood approach and tested by bootstrapping (1,000 replicates).

For Phanerochaetaceae family, two major and well supported groups were identified. One of the group represented all P. gigantea and Phanerochaete chrysosporium Burds. partial sequences (97–99% bootstrap support). The second group included: Candelabrochaete sp., Antrodiella Americana Ryvarden & Gilb. and Antrodiella faginea Vampola & Pouzar (48% bootstrap support). Phylogenetic analysis of partial sequences of EF1α gene for P. gigantea and selected Basidiomycota species, showed separate clades for following orders: Polyporales, Agaricales, Boletales, Atheliales, Russulales and Sebacinales (**Figure 3**).

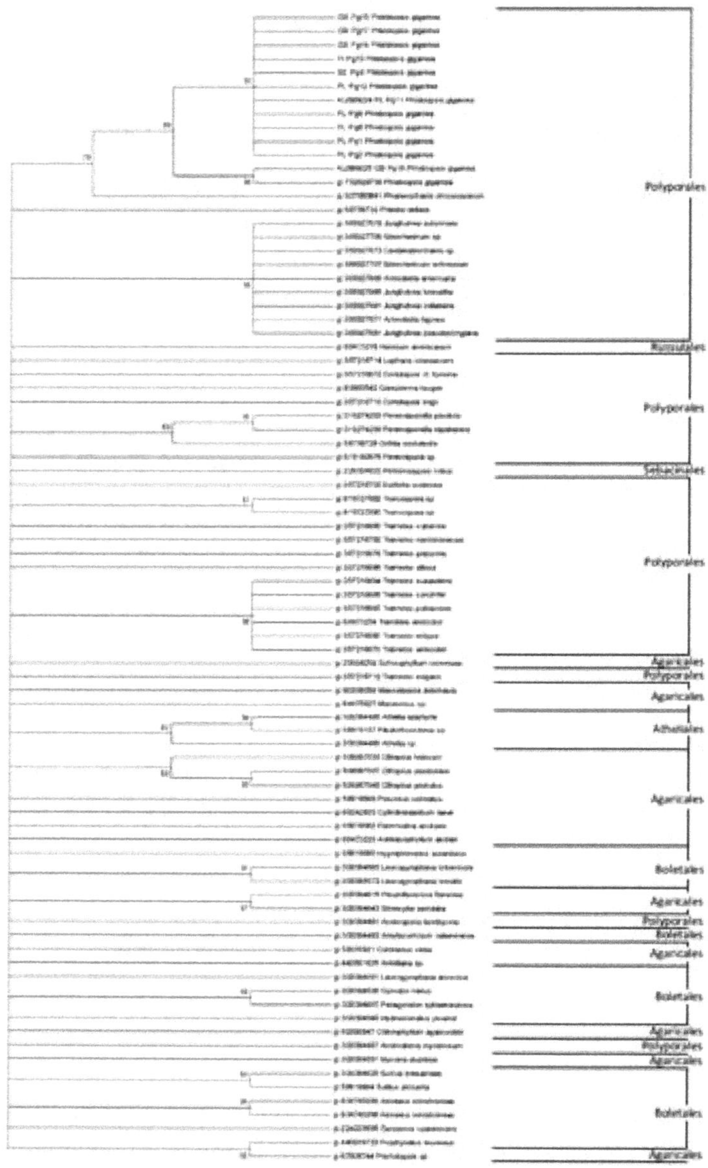

Figure 3. Phylogenetic tree for Phlebiopsis gigantea isolates and selected Basidiomycota, based on the EF1α partial gene sequences. The tree was reconstructed using the maximum likelihood approach and tested by bootstrapping (1000 replicates).

4. DISCUSSION

In this study, DNA partial sequence for the EF1α gene of Phlebiopsis gigantea was successfully identified and analyzed. All P. gigantea isolates yielded PCR products of similar size, indicating that the amplified partial gene region does not include large indels, and therefore is a suitable choice for phylogenetic studies. The aligned sequences showed considerable homology among P. gigantea but various species-specific nucleotide substitutions and indels were observed among all Basidiomycota species. No sequence variation was observed among all P. gigantea except of one isolate (GB Pg 16) from Great Britain and partial sequence of P. gigantea genome (gi: 752829739).

The remarkable variation was observed between different strains. This is consistent with previous studies employing ITS region and genetic fingerprinting using random amplified microsatellite (RAMS) markers in taxonomic studies of P. gigantea [**21**].

For the analyzed P. gigantea EF1α partial gene the occurrence of exons and introns was identified. Introns represented the so-called 3rd introns group, spliced during the maturation of RNA with the participation of spliceosome [**38**]. The size of introns ranged from 52 to 58 bp, at the average length of the gene from 0.75 to 1,000 bp of gene.

Among representatives of Basidiomycota six introns in gene are observed average [**39**]. Compared to higher Eukaryota, introns occurring in fungi are relatively short [**40**]. In the case of the analyzed gene, introns size corresponds to the average size of introns identified in representatives of the Fungi kingdom (50–200 bp) [**41**]. The length of introns of EF1α gene was slightly smaller than the average for the Puccinia graminis (0.65 kb, NCBI accession number: X73529.1) and Neurospora crassa gene (0.81 kb) [**42**].

Despite the randomness and variability of the structure of introns, in the case of I_5 of EF1α partial gene, a larger number of segregating sites (six) was identified. In the sequences of other introns this number were from 1 to 3. Furthermore, the characteristic feature of the identified intron sequences was the presence of the sequence GT on the donor side (5') and the sequence AG at the acceptor side (3'). They are essential in the process of identifying and splicing by spliceosome. Identified GT and AG sequences are considered as the most commonly occurring canonical dinucleotide fragments, respectively starting and ending the introns [**43**].

Intraspecies analysis of EF1α partial gene showed small variability within the coding and non-coding regions. Introns of analyzed EF1α partial gene were well conserved among all tested isolates of P. gigantea. The evolution of the structure of all introns and E_2–E_6 exon regions of EF1α partial gene was neutral. Isolates representing P. gigantea from Poland, Great Britain, Finland and Sweden had similar partial sequences of EF1α gene and were grouped together. The only British isolate (GB Pg 16 = FOC PG B 20/5) of this fungus showed coupled single mutations like single nucleotide polymorphisms (SNPs) and was very similar to P. gigantea (gi: 752829739) with identity 99% and query cover 100%.

Results of the study using DNA-RAMS markers indicated genetic similarity among isolates collected in Finland and Great Britain [**10**]. Additional studies using the previously mentioned markers showed that Polish isolate (PL 12) of P. gigantea was genetically analogous to FC 16 from Great Britain [**11**]. Vainio and Hantula [**21**] also showed that European and North American ITS/A alleles of P. gigantea were identical, while ITS/C alleles were different. The authors mentioned that the analysis of molecular variation and neighbor joining analysis using 28 RAMS markers revealed a considerable degree of differentiation between Europe and North America [**21**].

The main advantage of this study is the phylogenetic analysis of EF1α partial DNA sequence data for P. gigantea in comparing it to species belonging to the same family—Phanerochaetaceae and selected Basidiomycota species. Phylogenetic trees showed that P. gigantea is closely related to Phanerochaete chrysosporium. EF1α is also the first protein-coding gene and first single-copy gene used for phylogenetic analysis of P. gigantea. The whole genome sequence of P. gigantea (gi: 752829739, accession number: AZAG01000080) was published by Hori et al. [**26**]. The whole genome sequence size is approximately 30 Mbp and number of predicted genes (11,891). Sequence data from the majority of isolates belonging to the different species showed unique species-specific substitutions, allowing the isolates to be differentiated into clades representing the species.

The results of this study demonstrate that the EF1α region is useful for phylogenetic analysis and classification of Polyporales species. This is a large and taxonomically difficult order, which include several genera, for example: Phlebiopsis, Phanerochaete, Phlebia, Junghuhnia, Steccherinum, Androniella, Ganoderma, Coriolopsis, Perenniporiella and Trametes.

5. CONCISE SUMMARY

Our computational research demonstrated a correlation between P. gigantea and specific Basidiomycota species, as determined by analyzing the partial DNA sequence of elongation factor 1-alpha. The EF1α gene sequence of P. gigantea isolates from Poland, Finland, Sweden, and Great Britain constituted a monophyletic group, with the exception of one British strain GB Pg 16 (FOC PG B 20/5). Our study shown that the partial sequence of the EF1α gene is valuable for conducting phylogenetic analysis of the Phanerochaetaceae family.

6. AT A GLANCE

Phlebiopsis gigantea (Fr.) Jülich has effectively served as a biocontrol agent against Heterobasidion annosum (Fr.) Bref., a significant pathogen that affects pine and spruce trees. The species P. gigantea has been well-known for a considerable period of time. However, our comprehension of the correlation between different isolates of this fungus has significantly advanced due to the utilization of DNA sequence comparisons. This work aimed to establish the links between P. gigantea and certain species of Basidiomycota. This was achieved by analyzing the partial DNA sequence of the elongation factor 1-alpha (EF1α) and utilizing in silico data. This study comprised a total of 12 isolates, which were the most representative of P. gigantea. These isolates had various geographic ranges and hosts. The phylogenetic trees constructed using sequences obtained in this investigation placed the European taxa of P. gigantea and a partial sequence of the genome deposited in the NCBI database in a well-supported clade. This clade was shown to be basal to the other strains included in the study. The P. gigantea isolates from Poland, Finland, Sweden, Great Britain, and a partial genome sequencing established a monophyletic clade. Within this group, isolates of P. gigantea formed two distinct subclades, indicating their partial divergence, similar to the presence of two single nucleotide polymorphisms (SNPs) that differentiate them from the other isolates. Furthermore, the links between introns and exons among P. gigantea isolates were clarified. The findings derived from analyzing the EF1α area are expected to be valuable in identifying more effective P. gigantea isolates for controlling forest tree root diseases.

REFERENCES

1. Sierota, Z. Heterobasidion root rot in forests on former agricultural lands in Poland: Scale of threat and prevention. Sci. Res. Essays 2013, 8, 2298–2305. [**Google Scholar**]
2. Żółciak, A.; Sikora, K.; Nowakowska, J.A.; Małecka, M.; Borys, M.; Tereba, T.; Sierota, Z. Antrodia gossypium, Phlebiopsis gigantea and Heterobasidion parviporum: In vitro growth and Norway spruce wood block decay. Biocontrol Sci. Technol. 2016, 26, 1706–1718. [**Google Scholar**] [**CrossRef**]
3. Żółciak, A.; Sierota, Z.; Małecka, M. Characterization of some Phlebiopsis gigantea isolates with respect to enzymatic activity and decay of Norway spruce wood. Biocontrol Sci. Technol. 2012, 22, 777–790. [**Google Scholar**] [**CrossRef**]
4. Żółciak, A.; Sikora, K.; Wrzosek, M.; Damszel, M.; Sierota, Z. Why Does Phlebiopsis gigantea not Always Inhibit Root and Butt Rot in Conifers? Forests 2020, 11, 129. [**Google Scholar**] [**CrossRef**] [**Green Version**]
5. Sun, H.; Paulin, L.; Alatalo, E.; Asiegbu, F.O. Response of living tissues of Pinus sylvestris to the saprotrophic biocontrol fungus Phlebiopsis gigantea. Tree Physiol. 2011, 31, 438–451. [**Google Scholar**] [**CrossRef**]
6. Sun, H.; Terhonen, E.; Koskinen, K.; Paulin, L.; Kasanen, R.; Asiegbu, F.O. The impacts of treatment with biocontrol fungus (Phlebiopsis gigantea) on bacterial diversity of Norway spruce stumps. Biological Control 2013, 64, 238–246. [**Google Scholar**] [**CrossRef**]
7. Mgbeahuruike, A.C.; Karlsson, M.; Asiegbu, F.O. Differential expression of two hydrophobin genes (Pgh1 and Pgh2) from the biological control agent Phlebiopsis gigantea. Fungal Biol. 2012, 116, 620–629. [**Google Scholar**] [**CrossRef**]
8. Mgbeahuruike, A.C.; Kohler, A.; Asiegbu, F.O. Expression analysis of the impact of culture filtrates from the biocontrol agent, Phlebiopsis gigantea on the conifer pathogen,

Heterobasidion annosum s.s. Transcriptome. Microb. Ecol. 2013, 66, 669–681. [**Google Scholar**] [**CrossRef**]

9. Garbelotto, M.; Gonthier, P. Biology, epidemiology, and control of Heterobasidion species worldwide. Annu. Rev. Phytopathol. 2013, 51, 39–59. [**Google Scholar**] [**CrossRef**] [**Green Version**]
10. Sierota, Z.; Nowakowska, J.A.; Sikora, K.; Wrzosek, M.; Żółciak, A.; Małecka, M. What is important in selecting Phlebiopsis gigantea strain for commercial use? J. Agric. Sci. Technol. B 2015, 5, 55–64. [**Google Scholar**]
11. Sierota, Z.; Nowakowska, J.A.; Sikora, K.; Wrzosek, M.; Żółciak, A.; Małecka, M. Genetic variation among Phlebiopsis gigantea strains determined by Random Amplified Microsatellite Markers. Balt. For. 2015, 21, 178–183. [**Google Scholar**]
12. Blomquist, M.; Herrera, S.L.; Hofmann, J.; Beram, R.C.; Cleary, M.; Rönnberg, J. Size matters but is big always better? Effectiveness of urea and Phlebiopsis gigantea as treatment against Heterobasidion on Picea abies stumps of variable size. For. Ecol. Manag. 2020, 462, 117998. [**Google Scholar**] [**CrossRef**]
13. Parmasto, E.; Hallenberg, N. A taxonomic study of phlebioid fungi (Basidiomycota). Nord. J. Bot. 2000, 20, 105–118. [**Google Scholar**] [**CrossRef**]
14. Grillo, R.; Hantula, J.; Korhonen, K. Interfertility between North American and European strains of Phlebiopsis gigantea. For. Pathol. 2005, 35, 173–182. [**Google Scholar**] [**CrossRef**]
15. Żółciak, A. Inokulacja pniaków sosnowych preparatami biologicznymi z Phlebiopsis gigantea/Scots pine stumps inoculation with Phlebiopsis gigantea biological preparations. Leśne Pr. Badaw. 2007, 2, 77–94. (In Polish) [**Google Scholar**]
16. Kozlakidis, Z.; Hacker, C.V.; Bradley, D.; Jamal, A.; Phoon, X.; Webber, J.; Brasier, C.M.; Buck, K.W.; Coutts, R.H.A. Molecular characterization of two novel double-stranded RNA elements from Phlebiopsis gigantea. Virus Genes 2009, 39, 132–136. [**Google Scholar**] [**CrossRef**]
17. Menkis, A.; Burokienė, D.; Gaitnieks, T.; Uotila, A.; Johannesson, H.; Rosling, A.; Finlay, R.D.; Stenlid, J.; Vasaitis, R. Occurrence and impact of the root-rot biocontrol agent Phlebiopsis gigantea on soil fungal communities in Picea abies forests of Northern Europe. FEMS Microbiol. Ecol. 2012, 81, 438–445. [**Google Scholar**] [**CrossRef**] [**Green Version**]
18. Rehner, S. Primers for Elongation Factor 1-α (EF1-α); Insect Biocontrol Laboratory USDA, ARS, PSI: Beltsville, MD, USA, 2001; p. 4. 1p, Available online: **http://ocid.NACSE.ORG/research/deephyphae/EF1primer.pdf** (accessed on 20 May 2020).
19. Stępień, Ł.; Koczyk, G.; Waśkiewicz, A. Genetic and phenotypic variation of Fusarium proliferatum isolates from different host species. J. Appl. Genet. 2011, 52, 487–496. [**Google Scholar**] [**CrossRef**] [**Green Version**]
20. Schoch, C.L.; Seifert, K.A.; Huhndorf, S.; Robert, V.; Spouge, J.L.; Levesque, C.A.; Chen, W.; Fungal Barcoding Consortium. Nuclear ribosomal internal transcribed spacer (ITS) region as a universal DNA barcode marker for Fungi. Proc. Natl. Acad. Sci. USA 2012, 109, 6241–6246. [**Google Scholar**] [**CrossRef**] [**Green Version**]
21. Vainio, E.J.; Hantula, J. Genetic differentiation between European and North American populations of Phlebiopsis gigantea. Mycologia 2000, 92, 436–446. [**Google Scholar**] [**CrossRef**]
22. Liu, A.Z.; Samils, N.; Higgins, B.; Stenlid, J.; Slippers, B.; Nairn, C.J.; Covert, S.F. Microsatellite markers for the wood decay fungus Phlebiopsis gigantea. Conserv. Genet. 2009, 10, 1529–1532. [**Google Scholar**] [**CrossRef**]

23. Rehner, S.A.; Buckley, E.A. Beauveria phylogeny inferred from nuclear ITS and EF1-α sequences: Evidence for cryptic diversification and links to Cordyceps teleomorphs. Mycologia 2005, 97, 84–98. **[Google Scholar]** **[CrossRef]** **[PubMed]**

24. Matheny, P.B.; Wang, Z.; Binder, M.; Curtis, J.M.; Lim, Y.W.; Nilsson, R.H.; Hughes, K.W.; Hofstetter, V.; Ammirati, J.F.; Schoch, C.L.; et al. Contributions of rpb2 and tef1 to the phylogeny of mushrooms and allies (Basidiomycota, Fungi). Mol. Phylogenet. Evol. 2007, 43, 430–451. **[Google Scholar]** **[CrossRef]** **[PubMed]**

25. Roger, A.J.; Sandblom, O.; Doolittle, W.F.; Philippe, H. An evaluation of elongation factor 1 alpha as a phylogenetic marker for Eukaryotes. Mol. Biol. Evol. 1999, 16, 218–233. **[Google Scholar]** **[CrossRef]** **[Green Version]**

26. Hori, C.; Ishida, T.; Igarashi, K.; Samejima, M.; Suzuki, H.; Master, E.; Ferreira, P.; Ruiz-Dueñas, F.J.; Held, B.; Canessa, P.; et al. Analysis of the Phlebiopsis gigantea genome, transcriptome and secretome provides insight into its pioneer colonization strategies of wood. PLoS Genet. 2014, 10, e1004759. **[Google Scholar]** **[CrossRef]** **[Green Version]**

27. Sierota, Z.; Wrzosek, M.; Małecka, M.; Zółciak, A. Decay indices for evaluating wood decomposition activity. Biocontrol Sci. Technol. 2016, 26, 163–173. **[Google Scholar]** **[CrossRef]**

28. Grossbard, E. Antibiotic Production by Fungi on Organic Manures and in Soil. J. Gen. Microbiol. 1952, 6, 295–310. **[Google Scholar]** **[CrossRef]** **[Green Version]**

29. Schardl, C.L.; Craven, K.D. Interspecific hybridization in plant-associated Fungi and Oomycetes: A review. Mol. Ecol. 2003, 12, 2861–2873. **[Google Scholar]** **[CrossRef]**

30. Kwaśna, H.; Chełkowski, J.; Zajkowski, P. Grzyby (Mycota), Grzyby niedoskonałe (Deuteromycetes), Strzępczakowe (Hyphomycetales), Gruzełkowate (Tuberculariaceae), Sierpik (Fusarium); Instytut Botaniki PAN: Warszawa-Kraków, Poland, 1991; Volume T.22, p. 136. (In Polish) **[Google Scholar]**

31. Tamura, K.; Peterson, D.; Peterson, N.; Stecher, G.; Nei, M.; Kumar, S. MEGA5: Molecular evolutionary genetics analysis using maximum likelihood, evolutionary distance, and maximum parsimony methods. Mol. Biol. Evol. 2011, 28, 2731–2739. **[Google Scholar]** **[CrossRef]** **[Green Version]**

32. Larkin, M.A.; Blackshields, G.; Brown, N.P.; Chenna, R.; McGettigan, P.A.; McWilliam, H.; Valentin, F.; Wallace, I.M.; Wilm, A.; Lopez, R.; et al. Clustal W and Clustal X version 2.0. Bioinformatics 2007, 23, 2947–2948. **[Google Scholar]** **[CrossRef]** **[Green Version]**

33. Yu, J.; Hu, S.; Wang, J.; Wong, G.K.; Li, S.; Liu, B.; Deng, Y.; Dai, L.; Zhou, Y.; Zhang, X.; et al. A draft sequence of the rice genome (Oryza sativa L. ssp. indica). Science 2002, 296, 79–92. **[Google Scholar]** **[CrossRef]** **[PubMed]**

34. Solovyev, V.; Kosarev, P.; Seledsov, I.; Vorobyev, D. Automatic annotation of eukaryotic genes, pseudogenes and promoters. Genome Biol. 2006, 7 (Suppl. 1), 10.1-1012. **[Google Scholar]** **[CrossRef]** **[PubMed]** **[Green Version]**

35. Tajima, F. Statistical method for testing the neutral mutation hypothesis by DNA polymorphism. Genetics 1989, 123, 585–595. **[Google Scholar]** **[PubMed]**

36. Nei, M.; Kumar, S. Molecular Evolution and Phylogenetics; Oxford University Press: New York, NY, USA, 2000; p. 333. **[Google Scholar]**

37. Felsenstein, J. Confidence limits on phylogenies: An approach using the bootstrap. Evolution 1985, 39, 783–791. **[Google Scholar]** **[CrossRef]**

38. Roy, S.W.; Gilbert, W. The evolution of spliceosomal introns: Patterns, puzzles and progress. Nat. Rev. Genet. 2006, 7, 211–221. [**Google Scholar**]

39. Loftus, B.J.; Fung, E.; Roncaglia, P.; Rowley, D.; Amedeo, P.; Bruno, D.; Vamathevan, J.; Miranda, M.; Anderson, I.J.; Fraser, J.A.; et al. The genome of the basidiomycetous yeast and human pathogen Cryptococcus neoformans. Science 2005, 307, 1321–1324. [**Google Scholar**] [**CrossRef**] [**Green Version**]

40. Deutsch, M.; Long, M. Intron-exon structures of eukaryotic model organisms. Nucleic Acids Res. 1999, 27, 3219–3228. [**Google Scholar**]

41. Wakuliński, W. Ogólna charakterystyka grzybów i ich cechy jako patogenów roślin/General characteristics of fungi and their traits as plant pathogens. In Fitopatologia. T. 1, Podstawy fitopatologii; Kryczyński, S., Weber, Z., Eds.; Wyd. zbiorowe. PWRiL: Poznań, Poland, 2010; pp. 206–244. (In Polish) [**Google Scholar**]

42. Stajich, J.E.; Dietrich, F.S.; Roy, S.W. Comparative genomic analysis of fungal genomes reveals intron-rich ancestors. Genome Biol. 2007, 8, R223. [**Google Scholar**] [**CrossRef**] [**Green Version**]

43. Schwartz, S.H.; Silva, J.; Burstein, D.; Pupko, T.; Eyras, E.; Ast, G. Large-scale comparative analysis of splicing signals and their corresponding splicing factors in eukaryotes. Genome Res. 2008, 18, 88–103. [**Google Scholar**] [**CrossRef**]

21 THE DETRIMENTAL IMPACT OF ALIEN INVASIVE PATHOGENS AND PESTS ON TREES, FORESTS, AND PLANTATIONS: PATHWAYS, GLOBAL CONSEQUENCES AND MANAGEMENT

1. INTRODUCTIONAN OVERVIEW

Forests play a crucial role in providing income and well-being for human populations around the world. However, the global economic and ecological sustainability of forests is seriously threatened by biological incursions. Biological invasions are believed to produce the second highest amount of harm, after habitat destruction/fragmentation [1]. Each year, a significant portion of the world's forests is lost or severely harmed by invasive alien pathogens and pests (IAPPs) [2].

The rapid proliferation of invasive alien plant species (IAPPs) in new environments is a multifaceted phenomena influenced by various causes. At present, the primary factor is the escalation of trade, particularly in the realm of plants, on a worldwide level [3]. The economic and social changes that have occurred in recent decades have directly impacted the entrance of detrimental species, and this impact is ongoing. Specifically, advancements in transportation vehicles have facilitated the more efficient and rapid delivery of plants, foodstuffs, and plant components [4]. Another significant factor is the substantial rise in human mobility, whether for employment or tourism. It is not unusual for foreign organisms to be transported from one country to another through this means, either unintentionally (such as through contaminated seeds or dirt attached to footwear or clothing) or intentionally (for instance, through seeds, bulbs, fruits, or souvenirs made with plant materials) [5,6].

After the IAPPs have expanded into a new area, additional factors like human-caused disruption can promote their successful establishment. The modification of habitats and the transformation of natural ecosystems, such as the installation of roads, dams, power lines, and buildings, or the conversion of land for agriculture, deforestation, and changes in land use, all contribute to the creation of favorable conditions for the invasion of species [7,8]. More precisely, these adjustments have the following effects: they cause a release of resources, impact the relationships between species, change the substrates, and transform the physical environment [9]. The correlation between urbanization, tourism, and disturbed ecosystems, such as coastal and island areas, lakes, rivers, and periurban woods, is not accidental. These habitats are also the ones with the highest presence of alien species [10].

Climate change is another contributing element to IAPP infestations. The ongoing climate trend of global warming directly impacts IAPPs by establishing favorable conditions for their colonization in formerly inhospitable regions [11]. Moreover, elevated temperatures improve the adaptability of the thermophilic IAPPs, promoting their reproductive and dispersal capabilities. Additionally, milder climates result in an increased frequency of reproductive cycles per year and a higher survival rate throughout the winter season [12]. Climate changes indirectly support IAPPs by increasing the vulnerability of plants. Adverse climatic conditions like prolonged drought, extreme weather events such as rainstorms, windstorms, hail, and severe flooding, negatively affect plant health. This leads to reduced plant growth and vigor, changes in phenology, disruptions in physiological processes, and the creation of wounds that make plants more susceptible to IAPP attacks [13,14].

The significant effects of forest invasive alien plant species (IAPPs) include: decreased primary production; modifications to evolutionary processes in both host populations and organisms closely related to the invaders; alterations to ecological relationships within tree communities (such as changes to mutualistic or antagonistic symbiotic relationships between tree species and their animal or microbial symbionts); reduced or lost ecosystem services (such as water filtration and supply, climate regulation, erosion prevention, nutrient cycling, carbon storage, and habitat provision); and the extinction of native species [12,15].

Many countries' governments are focusing on the issue of introducing dangerous organisms to agriculture and forestry. As a result, phytosanitary protection has become a crucial aspect of their agricultural and forestry strategies [16]. To address this escalating trend, numerous governments have implemented more rigorous methods to promptly identify prohibited viruses and pests, and to eliminate detrimental quarantine species found within their borders [17]. Moreover, intergovernmental regional plant protection groups endeavor to address the issue at a regional level, with the goal of guaranteeing sufficient levels of plant protection through a synchronized set of phytosanitary measures [18].

2. IAPP PATHWAYS

The movement of IAPPs to novel locations can occur naturally or through human-mediated processes. Natural movement, even over long distances, always occurs along a trajectory that follows a regular, expected colonization pattern [**10**]. The populations that colonize a new environment normally originate from nearby areas, such as neighbouring islands or the adjacent mainland, and, in any case, they follow a path that can be predicted according to existing knowledge about their biology, the biology of their vectors, the prevailing winds and water currents (in particular for oomycetes), as well as their dispersal capabilities (migratory capacity) [**10**]. Human-mediated transport, on the other hand, takes place on a much larger scale [**19**], since a given species can be taken from one point on the planet and released on the opposite side of the globe. Such transportation both breaks up natural biogeographic barriers and disrupts natural evolutionary processes, constituting a concrete threat to the planet's biodiversity [**20**]. Such anthropogenic conveyance of agricultural and forest pests has increased immensely in terms of frequency, geographical breadth, and number of species involved.

Among the human-dependent causes of the introduction of forest IAPPs, postal traffic, tourist activities, and, of course, global trade play a leading role [**6**]. The movement of people in general, either international or domestically, influences the amount of interception. Liebhold et al. [**21**] has quantified this and showed how the number of insect pests intercepted in baggage was positively correlated to the number of travellers entering the U.S. As regards trade-related transport, the main pathways are germplasm (tree seeds, cuttings, and plants for planting) and wood (logs, firewood, processed wood, as well as packaging material, etc.). According to Meurisse et al. [**6**], live plants and wood packaging material are the main source for movement of forest IAPPs beyond their range limits into new, uncontaminated forest areas, based on the data collected in the ports of entry in the U.S. and New Zealand.

Live plants and wood packaging material have also been confirmed as the main pathways for IAPPs in Europe based on interceptions at ports of entry during the last 5 years (**Figure 1**). The European Union Notification System for Plant Health Interceptions (Europhyt) (**https://ec.europa.eu/food/plants/plant-health-and-biosecurity/europhyt/interceptions_en**, accessed on 1 June 2021) records interceptions of new pests and diseases made by EU Member States [**22**]. Among the main intercepted commodities in this database, there are the following categories: "plants", which includes plants for planting, scions, cuttings, leaves, flowers, branches with foliage; "wood-packaging material", such as wooden crates, wood pallets, wooden packaging material, dunnage; "wood and bark", including roundwood, bark, and processed wood; "produce", such as fruits, vegetables, seeds;

"soil", including soil and growing media. "Plants", in particular those coming from Asia and Africa, constitute the most important pathway of forest IAPPs into Europe. The "plants" pathway is followed by the "wood-packaging material", again mainly coming from Asia. Although wood has historically been one of the most common introduction pathways, Europhyt data relating to the 2016–2020 period show that interceptions of harmful organisms on this commodity have reduced significantly, while "wood-packaging material" has moved up to second place. In Europe, there is a third major pathway: "produce", which is an important entry route mainly for agricultural IAPPs (**Figure 1**).

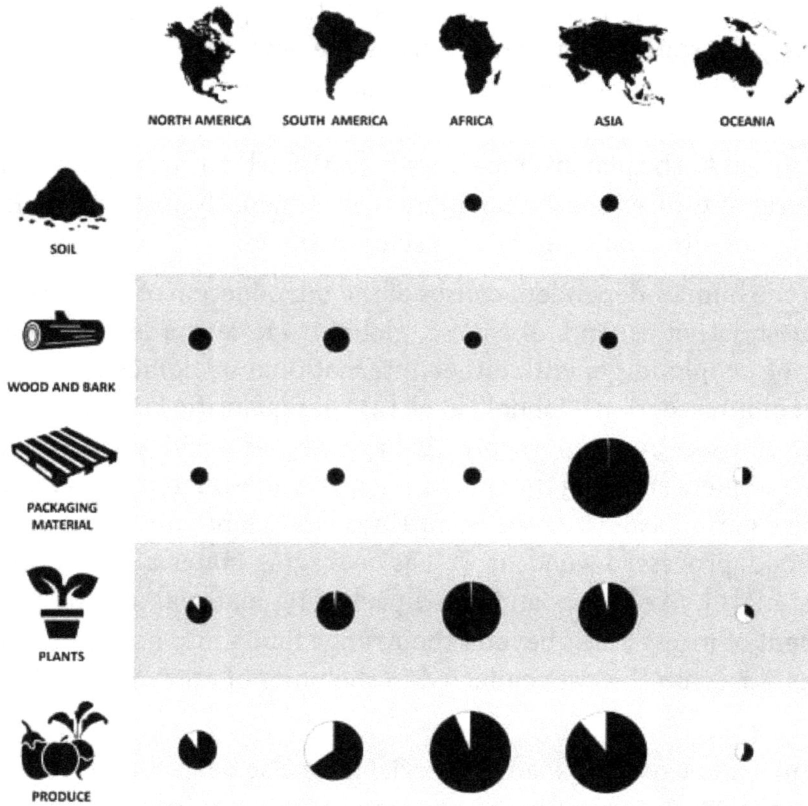

Figure 1. The frequency of fungus and insect interceptions at ports of entry in Europe in the 2016–2020 period (data obtained from reports of Europhyt). Circle sizes indicate the relative amount of interceptions of IAPPs (fungi and insects) within each transport pathway according to their continents of origin. The black sector of the circle denotes the relative frequency of insect interceptions; the white indicates fungi interceptions. Legend: soil = soil and growing media; wood and bark = roundwood, bark, and processed wood; packaging material = wooden crates, wood pallets, wooden packaging material, dunnage; plants = plants for planting, scions, cuttings, leaves, flowers, branches with foliage; produce = fruits, vegetables, seeds.

One feature of imports which is overlooked is "soil". The scarce or absent interception of alien organisms in the "soil" (**Figure 1**) is to be attributed to the very nature of this material, which hampers detection. This is because harmful IAPPs present in the soil remain hidden inside this matrix [**5**]. Asia was found to be the main source of alien plant insects, followed by Africa and, to a lesser extent, South America (**Figure 1**). On the other hand, fungal pathogens were intercepted on commodities coming mainly from South America, and secondly, from Asia. However, it should also be considered that EU phytosanitary rules impose a ban on different commodities imported to EU, a restriction that could impact the rate of interceptions on imports from different continents.

A fact that immediately stands out, according to McCullough et al. [**23**], is that insect interceptions largely prevail over fungal interceptions. This essentially depends on the fact that many fungal pathogens possess a latent phase during which they are not visible on plant surfaces [**24,25**]. **Figure 1** shows how fungal infections become more frequent on fresh plant produce. This is because the herbaceous (fleshy) consistency of fresh plant produce makes this commodity more easily colonizable and, as a consequence, disease symptoms and/or signs become noticeable in a shorter time [**23**].

The higher frequency of insect interceptions at ports of entry compared to fungal pathogens, corresponds also to a higher number of insects established in new areas (**Figure 2**). This is because fungi, as microorganisms, are generally not visible (the fungal fruiting bodies are not always present), thus they remain less noticeable and less detected. However, even considering only the species with a high impact, whose effects have been tangible in terms of damage, insects continue to be reported in the highest numbers, as results from the European Alien Species Information Network indicate (EASIN; **http://easin.jrc.ec.europa.eu**, accessed on 10 June 2021).

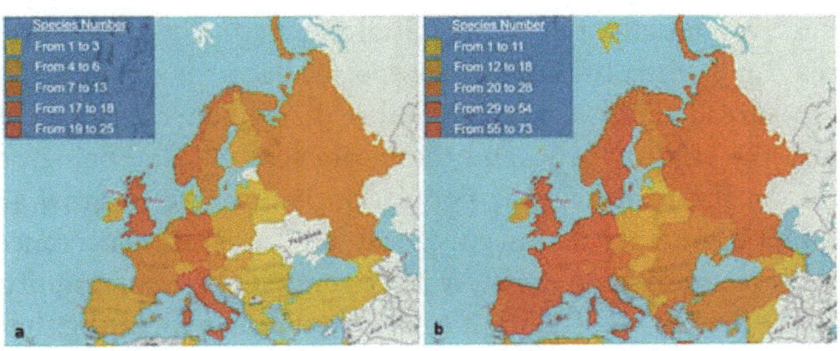

Figure 2. EU geographical distribution of invasive alien species of fungi (a) and insects (b) introduced into terrestrial and fresh-water environments from 1999 to 2021 (source: EASIN—European Alien Species Information Network). Fresh-water environments were also surveyed to include water moulds (Oomycetes).

The European climate turns out not to be a key factor in determining the abundance of IAPPs introduced. In fact, high numbers of new reported infestations are found in countries from both southern and northern Europe (**Figure 2**). Instead, other factors, such as trade volume, the presence of roads, and human population density are all correlated with the introduction of alien species to Europe [**26**].

3. FACTORS DRIVING THE INVASION PROCESS

The chance of newly introduced harmful organisms becoming invasive in new geographic areas is always uncertain [**10**]. Indeed, few of these survive either the chronic or occasional forces they encounter in the new environment. Only a very small percentage of them ever become naturalized, and even a smaller fraction ends up becoming invasive [**27**]. In fact, all non-native species must survive three distinct phases, not always clearly distinguishable, before being able to inflict ecological and economic damage: (1) transport: individuals are taken from their area of origin, transported to a new area, and released into the recipient environment; (2) establishment: the newcomers settle in the non-native environment and implant a reproductive population—among the set of the transported individuals, only a few are capable of this; (3) spread: colonizers grow in abundance and expand their geographical range. These phases are mandatory steps in the invasion process, and the advancement from one to the next one involves overcoming hefty ecological barriers [**28**]. These hindrances, in fact, allow only a small percentage of the introduced species to advance to the next phase (generally between 5% and 20%, with an average of roughly 10%). Based on this assumption, Williamson [**29**] introduced the so-called "10 rule", according to which only one out of 10 new introductions would be successful in overcoming any single barrier. This approach has made it possible to analyse the various opportunities IAPPs have to successfully establish into new areas. Invasions, for example, always start with a limited number of individuals being transported and released into a novel location; here, they must be able to survive and give rise to an initial population capable of persisting in this new area. Only when the newcomer species becomes numerically abundant and begins to spread rapidly may "an outbreak" be generated, causing damage to agriculture, the environment, and the local economy, and being perceived as "invasive". If, on the contrary, the population is not able to grow and expand, it remains small in number and is only local (an "endemic occurrence") [**30,31**]. However, there is an ongoing debate within the scientific community regarding the reliability of the "10 rule", since it can lead to underestimating the impact the introduced organisms can have in new introduction areas [**32**].

An alien species may have different outcomes in distinct sites of a same region [**33**]. The first arrivals may survive, reproduce, and spread to become invasive at some locations, while they may succumb at others, even if all individuals of that species share the same traits: e.g., aggressiveness, competitiveness, frugality, polyphagy, tolerance to adverse environmental conditions [**34**]. In fact, the physical and biological external forces, collectively known as "environmental resistance", which the alien species has to face, vary from site to site, ultimately determining their fate [**35**]. Among the most common physical forces the alien species has to withstand are: the availability of nutrients, water, oxygen, light, space, shelter, or protection (e.g., the inner tissues of a plant for a fungal endophyte). Biological forces that can effectively hinder the invasion, instead, are: competition, predation, parasitism, and diseases, as well as other interactions with the resident community, including anthropogenic perturbation [**36**].

Evidence exists that IAPPs perform better than native pathogens and pests in the invaded community, and this can be attributed to various factors. One key reason for their successful infiltration is a higher host susceptibility in the new non-native range, due to the absence of coevolution with the indigenous hosts, the newcomer IAPPs can grow very rapidly and inflict greater damage, in some instances generating overwhelming outbreaks [**37**]. Another cause is that in the new environment the IAPP may also jump onto new host(s), which may even be taxonomically unrelated to its native hosts [**38**]. This is a fairly common phenomenon, whose frequency probably depends on the presence in the introduction area of native plants taxonomically close to the original host plant [**39**]. Another factor favouring IAPP outbreaks is the scarcity of natural enemies in the new range compared with their native range. According to this "enemy release hypothesis", IAPPs leave behind all those natural enemies that would naturally keep their population under control in the area of origin. In the new environment, moreover, it encounters very few natural enemies able to use it as a trophic resource [**40,41**]. All these factors enable the IAPPs to exploit their enormous reproductive potential and to spread rapidly on a large scale, even displacing native species [**42**].

4. INVASIVE ALIEN PATHOGENS AND PESTS HARMING FORESTS

The harm caused by the introduction of alien forest pathogens and pests is often difficult to fully quantify due to the multiple ecosystem services forests offer, and because each invasion event must be analysed within the specific context (forest, nursery, district, country, region) in which it occurs [**43,44**]. The damage may not only be significant at the economic, ecological and landscape level, but it might also generate negative repercussions in related sectors, such as in marketing and tourism [**45**].

4.1. A Retrospective Look

The forest and rural landscapes of various areas of the planet were transformed in the last century by the introduction of dangerous non-native forest pathogens

[**46**]. Between the 1920s and 1940s, the agent behind Dutch elm disease (DED), Ophiostoma ulmi (Buisman) Nannf, spread pervasively throughout Europe and North America on native elm species [**47**]. The invasion by this vascular pathogen, native to Asia, soon turned into a true pandemic [**48**]. Moreover, the several variants the fungus spawned over time, including the more aggressive subspecies americana of Ophiostoma novo-ulmi Brasier, were introduced to Europe from North America via infected logs [**49**]. As clear evidence of the fungus' high evolutionary potential and destructiveness, it eventually led to the almost total disappearance of the European field elm (Ulmus minor Mill.) from many countryside and urban districts of Europe [**50**].

Similarly, iconic tree species of the genus Castanea, such as the American chestnut Castanea dentata (Marsh.) Borkh. and the sweet chestnut Castanea sativa Mill., were, in the last two centuries, wiped out from forest ecosystems and orchards due to the combined, sometimes synchronous, attack of the two fearsome oomycete pathogens, Phytophthora cambivora (Petri) Buisman and Phytophthora cinnamomi Rands [**51**], as well as of the ascomycete chestnut blight agent Cryphonectria parasitica (Murrill) M.E. Barr [**52**]. In many European countries, the epidemic spread of chestnut blight led to a severe loss of chestnut cultivation. The disease devastated coppices and orchards, with tragic effects on the already poor economy of many farming communities, for which the edible chestnut fruit was a primary source of food, causing famine and emigration [**12**].

Around the 1930s, another harmful pathogen, Ceratocystis platani (J.M. Walter) Engelbr. and T.C. Harr., the agent behind blue stain canker in plane trees or sycamores (Platanus occidentalis L.), spread epidemically first in the suburbs of Philadelphia, and then in many cities in the south-eastern U.S. [**53**]. Subsequently introduced into the Mediterranean basin either during World War II or soon after, in just a few decades this vascular pathogen gradually killed a multitude of monumental P. occidentalis and Platanus orientalis L., and their hybrid Platanus x acerifolia (Aiton) Willd., trees in city parks, squares, and boulevards [**54**]. The high genetic uniformity of host populations (most urban trees were clones) and the lack of knowledge about the high infectivity of fungus propagules, which can remain viable at length on pruning tools, favoured the outbreaks of blue stain canker [**55**].

Seiridium cardinale (W.W. Wagener) B. Sutton and I.A.S. Gibson, the agent behind cypress blight, is an anamorphic fungus native to the west coast of the U.S. (California), where it was first recorded on Monterey cypress (Cupressus macrocarpa Hartw. ex Gordon) in 1927 [**56**]. Despite lacking sexual reproduction, the pathogen was able to spread worldwide epidemically in a short time [**57**]. Accidentally introduced into Mediterranean Europe during the mid-1900s [**58,59**], in a few decades it killed planted and ornamental cypress trees (Cupressus sempervirens

L.) in stands, windbreaks, parks, villas, cemeteries, and treelines. The destructive cypress blight outbreaks deeply altered the Mediterranean landscape, causing a true ecological disaster [**60**]. Cypress is, in fact, an iconic part of the Mediterranean countryside, so its disappearance also impacted local economies and tourism [**57**].

Alien insects can pose a direct danger to plants, but they can also act as vectors for phytopathogenic microorganisms [**61**]. In addition, native insects sometimes also play an important role in the spread of exotic diseases. For example, the bark beetle, Phloeosinus aubei (Perris), native to the Caucasus, Asia Minor, and the Mediterranean area, contributed to the increase in damage caused by the exotic S. cardinale by carrying its fungal spores; thus the beetle facilitated the fungus' colonization of host trees. In addition, over the last few decades, this bark beetle has continued to spread into Central Europe, where it is considered an alien pest [**62**]. Similarly, Scolytus multistriatus (Marsham) (Coleoptera: Curculionidae), with a Palearctic native range (from western Europe to Russia and northeast Africa) [**63**], is the main vector of D.E.D., not only in its native range but also in introduced areas, such as North America, where, before the 1930s, 50%–75% of the elm population in the north-eastern part of the continent were killed by this vascular pathogen [**64**].

Insects harmful to forest trees are also included among the world's worst 100 IASs [**65**]; those of forest concern include the gypsy moth, Lymantria dispar L. (Lepidoptera, Lymantriidae), and the cypress aphid, Cinara cupressi (Buckton) (Hemiptera, Aphididae). The gypsy moth, native to Europe and Asia, was introduced near Boston, Maine, in the late 1860s, and currently is listed as one of the most destructive invasive pests, ranking third among the costliest IASs in the world [**66**]. Its larvae cause serious defoliation of many tree and shrub species, which may ultimately die or become susceptible to subsequent infestation by secondary pests. From 1924 to 2013, in the north-eastern and mid-western U.S., over 37 million hectares were defoliated [**67**], with a high impact on timber, costs and losses to urban and suburban forests, as well as damage to the recreational sector [**68,69**]. In addition, gypsy moth outbreaks have contributed to the decline of oak forests in eastern North America [**70**]. The cypress aphid is also a highly invasive insect; probably native to North America and Syria, since the 1960s it has become newly established in parts of Europe, Africa, South America, and the Middle East [**71**]. The cypress aphid, which sucks sap from twigs, causes yellowing to browning of the foliage. Depending on the severity and duration of the infestation, trees may suffer more or less severe damage, or even death in the worst cases [**72**]. In recent decades, sudden and explosive outbreaks of the cypress aphid have occurred in many countries (some African countries, Italy, Jordan, Yemen, Mauritius, and Colombia). As a consequence, populations of the host trees (Cupressus, Juniperus, Widdringtonia, and other Cupressaceae) have been decimated, both in commercial and ornamental plantings or native stands [**73**].

4.2. Emerging Invaders

Among the many currently emerging threats to forest ecosystems, oomycete Phytophthora species are at the pinnacle. For example, Phytophthora ramorum Werres, De Cock et Man in 't Veld, a polyphagous, invasive pathogen originally from the laurosilva forests of eastern Indochina and Japan [**74**], attacks a multitude of shrub ornamentals and tree species. After causing extensive tree mortality in oak (Quercus spp.) and tanoak Notholithocarpus densiflorus (Hook. & Arn.) Manos, Cannon and S. H. Oh) forests in California (for this it was called the agent of "sudden oak death") [**75**], this destructive oomycete has been more recently killing off plantations of Japanese larch [Larix kaempferi (Lamb.) Carr.] in the UK, where it was renamed the agent behind "sudden larch death" [**76**]. Furthermore, the infamous P. cinnamomi, the causal agent of ink disease in chestnuts, already mentioned above, continues to destroy woody plants in natural environments, plantations and urban forests, as well as in parks, horticultural nurseries, and urban ornamental plantings around the world, to the point of earning the nickname of "biological bulldozer" [**77**]. P. lateralis Tucker and Milbrath is considered to be of Asian origin, such as the several Chamaecyparis species that, due to the similar native range, exhibit some tolerance or resistance to infection. It is also a harmful pathogen, responsible for serious root and aerial infections [**78**]. This oomycete shows considerable host specificity, with Port-Orford cedar (Chamaecyparis lawsoniana (Murray) Parl.) its main host, but it has also caused epidemic damage to Taxus brevifolia, Thuja, and other Chamaecyparis species growing in proximity to infected C. lawsoniana in forests, plantations and in the landscape [**79**].

Rust fungi also pose a constant threat to many forest ecosystems, mainly because of their high dispersal ability over long distances. It is therefore no coincidence that rust fungi are the largest group of pathogenic fungi in the EPPO (European and Mediterranean Plant Protection Organization) A1 List of Pests Recommended for Regulation as Quarantine Pests (EPPO, **https://www.eppo.int/ACTIVITIES/plant_quarantine/A1_list**, accessed on 30 June 2021). One invasive rust that has been spreading pervasively throughout Europe in recent years is the fungus Melampsoridium hiratsukanum S. Ito ex Hirats. f., of Asian origin [**80**]. This rust is currently causing heavy defoliation, stunted growth and death to riparian Alnus incana (L.) Moench trees in the eastern Alps, threatening the stability of some riverbanks. In this mountain range the fungus has also found Larix europea L. to be a suitable alternate host on which to complete its life cycle [**81**].

Moreover, a highly destructive fungal disease of ash trees (Fraxinus spp.) is at present causing significant damage to ash populations throughout the European continent. The causative agent, Hymenoscyphus fraxineus (T. Kowalski) Baral, Queloz and Hosoya (anamorph: Chalara fraxinea T. Kowalski), kills young and coppiced ash trees in a very short time, whereas older trees can resist its irritative action for some time, until they are exposed to high inoculation pressure [**82**]. The resultant ash dieback is seriously impacting woodland biodiversity and ecology, reducing landscape aesthetics and gravely damaging hardwood industries, with heavy economic losses [**83**].

Yet another tree staple, pine stands, are currently being impacted worldwide by the destructive "Pine Pitch canker", caused by Fusarium circinatum Nirenberg and O'Donnell, (teleomorph: Gibberella circinata) [**84,85**], as well as by some highly damaging foliar pathogens, such as Lecanosticta acicola (Thümen) H. Sydow [**86**], agent of Lecanosticta needle blight (LNB) and the related fungi, agents of Dothistroma needle blight (DNB) or of "red band needle blight", Dothistroma septosporum (Dorog.) Morelet (syn. Scirrhia pini Funk and Parker, teleomorph: Mycosphaerella pini Rostr.) and Dothistroma pini Hulbary (teleomorph: unknown) [**87,88,89**]. F. circinatum, considered native to Mexico, has spread in the last half century to many pine growing areas, causing severe damage to artificial plantations in the Americas, the Far East (Japan and the Republic of Korea), and in southern Africa [**90**]. DNB causes premature needle abscission, stunted growth, and, in some cases, tree mortality [**91**]. The disease represents a serious threat to both natural forest ecosystems and exotic pine plantations on all continents (it occurs everywhere except Antarctica) [**92**]. The disease has been primarily associated with exotic plantations of Monterey pine (Pinus radiata D. Don) but, starting from the 1990s, it has caused serious damage in stands of both planted and native pine species, e.g., in Pinus contorta subsp. latifolia stands in Canada and in P. nigra subsp. laricio stands in the Mediterranean region [**8,93**].

Some invasive pests involve symbiotic relationships between insects and fungi; in this case, the whole complex (the insect and its fungal symbiont) is introduced into the new environment. For example, the complex of Geosmithia morbida Kolařik (Ascomycota, Hypocreales, Bionectriaceae) and its vector, the bark beetle Pityophthorus juglandis Blackman (Coleoptera, Curculionidae, Scolytinae) is responsible for Thousand Cankers Disease (TCD) in walnut trees. First described in North America, TCD causes progressive crown decline, with the fungus colonizing the copious feeding and reproductive galleries made by its symbiont, the phloem-boring bark beetle. This results in numerous, coalescent cankers that girdle branches and trunks, ultimately leading to the tree's death [**94**]. This insect–fungus complex has been recorded in Italy, where in northern and central regions several plantations were found to be being attacked [**95,96**]. Another example is the polyphagous shot hole borer, Euwallacea fornicatus (Eichhoff) (Coleoptera: Curculionidae), an ambrosia beetle native to Asia that was accidentally introduced into Central and North America, Israel, and South Africa. This emerging tree pest carries the spores of three ambrosial fungi within specific body structures (mycangia), which are the only trophic source for insect adults and larvae. One of these fungi, Neocosmospora (= Fusarium) euwallaceae (S. Freeman, Z. Mendel, T. Aoki and O'Donnell) Sand.-Den., L. Lombard and Crous, a moderately virulent pathogen, is responsible for causing dieback in a range of susceptible host trees, or even mortality, depending on the level of infestation. This insect–fungus complex has caused significant damage to orchards and ecosystems around the world, even attacking important forest species [**97**].

Other insect species of great impact are the Asian long-horned beetle, Anoplophora glabripennis (Motschulsky) (Coleoptera: Cerambycidae), and the citrus long-horned beetle, A. chinensis Forster (Coleoptera: Cerambycidae). These are xylophagous insects native to Asia, whose big larvae can kill healthy broadleaf trees. The first record of an established population of A. glabripennis outside its native range was in 1996 [**98**], while that of A. chinensis was in 1999; both occurred in North America [**99**]. In Europe, eradication was successful at the sites where A. chinensis was discovered shortly after its introduction, proving how early detection is key to prevent the lasting establishment of the pest and to implement effective control measures [**100**]. Furthermore, the importance of citizen involvement in early detection was highlighted in 2008, when every new outbreak was first reported by members of the public [**101**].

However, when an IAPP species is not considered a major pest in its native range, it is unlikely to be included on IAS watch lists. Consequently, specific early detection surveys for that pest are not in place. Furthermore, it is difficult to predict the invasiveness of a non-native species in a new environment. Such was the case not only for pathogenic fungi, such as the ash dieback agent H. fraxineus, but also for several insect species, such as, for example, the emerald ash borer, Agrilus planipennis (Fairmaire) (Coleoptera: Buprestidae), a phloem-feeding beetle native to Asia, where it was only marginally studied as a minor pest. Once accidentally introduced (in 2002) in the United States, however, it became the most destructive forest pest ever recorded in North America [**102**]. Millions of ash trees have already died because of A. planipennis; the infestations have spread over 35 US States and two Canadian Provinces, as of 2020 [**103**]. The emerald ash borer has also reached European Russia, being first recorded in Moscow in 2003; it is now already present in nine Russian regions [**104**]. In 2019, it was also confirmed for the first time in Ukraine, revealing how the beetle is continuing to expand its range towards Europe [**105**]. Because of what already happened in North America, the beetle is being addressed in Europe with more groundwork, trying to develop all the necessary measures to contain the problem [**106**]. However, it has yet to be determined whether the European ash (Fraxinus excelsior L.) is going to be susceptible to severe damage; in fact, most of the infested trees in Russia are in artificial urban settings of F. pennsylvanica Marshall [**104**].

5. MITIGATING THREATS TO FOREST HEALTH: POSSIBLE SOLUTIONS AND ACTION STRATEGIES

In natural forests, the better approach is to actively maintain ecological integrity to ensure forest resilience, whereas in forest plantations the restoration of increasing levels of biodiversity should be favoured as the first line of defence against invasive pests and pathogens [**107**]. In this last case, the careful choice of appropriate plant material is always fundamental, specifically, avoiding the use of species unsuited to site conditions, as well as limiting the number of exotic plant species [**108**], since in the last century, non-native tree species, widely used as ornamentals or in afforestation programmes, acted as "Trojan horses" for many harmful diseases and pests [**8**]. Clear examples are the accidental introductions of P. ramorum into North America with nursery plant material [**75**], of Teratosphaeria (Mycosphaerella) nubilosa (Cooke) Crous and U. Braun, a causal agent of Mycosphaerella leaf disease of eucalypt, into South America (Uruguay) with propagation material of this exotic tree species [**109**], of Rhynchophorus ferrugineus (Olivier), harmful to palm trees widely diffused in South Europe [**110**], and of Ophelimus maskelli (Ashmead) (Hymenoptera: Eulophidae), which damages eucalypt plants in the Mediterranean Basin [**111**]. Furthermore, in forestry it is always advisable to avoid genetic uniformity, such as clonal plantations, as in poplar cultivation. Mixed forests, instead, are generally less prone to diseases and pests than monotypic stands, with nonhost trees creating effective physical and chemical barriers to the spread of pathogens' inoculum, as well as hindering host searches by pests. In addition, higher biodiversity favours natural enemies and competitors which, interacting with IAPPs, contribute to keeping the invaders' population densities low [**107**].

Preventing the introduction of dangerous IAPPs into new areas is the first step in their management, so this has become an important priority in many countries. Preventing introductions is carried out through effective surveillance activities, namely phytosanitary inspections at ports of entry (ports, airports, and custom barriers), as well as at vulnerable sites, i.e., those places where the plant material is in transit or processed, such as nurseries, loading and sorting stations, and wood processing companies [**112,113**]. Considering the enormous volume of commerce at present, and the incessant mobility of humans for work, tourism, etc., many IAPPs elude preventative measures, it being practically impossible to intercept all the species that transit on all commodities [**114,115**].

Careful prediction of potential invasiveness of harmful pests and diseases can aid interceptions [**116**]. Important predictive tools are those provided by international organizations such as the CABI (Commonwealth Agricultural Bureaux International) and the EFSA (European Food Safety Authority), as well as various Regional Plant Protection Organizations such as the EPPO, the NAPPO (North

American Plant Protection Organization) and the APPPC (Asia and Pacific Plant Protection Commission), which were created to develop international strategies to fight the introduction and spread of harmful pests and diseases and to promote safe and effective pest control methods. In fact, they provide constantly updated lists of organisms considered as threats to agriculture, forestry, and the environment and that are most likely to be introduced into Europe and North America. These agencies' published assessments of invasion risks are important tools since they focus the efforts of inspection services, plant health scientists, and all the stakeholders affected by imported pests and pathogens [**117**]. Another important early warning system of new and emerging pest and pathogen risks is sentinel plants [**118**]. These can be: (a) plantings made with young, frequently exported native plants which are used to identify pests and diseases that may be introduced to import countries through the trade in live plants; and (b) plantings made of non-indigenous young or mature plants in import countries whose monitoring could give useful information about the potential damage of non-native pathogens and pests if they were to arrive and become established [**119**].

A primary obstacle to managing and mitigating alien forest pests is correctly identifying the agents of damage [**117**]. Traditional identification methods, essentially based on the analysis of macro- and/or micro-morphological characters, have inherent limitations. Considerable experience is necessary to induce the formation of fungal distinguishing structures in vitro and substantial expertise is needed for their microscopic observation and identification; moreover, special growth media (i.e., differential or selective substrates) may not be available for certain, fastidious microorganisms. In any case, these procedures are not applicable for unculturable microbes such as biotrophic fungi [**120**]. The above difficulties involve the fact that some taxa, for example, are not easily identified through morphological characters, so this entails inspectors with specific expertise [**114**]. In addition, some insects have juvenile stages that are indistinguishable by classical diagnostic methods. Furthermore, phytosanitary inspectors may face plant material that has been attacked, but which has already been abandoned by its phytophages; therefore, they have to track the pest only on the basis of the traces left behind, which becomes a more difficult task. It is also true, on the other hand, that the reproductive structures of many forest pathogens can be observed on infected plant material, and these are of a great diagnostic value [**81**]. Furthermore, when the reproductive structures (e.g., fruiting bodies) are not present on the host surface, they can still be obtained in the laboratory by using a moist chamber; alternatively, the isolation from infected plant material into pure culture can permit the growth and identification of the causal agent.

Many of the difficulties inherent in traditional morphological identification can nowadays be overcome by employing molecular techniques. Furthermore, for an effective inspection system which can cover the ever-increasing volume of import–export material, simple, timely, and repeatable methods have to be developed. Modern molecular diagnostic tools, which detect DNA polymorphisms, ensure high accuracy and diagnostic sensitivity, allowing the detection of even minimal traces of nucleic acids, such as those remaining in the frass of xylophagous insects [**113**]. Some of these methods have the advantage of being able to process a large number of samples in a short time [**121**]. For example, loop-mediated isothermal amplification (LAMP) is a nucleic acid amplification method that offers rapid, accurate, and cost-effective diagnosis of diseases and pests [**122,123,124**]. The simplicity and the portability of the equipment also make it a tool of choice for carrying out tests in situ, making inspection greatly easier [**8**]. Molecular methods are therefore of great help when it is necessary to focus attention on clearly defined species, for example for delimiting surveys at vulnerable sites known or suspected to harbour a given pathogen or pest, while conventional (e.g., morphology-based) methods continue to retain their high validity for the control of the territory for the occurrence of unknown IAPPs. In other words, morphological and molecular methods should go hand-in-hand and complement each other, with DNA-based techniques representing versatile tools to support and extend more classical species identification and taxonomy [**125**].

Novel diagnostic methods are also useful in monitoring the dynamics and ecology of a given IAPP after its introduction. These techniques are in fact also highly effective in: detecting early disease and insect foci; tracing and quantifying pathogens' inoculum (especially airborne inoculum) and insect populations; and locating infection and infestation reservoirs (e.g. storage facilities) that sustain invasion events [**126**]. National and local inspection services also make use of innovative strategies, including remote sensing technology (GIS and unmanned aerial vehicles like drones) [**127,128**] and detection dogs, whose keen smell can locate biological targets [**129**]. This last method was used for the first time for gypsy moths and later for other alien and invasive insect species [**130**].

The mobilisation of citizens as additional observers and reporters (citizen science) has become increasingly utilized for the early detection of unwanted IAPPs. Active monitoring by citizens, when extensively distributed throughout the territory, has in fact become an important part of biosecurity efforts to prevent IAPPs' establishment and spread [**131**]. However, to obtain the consent and collaboration of citizens, they must be properly informed about the risks and dangers associated with non-native pathogens and pests for native plants, also because citizens have sometimes opposed management interventions aimed at containing and eradicating harmful alien organisms [**132**]. Well-informed citizens, besides supporting the accurate monitoring of the territory by inspecting sites and trees, are also a fundamental resource for collecting, categorizing, transcribing, and analysing data [**133**]. Citizen science is possible today due to a growing number of citizens with ecological awareness and knowledge eager to contribute as volunteers [**134**]. Some examples of citizen science projects include Observatree (**https://www.observatree.org.uk**, accessed on 31 July 2021) in the UK and LIFE ARTEMIS (**https://www.tujerodne-vrste.info/en/**, accessed on 31 July 2021) in Slovenia; in addition there are several citizen science networks, such as the EASIN (**https://easin.jrc.ec.europa.eu/easin/CitizenScience/BecomeACitizen**, accessed on 31 July 2021), which is an initiative of the European Commission that aims to connect citizens, scientists and policymakers in an effort to manage harmful alien invasive species.

Once an IAPP has been detected within an area of concern, the next move is to try to eradicate invasion foci [**135,136**]. Quick action while IAPP populations are still small is the most efficient way to prevent ecological or economic harm. With forest IAPPs the only control options are mechanical approaches (such as sanitation cutting) [**137**], biological control, and the use of semiochemicals (e.g., mass trapping) [**10**]. While eradication of alien forest insects has been effective in some cases [**138**], the eradication of forest pathogens has always been incredibly challenging, with most of the successful interventions taking place in controlled environments (greenhouses and tree nurseries) or in urban green spaces (e.g., parks, trees, and gardens) [**139**].

After an IAPP population has newly established in a forest, biological control is one of the main measures to achieve self-sustaining long-term control [**140**]. In some cases, biological control has managed to restore more sustainable pest population dynamics, succeeding in keeping the population density of the invasive species below the damage thresholds [**141,142,143**]. For instance, the parasitoid Torymus sinensis Kamijo (Hymenoptera: Torymidae) proved to be a good control agent of the Asian chestnut gall wasp D. kuriphilus in several areas of the world, significantly reducing the pest populations with a self-sustaining effect [**144**]. Similarly, in North European conifer forests, the fungus Phlebiopsis gigantea (Fr.) Jul. has been successfully employed in the biological control of the root and butt rot pathogen Heterobasidion annosum (Fr.) Bref. [**145,146**]. Commercial formulations based on this biocontrol agent have even since been developed and are commercially available under the names of 'PG suspension™' in the UK, 'Rotstop®' in Finland, and PG IBL in Poland [**146,147**].

Biological control is not, however, a universal remedy; it is not always effective and several negative effects may occur. This variability in the efficacy of biological control in forest ecosystems is a frequent drawback, mainly due to the complex and multi-partite interactions between the tree, the IAPP, the biocontrol agent, and the other biotic factors [**148**], all of which are, in turn, impacted by the physical environment and human interference. Biological control in heterogeneous forest ecosystems, therefore, in order to be successful, should take into consideration the multiple biological factors involved: including the environmental conditions (e.g., soil properties, microclimate, etc.), type of forest stand, and silvicultural management [**149**]. Controlling IAPPs by introducing their natural enemies, many of which are exotic themselves, is a controversial issue whose effects may be viewed as "deliberate ecological invasions". Retrospective analyses of several biological control programs around the globe are now available, thus providing quantitative data on negative effects: parasitism/predation on non-targets; competition with native natural enemies; intraguild predation; vectoring of pathogens [**150**]; and risk of hybridization with related native species [**151**]. Due to the increasing awareness of the related environmental and economic risks, many countries have implemented regulations for the release of biological control agents [**152**].

For the management of IAPP invasions, more stringent rules and policies are needed to protect plant health and to ensure a safer trade [**153**]. The need for such measures has long been recognised: at a regional level Plant Protection Organizations are trying to enforce phytosanitary policies, regulations and technical recommendations [**154**]. For example, the defence of the European territory and its plants has been recently tackled with the adoption of new rules (Regulation (EU) 2016/2031), which improve the European phytosanitary regime through more effective measures. A global effort is also required, involving all the countries and policymakers that in various ways may contribute to counteracting the problem. Specifically, international politics should work towards: consolidating phytosanitary coordination among the countries; harmonizing phytosanitary measures; eliminating the gaps between international and national laws; filling legislative gaps by implementing preventive and/or corrective legal actions; promoting more incisive and restrictive actions at national, regional, and international levels; and establishing legal responsibility frameworks to sanction countries that do not respect international regulations [**117**].

6. ROLE OF RESEARCH

The introduction of IAPPs into new territories is a current environmental issue and a dominant theme of scientific research. Phytosanitary protection has become one of the priorities of many nations, as proven by the numerous funded research projects on this subject [**155**]. Information about the taxonomy, biology, and ecology of IAPPs during the onset of an invasion process is often lacking, especially as regards their reproduction and dispersal abilities into areas where they have been newly introduced; this knowledge gap may favour their establishment and uncontrolled spread [**156**]. Plant health scientists, besides being involved in the diagnosis of IAPPs, which, often being new, are sometimes little known, are also concerned with their taxonomic positioning, as well as with the characterization of those subspecific entities that are difficult to determine by traditional approaches [**157**]. The development of diagnostic methods is especially important for those IAPPs endowed with a latent phase, because latency increases the risk of their spreading undetected. However, research also needs to be performed in other, fundamental aspects of the IAPPs' life history strategies: their demographic history (gene flow, migration); mode of reproduction (i.e., sexual vs. asexual); recombinational events outside the sexual cycle, such as parasexuality and horizontal gene transfer, affecting pathogen virulence [**158**]; occurrence of hybridization events between related taxa [**159**]; microbial/pest virulence/aggressiveness versus host resistance. A constant threat is posed to forests by the possible arrival and co-occurrence of new mating types and haplotypes. While the coexistence of different mating types increases the chances for sexual recombination between genetically divergent lineages, continual haplotype introgression can give rise to the emergence of new pathotypes with heightened virulence. These natural events, which have become more frequent as a result of international trade and climate change, are strongly feared in many pest–plant interactions. The evolutionary and pathogenetic potential of P. ramorum, H. fraxineus, as well as of the walnut twig beetle P. juglandis, vector of TCD, to cite but a few examples, strongly depend on these mechanisms [**78,137,160**]. Research can thus dramatically advance our understanding of the evolution of pathogenesis and aggressiveness of pathogens and pests and of the epidemic risks associated with these phenomena.

This knowledge is crucial for enabling stakeholders (landowners, foresters, phytosanitary inspection services, and policymakers) to implement adequate management strategies. In addition, accurate information from the research community would also permit valid, standardized phytosanitary certification, which could be adopted in partecipating countries and support inspections of feared IAPP introduction, including in the exporting country. On an open market, which allows the free movement of plant material, the certification of propagation material is of primary importance to avoid the spread of diseases and pests into new areas [**161**].

7. CONCISE SUMMARY

The issue of forest invasions by invasive alien plant species (IAPPs) is vast and intricate. This issue pertains to nations, global organizations, and agencies, and it impacts economic interests. It encompasses a wide range of activities that span from local to global levels [155,162]. At the local level, it is necessary to address the problem by focusing efforts on both preventing future introductions and eliminating early invasion foci. Immediate action should be made in the area of trade-related travel, which is the primary source of forest invasive alien plant pests (IAPPs), especially those associated with germplasm and wood-packaging material [6]. Implementing surveillance campaigns on these channels would thus aid in the prevention, or at the very least, reduction, of the dangers associated with introducing forest IAPPs. Furthermore, the monitoring of forest health has now become of paramount significance. Inspecting forest ecosystems is tough due to the wide expanse of many wooded areas and the complexity of their different compartments, such as soil, water, live plants, snags, and logs. Well-trained professionals are required for this task [129]. In addition, as the number of IAPPs being introduced continues to rise, it is crucial to maintain ongoing monitoring efforts. This monitoring should encompass various activities, including extensive surveillance of large areas, intensive monitoring of vulnerable locations such as loading stations, nurseries, and import-export material checking points [112]. Additionally, the use of sentinel plants [118] and the implementation of early warning systems supported by both traditional and innovative molecular diagnostic tools [123] are essential. Furthermore, involving volunteers in early detection through citizen science initiatives [131] is recommended. Research has long been crucial in pest management and remains more important, particularly when addressing IAPPs. In this paper, we have highlighted that the scientific community is studying various aspects of IAPPs. These include their accurate classification, diagnosis methods, analysis of genetic exchange between non-native and native species, and their ability to adapt to new environments. The provision of scientific support is essential for government agencies to establish effective control measures [100]. Additionally, lawmakers responsible for enacting laws and regulations [117] should be regularly informed on scientific achievements in this sector.

8. AT A GLANCE

The global condition of forests is influenced by several invasive alien pathogens and pests (IAPPs) that result in substantial damage, leading to major economic losses and environmental changes. Past occurrences of destructive tree viruses and pests have caused significant damage to our forests, natural landscapes, and cityscapes. Furthermore, these threats persist and continue to pose a serious danger. Human activities, particularly global trade, are the primary catalyst for the spread of pathogens and pests. This enables invasive species to surpass their natural distribution boundaries. Natural transport follows a consistent colonization pattern based on the organism's ability to disperse, while human-mediated movement occurs on a larger and less predictable scale. For a virus or pest species to become invasive in a new territory, it must successfully navigate through several stages, also known as barriers, which significantly influence the outcome of the invasion. Timely identification is essential for achieving effective elimination and control. While advanced diagnostic techniques exist for disease and pest surveillance and monitoring, there are few control and mitigation alternatives available for forestry. Among these possibilities, biological management is commonly utilized. Given that the invasion of viruses and pests poses a significant economic and ecological challenge that extends beyond national boundaries, it is imperative for governments to support and implement all essential measures for prevention and mitigation. In order to achieve this goal, it is crucial to establish and coordinate actions among countries. This is necessary to avoid the entry of new invasive alien plant species (IAPPs) and to reduce their potential spread at a local level. Research is essential for: (i) creating efficient and expedient diagnostic techniques; (ii) studying the epidemiology and ecology of IAPPs in recently imported regions; and (iii) assisting policymakers in enforcing quarantine legislation.

REFERENCES

1. Wilcove, D.S.; Rothstein, D.; Dubow, J.; Phillips, A.; Losos, E. Quantifying threats to imperiled species in the United States. BioScience 1998, 48, 607–615. [**Google Scholar**] [**CrossRef**] [**Green Version**]
2. Roy, B.A.; Alexander, H.M.; Davidson, J.; Campbell, F.T.; Burdon, J.J.; Sniezko, R.; Brasier, C. Increasing forest loss worldwide from invasive pests requires new trade regulations. Front. Ecol. Environ. 2014, 12, 457–465. [**Google Scholar**] [**CrossRef**] [**Green Version**]
3. Chapman, D.; Purse, B.V.; Roy, H.E.; Bullock, J.M. Global trade networks determine the distribution of invasive non-native species. Glob. Ecol. Biogeogr. 2017, 26, 907–917. [**Google Scholar**] [**CrossRef**]
4. Ascensão, F.; Capinha, C. Aliens on the move: Transportation networks and non-native species. In Railway Ecology; Borda-de-Água, L., Barrientos, R., Beja, P., Pereira, H.M., Eds.; Springer: Cham, Switzerland, 2017; pp. 293–297. ISBN 978-3-319-57496-7. [**Google Scholar**] [**CrossRef**] [**Green Version**]
5. McNeill, M.; Phillips, C.; Young, S.; Shah, F.; Aalders, L.; Bell, N.; Gerard, E.; Littlejohn, R. Transportation of nonindigenous species via soil on international aircraft passengers' footwear. Biol. Invasions 2011, 13, 2799–2815. [**Google Scholar**] [**CrossRef**]
6. Meurisse, N.; Rassati, D.; Hurley, B.P.; Brockerhoff, E.G.; Haack, R.A. Common pathways by which non-native forest insects move internationally and domestically. J. Pest. Sci. 2019, 92, 13–27. [**Google Scholar**] [**CrossRef**] [**Green Version**]
7. Ginetti, B.; Moricca, S.; Squires, J.N.; Cooke, D.E.L.; Ragazzi, A.; Jung, T. Phytophthora acerina sp. nov., a new species causing bleeding cankers and dieback of Acer pseudoplatanus trees in planted forests in Northern Italy. Plant Pathol. 2014, 63, 858–876. [**Google Scholar**] [**CrossRef**] [**Green Version**]
8. Aglietti, C.; Benigno, A.; Scali, E.; Capretti, P.; Ghelardini, L.; Moricca, S. Molecular-based reappraisal of a historical record of Dothistroma needle blight in the centre of the Mediterranean region. Forests 2021, 12, 983. [**Google Scholar**] [**CrossRef**]
9. Hobbs, R.J.; Huenneke, L.F. Disturbance, diversity and invasion: Implications for conservation. Conserv. Biol. 1992, 6, 324–337. [**Google Scholar**] [**CrossRef**] [**Green Version**]
10. Lockwood, J.L.; Hoopes, M.F.; Marchetti, M.P. Invasion Ecology; Blackwell Publishing: Oxford, UK, 2007; 304p. [**Google Scholar**]

11. Jeschke, J.M.; Strayer, D.L. Invasion success of vertebrates in Europe and North America. Proc. Natl. Acad. Sci. USA 2005, 102, 7198–7202. **[Google Scholar]** **[CrossRef]** **[Green Version]**

12. Moricca, S.; Bracalini, M.; Croci, F.; Corsinovi, S.; Tiberi, R.; Ragazzi, A.; Panzavolta, T. Biotic factors affecting ecosystem services in urban and peri-urban forests in Italy: The role of introduced and impending pathogens and pests. Forests 2018, 9, 65. **[Google Scholar]** **[CrossRef]** **[Green Version]**

13. Panzavolta, T.; Panichi, A.; Bracalini, M.; Croci, F.; Ginetti, B.; Ragazzi, A.; Tiberi, R.; Moricca, S. Dispersal and propagule pressure of Botryosphaeriaceae species in a declining oak stand is affected by insect vectors. Forests 2017, 8, 228. **[Google Scholar]** **[CrossRef]** **[Green Version]**

14. Jactel, H.; Koricheva, J.; Castagneyrol, B. Responses of forest insect pests to climate change: Not so simple. Curr. Opin. Insect Sci. 2019, 35, 103–108. **[Google Scholar]** **[CrossRef]**

15. Freer-Smith, P.H.; Webber, J.F. Tree pests and diseases: The threat to biodiversity and the delivery of ecosystem services. Biodivers. Conserv. 2015, 26, 3167–3181. **[Google Scholar]** **[CrossRef]**

16. Shine, C.; Williams, N.; Gundling, L. A Guide to Designing Legal and Institutional Frameworks on Alien Invasive Species; IUCN: Gland, Switzerland, 2000; 152p. **[Google Scholar]**

17. Hurley, B.P.; Garnas, J.; Wingfield, M.J.; Branco, M.; Richardson, D.M.; Slippers, B. Increasing numbers and intercontinental spread of invasive insects on eucalypts. Biol. Invasions 2016, 18, 921–933. **[Google Scholar]** **[CrossRef]** **[Green Version]**

18. Petter, F.; Suffert, M.; McMullen, M.; Griessinger, D.; Roy, A.S. Seed-borne pests and phytosanitary issues: The role of EPPO. In Global Perspectives on the Health of Seeds and Plant Propagation Material; Gullino, M., Munkvold, G., Eds.; Springer: Dordrecht, The Netherlands, 2014. **[Google Scholar]** **[CrossRef]**

19. Banks, N.C.; Paini, D.R.; Bayliss, K.L.; Hodda, M. The role of global trade and transport network topology in the human-mediated dispersal of alien species. Ecol. Lett. 2015, 18, 188–199. **[Google Scholar]** **[CrossRef]** **[Green Version]**

20. Eldredge, N. Life on Earth: An Encyclopedia of Biodiversity, Ecology, and Evolution; ABC-CLIO Inc.: Santa Barbara, CA, USA, 2002; 800p. **[Google Scholar]**

21. Liebhold, A.M.; Work, T.T.; McCullough, D.G.; Cavey, J.F. Airline baggage as a pathway for alien insect species entering the United States. Am. Entomol. 2006, 52, 48–54. **[Google Scholar]** **[CrossRef]** **[Green Version]**

22. Europhyt. Interceptions of Harmful Organisms in Imported Plants and Other Objects, Annual and Monthly Interception. Available online: **https://ec.europa.eu/food/plants/plant-health-and-biosecurity/european-union-notification-system-plant-health-interceptions-2_en** (accessed on 26 July 2021).

23. McCullough, D.G.; Work, T.T.; Cavey, J.F.; Liebhold, A.M.; Marshall, D. Interceptions of nonindigenous plant pests at US ports of entry and border crossings over a 17-year period. Biol. Invasions 2006, 8, 611–630. **[Google Scholar]** **[CrossRef]**

24. Boddy, L.; Griffith, G.S. Role of endophytes and latent invasion in the development of decay communities in sapwood of angiospermous trees. Sydowia 1989, 41, 41–73. **[Google Scholar]**

25. Précigout, P.A.; Claessen, D.; Makowski, D.; Robert, C. Does the latent period of leaf fungal pathogens reflect their trophic type? A meta-analysis of biotrophs, hemibiotrophs, and necrotrophs. Phytopathology 2020, 110, 345–361. [**Google Scholar**] [**CrossRef**] [**Green Version**]
26. Roques, A. Alien forest insects in a warmer world and a globalised economy: Impacts of changes in trade, tourism, and climate on forest biosecurity. N. Zeal. J. For. Sci. 2010, 40, 77–94. [**Google Scholar**]
27. Mack, R.N.; Simberloff, D.; Lonsdale, W.M.; Evans, H.; Clout, M.; Bazzaz, F.A. Biotic invasions: Causes, epidemiology, global consequences, and control. Ecol. Applic. 2000, 10, 689–710. [**Google Scholar**] [**CrossRef**]
28. Williamson, M.; Fitter, A. The varying success of invaders. Ecology 1996, 77, 1661–1665. [**Google Scholar**] [**CrossRef**]
29. Williamson, M. Biological Invasions; Chapman & Hall: London, UK, 1996; 244p. [**Google Scholar**]
30. Liebhold, A.M.; MacDonald, W.L.; Bergdahl, D.; Mastro, V.C. Invasion by Exotic Forest Pests: A Threat to Forest Ecosystems. For. Sci. 1995, 41, a0001–z0001. [**Google Scholar**] [**CrossRef**] [**Green Version**]
31. Moricca, S.; Linaldeddu, B.T.; Ginetti, B.; Scanu, B.; Franceschini, A.; Ragazzi, A. Endemic and emerging pathogens threatening cork oak trees: Management options for conserving a unique forest ecosystem. Plant Dis. 2016, 100, 2184–2193. [**Google Scholar**] [**CrossRef**] [**Green Version**]
32. Jarić, I.; Cvijanović, G. The Tens Rule in invasion biology: Measure of a true impact or our lack of knowledge and understanding? Enviro. Manag. 2012, 50, 979–981. [**Google Scholar**] [**CrossRef**]
33. Lovell, R.S.L.; Blackburn, T.M.; Dyer, E.E.; Pigot, A.L. Environmental resistance predicts the spread of alien species. Nat. Ecol. Evol. 2021, 5, 322–329. [**Google Scholar**] [**CrossRef**] [**PubMed**]
34. Philibert, A.; Desprez-Loustau, M.L.; Fabre, B.; Frey, P.; Halkett, F.; Husson, C.; Lung-Escarmant, B.; Marçais, B.; Robin, C.; Vacher, C.; et al. Predicting invasion success of forest pathogenic fungi from species traits. J. Appl. Ecol. 2011, 48, 1381–1390. [**Google Scholar**] [**CrossRef**]
35. Warren, R.J.; Candeias, M.; Lafferty, A.; Chick, L.D. Regional-scale environmental resistance to non-native ant invasion. Biol. Invasions 2020, 22, 813–825. [**Google Scholar**] [**CrossRef**]
36. Zenni, R.D.; Nuñez, M.A. The elephant in the room: The role of failed invasions in understanding invasion biology. Oikos 2013, 122, 801–815. [**Google Scholar**] [**CrossRef**]
37. Morrison, W.E.; Hay, M.E. Herbivore preference for native vs. exotic plants: Generalist herbivores from multiple continents prefer exotic plants that are evolutionarily naïve. PLoS ONE 2011, 6, e17227. [**Google Scholar**] [**CrossRef**] [**Green Version**]
38. Gladieux, P.; Feurtey, A.; Hood, M.E.; Snirc, A.; Clavels, J.; Dutech, C.; Roy, M.; Giraud, T. The population biology of fungal invasions. Mol. Ecol. 2015, 24, 1969–1986. [**Google Scholar**] [**CrossRef**]
39. Roques, A.; Shi, J.; Auger-Rozenberg, M.-A.; Ren, L.; Augustin, S.; Luo, Y. Are invasive patterns of non-native insects related to woody plants differing between Europe and China? Front. For. Glob. Chang. 2020, 2, 91. [**Google Scholar**] [**CrossRef**] [**Green Version**]

40. Keane, R.M.; Crawley, M.J. Exotic plant invasions and the enemy release hypothesis. Trends Ecol. Evol. 2010, 17, 164–170. **[Google Scholar]** **[CrossRef]**

41. Wolfe, L.M. Why alien invaders succeed: Support for the escape-from-enemy hypothesis. Am. Nat. 2002, 160, 705–711. **[Google Scholar]** **[CrossRef]** **[PubMed]**

42. Kenis, M.; Auger-Rozenberg, M.A.; Roques, A.; Timms, L.; Péré, C.; Cock, M.J.W.; Settele, J.; Augustin, S.; Lopez-Vaamonde, C. Ecological effects of invasive alien insects. Biol. Invasions 2009, 11, 21–45. **[Google Scholar]** **[CrossRef]**

43. Sache, I.; Roy, A.S.; Suffert, F.; Desprez-Loustau, M.L. Invasive plant pathogens in Europe. In Biological Invasions: Economic and Environmental Costs of Alien Plant, Animal, and Microbe Species, 2nd ed.; Pimentel, D., Ed.; CRC Press-Taylor and Francis Group: Boca Raton, FL, USA, 2011. **[Google Scholar]** **[CrossRef]**

44. Lovett, G.M.; Weiss, M.; Liebhold, A.M.; Holmes, T.P.; Leung, B.; Lambert, K.F.; Orwig, D.A.; Campbell, F.T.; Rosenth, J.; McCulloug, D.G.; et al. Nonnative forest insects and pathogens in the United States: Impacts and policy options. Ecol. Appl. 2016, 26, 1437–1455. **[Google Scholar]** **[CrossRef]** **[PubMed]**

45. Hulme, P.E.; Bacher, S.; Kenis, M.; Klotz, S.; Kühn, I.; Minchin, D.; Nentwig, W.; Olenin, S.; Panov, V.; Pergl, J.; et al. Grasping at the routes of biological invasions: A framework for integrating pathways into policy. J. Appl. Ecol. 2008, 45, 323–341. **[Google Scholar]** **[CrossRef]**

46. Moore, B.A. Alien Invasive Species: Impacts on Forests and Forestry. A Review; Forest Health and Biosecurity Working Paper 8. Forest Resources Development Service Working Paper FBS/8E; Forest Resources Division FAO: Rome, Italy, 2005. **[Google Scholar]**

47. Brasier, C.M. China and the origins of Dutch elm disease: An appraisal. Plant Pathol. 1990, 39, 5–16. **[Google Scholar]** **[CrossRef]**

48. Brasier, C.M.; Kirk, S.A. Designation of the EAN and NAN races of Ophiostoma novo-ulmi as subspecies. Mycol. Res. 2001, 105, 547–554. **[Google Scholar]** **[CrossRef]**

49. Brasier, C.M.; Gibbs, J.N. Origin of the Dutch elm disease Epidemic in Britain. Nature 1973, 242, 607–609. **[Google Scholar]** **[CrossRef]**

50. Brasier, C.M. Intercontinental spread and continuing evolution of the Dutch elm disease pathogens. In The Elms; Dunn, C.P., Ed.; Springer: Boston, MA, USA, 2000; pp. 61–72. **[Google Scholar]**

51. Vettraino, A.M.; Morel, O.; Perlerou, C.; Robin, C.; Diamandis, S.; Vannini, A. Occurrence and distribution of Phytophthora species in European chestnut stands and their association with ink disease and crown decline. Eur. J. Plant Pathol. 2005, 111, 169–180. **[Google Scholar]** **[CrossRef]**

52. Milgroom, M.G.; Wang, K.; Zhou, Y.; Lipari, S.E.; Kaneko, S. Intercontinental population structure of the chestnut blight fungus, Cryphonectria parasitica. Mycologia 1996, 88, 179–190. **[Google Scholar]** **[CrossRef]**

53. Panconesi, A. Canker stain of plane trees: A serious danger to urban plantings in Europe. J. Plant Pathol. 1999, 81, 3–15. **[Google Scholar]**

54. Moricca, S.; Panconesi, A. Canker stain of plane-trees: A serious threat to North-European urban plantations. Mitt. Biol. Bundesanst. Land- Forstwirtsch. Berl.-Dahl. 2000, 370, 97–100. **[Google Scholar]**

55. Panconesi, A.; Moricca, S.; Dellavalle, I.; Torraca, G. The epidemiology of canker stain of Plane tree and its spread from urban plantings to spontaneous groves and natural forests. In Proceedings of the Second International Symposium on Plant Health in Urban Horticulture, Berlin, Germany, 27–29 August 2003; pp. 84–91. [**Google Scholar**]

56. Wagener, W.W. Diseases of Cupressus. In Proceedings of the FAO-IUFRO Symposium on Internationally Dangerous Forest Diseases and Insects, Oxford, UK, 20–29 July 1964; pp. 17–24. [**Google Scholar**]

57. Graniti, A. Cypress canker: A pandemic in progress. Annu. Rev. Phytopathol. 1998, 36, 91–118. [**Google Scholar**] [**CrossRef**] [**PubMed**]

58. Sutton, B.C.; Gibson, I.A.S. Seiridium cardinale. CMI Description of Pathogenic Fungi and Bacteria No. 326; Commonwealth Mycological Institute: Kew, UK, 1972. [**Google Scholar**]

59. Grasso, V.; Ponchet, J. Historique, distribution géographique et hôtes du Coryneum cardinale. Wag. See Ref. 1979, 31, 119–126. [**Google Scholar**]

60. Graniti, A. Seiridium blight of cypress—Another ecological disaster? Plant Dis. 1993, 77, 544. [**Google Scholar**]

61. Panzavolta, T.; Panichi, A.; Bracalini, M.; Croci, F.; Benigno, A.; Ragazzi, A.; Tiberi, R.; Moricca, S. Tree pathogens and their insect-mediated transport: Implications for oak tree die-off in a natural park area. Glob. Ecol. Conserv. 2018, 15, e00437. [**Google Scholar**] [**CrossRef**]

62. Fiala, T.; Holuša, J. Occurrence of the invasive bark beetle Phloeosinus aubei on common juniper trees in the Czech Republic. Forests 2019, 10, 12. [**Google Scholar**] [**CrossRef**] [**Green Version**]

63. Michalski, J. Revision of the Palearctic Species of the Genus Scolytus Geoffroy (Coleoptera, Scolytidae); Panstwowe Wydawnictwo Naukowe: Warsaw, Poland, 1973; 214p. [**Google Scholar**]

64. Bloomfield, H. Elms for always. Am. For. 1979, 85, 24–26, 48, 50. [**Google Scholar**]

65. Lowe, S.; Browne, M.; Boudjelas, S.; De Poorter, M. 100 of the World's Worst Invasive Alien Species A Selection from the Global Invasive Species Database; Invasive Species Specialist Group: Auckland, New Zealand, 2000; 12p. [**Google Scholar**]

66. Bradshaw, C.; Leroy, B.; Bellard, C.; Roiz, D.; Albert, C.; Fournier, A.; Barbet-Massin, M.; Salles, J.M.; Simard, F.; Courchamp, F. Massive yet grossly underestimated global costs of invasive insects. Nat. Commun. 2016, 7, 12986. [**Google Scholar**] [**CrossRef**]

67. Montgomery, M.E.; Wallner, W.E. The gypsy moth. A westward migrant. In Dynamics of Forest Insect Populations: Patterns, Causes, Implications; Berryman, A.A., Ed.; Plenum Press: New York, NY, USA, 1988; pp. 353–376. [**Google Scholar**]

68. Leuschner, W.A.; Young, J.A.; Waldon, S.A.; Ravlin, F.W. Potential benefits of slowing the gypsy moth's spread. South. J. Appl. For. 1996, 20, 65–73. [**Google Scholar**] [**CrossRef**] [**Green Version**]

69. Bigsby, K.M.; Ambrose, M.J.; Tobin, P.C.; Sills, E.O. The cost of gypsy moth sex in the city. Urban For. Urban Green. 2014, 13, 459–468. [**Google Scholar**] [**CrossRef**]

70. Morin, R.S.; Liebhold, A.M.; Gottschalk, K.W.; Acciavatti, R.; Twardus, D.; White, R.; Horsley, S.; Smith, W. Forest Health Conditions on the Allegheny National Forest (1989–1999): Analysis of Forest Health Monitoring Surveys. NA-TP-04-01; Department of Agriculture,

Forest Service, Northeastern Area, State and Private Forestry: Newton Square, PA, USA, 2001. [**Google Scholar**]

71. CABI. Available online: **https://www.cabi.org/isc/datasheet/13579** (accessed on 26 July 2021).
72. O'Neil, C. Cypress Aphid, Cinara cupressi; The Entomology and Forest Resources Digital Information Work Group, College of Agricultural and Environmental Sciences and Warnell School of Forest Resources, The University of Georgia: Tifton, GA, USA, 1998; Available online: **www.afae.org/html/98-202.html** (accessed on 27 July 2021).
73. Watson, G.W.; Voegtlin, D.J.; Murphy, S.T.; Foottit, R.G. Biogeography of the Cinara cupressi complex (Hemiptera: Aphididae) on Cupressaceae, with description of a pest species introduced into Africa. Bull. Entomol. Res. 1999, 89, 271–283. [**Google Scholar**] [**CrossRef**]
74. Jung, T.; Horta Jung, M.; Webber, J.F.; Kageyama, K.; Hieno, A.; Masuya, H.; Uematsu, S.; Pérez-Sierra, A.; Harris, A.R.; Forster, J.; et al. The destructive tree pathogen Phytophthora ramorum originates from the laurosilva forests of East Asia. J. Fungi 2021, 7, 226. [**Google Scholar**] [**CrossRef**] [**PubMed**]
75. Rizzo, D.M.; Garbelotto, M.; Davidson, J.M.; Slaughter, G.W.; Koike, S.T. Phytophthora ramorum as the cause of extensive mortality of Quercus spp. and Lithocarpus densiflorus in California. Plant Dis. 2002, 86, 205–214. [**Google Scholar**] [**CrossRef**] [**PubMed**] [**Green Version**]
76. Brasier, C.M.; Webber, J.L. Sudden larch death. Nature 2010, 466, 824–825. [**Google Scholar**] [**CrossRef**]
77. Shearer, B.; Crane, C.; Cochrane, A. A thief of time: Phytophthora cinnamomi and threatened flora. Austr. Plant Conserv. 2005, 13, 14–15. [**Google Scholar**] [**CrossRef**]
78. Hansen, E.M. Phytophthora species emerging as pathogens of forest trees. Curr. For. Rep. 2015, 1, 16–24. [**Google Scholar**] [**CrossRef**] [**Green Version**]
79. Green, S.; Brasier, C.M.; Schlenzig, A.; McCracken, A.; MacAskill, G.A.; Wilson, M.; Webber, J.F. The destructive invasive pathogen Phytophthora lateralis found on Chamaecyparis lawsoniana across the UK. For. Pathol. 2013, 43, 19–28. [**Google Scholar**] [**CrossRef**]
80. Hantula, J.; Scholler, M. NOBANIS—Invasive Alien Species Fact Sheet—Melampsoridium hiratsukanum. Database of the European Network on Invasive Alien Species. NOBANIS. 2013. Available online: **https://www.nobanis.org/fact-sheets/** (accessed on 23 March 2021).
81. Moricca, S.; Benigno, A.; Oliveira Longa, C.M.; Cacciola, S.O.; Maresi, G. First documentation of life cycle completion of the alien rust pathogen Melampsoridium hiratsukanum in the Eastern Alps proves its successful establishment in this mountain range. J. Fungi 2021, 7, 617. [**Google Scholar**] [**CrossRef**] [**PubMed**]
82. McKinney, L.V.; Nielsen, L.R.; Collinge, D.B.; Thomsen, I.M.; Hansen, J.K.; Kjaer, E.D. The ash dieback crisis: Genetic variation in resistance can prove a long-term solution. Plant Pathol. 2014, 63, 485–499. [**Google Scholar**] [**CrossRef**]
83. Hill, L.; Jones, G.; Atkinson, N.; Hector, A.; Hemery, G.; Brown, N. The £15 billion cost of ash dieback in Britain. Curr. Biol. 2019, 29, R315–R316. [**Google Scholar**] [**CrossRef**] [**PubMed**]
84. Nirenberg, H.I.; O'Donnell, K. New Fusarium species and combinations within the Gibberella fujikuroi species complex. Mycologia 1998, 90, 434–458. [**Google Scholar**] [**CrossRef**]

85. Wingfield, M.J.; Hammerbacher, A.; Ganley, R.J.; Steenkamp, E.T.; Gordon, T.R.; Wingfield, B.D. Pitch canker caused by Fusarium circinatum—A growing threat to pine plantations and forests worldwide. Australas. Plant Pathol. 2008, 37, 319–334. **[Google Scholar]** **[CrossRef]**

86. Pehl, L.; Cech, T.L.; Ioos, R. Lecanosticta acicola (formerly Mycosphaerella dearnessii), Dothistroma septosporum (formerly Mycosphaerella pini) and Dothistroma pini. EPPO Bull. 2015, 45, 163–182. **[Google Scholar]**

87. Barnes, I.; Crous, P.W.; Wingfield, B.D.; Wingfield, M.J. Multigene phylogenies reveal that red band needle blight of Pinus is caused by two distinct species of Dothistroma, D. septosporum and D. pini. Stud. Mycol. 2004, 50, 551–566. **[Google Scholar]**

88. Watt, M.S.; Kriticos, D.J.; Alcaraz, S.; Brown, A.V.; Leriche, A. The hosts and potential geographic range of Dothistroma needle blight. For. Ecol. Manag. 2009, 257, 1505–1519. **[Google Scholar]** **[CrossRef]**

89. Ghelardini, L.; Aglietti, C.; Loria, F.; Cerboneschi, M.; Gionni, A.; Goti, E.; Maresi, G.; Moricca, S.; Marchi, G. Dothistroma Needle Blight in protected pine forests in Italy. Manag. Biol. Invasions 2020, 11, 689–702. **[Google Scholar]** **[CrossRef]**

90. Wingfield, M.J.; Coutinho, T.A.; Roux, J.; Wingfield, B.D. The future of exotic plantation forestry in the tropics and southern hemisphere: Lessons from pitch canker. S. Afr. For. J. 2002, 195, 79–82. **[Google Scholar]** **[CrossRef]**

91. Bulman, L.S.; Bradshaw, R.E.; Fraser, S.; Martn-Garca, J.; Barnes, I.; Musolin, D.L.; La Porta, N.; Woods, A.J.; Diez, J.J.; Koltay, A.; et al. A worldwide perspective on the management and control of Dothistroma needle blight. Forest Pathol. 2016, 46, 472–488. **[Google Scholar]** **[CrossRef]** **[Green Version]**

92. Gibson, I.A.S. Impact and control of Dothistroma blight of pines. Eur. J. For. Pathol. 1974, 4, 89–100. **[Google Scholar]** **[CrossRef]**

93. Villebonne, D.; Maugard, F. Rapid development of Dothistroma needle blight (Scirrhia pini) on Corsican pine (Pinus nigra subsp. laricio) in France. In La Sante des Forets, Annual Report 1998; Les Cahiers du Département de la santé des forêts (DSF); DERF: Paris, France, 1999. **[Google Scholar]**

94. Newton, L.P.; Fowler, G.; Neeley, A.D.; Schall, R.A.; Takeuchi, Y. Pathway Assessment: Geosmithia sp. and Pityophthorus juglandis Blackman Movement from the Western into the Eastern United States; U.S. Department of Agriculture Animal and Plant Health Inspection Service: Raleigh, NC, USA, 2009; 50p.

95. Montecchio, L.; Faccoli, M. First record of thousand cankers disease Geosmithia morbida and walnut twig beetle Pityophthorus juglandis on Juglans nigra in Europe. Plant Dis. 2014, 98, 696. **[Google Scholar]** **[CrossRef]**

96. Moricca, S.; Bracalini, M.; Benigno, A.; Ginetti, B.; Pelleri, F.; Panzavolta, T. Disease Note. Thousand cankers disease caused by Geosmithia morbida and its insect vector Pityophthorus juglandis first reported on Juglans nigra in Tuscany, Central Italy. Plant Dis. 2019, 103, 369. **[Google Scholar]** **[CrossRef]**

97. Umeda, C.; Eskalen, A.; Paine, T.D. Polyphagous Shot Hole Borer and Fusarium Dieback in California. In Insects and Diseases of Mediterranean Forest Systems; Paine, T., Lieutier, F., Eds.; Springer: Cham, Switzerland, 2016. **[Google Scholar]** **[CrossRef]**

98. Haack, R.A.; Cavey, J.F.; Hoebeke, E.R.; Law, K. Anoplophora glabripennis: A new tree-infesting exotic cerambycid invades New York. Mich. Entomol. Soc. Newsl. 1996, 41, 1–3. [**Google Scholar**]

99. EPPO. Anoplophora chinensis Found under Glasshouse in Georgia (US). Available online: **https://gd.eppo.int/reporting/article-3421** (accessed on 26 July 2021).

100. Hérard, F.; Maspero, M. History of discoveries and management of the citrus longhorned beetle, Anoplophora chinensis, in Europe. J. Pest Sci. 2019, 92, 117–130. [**Google Scholar**] [**CrossRef**]

101. Haack, R.A.; Hérard, F.; Sun, J.; Turgeon, J.J. Managing invasive populations of Asian longhorned beetle and citrus longhorned beetle: A worldwide perspective. Annu. Rev. Entomol. 2010, 55, 521–546. [**Google Scholar**] [**CrossRef**] [**PubMed**] [**Green Version**]

102. Siegert, N.W.; Mccullough, D.G.; Liebhold, A.M.; Telewski, F.W. Dendrochronological reconstruction of the epicentre and early spread of emerald ash borer in North America. Divers. Distrib. 2014, 20, 847–858. [**Google Scholar**] [**CrossRef**]

103. EPPO. Agrilus Planipennis. EPPO Datasheets on Pests Recommended for Regulation. Available online: **https://gd.eppo.int** (accessed on 26 July 2021).

104. Orlova-Bienkowskaja, M.J.; Drogvalenko, A.N.; Zabaluev, I.A.; Sazhnev, A.S.; Peregudova, E.Y.; Mazurov, S.G.; Komarov, E.V.; Struchaev, V.V.; Martynov, V.V.; Nikulina, T.V.; et al. Current range of Agrilus planipennis Fairmaire, an alien pest of ash trees, in European Russia and Ukraine. Ann. For. Sci. 2020, 77, 29. [**Google Scholar**] [**CrossRef**]

105. Drogvalenko, A.N.; Orlova-Bienkowskaja, M.J.; Bieńkowski, A.O. Record of the emerald ash borer (Agrilus planipennis) in Ukraine is confirmed. Insects 2019, 10, 338. [**Google Scholar**] [**CrossRef**] [**Green Version**]

106. Marzano, M.; Hall, C.; Dandy, N.; Fisher, C.L.; Diss-Torrance, A.; Haight, R.G. Lessons from the frontline: Exploring how stakeholders may respond to emerald ash borer management in Europe. Forests 2020, 11, 617. [**Google Scholar**] [**CrossRef**]

107. Thompson, I.; Mackey, B.; McNulty, S.; Mosseler, A. Forest Resilience, Biodiversity, and Climate Change: A Synthesis of the Biodiversity/Resilience/Stability Relationship in Forest Ecosystems; Technical Series no. 43; Secretariat of the Convention on Biological Diversity: Montreal, QC, Canada, 2009; 67p. [**Google Scholar**]

108. Rytter, L.; Johansson, K.; Karlsson, B.; Stener, L.-G. Tree species, genetics and regeneration for bioenergy feedstock in Northern Europe. In Forest Bioenergy Production, Management, Carbon Sequestration and Adaptation; Kellomäki, S., Kilpeläinen, A., Alam, A., Eds.; Springer: New York, NY, USA, 2013; pp. 7–37. [**Google Scholar**]

109. Pérez, G.; Hunter, G.C.; Slippers, B.; Pérez, C.; Wingfield, B.D.; Wingfield, M.J. Teratosphaeria (Mycosphaerella) nubilosa, the causal agent of Mycosphaerella leaf disease (MLD), recently introduced into Uruguay. Eur. J. Plant Pathol. 2009, 125, 109–118. [**Google Scholar**] [**CrossRef**]

110. European Commission. The Insect Killing Our Palm Trees. EU Efforts to Stop the Red Palm Weevil; Office for Official Publications of the European Communities: Luxembourg, 2011. [**Google Scholar**]

111. Burks, R.A.; Mottern, J.L.; Waterworth, R.; Paine, T.D. First report of the Eucalyptus gall wasp, Ophelimus maskelli (Hymenoptera: Eulophidae), an invasive pest on Eucalyptus, from the Western Hemisphere. Zootaxa 2015, 3926, 448–450. [**Google Scholar**] [**CrossRef**] [**PubMed**] [**Green Version**]

112. Wittenberg, R.; Cock, M.J.W. Best practices for the prevention and management of invasive alien species. In Invasive Alien Species: A New Synthesis; Mooney, H.A., Mack, R.N., McNeely, J.A., Neville, L.E., Scheiand, P.J., Waage, J.K., Eds.; Island Press: Washington, DC, USA, 2005; pp. 209–232. [**Google Scholar**]

113. Rizzo, D.; Moricca, S.; Bracalini, M.; Benigno, A.; Bernardo, U.; Luchi, N.; Da Lio, D.; Nugnes, F.; Cappellini, G.; Salemi, C.; et al. Rapid detection of Pityophthorus juglandis (Blackman) (Coleoptera, Curculionidae) with the Loop-Mediated Isothermal Amplification (LAMP) method. Plants 2021, 10, 1048. [**Google Scholar**] [**CrossRef**] [**PubMed**]

114. Piel, F.; Gilbert, M.; De Canniere, C.; Gregoire, J.C. Coniferous round wood imports from Russia and Baltic countries to Belgium. A pathway analysis for assessing risks of exotic pest insect introductions. Divers. Distrib. 2008, 14, 318–328. [**Google Scholar**] [**CrossRef**]

115. Matthews, J.; Beringen, R.; Creemers, R.; Hollander, H.; Van Kessel, N.; Van der Kleef, H.; van der Velde, G.; Verbrugge, L.N.H.; Leuven, R.S.E.W. Horizon Scanning for New Invasive Non-Native Species in the Netherlands; Department of Environmental Science, Institute for Water and Wetland Research, Faculty of Science, Radboud University Nijmegen: Nijmegen, The Netherlands, 2014; 117p. [**Google Scholar**]

116. Mack, R.N.; Barrett, S.C.H.; deFur, P.L.; MacDonald, W.L.; Madden, L.V.; Marshall, D.S.; McCullough, D.G.; McEvoy, P.B.; Nyrop, J.P.; Reichard, S.E.H.; et al. Predicting Invasions of Nonindigenous Plants and Plant Pests; National Academy of Sciences: Washington DC, USA, 2002; 194p. [**Google Scholar**]

117. Shine, C.; Kettunen, M.; Genovesi, P.; Essl, F.; Gollasch, S.; Rabitsch, W.; Scalera, R.; Starfinger, U.; Ten Brink, P. Assessment to Support Continued Development of the EU Strategy to Combat Invasive Alien Species; Institute for European Environmental Policy (IEEP): Brussels, Belgium, 2010; 298p. [**Google Scholar**]

118. Kenis, M.; Li, H.; Fan, J.T.; Courtial, B.; Auger-Rozenberg, M.A.; Yart, A.; Eschen, R.; Roques, A. Sentinel nurseries to assess the phytosanitary risks from insect pests on importations of live plants. Sci. Rep. 2018, 8, 11217. [**Google Scholar**] [**CrossRef**]

119. Vettraino, A.M.; Li, H.-M.; Eschen, R.; Morales-Rodriguez, C.; Vannini, A. The sentinel tree nursery as an early warning system for pathway risk assessment: Fungal pathogens associated with Chinese woody plants commonly shipped to Europe. PLoS ONE 2017, 12, e0188800. [**Google Scholar**] [**CrossRef**] [**Green Version**]

120. McCartney, H.A.; Foster, S.J.; Fraaije, B.A.; Ward, E. Molecular diagnostics for fungal plant pathogens. Pest. Manag. Sci. 2003, 59, 129–142. [**Google Scholar**] [**CrossRef**]

121. Rizzo, D.; Da Lio, D.; Bartolini, L.; Cappellini, G.; Bruscoli, T.; Bracalini, M.; Benigno, A.; Salemi, C.; Del Nista, D.; Aronadio, A.; et al. A duplex real-time PCR with probe for simultaneous detection of Geosmithia morbida and its vector Pityophthorus juglandis. PLoS ONE 2020, 15, e0241109. [**Google Scholar**] [**CrossRef**]

122. Fekrat, L.; Aghl, M.Z.; Tahan, V. Application of the LAMP assay as a diagnostic technique for rapid identification of Thrips tabaci (Thysanoptera: Thripidae). J. Econ. Entomol. 2015, 108, 1337–1343. [**Google Scholar**] [**CrossRef**] [**PubMed**]

123. Blaser, S.; Diem, H.; Von Felten, A.; Gueuning, M.; Andreou, M.; Boonham, N.; Tomlinson, J.; Müller, P.; Utzinger, J.; E Frey, J.; et al. From laboratory to point of entry: Development and implementation of a loop-mediated isothermal amplification (LAMP)-based genetic identification system to prevent introduction of quarantine insect species. Pest Manag. Sci. 2018, 74, 1504–1512. [**Google Scholar**] [**CrossRef**] [**Green Version**]

124. Rizzo, D.; Luchi, N.; Da Lio, D.; Bartolini, L.; Nugnes, F.; Cappellini, G.; Bruscoli, T.; Salemi, C.; Griffo, R.V.; Garonna, A.P.; et al. Development of a loop-mediated isothermal amplification (LAMP) assay for the identification of the invasive wood borer Aromia bungii (Coleoptera: Cerambycidae) from frass. Biotech 2021, 11, 1–12. **[Google Scholar]** **[CrossRef]**

125. Tordoni, E.; Ametrano, C.G.; Banchi, E.; Ongaro, S.; Pallavicini, A.; Bacaro, G.; Muggia, L. Integrated eDNA metabarcoding and morphological analyses assess spatio-temporal patterns of airborne fungal spores. Ecol. Indic. 2021, 121, 107032. **[Google Scholar]** **[CrossRef]**

126. Martin, R.R.; James, D.; Lévesque, C.A. Impacts of molecular diagnostic technologies on plant disease management. Annu. Rev. Phytopathol. 2000, 38, 207–239. **[Google Scholar]** **[CrossRef]** **[Green Version]**

127. Li, M.; Liu, M.; Liu, M.; Ju, Y. Monitoring exotic forest pest based on high-resolution remote sensing image and CART model. In Proceedings of the 3rd International Congress on Image and Signal Processing, Yantai, China, 16–18 October 2010; Tan, Z.H., Wan, Y., Xiang, T., Song, Y., Eds.; Institute of Electrical and Electronics Engineers (IEEE): Piscataway, NJ, USA; pp. 2203–2206. **[Google Scholar]** **[CrossRef]**

128. Hall, R.; Castilla, G.; White, J.; Cooke, B.; Skakun, R. Remote sensing of forest pest damage: A review and lessons learned from a Canadian perspective. Can. Entomol. 2016, 148, S296–S356. **[Google Scholar]** **[CrossRef]**

129. Hoyer-Tomiczek, U.; Sauseng, G.; Hoch, G. Scent detection dogs for the Asian longhorn beetle, Anoplophora glabripennis. EPPO Bull. 2016, 46, 148–155. **[Google Scholar]** **[CrossRef]**

130. Mosconi, F.; Campanaro, A.; Carpaneto, G.M.; Chiari, S.; Hardersen, S.; Mancini, E.; Maurizi, E.; Sabatelli, S.; Zauli, A.; Mason, F.; et al. Training of a dog for the monitoring of Osmoderma eremita. Nat. Conserv. 2017, 20, 237–264. **[Google Scholar]** **[CrossRef]** **[Green Version]**

131. Brown, N.; van den Bosch, F.; Parnell, S.; Denman, S. Integrating regulatory surveys and citizen science to map outbreaks of forest diseases: Acute oak decline in England and Wales. Proc. Royal Soc. B 2017, 284, 20170547. **[Google Scholar]** **[CrossRef]** **[Green Version]**

132. Panzavolta, T.; Bracalini, M.; Tiberi, R. Impatto delle invasioni biologiche sul paesaggio urbano. In Paesaggi Abitati: Dalla Percezione al Sistema Complesso; Bolletti, S., Puma, P., Eds.; Edifir Edizioni Firenze s.r.l.: Florence, Italy, 2021; pp. 73–79. **[Google Scholar]**

133. Bonney, R.; Shirk, J.L.; Phillips, T.B.; Wiggins, A.; Ballard, H.L.; Miller-Rushing, A.J.; Parrish, J.K. Next steps for citizen science. Science 2014, 343, 1436–1437. **[Google Scholar]** **[CrossRef]**

134. Bonney, R.; Phillips, T.B.; Ballard, H.L.; Enck, J.W. Can citizen science enhance public understanding of science? Public Underst. Sci. 2016, 25, 2–16. **[Google Scholar]** **[CrossRef]**

135. Suckling, D.M.; Stringer, L.D.; Stephens, A.E.A.; Woods, B.; Williams, D.G.; Baker, G.; El-Sayed, A.M. From integrated pest management to integrated pest eradication: Technologies and future needs. Pest Manag. Sci. 2013, 70, 179–189. **[Google Scholar]** **[CrossRef]**

136. Pluess, T.; Jarosik, V.; Pysek, P.; Cannon, R.; Pergl, J.; Breukers, A.; Bacher, S. Which factors affect the success or failure of eradication campaigns against alien species? PLoS ONE 2012, 7, e48157. **[Google Scholar]** **[CrossRef]**

137. Moricca, S.; Bracalini, M.; Benigno, A.; Panzavolta, T. Observations on the non-native thousand cankers disease of walnut in Europe's southernmost outbreak. Global Ecol. Conserv. 2020, 23, e01159. **[Google Scholar]** **[CrossRef]**

138. Liebhold, A.M.; Kean, J.M. Eradication and containment of non-native forest insects: Successes and failures. J. Pest Sci. 2019, 92, 83–91. **[Google Scholar]** **[CrossRef]**

139. Vainio, E.J.; Bezos, D.; Bragança, H.; Cleary, M.; Fourie, G.; Georgieva, M.; Ghelardini, L.; Hannunen, S.; Ioos, R.; Martín-García, J.; et al. Sampling and detection strategies for the pine pitch canker (PPC) disease pathogen Fusarium circinatum in Europe. Forests 2019, 10, 723. **[Google Scholar]** **[CrossRef]** **[Green Version]**

140. Ramsfield, T.D.; Bentz, B.J.; Faccoli, M.; Jactel, H.; Brockerhoff, E.G. Forest health in a changing world: Effects of globalization and climate change on forest insect and pathogen impacts. Forestry 2016, 89, 245–252. **[Google Scholar]** **[CrossRef]**

141. Caltagirone, L.E.; Doutt, R.L. The history of the vedalia beetle importation to California and its impact on the development of biological control. Annu. Rev. Entomol. 1989, 34, 1–16. **[Google Scholar]** **[CrossRef]**

142. McEvoy, P.; Cox, C.; Coombs, E. Successful biological control of ragwort, Senecio jacobaea, by introduced insects in Oregon. Ecol. Appl. 1991, 1, 430–442. **[Google Scholar]** **[CrossRef]**

143. Radcliffe, E.B.; Flanders, K.L. Biological control of alfalfa weevil in North America. Integr. Pest Manag. Rev. 1998, 3, 225–242. **[Google Scholar]** **[CrossRef]**

144. Ferracini, C.; Ferrari, E.; Saladini, M.A.; Pontini, M.; Corradetti, M.; Alma, A. Non-target host risk assessment for the parasitoid Torymus sinensis. BioControl 2015, 60, 583–594. **[Google Scholar]** **[CrossRef]**

145. Vasiliauskas, R.; Larsson, E.; Larsson, K.-H.; Stenlid, J. Persistence and long-term impact of Rotstop biological control agent on mycodiversity in Picea abies stumps. Biol. Control. 2005, 32, 295–304. **[Google Scholar]** **[CrossRef]**

146. Tubby, K.V.; Scott, D.; Webber, J.F. Relationship between stump treatment coverage using the biological control product PG Suspension, and control of Heterobasidion annosum on Corsican pine, Pinus nigra ssp. Laricio. For. Path. 2008, 38, 37–46. **[Google Scholar]** **[CrossRef]**

147. Korhonen, K.; Stenlid, J. Biology of Heterobasidion annosum. In Heterobasidion annosum. Biology, Ecology, Impact and Control; Woodward, S., Stenlid, J., Karjalainen, R., Hüttermann, A., Eds.; CAB International: Wallingford, UK, 1998; pp. 43–70. **[Google Scholar]**

148. Slippers, B.; Hurley, B.P.; Wingfield, M.J. Sirex woodwasp: A model for evolving management paradigms of invasive forest pests. Annu. Rev. Entomol. 2015, 60, 601–619. **[Google Scholar]** **[CrossRef]** **[PubMed]** **[Green Version]**

149. Moricca, S.; Ragazzi, A. Biological and integrated means to control rust diseases. In Integrated Management of Diseases Caused by Fungi, Phytoplasma and Bacteria; Ciancio, A., Mukerji, K.J., Eds.; Springer: Berlin, Germany, 2008; pp. 303–329. **[Google Scholar]**

150. Van Lenteren, J.C.; Bale, J.; Bigler, F.; Hokkanen, H.M.T.; Loomans, A.J.M. Assessing risks of releasing exotic biological control agents of arthropod pests. Annu. Rev. Entomol. 2006, 51, 609–634. **[Google Scholar]** **[CrossRef]** **[Green Version]**

151. Yara, K.; Sasawaki, T.; Kunimi, Y. Hybridization between introduced Torymus sinensis (Hymenoptera: Torymidae) and indigenous T. beneficus (late-spring strain), parasitoids of the asian chestnut gall wasp Dryocosmus kuriphilus (Hymenoptera: Cynipidae). Biol. Control. 2010, 54, 14–18. [**Google Scholar**] [**CrossRef**]

152. Messing, R.; Brodeur, J. Current challenges to the implementation of classical biological control. BioControl 2018, 63, 1–9. [**Google Scholar**] [**CrossRef**]

153. ec.europa.eu. Available online: **https://ec.europa.eu/commission/presscorner/detail/en/QANDA_19_6710** (accessed on 6 September 2021).

154. IPPC. Establishing a National Plant Protection Organization; A guide to understand the principal requirements for establishing an organization to protect national plant resources from pests; FAO: Rome, Italy, 2015. [**Google Scholar**]

155. Scalera, R. How much is Europe spending on invasive alien species? Biol. Invasions 2010, 12, 173–177. [**Google Scholar**] [**CrossRef**]

156. Holdenreider, O.; Pautasso, M.; Weisberg, P.J.; Lonsdale, D. Tree diseases and landscape processes: The challenge of landscape pathology. Trends. Ecol. Evol. 2004, 19, 446–452. [**Google Scholar**] [**CrossRef**]

157. Douglas, H.; Dang, B.T.; Gill, B.D.; Huber, J.; Mason, P.F.; Parker, D.J.; Sinclair, B.J. The importance of taxonomy in responses to invasive alien species. Biodiversity 2009, 10, 92–99. [**Google Scholar**] [**CrossRef**]

158. Gluck-Thaler, E.; Jason, C.; Slot, J.C. Dimensions of horizontal gene transfer in eukaryotic microbial pathogens. PLoS Pathog. 2015, 11, e1005156. [**Google Scholar**] [**CrossRef**] [**Green Version**]

159. Brasier, C.M. The rise of the hybrid fungi. Nature 2000, 405, 134–135. [**Google Scholar**] [**CrossRef**] [**PubMed**]

160. McMullan, M.; Rafiqi, M.; Kaithakottil, G.; Clavijo, B.J.; Bilham, L.; Orton, E.; Percival-Alwyn, L.; Ward, B.J.; Edwards, A.; Saunders, D.G.O.; et al. The ash dieback invasion of Europe was founded by two genetically divergent individuals. Nat. Ecol. Evol. 2018, 2, 1000–1008. [**Google Scholar**] [**CrossRef**] [**Green Version**]

161. Albanese, G.; Saponari, M.; Faggioli, F. Phytosanitary certification. In Olive Germplasm—The Olive Cultivation, Table and Olive Oil Industry in Italy; Muzzalupo, I., Ed.; InTech Publisher: Rijeka, Croatia, 2012; pp. 107–132. ISBN 9789535108849. [**Google Scholar**]

162. Ricciardi, A.; Hoopes, M.F.; Marchetti, M.P.; Lockwood, J.L. Progress toward understanding the ecological impacts of non-native species. Ecol. Monogr. 2013, 83, 263–282. [**Google Scholar**] [**CrossRef**] [**Green Version**]

22 AN UPDATE ON IMPORTANT PHYTOPLASMA DISEASES AFFECTING FOREST TREES

1. INTRODUCTION: AN OVERVIEW

Phytoplasmas are a vast collection of bacteria that lack cell walls and are closely linked to diseases in over a thousand plant species globally, including commercially and ecologically significant forest trees. Within the plant host, these organisms establish colonies in the phloem components and are naturally spread by insects that feed on phloem, specifically homopterans [1]. Phytoplasmas induce either distinct symptoms (virescence, phyllody, witches' brooms) or nonspecific symptoms (yellowing, reddening, diminished growth, die-back, and decline). Nevertheless, the manifestation of symptoms might greatly differ for a specific mix of plant and host.

Phytoplasmas belong to the family Mollicutes and are closely linked to acholeplasmas. They are currently classified as members of the tentative genus 'Candidatus Phytoplasma' mostly based on 16S rDNA sequence analysis [2,3]. A total of 34 phytoplasma groups, also known as 16Sr groups, were identified by the use of restriction fragment length polymorphism (RFLP) analysis. This investigation involved examining the PCR-amplified rDNA sequences within the specified genus [3,4]. Furthermore, the utilization of multi-locus sequence typing (MLST) with less-conserved genes has demonstrated its efficacy as a valuable method for distinguishing genetically similar strains that may differ in terms of pathology or epidemiology. Accurate identification of these strains is crucial for conducting epidemiological research [5].

2. PHYTOPLASMA DISEASES OF FOREST TREES

2.1. ELM

Various Ulmus (elm) species and hybrids are affected by elm yellows (EY), which is a disease common in North America and Europe [**6**]. This disease was formerly named elm phloem necrosis because the discoloration and death (= necrosis) of phloem tissue are the most characteristic symptoms in highly susceptible hosts, e.g., Ulmus americana (American or white elm) [7]. EY was first reported in Ohio in 1938 [7]. However, there is evidence that it was present there as well as in neighboring states long before this time [**8,9**]. Once established in the midwestern states, EY spread into eastern states and southeastern Ontario [**10**].

It is known for causing destructive epidemics in North America, which killed thousands of elm trees. Until the 1980s, EY was considered a typical North American disease. Conti et al. [**11**] first reported the occurrence of this disease in Italy, although it was observed in Italy since at least 1918 (for a review, see [**6**]). Following this finding, phytoplasma diseases of elm have also been been recorded in several other European countries [**6,12,13**]. The detection of EY in Europe was first made on the basis of symptoms [**11,14**]. Later, molecular studies using mainly RFLP and sequence analyses of PCR-amplified rDNA showed that the phytoplasma diseases of elm in Europe and North America are caused by the same organism, the EY agent [**15,16,17,18**]. This pathogen was taxonomically delineated as 'Candidatus (Ca.) Phytoplasma ulmi' [**2,19**].

Taxonomically, the EY agent 'Ca. Phytoplasma ulmi' is assigned to the EY phytoplasma group or 16SrV group, subgroup 16SrV-A [**19**]. Other members of this group are phytoplasmas causing mainly diseases of trees and shrubs in the northern hemisphere [**6**]. An updated picture of the phylogenetic relationships among phytoplasmas associated with diseases of forest trees reviewed in this paper is shown in **Figure 1**. In contrast to 16S rRNA gene, genetic diversity within the EY pathogen have been observed in rplV, rpsC, secY, map, groEL, and imp gene sequences [**19,20,21,22,23,24**].

A phytoplasma of the clover proliferation group (=16SrVI group), subgroup 16SrVI-C, the Illinois elm yellows (ILEY) phytoplasma, was detected in American elm trees showing EY symptoms in Illinois, USA [**26**]. However, some trees proved to be doubly infected with the ILEY phytoplasma and a phytoplasma of the aster yellows (AY) group (16SrI group) [**26**]. Diseased elm trees doubly or multiply infected with the EY agent and a phytoplasma of the 16SrI and/or the stolbur group (16SrXII group) were also recorded in Italy [**27**]. However, in these trees, the EY agent was predominant, while the other phytoplasmas were present at low titer. A few diseased elm trees showing yellowing symptoms, singly infected with 16SrI and 16SrXII-A phytoplasmas, were found in Croatia [**23**], whereas diseased elm trees in China harbored 16SrV-B and 16SrI-B phytoplasmas [**28**]. A 'Ca. Phytoplasma fragariae'-related strain (16SrXII-E subgroup) was detected in five symptomless elm trees in Belgium, whereas 16SrV-C phytoplasmas were detected in two elm trees in Germany with the absence of yellows disease symptoms [**13,29**].

Figure 1. Phylogenetic dendrogram generated by using the neighbor-joining method software MEGA, version X [25] with 16S rDNA sequences from phytoplasmas detected in naturally infected elm, alder, ash, conifer, sandal and eucalyptus trees (in bold type), and a number of reference phytoplasmas. Acholeplasma laidlawii was used as the outgroup. Bar represents a phylogenetic distance of 2%. GenBank accession number is given for each phytoplasma. Bootstrap values are shown on branches of the dendrogram.

EY symptoms may differ among elm species. In Northern American species, e.g., U. americana, U. rubra (red or slippery elm), U. alata (winged elm), U. serotina (September elm), and U. crassifolia (cedar elm), the main symptoms are yellowing, premature casting of the leaves, early bud break, brown discoloration of the phloem tissue, and tree death. Red elm often shows witches' brooms. Discolored phloem tissue of American, winged, September and cedar elms usually have a characteristic odor of oil of wintergreen (methyl salicylate). In U. minor (syn.: U. carpinifolia, European field elm), witches' brooms are predominant. For this reason, the disease occurring in Europe is often called elm witches' broom. Other symptoms are yellowing, stunting, and small leaves (**Figure 2**).

Figure 2. Small leaves, yellowing, and witches' broom on European field elm (Ulmus minor) shoot affected by elm yellows disease. Left, healthy shoot.

Brooming and stunting are also typical symptoms of EY-affected U. glabra (Scots or wych elm) and U. parvifolia (Chinese elm) [**14,15,30**]. Yellowing or reddening, reduced vigor, witches' brooms, and die-back but not phloem discoloration are known to occur in several other European and Asian elm genotypes, including U. pumila (Siberian elm), U. chenmoui (Chenmou elm), U. villosa (cherry bark elm), U. laevis (European white elm), U. wallichiana (Himalayan elm), U. wilsoniana (Wilson elm), U. japonica (Japanese elm), and U. × hollandica (Dutch elm) [**11,21,22,27,31,32**]. In diseased American elm trees, an abnormal callose deposition on the sieve plates is usually followed by the collapse of sieve tube elements. Thereby, phloem transport is severely compromised. Starch accumulates in the aerial parts of affected trees while roots starve and die. The sieve tubes of affected U. americana and U. rubra trees are so sensitive to 'Ca. Phytoplasma ulmi' that they suddenly collapse without allowing the pathogen to reach a high titer [**14,33,34**]. In contrast, sieve tubes of U. minor and U. parvifolia trees as well as of other European and Asian elm genotypes being rather tolerant to the EY pathogen are only slightly damaged and allow it to reach a quite high titer [**14,17,35,36**]. 'Ca. Phytoplasma ulmi' also causes stomatal closure and elevated diffusive resistance, which is associated with high water potential, in leaves of U. americana and U. rubra trees [**10**]. There are also several reports on the presence of 'Ca. Phytoplasma ulmi' in non-symptomatic trees belonging to some European and Asian elm genotypes [**13,23,27,29,30,31,32,37**]. In China, elm trees of unspecified species showing yellowing and witches' brooms were found to be infected by 16SrV-B and 16SrI-B phytoplasmas, respectively [**28**].

EY is common in eastern North American states where several severe epidemics spread at rates of 5 to 8 km per year in some areas and destroyed a large number of native elm tree [**30**]. In some states, EY epidemics occurred together with the Dutch elm disease and exacerbated the latter by providing additional breeding material for elm bark beetles, which are the insects responsible for the spread of the Dutch elm disease [**30**]. In Italy, significant EY epidemics have been observed in some experimental fields established during the 1980s to test the adaptability of a number of elm species and various hybrid clones to local environmental conditions [**31**]. U. americana, U. villosa, U. parvifolia, U. japonica, U. parvifolia × U. wallichiana (= clone P628), (U. wallichiana × (U. × hollandica 'Vegeta' × U. minor) (= clone 793), [(U. wallichiana × U. minor)] × U. laciniata) × U. laciniata 'Nikkoensis' open-pollinated (= clone 1094), and (U. wallichiana × U. minor) × U. laciniata open-pollinated (= clone 1098) were severely affected [**31**]. In one of the experimental field, 30% of trees were infected five years after the first evidence of the disease, reaching nearly 80% within fourteen years. EY epidemics observed in southern and northern Italy on European field elm and Siberian elm showed a disease incidence greater than 80% [**6,11,31,37**]. A detection rate of 75% of EY phytoplasma in European white elm trees has been recorded in Croatia using nested PCR assays. More than half of the infected trees were symptomless [**23**]. In Croatia, 75% of the European white elm trees that were examined by nested PCR assays proved to be infected by 'Ca. Phytoplasma ulmi'. Most of the examined trees were non-symptomatic [**23**]. In a recent survey performed in Germany on a nationwide scale by using real-time PCR, 'Ca. Phytoplasma ulmi' was detected in almost 28% of elm trees examined. Among the elm species examined, the highest detection rate was recorded in the European white elm, followed by the Scots elm and European field elm. EY-specific symptoms were rarely observed only in infected Scots and European field elm trees [**13**]. In Illinois, a severe EY outbreak associated to ILEY phytoplasma occurred in the 1990s and caused the death of more than one thousand American elms [**26**]. No epidemics are known to occur in China where single disease occurrences were recorded [**28**].

The leafhopper Scaphoideus luteolus is known to transmit 'Ca. Phytoplasma ulmi' in North America. However, other vectors are most probably involved in the transmission process since numerous homopteran insects were captured on elm, whereas S. luteolus is not present in some EY-infested areas (for references see [**6,30**]). In New York State, of the various homopteran insects collected in EY-infested areas and examined for their ability to transmit 'Ca. Phytoplasma ulmi', single transmissions were obtained for the leafhopper Allygus atomarious and the spittlebug Philaenus spumarius [**10**]. In addition, 'Ca. Phytoplasma ulmi' was detected by real-time PCR in several leafhoppers belonging to Allygus, Colladonus, Empoasca, Erythronneura, Graphocephala, Homalodisca, Orientus, Scaphoideus, and Typhlocyba, which were collected in the University Park Campus of Pennsylvania State University, USA [**38**].

However, it has not been shown whether these leafhoppers transmit the pathogen. Work by Rosa et al. [**39**] revealed that three out of 30 American elm seedlings exposed to individuals of Lepyronia quadrangularis, P. spumarius, and Latalus sp., captured from an EY-infected red elm tree in the Pennsylvania State University campus, were infected by 'Ca. Phytoplasma ulmi'. S. luteolus is not present in Europe, whereas Carraro et al. [**37**] provided evidence that Macropsis mendax is vectoring 'Ca. Phytoplasma ulmi' in northern Italy. There are no reports on the role of M. mendax in the natural transmission of 'Ca. Phytoplasma ulmi' in other European countries as well as on its transmission efficiency. In Germany, the finding of naturally infected elm trees at elevations beyond 750 m suggests the presence of a vector different from M. mendax, since this insect occurs only to an altitude of 400 m [**13**].

'Ca. Phytoplasma ulmi' was also detected by PCR in individuals of Hyalesthes luteipes, and Iassus scutellaris, Allygidius furcatus, and Cixus sp., captured from EY-affected elm trees in Serbia and France, respectively [**40,41**]. However, no transmission experiments were carried out in both countries. 'Ca. Phytoplasma ulmi' seems to be transmitted only among elm trees, because other plants growing near EY-infected elm trees have so far not been found infected by the pathogen. It is also transmitted by natural root grafts. Although in nature this pathogen preferentially infects Ulmus plants, it was also detected in Zelkova serrata (Japanese zelkova) and Ugni molinae (murta) plants showing symptoms of yellowing and witches' brooms in central Italy and Chile, respectively [**42,43**].

Several elm genotypes that are resistant to the Dutch elm disease have been examined for EY resistance or tolerance by graft-inoculation experiments. The inoculated trees of these genotypes greatly differed in their response to 'Ca. Phytoplasma ulmi' [**44**]. Diseased trees of 'Frontier', 'Pathfinder', and 'Patriot' showed foliar yellowing and reddening, witches' brooms, reduced terminal growth, and stunting. Since phloem necrosis and death were not observed, these elms may be rated as tolerant. Trees of 'Pioneer' were considerably more affected; most of them became infected and died. Only two out of 20 inoculated 'Prospector' trees became infected, one of which died, whereas none of the inoculated 'Homestead' trees was infected. The latter trees showed localized phloem necrosis as a defense reaction that prevented spreading of the pathogen, suggesting thus resistance of

'Homestead' [**44**]. Trees of 'American Liberty' proved to be highly susceptible to 'Ca. Phytoplasma ulmi', which is a finding that is confirmed also under natural infection conditions [**30**]. The clonal cultivars Independence, New Harmony, Valley Forge, and the triploid putative hybrid cultivar Jefferson were susceptible as well [**45**]. Since EY is lethal to North American species, while some Eurasian genotypes are tolerant or resistant, it seems that the disease is of European origin [**30**]. Tolerance or resistance is usually expected in regions where the pathogen and its natural hosts co-evolved.

Currently, PCR assays are most widely used for the detection of 'Ca. Phytoplasma ulmi'. Primers differing in specificity have been developed and directed to either ribosomal or non-ribosomal DNA sequences. The sensitivity of detection can be increased by the use of nested PCR. For a review on primer sequences and primer combinations for the detection of 'Ca. Phytoplasma ulmi' and other phytoplasmas infecting elm, see [**6,46,47**]. A real-time PCR (qPCR) assay using nonribosomal primers in combination with TaqMan minor-groove-binder (MGB) probe chemistry was developed for the specific detection of 'Ca. Phytoplasma ulmi'. This assay allowed detection of the pathogen in affected trees and several leafhoppers, but it did not show cross reactivity with the DNA of closely related phytoplasmas belonging to other 16SrV subgroups [**38**]. In Germany, 'Ca. Phytoplasma ulmi' was nationwide surveyed by using a universal 16Sr DNA-based qPCR assay and a newly developed 'Ca. Phytoplasma ulmi'-specific qPCR Taq-Man assay directed to 16S–23S spacer region sequences. Both assays enabled the detection of 'Ca. Phytoplasma ulmi' in the same number of elm trees (1801 trees out of 6486 trees examined), although the pathogen specific assay showed a slightly lower sensitivity as shown by different cycle threshold (Ct) values. In addition, the universal phytoplasma assay enabled the identification of two elm trees each infected by a different 16SrV-C phytoplasma strain [**13**]. Scots elm trees showed a higher phytoplasma titer in comparison to the other two elm species examined. However, the phytoplasma titer was not correlated to the presence of EY symptoms [**13**]. In most trees, a high titer of 'Ca. Phytoplasma ulmi' was identified, as evidenced by a Ct value ≤ 28, which corresponds to a titer of 10^6 per gram of phloem tissue at least. In addition, all parts of an infected Scots elm tree such as roots, trunk at both ground and aerial levels, branches, leaves, and buds from the top and bottom of the tree were heavily colonized by 'Ca. Phytoplasma ulmi' throughout the year. The survey also revealed that the pathogen was not homogeneously distributed throughout Germany. Sites with a high number of infected trees occurred in east, south, and central Germany, whereas only a few sites with low infection rates were found in the remaining parts of the country. Infection rates did not vary considerably among sites located at different altitude levels [**13**]. On the basis of groEL and imp gene sequences, 29 and 74 genotypes, respectively, have been identified among 'Ca. Phytoplasma ulmi' strains infecting elms in Germany [**24**]. In phylogenetic analysis, genotypes obtained from Scots elm trees clustered together, while those from European white elm and European field elm trees did not. A regional distribution pattern was evident only for a few genotypes [**24**].

2.2. Alder

Alder yellows (ALY) is a phytoplasma disease affecting several Alnus (alder) species. This disease has been described in several European countries such as Germany, Italy, Austria, France, Switzerland, Hungary, Serbia, Slovenia, Lithuania, Serbia, Montenegro, Macedonia, and Poland (for reviews, see [**48,49**]). ALY has also been recorded in the Washington State (USA) and Ontario, Canada, on A. rubra (red alder) [**16,50,51**]. However, molecular identification and classification of the phytoplasma(s) infecting A. rubra in the Washington State have not yet been reported.

Figure 3. Pronounced shoot proliferation at the base of trunk of a several-year-old alder (Alnus glutinosa) tree affected by alder yellows disease.

Symptoms of ALY are yellowing, premature autumn coloration, sparse foliage, small leaves, die-back, and decline. Shoot proliferation may occur at the base of trunk of affected trees (**Figure 3**). Latent infections have been observed as well [**50,52,53,54**]. Lederer and Seemüller [**50**] examined 500 A. glutinosa (European alder) trees for phytoplasma infections using DAPI (4'-6-diamidino-2-phenylindole) fluorescence methods. All trees older than 5 years tested positive for phytoplasma infections. However, the majority of infected trees were non-symptomatic. Usually, non-symptomatic trees revealed a higher colonization density than symptomatic ones. The colonization density was always higher in leaves and young twigs than in other parts of the trees. The phytoplasma distribution within the affected phloem tissue was uneven, with some sieve tubes filled by phytoplasmas, while others completely empty or containing only a few phytoplasmas. Similar findings were obtained by examining A. incana trees [**50**]. Under natural infection conditions, high detection rates ranging from 85 to 100% of ALY phytoplasma infections in A. glutinosa were also recorded by PCR [**53,55,56,57,58**]. No phytoplasma infections have been recorded in A. viridis either by DAPI fluorescence methods or PCR assays [**50,57**]. This lack of infections may be ascribed to the host preference or unfavorable conditions of the insect vector(s).

The ALY phytoplasma is taxonomically assigned to subgroup 16SrV-C within the EY group [**19**]. It has a 99.7%–100% 16S rDNA sequence similarity with other members of the mentioned subgroup such as flavescence dorée (FD), Palatinate grapevine yellows (PGY), and spartium witches' broom (SpaWB) agents and phytoplasmas infecting Clematis vitalba in Italy and the Balkans, and hemp dogbane (Apocynum cannabinum) in New York state, respectively [**19,59,60**]. ALY phytoplasma strains originating from different parts of Europe are identical at the 16S rDNA and 16S–23S rDNA spacer region sequence level. However, studies through ribosomal protein gene and nonribosomal loci analyses showed a considerable molecular variability among them. Some strains proved to be identical or near identical to either FD phytoplasma strains or PGY phytoplasma strains [**19,20,49,53,55,56,61,62**]. Many different strains of the ALY phytoplasma have been identified through map-based gene analysis in singly and multiply infected alder trees throughout Europe. These strains, phylogenetically, do not form a homogeneous group but are scattered in various clusters that comprise FD, Palatinate grapevine yellows (PGY), and SpaWB phytoplasma strains. Some ALY phytoplasma strains were more closely related to FD and/or PGY phytoplasma strains than to other ALY phytoplasma strains. In addition, clustering was not linked to the geographic origin of the strains [**20,53,55,61**]. Phylogenetic analyses of the concatenated tuf, rplV-rpsC, rplF-rplR, map, and uvrB-degV gene sequences showed that strains infecting alder trees in Canada (two strains) form together with strain HD1 of hemp dogbane (Apocynum cannabinum) phytoplasma a monophyletic discrete genetic cluster clearly separated from the cluster comprising ALY phytoplasma strains from Europe [**51**].

The ALY phytoplasma is naturally spread by the leafhopper Oncopsis alni [**52**], which is very common on alder trees and is characterized by an oligophagous feeding behavior. The distribution of this leafhopper in Europe corresponds to that of ALY [**52**]. ALY phytoplasma infections were also recorded in individuals of the psyllid Psylla alni and leafhoppers Orientus ishidae, Allygus mixtus, and Allygus modestus, which were all collected from ALY-infected alder trees. However, all attempts to transmit the ALY phytoplasma to healthy alder plants using ALY phytoplasma-infested P. alni psyllids failed, whereas successful transmission was obtained with ALY phytoplasma-infested O. ishidae, A. mixtus, and A. modestus leafhoppers [**52,58,63**]. It is well known that ALY phytoplasma strains are also responsible of the Palatinate grapevine yellows (PGY) disease in Germany. Spreading of these strains from alder to grapevine occurs through O. alni, which feeds occasionally on grapevine [**64,65,66**]. Thus, infected alder trees growing in proximity of vineyards may act as pathogen reservoirs. In addition, due to the close phylogenetic relationship of ALY phytoplasma with the FD agent, diseased alder trees may also function as pathogen reservoirs in areas where FD disease occurs. In these cases, some ALY phytoplasma strains may be transmitted to grapevine by occasional grapevine-feeding vectors leading to FD disease, which is then spread from grapevine to grapevine by Scaphoideus titanus, the strictly ampelophagous vector of FD phytoplasma strains. S. titanus is not known to occur in the Palatinate [**20,55,58,61**].

A recent study by Jurga and Zwolińska [**49**] has shown that diseased common mugwort (Artemisia vulgaris) plants showing witches'-broom symptoms, grown next to an ALY-affected alder (A. glutinosa) tree in Southern Poland, were infected by a 16SrV-C subgroup phytoplasma that had the same 16S rDNA sequences of the phytoplasma present in ALY-affected alder trees. Therefore, it is possible that common mugwort serves as alternative host of the ALY phytoplasma in eastern Europe [**49**].

As mentioned above, in nature, a high percentage of alder trees infected by ALY agent do not develop obvious symptoms. The reasons were elucidated by Berges and Seemüller [**67**] through graft-inoculation of healthy alder plants with icoculum from both differently affected and symptomless alder trees and observation of symptom development for five years. It resulted that the severity of symptoms shown by inoculated plants corresponded to that of plants from which icoculum was taken. Thus, it was concluded that ALY phytoplasma may affect alder by inducing severe symptoms, but avirulent strains occurring within this phytoplasma are responsible for the latent infections that are widespread in Europe. It is also possible that avirulent strains mediate cross-protection and suppress severe strains, since infected alder trees remain symptomless over time in spite of the presence of an abundant and efficient insect vector such as O. alni. Differences in strain virulence are also known for other phytoplasmas, including those causing apple proliferation, European stone fruit yellows, ash yellows and aster yellows diseases, whereas antagonistic interactions among different strains of the same pathogen have been observed in many instances [**68,69,70**]. On the other hand, it is also possible that symbiotic microorganisms of alder such as Frankia alni may compensate for the detrimental effects of phytoplasma infections [**54**]. However, it is unknown whether these microorganisms are able to colonize systemically the phloem tissue of alder trees.

2.3. Ash

Ash yellows (AshY) is a prominent phytoplasma disease of Fraxinus (ash) species known to occur widely in North America [**71**]. This disease is caused by the AshY agent 'Ca. Phytoplasma fraxini', which is a pathogen of the AshY group or 16SrVII group, subgroup 16SrVII-A [**72**]. Although AshY may have occurred in the northeastern USA states since the 1930s, it was not considered as a major forest disease until the early 1980s. This late recognition was due to the difficulties met in identifying phytoplasma diseases and to the inconsistency with which infected trees show specific symptoms [**71**]. In North America, AshY occurs in twelve ash species and numerous intraspecific and interspecific ash genotypes [**71**]. Most of the affected species are native to North America. A 16SrI phytoplasma has been identified in yellows-diseased trees of F. excelsior (European ash) in Poland and F. uhdei (Urapan ash) in Colombia [**73,74**]. In the latter country, other studies showed that F. uhdei trees showing typical AshY symptoms were infected by 'Ca. Phytoplasma fraxini' [**75,76**]. In France, ash trees showing die-back symptoms were infected with 'Ca. Phytoplasma fragariae' (subgroup 16SrXII-E) [**77**].

AshY causes reduced growth, loss of apical dominance, suppressed root development, premature flowering, shoot proliferation, and witches' brooms (**Figure 4**). Light green and deformed leaves are also common. Highly susceptible taxa show mainly die-back, shoot proliferation, loss of apical dominance, and premature death. Short roots and necrosis of rootlets that precedes wilting and death are predominant in diseased F. americana (white ash) trees. Infected ash trees are highly sensitive to frost damages, which appear as cracks of the bark at the base of the trunk [**10**]. Histological symptoms which are more readily observed in the roots than in the aerial parts of affected plants, include autofluorescent sieve tubes and pathological sclerenchyma. Autofluorescent sieve tubes are sieve tubes that fluoresce without the addition of any reagent when exposed to UV radiation. Such sieve tubes are presumed to be non-functional and often collapse. Pathological sclerenchyma is a distinct parenchyma tissue adjacent to infected sieve tubes, which is characterized by prominently thickened and lignified cell walls, which greatly differs from the sclerotic parenchyma that is often seen in healthy phloem. Stomatal closure and high diffusive resistance are also known to occur in the leaves of affected white ash trees [**10**].

Witches' brooms are regarded as specific symptoms of the disease, whereas other symptoms are non-specific, since they can be induced by several other biotic and abiotic factors. A disease incidence of ≥50% has been observed in some F. americana stands in northern USA states and in a F. velutina (velvet ash) stand in Utah. Incidence ranging from of 3 to 27% was recorded in F. pennsylvanica (green ash) in Iowa and Wisconsin (for review, see [**71**]). 'Ca. Phytoplasma fraxini' was also detected in non-symptomatic ash trees [**71,78**]. In Colombia, more than 53,000 F. uhdei trees showing typical AshY symptoms were recorded in several areas of the Andes; many of them died.

Figure 4. Witches' brooms at the root collar of an ash yellows-affected velvet ash (Fraxinus velutina) tree (courtesy W.A. Sinclair, Department of Plant Pathology, Cornell University, Ithaca, NY, USA).

Phytoplasma infections were detected in all symptomatic trees examined (100 trees) either through DAPI fluorescence methods or PCR assays. Phytoplasma infections were strictly related to the presence of symptoms [**76**]. The phytoplasma identified in diseased F. uhdei trees, the AshY agent 'Ca. Phytoplasma fraxini', was also transmitted to Catharanthus roseus (periwinkle) through dodder (Cuscuta sp.) [**76**]. In a study conducted in France to test the presence of the ascomycete fungus Hymenoscyphus fraxineus in ash trees displaying symptoms of die-back, a relatively large proportion of symptomatic trees (27%) proved to be not infected by such fungal pathogens [**79**]. This finding prompted a further study aimed at verifying the presence of phytoplasma infections in affected trees [**77**]. 'Ca. Phytoplasma fragariae' was identified in 5% of the examined ash trees (13 out of 260 trees). Four phytoplasma-positive trees were also infected by the fungus H. fraxineus. However, further data are necessary to validate the role of phytoplasma infections in the die-back disease affecting ash in France.

Sinclair and Griffiths [**68**] investigated the virulence of 12 'Ca. Phytoplasma fraxini' strains, originating from different areas of North America, by graft-inoculating F. pennsylvanica seedlings and periwinkle plants and monitoring symptom development under greenhouse conditions. In graft-inoculated plants, the examined strains caused symptoms that ranged considerably from imperceptible to severe. The same study [**68**] revealed the occurrence of interference among strains of the same taxon. Strains of 'Ca. Phytoplasma fraxini' differing in virulence have also been recorded under natural infection conditions [**80**]. These strains greatly differed in aggressiveness as indicated by the different rates of growth suppression, reduction of foliar greenness, and frequency of witches' brooms that they caused. However, specific interactions between strain and cultivar were not discovered [**80**]. Several investigations have shown that F. pennsylvanica and F. velutina are more tolerant than F. americana to 'Ca. Phytoplasma fraxini' and that heritable intraspecific variation in tolerance occurs in F. pennsylvanica and is supposed to be present also in other ash species [**81,82**].

In graft-inoculation trials, the shoot growth of F. americana and F. pennsylvanica was suppressed earlier than that of F. velutina. In addition, the highest growth losses in height, stem diameter, and root volume were recorded for F. americana followed in decreasing order by F. pennsylvanica and F. velutina. The growth of diseased F. velutina scions on F. americana rootstocks was severely suppressed in comparison with that of diseased own-rooted F. velutina. On the other hand, diseased F. americana scions grafted on F. velutina rootstocks registered a significantly lower growth suppression than diseased own-rooted F. americana. When witches' brooms arisen from F. americana trees were grafted on F. velutina rootstocks, they continued to maintain their original form but did not proliferate new shoots. These findings imply that although tolerant rootstocks may alleviate the impact of 'Ca. Phytoplasma fraxini' on scions, the management of AshY through tolerant genotypes requires tolerance in both scions and rootstocks [**82**]. Among the ash cultivars tested by graft inoculation to evaluate their responses to 'Ca. Phytoplasma fraxini', F. pennsylvanica cvs. Bergeson, Dakota Centennial, and Patmore and F. americana cv. Autumn Applause were least affected, i.e., they sustained less growth and foliar color depression [**80**].

There are no reports concerning insect vectors involved in the natural spread of AshY. A previous work showed that field-captured Paraphlepsius irroratus and P. spumarius individuals transmitted acquired phytoplasmas to ash seedlings [**10**]. However, the molecular identities of these phytoplasmas were not determined, and further experiments using progenies of these insects failed to transmit 'Ca. Phytoplasma fraxini'. In the work by Hill and Sinclair [**83**], 33 taxa of leafhoppers were collected in two locations where AshY incidence was high, in New York State, and tested by PCR assays for phytoplasma infections. The most abundant genus was Scaphoideus. 'Ca. Phytoplasma fraxini' infections were recorded in some individuals of Scaphoideus spp. Therefore, these insects should be further examined to prove their transmission ability [**83**].

2.4. Conifers

After a few findings that were not confirmed by other studies on the presence of phytoplasma infections in various gymnosperms using electron microscope observations, phytoplasmas were detected and molecularly identified in several conifers over the last few years. Schneider et al. [**84**] reported on the presence of a previously undescribed taxon, 'Ca. Phytoplasma pini', in Pinus sylvestris (Scots pine) and P. halepensis (Aleppo pine) trees in Germany and Spain. Affected trees showed symptoms of yellowing, stunting, needle dwarfing, and shoot proliferation. In addition, affected P. sylvestris trees were characterized by the presence of ball-like structures in the canopy, resulting in the combination of proliferation of shoots with dwarfed needles. 'Ca. Phytoplasma pini' was detected in symptomatic and non-symptomatic parts of P. sylvestris and P. halepensis trees as well as in some neighboring non-symptomatic trees of both species.

Following the finding of Schneider et al. [**84**], 'Ca. Phytoplasma pini' and related strains were identified in several other Pinus species as well as in Abies procera (noble fir), Tsuga canadensis (Canadian hemlock), and Picea pungens (Colorado blu spruce) in Poland, the Czech Republic, Croatia, Lithuania, and Maryland (USA). In all affected species, the symptoms were similar to those observed in Germany and Spain, whereas ball-like structures were recorded other than in P. sylvestris also on P. mugo (mountain pine) in Lithuania (**Figure 5**). Abnormal shoot branching resulting in typical witches' brooms were shown by pine trees in Maryland [**85,86,87,88,89,90**]. Furthermore, 'Ca. Phytoplasma pini' was identified in China

[**91**] in Taxodium distichum var. imbricarium (pond cypress) showing symptoms of shoot proliferation, little leaf, and leaf necrosis.

Figure 5. Phytoplasma-infected Scots pine (Pinus sylvestris) (left) and mountain pine (P. mugo) (right) trees showing ball-like structures.

'Ca. Phytoplasma pini' is a member of the pine shoot proliferation group or 16SrXXI group, subgroup 16SrXXI-A. A 'Ca. Phytoplasma pini'-related strain detected in a witches'-broom affected pine tree in Maryland, strain MDPP, was assigned to a new subgroup, designated as 16SrXXI-B within the group 16SrXXI, on the basis of virtual RFLP pattern similarity coefficient values [**90**].

A phytoplasma of the X-disease group (16SrIII group) was detected in Cupressus sp. (cypress) trees showing witches'-brooms, stunting, and fasciation in Italy [**92**] and in Picea abies (Norway spruce) and P. glauca (white spruce) trees with symptoms similar to those associated with 'Ca. Phytoplasma pini' in Poland [**86**]. 16SrI phytoplasmas were identified in diseased P. pungens and Larix sp. (larch) trees in Poland and Ukraine, respectively [**86,93**]. The main symptoms observed in diseased larch trees were dwarfing and the proliferation of needles. Therefore, the name "larch dwarfed needle proliferation (LDNP)" was given to the pathogen identified in this species [**93**]. Valiunas et al. [**89**] examined 300 P. sylvestris and P. mugo trees in Lithuania for phytoplasma infections using PCR assays. Of the examined trees, 80% proved to be phytoplasma-positive, of which 98% were infected by Ca. Phytoplasma pini-related strains, whereas the remaining trees harbored a subgroup 16SrI-A phytoplasma. A 'Ca. Phytoplasma phoenicium'-related strain, subgroup 16SrIX-F, was found in diseased Juniperus occidentalis (western juniper) plants in Oregon (USA) showing symptoms of shoot proliferation, little leaves, shortened internodes, and ball-like structures. The percentage of affected plants was about 1% [**94**]. In India, yellows-diseased trees of Araucaria heterophylla (Norfolk Island pine) were infected with a 'Ca. Phytoplasma trifolii'-related strain, which is a member of the clover proliferation group or 16SrVI group. The incidence of the disease was about 15% [**95**].

Phytoplasma titer in gymnosperms is usually so low that infections could be detected only through nested PCR assays. In addition, the mechanism by which phytoplasmas colonize systemically the affected gymnosperm plants is unknown, since pore sizes in the sieve cells of these plants being too small may hinder phytoplasma movement. Valiunas et al. [**96**] developed direct PCR methods based on primer pairs directed to 16S rDNA and tuf gene sequences which allowed detecting 'Ca. Phytoplasma pini'-related strains infecting pine trees in Lithuania with great sensitivity and specificity.

The amplified sequences contain molecular markers which are suitable for distinguishing closely related strains of this pathogen. The mentioned PCR methods, which did not allow the amplification of DNA from the pine-infecting phytoplasma strain MDPP recorded in Maryland, are useful tools for detecting and identifying 'Ca. Phytoplasma pini' on large-scale studies. A draft genome sequence of the 'Ca. Phytoplasma pini'-related strain MDPP has recently been reported [**97**]. This sequence, which consists of 474,136 bases, will facilitate comparative genomic studies of phytoplasmas infecting conifer plants.

2.5. Sandal

Sandal spike (SSD) is a serious disease of Santalum album (sandal) that is widespread in India, especially in Karnataka, Tamil Nadu, and Kerala states [**98,99,100,101**]. The disease was first described in the late 1890s and was supposed to be of viral origin. Later, further investigations using fluorescence and electron microscopy and symptom remission achieved by tetracycline treatments proved its phytoplasmal etiology (for references, see [**48**]). SSD has spread gradually during the last decades, devastating large forest areas and threatening the sandal industry of southern India. The most characteristic symptom of SSD is small and extremely narrow leaves, which stand out stiffly from the shoots, giving them a spike-like appearance. Other symptoms include yellowing and decline. Severely affected trees die within a two- or three-year period (**Figure 6**). Disease incidences reaching up to 33 and 55% were recorded in southern Karnataka [**100,102**]. SSD is caused by 'Ca. Phytoplasma asteris', subgroup 16SrI-B [**98,103**]. In SSD-affected sandal trees in the Marayoor sandalwood reserve in Kerala, a new strain of the 16SrI group, which on the basis of the virtual RFLP pattern similarity coefficient values differed from the phytoplasmas of all 16SrI subgroups, has been identified. This strain represents a new 16SrI subgroup [**104**].

In addition, a survey conducted in the same reserve has shown the presence in SSD-affected sandal trees of a phytoplasma of the rice yellow dwarf group or 16SrXI group, subgroup 16SrXI-B (**Figure 1**). This phytoplasma was recorded either in singly infected trees or in doubly infected trees with 'Ca. Phytoplasma asteris' [**105**]. SSD phytoplasma was transmitted to periwinkle, which induced witches' brooms, and from this host back to sandal through dodder (Cuscuta subinclusa) [**106**]. However, attempts to transmit SSD phytoplasma from diseased sandal trees to Stachytarpheta indica, a weed common in sandal plantations, through dodder (C. subinclusa) failed [**100**]. SSD is naturally spread by the leafhopper Coelidia indica, which was formerly known as Jassus indicus [**1**].

Figure 6. Decline, small and narrow leaves in sandal spike-infected Santalum album (sandal). Right, healthy plant.

2.6. Eucalyptus

Eucalyptus little-leaf (ELL) is a phytoplasma disease affecting several eucalyptus (Eucalyptus spp.) species that was first observed in India in 1971 and was thought to be caused by the virus [**107**]. Later, ELL was observed by other Indian researchers, who discovered phytoplasma infections in diseased eucalyptus trees by using electron and light microscope observations [**48**]. The disease has also been recorded in other countries including Italy, China, Sudan, Iran, and Brazil [**48,108,109,110,111,112**].

The symptoms described from the various geographic areas and different Eucalyptus spp. are similar. They include small leaves, yellowing, and proliferation from axillary buds that confers a broom-like appearance. Diseased trees or their symptomatic parts do not set fruits, are stunted, and die-back. Severely affected trees decline (**Figure 7**). A 16SrV phytoplasma was identified in ELL-affected trees in Italy [**108**]. However, in the affected trees, the phytoplasma concentration was so low that infections could be identified only by nested PCR assays. In addition, 16SrI-B and 16SrI-C phytoplasmas were identified in diseased eucalyptus trees in Italy, whereas a 16SrI-B phytoplasma was recorded in diseased eucalyptus (E. camaldulensis) trees in Iran [**48,110**]. In the latter country, phytoplasmas of other taxonomic groups were recorded as well. Azimi et al. [**111**] reported on the detection of a 'Ca. Phytoplasma aurantifolia'-related strain, subgroup 16SrII-D in yellows-diseased E. camaldulensis trees in Ahvaz, southwestern Iran. The percentage of affected trees was approximately 5%. Baghaee-Ravari et al. [**112**] reported on the detection of a 'Ca. Phytoplasma solani'- related strain, subgroup 16SrXII-A in yellows-diseased E. camaldulensis trees in Fars and Khozestan provinces of Iran. This strain was detected through nested PCR assays in 14 out of 22 (approximately 64%) symptomatic trees examined. A 'Ca. Phytoplasma aurantifolia'-related strain, subgroup 16SrII-C, was detected in diseased E. urophylla trees in Brazil. Fourteen out of 22 (approximately 64%) symptomatic eucalyptus trees examined using nested PCR assays tested positive [**109**]. The identity of phytoplasmas infecting eucalyptus in India, China, and Sudan has never been determined by molecular methods.

Figure 7. Small leaves, yellowing, and shortened internodes on eucalyptus (Eucalyptus spp.) shoots affected by eucalyptus little-leaf. Healthy shoot on the left of the left panel.

3. DISCUSSION

With few exceptions, phytoplasma diseases of forest trees have been less extensively studied than those affecting fruit trees. One reason may be that until 10–15 years ago, the majority of the research carried out in forest pathology was mainly focused on fungal diseases and insect damages. A second reason is the technical difficulty in detecting and identifying phytoplasmas in forest trees due to the low phytoplasma titer and the presence of non-specific symptoms in affected plants. Research on the role of phytoplasmas as cause of diseases of forest trees has only in the last few years been intensified, after sensitive and specific detection methods, greatly based on PCR technology, became available. By using these methods, phytoplasma infections have been detected for the first time in several forest trees, highlighting that phytoplasma diseases may play in forest trees the same role as in fruit trees. Therefore, the number of phytoplasma diseases of forest trees and associated phytoplasmas has recently been increased considerably [**48**].

Many phytoplasma diseases of forest trees are each associated with taxonomically distinct phytoplasmas, which induce identical symptoms in a given plant and are present in the same areas or different countries. In addition, a given plant host may be singly or multiply infected with various phytoplasmas. This seems common in woody plants, which have more opportunities to be visited by insect vectors over time. Phytoplasmas belonging to various taxonomic groups and subgroups have been identified in naturally infected elm, ash, conifer, sandal, and eucalyptus trees (**Figure 1**).

In elm, in addition to the specific agent 'Ca. Phytoplasma ulmi' (16SrV group, subgroup 16SrV-A), phytoplasmas of the 16SrI group, subgroup 16SrI-B, 16SrVI group, subgroup 16SrVI-C, and 16SrXII group, subgroups 16SrXII-A and 16SrXII-E were identified. Subgroups 16SrV-B and 16SrV-C phytoplasmas were also detected in elm. While the pathological role of the EY agent 'Ca. Phytoplasma ulmi' has been well established by numerous graft-inoculation experiments, the pathogenicity of other phytoplasma taxa present in elm has to be verified. In addition, the current knowledge on the latter phytoplasmas are insufficient to draw conclusions that EY can be considered as a disease induced by various phytoplasmas. Only one phytoplasma taxon, the ALY agent (16SrV group, subgroup 16SrV-C), has been recorded in naturally infected alder trees. Strains of this pathogen are highly homogeneous at the rDNA sequence level, but they greatly differ on the basis of less-conserved gene sequences. In particular, many different strains of the ALY phytoplasma have been identified through map-based gene analysis in singly and multiply infected alder trees. These strains phylogenetically do not form a homogeneous group but are scattered among various clusters, which comprise strains of agriculturally important phytoplasmas such as FD and PGY agents.

In addition, the presence of avirulent strains occurring within this taxon may be responsible for the latent infections as has been shown by graft-inoculation experiments [**67**]. AshY-affected ash trees were found to harbor only 'Ca. Phytoplasma fraxini' (16SrVII group, subgroup 16SrVII-A), whereas ash trees with symptoms of decline and die-back were infected with phytoplasmas of the 16SrI group and 16SrXII group (subgroup 16SrII-E), respectively. The etiological role of the AshY agent 'Ca. Phytoplasma fraxini' has been clearly demonstrated by numerous graft-transmission experiments and field observations. However, more data are needed to clearly elucidate the role of 16SrI and 16SrII-E phytoplasmas in the decline and die-back diseases of ash. 'Ca. Phytoplasma fraxini'-infected ash trees have been recorded only in North America and Colombia. Phytoplasmas belonging to several taxonomic groups such as 16SrXXI (subgroups 16SrXXI-A and 16SrXXI-B), 16SrI (subgroups 16SrI-A and 16SrI-B), 16SrIII, 16SrVI, and 16SrIX (subgroup 16SrIX-F) have been detected in several yellows-diseased conifer species. Among these, the most frequently detected agent is 'Ca. Phytoplasma pini' (16SrXXI group, subgroup 16SrXXI-A). This pathogen seems particularly widespread in Europe. In sandal trees affected by SSD, in addition to the AY agent 'Ca. Phytoplasma asteris' (16SrI group, subgroup 16SrI-B), a new strain of the 16SrI group, which on the basis of virtual RFLP pattern similarity coefficient values differed from phytoplasmas of all 16SrI subgroups, as well as a 16SrXI phytoplasma have been identified [**104,105**]. Taxonomic groups of phytoplasmas infecting eucalyptus trees include 16SrV, 16SrI (subgroups 16SrI-B and 16SrI-C), 16SrII (subgroups 16SrII-C and 16SrII-D), and 16SrXII (subgroup 16SrXII-A).

4. CONCISE SUMMARY

Phytoplasma infections affecting forest trees have significant economic and ecological importance globally. In recent years, researchers have uncovered multiple phytoplasma taxa that are linked to different illnesses causing yellowing and decline in forest trees. These phytoplasma taxa can be found in trees that are infected with one, two, or multiple types of phytoplasma. Nevertheless, despite the extensive research conducted on phytoplasma diseases of forest trees, such as EY and AshY, there remains a dearth of information regarding insect vectors, plant host range, strain virulence, pathogenicity, and host tolerance and resistance. Understanding these characteristics forms the foundation for effective disease management. Specifically, additional investigation is necessary to clarify the function of phytoplasmas in trees that show no symptoms and how it relates to the absence of symptom manifestation. Furthermore, it is necessary to conduct further research on the causal involvement of several 'non-specific' phytoplasma species, which have been observed in forest trees but lack pathological study data.

5. AT A GLANCE

This chapter presents updated information on some significant phytoplasma diseases affecting forest trees, with a specific emphasis on the phytoplasma species linked with these diseases. Research on phytoplasma infections in forest trees has been less comprehensive compared to the study of same diseases in fruit trees. Research on the function of phytoplasmas as the cause of illnesses in forest trees has recently been expanded due to the availability of sensitive and precise detection technologies, primarily based on PCR technology. Different phytoplasma species have been found in naturally infected elm, ash, conifer, sandal, and eucalyptus trees. However, only one phytoplasma species has been observed in naturally infected alder trees. Nevertheless, there is limited data available on insect vectors, plant host range, strain virulence, pathogenicity, and host tolerance and resistance for the majority of the diseases that have been studied. Understanding these factors is essential for effective disease management. Specifically, additional research is necessary to elucidate the role of phytoplasmas in trees that show no symptoms. Furthermore, it is necessary to conduct further research on the causal function of several "non-specific" phytoplasma species, which have been observed in forest trees, but for which there is a lack of pathological study data.

REFERENCES

1. Weintraub, P.G.; Beanland, L. Insect vectors of phytoplasmas. Ann. Rev. Entomol. 2006, 51, 91–111. **[Google Scholar]** **[CrossRef]**
2. IRPCM Phytoplasma/Spiroplasma Working Team—Phytoplasma Taxonomy Group. 'Candidatus phytoplasma', a taxon for the wall-less, non-helical prokaryotes that colonize plant phloem and insects. Int. J. Syst. Evol. Microbiol. 2004, 54, 1243–1255. **[Google Scholar]** **[CrossRef]** **[PubMed]** **[Green Version]**
3. Martini, M.; Marcone, C.; Lee, I.-M.; Firrao, G. The family Acholeplasmataceae (including phytoplasmas). In The Prokaryotes—Firmicutes and Tenericutes, 4th ed.; Rosenberg, E., DeLong, E.F., Lory, S., Stackebrandt, E., Thompson, F., Eds.; Springer-Verlag: Berlin, Germany, 2014; Volume XXII, pp. 469–504. **[Google Scholar]**
4. Bertaccini, A.; Lee, I.-M. Phytoplasmas: An update. In Phytoplasmas: Plant Pathogenic Bacteria-I. Characterization and Epidemiology of Phytoplasma-Associated Diseases; Rao, G.P., Bertaccini, A., Fiore, N., Liefting, L., Eds.; Springer Nature: Singapore, 2018; pp. 1–29. **[Google Scholar]**
5. Zhao, Y.; Davis, R.E. Criteria for phytoplasma 16Sr group/subgroup delineation and the need of a platform for proper registration of new groups and subgroups. Int. J. Syst. Evol. Microbiol. 2016, 66, 2121–2123. **[Google Scholar]** **[CrossRef]** **[PubMed]**
6. Marcone, C. Elm yellows: A phytoplasma disease of concern in forest and landscape ecosystems. For. Pathol. 2017, 47, e12324. **[Google Scholar]** **[CrossRef]** **[Green Version]**
7. Swingle, R.U. A phloem necrosis of elm. Phytopathology 1938, 28, 757–759. **[Google Scholar]**
8. Forbes, S.A. What is the matter with the elms in Illinois? Ill. Agr. Exp. Sta. Bull. 1912, 154, 1–22. **[Google Scholar]**

9. Garman, H. The elms and their diseases. Kentucky Agr. Exp. Sta. Bull. 1899, 84, 51–75. [**Google Scholar**]

10. Matteoni, J.A.; Sinclair, W.A. Elm yellows and ash yellows. In Tree Mycoplasmas and Mycoplasma Diseases; Hiruki, C., Ed.; University of Alberta Press: Edmonton, AB, Canada, 1988; pp. 19–31. [**Google Scholar**]

11. Conti, M.; D'Agostino, G.; Mittempergher, L. A recent epiphytotic of elm yellows in Italy. In Proceedings of the 7th Congress of the Mediterranean Phytopathological Union, Granada, Spain, 20–27 September 1987; pp. 208–209. [**Google Scholar**]

12. De Jonghe, K.; Deeren, A.-M.; Goedefroit, T.; Ronse, A. First report of 'Candidatu Phytoplasma ulmi' on elm in Belgium. Plant Dis. 2019, 103, 1763. [**Google Scholar**] [**CrossRef**]

13. Schneider, B.; Kätzel, R.; Kube, M. Widespread occurrence of 'Candidatus Phytoplasma ulmi' in elm species in Germany. BMC Microbiol. 2020, 20, 74. [**Google Scholar**] [**CrossRef**] [**PubMed**]

14. Braun, E.J.; Sinclair, W.A. Phloem necrosis of elms: Symptoms and histopathological observations in tolerant hosts. Phytopathology 1979, 69, 354–358. [**Google Scholar**] [**CrossRef**]

15. Lee, I.-M.; Davis, R.E.; Sinclair, W.A.; De Witt, N.D.; Conti, M. Genetic relatedness of mycoplasmalike organisms detected in Ulmus spp. in the United States and Italy by means of DNA probes and polymerase chain reactions. Phytopathology 1993, 83, 829–833. [**Google Scholar**] [**CrossRef**]

16. Mäurer, R.; Seemüller, E.; Sinclair, W.A. Genetic relatedness of mycoplasmalike organisms affecting elm, alder, and ash in Europe and North America. Phytopathology 1993, 83, 971–976. [**Google Scholar**] [**CrossRef**]

17. Marcone, C.; Ragozzino, A.; Seemüller, E. Identification and characterization of the phytoplasma associated with elm yellows in southern Italy and its relatedness to other phytoplasmas of the elm yellows group. Eur. J. For. Pathol. 1997, 27, 45–54. [**Google Scholar**] [**CrossRef**]

18. Seemüller, E.; Marcone, C.; Lauer, U.; Ragozzino, A.; Göschl, M. Current status of molecular classification of the phytoplasmas. J. Plant Pathol. 1998, 80, 3–26. [**Google Scholar**]

19. Lee, I.-M.; Martini, M.; Marcone, C.; Zhu, S.F. Classification of phytoplasma strains in the elm yellows group (16SrV) and proposal of 'Candidatus Phytoplasma ulmi' for the phytoplasma associated with elm yellows. Int. J. Syst. Evol. Microbiol. 2004, 54, 337–347. [**Google Scholar**] [**CrossRef**] [**PubMed**] [**Green Version**]

20. Arnaud, G.; Malembic-Maher, S.; Salar, P.; Bonnet, P.; Maixner, M.; Marcone, C.; Boudon-Padieu, E.; Foissac, X. Multilocus sequence typing confirms the close genetic interrelatedness of three distinct flavescence dorée phytoplasma strain clusters and group 16SrV phytoplasmas infecting grapevine and alder in Europe. Appl. Environ. Microbiol. 2007, 73, 4001–4010. [**Google Scholar**] [**CrossRef**] [**Green Version**]

21. Jović, J.; Cvrković, T.; Mitrović, M.; Petrović, A.; Krnjajić, S.; Toševski, I. New strain of 'Candidatus Phytoplasma ulmi' infecting Ulmus minor and U. laevis in Serbia. Plant Pathol. 2008, 57, 1174. [**Google Scholar**] [**CrossRef**]

22. Jović, J.; Cvrković, T.; Mitrović, M.; Petrović, A.; Krstić, O.; Krnjajić, S.; Toševski, I. Multigene sequence data and genetic diversity among 'Candidatus Phytoplasma ulmi' strains infecting Ulmus spp. in Serbia. Plant Pathol. 2011, 60, 356–368. [**Google Scholar**] [**CrossRef**]

23. Katanić, Z.; Krstin, L.; Ježić, M.; Zebec, M.; Ćurković-Perica, M. Molecular characterization of elm yellows phytoplasmas in Croatia and their impact on Ulmus spp. Plant Pathol. 2016, 65, 1430–1440. **[Google Scholar]** **[CrossRef]**

24. Schneider, B.; Hüttel, B.; Zübert, C.; Kube, M. Genetic variation, phylogenetic relationship and spatial distribution of 'Candidatus Phytoplasma ulmi' strains in Germany. Sci. Rep. 2020, 10, 21864. **[Google Scholar]** **[CrossRef]**

25. Kumar, S.; Stecher, G.; Li, M.; Knyaz, C.; Tamura, K. MEGA X: Molecular evolutionary genetics analysis across computing platforms. Mol. Biol. Evol. 2018, 35, 1547–1549. **[Google Scholar]** **[CrossRef]** **[PubMed]**

26. Jacobs, K.A.; Lee, I.-M.; Griffiths, H.M.; Miller, F.D., Jr.; Bottner, K.D. A new member of the clover proliferation phytoplasma group (16SrVI) associated with elm yellows in Illinois. Plant Dis. 2003, 87, 241–246. **[Google Scholar]** **[CrossRef]** **[Green Version]**

27. Lee, I.-M.; Bertaccini, A.; Vibio, M.; Gundersen, D.E.; Davis, R.E.; Mittempergher, L.; Conti, M.; Gennari, F. Detection and characterization of phytoplasmas associated with disease in Ulmus and Rubus in northern and central Italy. Phytopathol. Mediterr. 1995, 34, 174–183. **[Google Scholar]**

28. Gao, R.; Wang, J.; Shao, Y.-H.; Li, X.-D.; Yang, B.-H.; Chang, W.C.; Zhao, W.-J.; Zhu, S.-F. Molecular identification of a phytoplasma associated with elm witches'-broom in China. For. Pathol. 2011, 41, 355–360. **[Google Scholar]** **[CrossRef]**

29. De Jonghe, K.; Goedefroit, T.; Deeren, A.-M.; Fauche, F.; Steyer, S. A phytoplasma survey reveals the presence of 'Candidatus Phytoplasma fragariae' in Ulmus spp. and Acer pseudoplatanus in Belgium. For. Pathol. 2020, 50, e12635. **[Google Scholar]**

30. Sinclair, W.A. Elm yellows in North America. In The Elms: Breeding, Conservation, and Disease Management; Dunn, C.P., Ed.; Kluwer Academic Publisher: Boston, MA, USA, 2000; pp. 121–136. **[Google Scholar]**

31. Mittempergher, L. Elm yellows in Europe. In The Elms: Breeding, Conservation, and Disease Management; Dunn, C.P., Ed.; Kluwer Academic Publisher: Boston, MA, USA, 2000; pp. 103–119. **[Google Scholar]**

32. Sfalanga, A.; Martini, M.; Surico, G.; Bertaccini, A. Involvement of phytoplasmas in a decline of Ulmus chenmoui in central Italy. For. Pathol. 2002, 32, 265–275. **[Google Scholar]** **[CrossRef]**

33. Braun, E.J.; Sinclair, W.A. Histopathology of phloem necrosis in Ulmus americana. Phytopathology 1976, 66, 598–607. **[Google Scholar]** **[CrossRef]**

34. Braun, E.J.; Sinclair, W.A. Translocation in phloem necrosis-diseased American elm seedlings. Phytopathology 1978, 68, 1733–1737. **[Google Scholar]** **[CrossRef]**

35. Pisi, A.; Marani, F.; Bertaccini, A. Mycoplasma-like organisms associated with Elm witches' broom symptoms. Phytopathol. Mediterr. 1981, 20, 189–191. **[Google Scholar]**

36. Marcone, C.; Ragozzino, A.; Firrao, G.; Locci, R. Detection of elm witches'-broom agent in Basilicata, Southern Italy. Phytopathol. Mediterr. 1994, 33, 194–199. **[Google Scholar]**

37. Carraro, L.; Ferrini, F.; Ermacora, P.; Loi, N.; Martini, M.; Osler, R. Macropsis mendax as a vector of elm yellows phytoplasma of Ulmus species. Plant Pathol. 2004, 53, 90–95. **[Google Scholar]** **[CrossRef]**

38. Herath, P.; Hoover, G.A.; Angelini, E.; Moorman, G.W. Detection of elm yellows phytoplasma in elms and insects using real-time PCR. Plant Dis. 2010, 94, 1355–1360. [**Google Scholar**] [**CrossRef**] [**Green Version**]

39. Rosa, C.; McCarthy, E.; Duong, K.; Hoover, G.; Moorman, G. First report of the spittlebug Lepyronia quadrangularis and the leafhopper Latalus sp. as vectors of the elm yellows associated phytoplasma, Candidatus Phytoplasma ulmi in North America. Plant Dis. 2014, 98, 154. [**Google Scholar**] [**CrossRef**] [**PubMed**]

40. Boudon-Padieu, E.; Larrue, J.; Clair, D.; Hourdel, A.; Jeanneau, A.; Sforza, R.; Collin, E. Detection and prophylaxis of elm yellows phytoplasma in France. Invest. Agrar. Sist. Recu. For. 2004, 13, 71–80. [**Google Scholar**]

41. Jović, J.; Cvrković, T.; Mitrović, M.; Petrović, A.; Krstić, O.; Krnjajić, S.; Toševski, I. Genetic variability among 'Candidatus Phytoplasma ulmi' strains infecting elms in Serbia and survey of potential vectors. In COST Action FA0807. Proceedings of the Current Status and Perspectives of Phytoplasma Disease Research and Management, Sitges, Spain, 1–2 February 2010; Bertaccini, A., Laviña, A., Torres, E., Eds.; IRTA: Cabrils, Spain, 2010; p. 18. [**Google Scholar**]

42. Romanazzi, G.; Murolo, S. 'Candidatus Phytoplasma ulmi' causing yellows in Zelkova serrata newly reported in Italy. Plant Pathol. 2008, 57, 1174. [**Google Scholar**] [**CrossRef**]

43. Arismendi, N.; Andrade, N.; Riegel, R.; Zamorano, A.; Fiore, N. Molecular identification of 16SrIII-J and 16SrV-A phytoplasmas in Ugni molinae and Amplicephalus curtulus. In Proceedings of the 20th Chilean Congress of Plant Pathology, Santiago, Chile, 19 November–1 December 2011; pp. 11–12. [**Google Scholar**]

44. Sinclair, W.A.; Townsend, A.M.; Griffiths, H.M.; Whitlow, T.H. Responses of six Eurasian Ulmus cultivars to a North American elm yellows phytoplasma. Plant Dis. 2000, 84, 1266–1270. [**Google Scholar**] [**CrossRef**] [**PubMed**] [**Green Version**]

45. Sinclair, W.A.; Townsend, A.M.; Sherald, J.L. Elm yellows phytoplasma lethal to Dutch elm disease-resistant Ulmus americana cultivars. Plant Dis. 2001, 85, 560. [**Google Scholar**] [**CrossRef**]

46. Bertaccini, A.; Paltrinieri, S.; Contaldo, N. Standard detection protocol: PCR and RFLP analyses based on 16S rRNA gene. In Phytoplasmas: Methods and Protocols, Methods in Molecular Biology; Musetti, R., Pagliari, L., Eds.; Springer Science+Business Media, LLC: New York, NY, USA, 2019; Volume 1875, pp. 83–95. [**Google Scholar**]

47. Martini, M.; Bottner-Parker, K.D.; Lee, I.-M. PCR-based sequence analysis on multiple genes other than 16S rRNA gene for differentiation of phytoplasmas. In Phytoplasmas: Methods and Protocols, Methods in Molecular Biology; Musetti, R., Pagliari, L., Eds.; Springer Science+Business Media, LLC: New York, NY, USA, 2019; Volume 1875, pp. 97–115. [**Google Scholar**]

48. Marcone, C. Current status of phytoplasma diseases of forest and landscape trees and shrubs. J. Plant Pathol. 2015, 97, 9–36. [**Google Scholar**]

49. Jurga, M.; Zwolińska, A. Artemisia vulgaris, a new host of 16SrV-C phytoplasma related strains infecting black alder in Poland. J. Phytopathol. 2020, 168, 659–667. [**Google Scholar**] [**CrossRef**]

50. Lederer, W.; Seemüller, E. Occurrence of mycoplasma-like organisms in diseased and non-symptomatic alder trees (Alnus spp.). Eur. J. For. Pathol. 1991, 21, 90–96. [**Google Scholar**] [**CrossRef**]

51. Foissac, X.; Salar, P.; Olivier, C.; Malembic-Maher, S. Proposal to establish a common taxon for 16SrV-C and 16SrV-D phytoplasmas and three genetic clades based on combined 16S rDNA signatures and variability of housekeeping genes. Phytopathogenic Mollicutes 2019, 9, 85–86. [**Google Scholar**] [**CrossRef**]

52. Maixner, M.; Reinert, W. Oncopsis alni (Schrank) (Auchenorrhyncha: Cicadellidae) as a vector of the alder yellows phytoplasma of Alnus glutinosa (L.) Gaertn. Eur. J. Plant Pathol. 1999, 105, 87–94. [**Google Scholar**] [**CrossRef**]

53. Holz, S.; Duduk, B.; Büttner, C.; Kube, M. Genetic variability of alder yellows phytoplasma in Alnus glutinosa in its natural Spreewald habitat. For. Pathol. 2016, 46, 11–21. [**Google Scholar**] [**CrossRef**]

54. Kube, M.; Furch, A.C.U. Symptomless phytoplasmosis of black alder (Alnus glutinosa): A model for host tolerance. Plant Physiol. Rep. 2019, 24, 550–554. [**Google Scholar**] [**CrossRef**]

55. Malembic-Maher, S.; Salar, P.; Carle, P.; Foissac, X. Ecology and taxonomy of flavescence doree phytoplasmas: The contribution of genetic diversity studies. In Progrès Agricole et Viticole, Hors Série, Proceedings of the Extended Abstracts of the 16th Meeting of ICVG, Dijon, France, 31 August–4 September 2009; Boudon-Padieu, E., Ed.; Le Progrès Agricole et Viticole: Montpellier, France, 2009; pp. 132–134. [**Google Scholar**]

56. Ember, I.; ACS, Z.; Salar, P.; Danet, J.-L.; Foissac, X.; Kölber, M.; Malembic-Maher, S. Survey and genetic diversity of phytoplasmas from the 16SrV-C and -D subgroups in Hungary. Bull. Insectol. 2011, 64, 33–34. [**Google Scholar**]

57. Marjanovic, M.; Stepanovic, J.; Rekanovic, E.; Kube, M.; Duduk, B. Alder yellows phytoplasmas in Alnus species in Serbia. Phytopathogenic Mollicutes 2019, 9, 57–58. [**Google Scholar**] [**CrossRef**]

58. Jarausch, B.; Biancu, S.; Lang, F.; Desqué, D.; Salar, P.; Jarausch, W.; Foissac, X.; Malembic-Maher, S.; Maixner, M. Study of the epidemiology of "flavescence dorée" (FD)-related phytoplasmas and potential vectors in a FD-free area. Phytopathogenic Mollicutes 2019, 9, 59–60. [**Google Scholar**] [**CrossRef**]

59. Filippin, L.; Jović, J.; Cvrković, T.; Forte, V.; Clair, D.; Toševski, I.; Boudon-Padieu, E.; Borgo, M.; Angelini, E. Molecular characteristics of phytoplasmas associated with Flavescence dorée in clematis and grapevine and preliminary results on the role of Dictyophora europaea as a vector. Plant Pathol. 2009, 58, 826–837. [**Google Scholar**] [**CrossRef**] [**Green Version**]

60. Malembic-Maher, S.; Salar, P.; Filippin, L.; Carle, P.; Angelini, E.; Foissac, X. Genetic diversity of European phytoplasmas of the 16SrV taxonomic group and proposal of 'Candidatus Phytoplasma rubi'. Int. J. Syst. Evol. Microbiol. 2011, 61, 2129–2134. [**Google Scholar**] [**CrossRef**] [**PubMed**]

61. Malembic-Maher, S.; Salar, P.; Vergnes, D.; Foissac, X. Detection and diversity of "flavescence doree"-related phytoplasmas in alders surrounding infected vineyards in Aquitaine (France). Bull. Insectol. 2007, 60, 329–330. [**Google Scholar**]

62. Mehle, N.; Rupar, M.; Seljak, G.; Ravnikar, M.; Dermastia, M. Molecular diversity of 'flavescence dorée' phytoplasma strains in Slovenia. Bull. Insectol. 2011, 64, 29–30. [**Google Scholar**]

63. Desqué, D.; Salar, P.; Danet, J.-L.; Lusseau, T.; Garcion, C.; Moreau, E.; Dubus, C.; Dureuil, J.; Delbac, L.; Binet, D.; et al. Impact of Orientus ishidae on "flavescence dorée" emergence in

the vineyards of riparian ecosystems. Phytopathogenic Mollicutes 2019, 9, 69–70. [**Google Scholar**] [**CrossRef**]

64. Maixner, M.; Rüdel, M.; Daire, X.; Boudon-Padieu, E. Diversity of grapevine yellows in Germany. Vitis 1995, 34, 235–236. [**Google Scholar**]
65. Maixner, M.; Reinert, W.; Darimont, H. Transmission of grapevine yellows by Oncopsis alni (Schrank) (Auchenorrhynca: Macropsinae). Vitis 2000, 39, 83–84. [**Google Scholar**]
66. Angelini, E.; Clair, D.; Borgo, M.; Bertaccini, A.; Boudon-Padieu, E. Flavescence dorée in France and Italy—Occurrence of closely related phytoplasma isolates and their near relationships to Palatinate grapevine yellows and an alder yellows phytoplasma. Vitis 2001, 40, 79–86. [**Google Scholar**]
67. Berges, R.; Seemüller, E. Impact of phytoplasma infection of common alder (Alnus glutinosa) depends on strain virulence. For. Pathol. 2002, 32, 357–363. [**Google Scholar**] [**CrossRef**]
68. Sinclair, W.A.; Griffiths, H.M. Variation in aggressiveness of ash yellows phytoplasmas. Plant Dis. 2000, 84, 282–288. [**Google Scholar**] [**CrossRef**] [**Green Version**]
69. Seemüller, E.; Kiss, E.; Sule, S.; Schneider, B. Multiple infection of apple trees by distinct strains of 'Candidatus Phytoplasma mali' and its pathological relevance. Phytopathology 2010, 100, 863–870. [**Google Scholar**] [**CrossRef**] [**Green Version**]
70. Rossi, M.; Vallino, M.; Galetto, L.; Marzachì, C. Competitive exclusion of flavescence dorée phytoplasma strains in Catharanthus roseus plants. Plants 2020, 9, 1594. [**Google Scholar**] [**CrossRef**]
71. Sinclair, W.A.; Griffiths, H.M.; Davis, R.E. Ash yellows and lilac witches'-broom: Phytoplasmal diseases of concern in forestry and horticulture. Plant Dis. 1996, 80, 468–475. [**Google Scholar**] [**CrossRef**]
72. Griffiths, H.M.; Sinclair, W.A.; Smart, C.D.; Davis, R.E. The phytoplasma associated with ash yellows and lilac witches'-broom: 'Candidatus Phytoplasma fraxini'. Int. J. Syst. Evol. Microbiol. 1999, 49, 1605–1614. [**Google Scholar**] [**CrossRef**]
73. Kamińska, M.; Berniak, H. 'Candidatus Phytoplasma asteris' in Fraxinus excelsior and its association with ash yellows newly reported in Poland. Plant Pathol. 2009, 58, 788. [**Google Scholar**] [**CrossRef**]
74. Perilla-Henao, L.M.; Dickinson, M.; Franco-Lara, L. First report of 'Candidatus Phytoplasma asteris' affecting woody hosts (Fraxinus udhei, Populus nigra, Pittosporum undulatum, and Croton spp.) in Colombia. Plant Dis. 2012, 96, 1372. [**Google Scholar**] [**CrossRef**] [**PubMed**]
75. Filgueira, J.J.; Franco-Lara, L.; Salcedo, J.E.; Gaitan, S.L.; Boa, E.R. Urapan (Fraxinus udhei) dieback, a new disease associated with a phytoplasma in Colombia. Plant Pathol. 2004, 53, 520. [**Google Scholar**] [**CrossRef**]
76. Filgueira, J.J.; Ávila, K.J.; López, K.R.; Mugno, P.A.; Zambrano, J.C.; Cruz, A.M.; Villamil-Garzón, L.A.; Rodríguez, A.M. Incidence and characterization of the presence of phytoplasmas in Fraxinus uhdei related with Ash Yellow disease in Colombia. J. Plant Dis. Prot. 2018, 125, 63–71. [**Google Scholar**] [**CrossRef**]
77. Loiseau, M.; Massot, M.; Cousseau-Suhard, P.; Foissac, X.; Piou, D.; Robin, C. Are phytoplasmas associated with dieback symptoms of Fraxinus in France? Phytopathogenic Mollicutes 2019, 9, 53–54. [**Google Scholar**] [**CrossRef**]

78. Sinclair, W.A.; Griffiths, H.M. Epidemiology of a slow-decline phytoplasmal disease: Ash yellows on old-fields sites in New York State. Phytopathology 1995, 85, 123–128. [**Google Scholar**] [**CrossRef**]

79. Ioos, R.; Kowalski, T.; Husson, C.; Holdenrieder, O. Rapid in planta detection of Chalara fraxinea by a real-time PCR assay using a dual-labelled probe. Eur. J. Plant Pathol. 2009, 125, 329–335. [**Google Scholar**] [**CrossRef**]

80. Sinclair, W.A.; Gleason, M.L.; Griffiths, H.M.; Iles, J.K.; Zriba, N.; Charlson, D.V.; Batzer, J.C.; Whitlow, T.H. Responses of 11 Fraxinus cultivars to ash yellows phytoplasma strains of differing aggressiveness. Plant Dis. 2000, 84, 725–730. [**Google Scholar**] [**CrossRef**]

81. Sinclair, W.A.; Whitlow, T.H.; Griffiths, H.M. Heritable tolerance of ash yellows phytoplasmas in green ash. Can. J. For. Res. 1997, 27, 1928–1935. [**Google Scholar**] [**CrossRef**]

82. Sinclair, W.A.; Griffiths, H.M.; Whitlow, T.H. Comparisons of tolerance of ash yellows phytoplasmas in Fraxinus species and rootstock-scion combinations. Plant Dis. 1997, 81, 395–398. [**Google Scholar**] [**CrossRef**] [**Green Version**]

83. Hill, G.T.; Sinclair, W.A. Taxa of leafhoppers carrying phytoplasmas at sites of ash yellows occurrence in New York State. Plant Dis. 2000, 84, 134–138. [**Google Scholar**] [**CrossRef**] [**PubMed**]

84. Schneider, B.; Torres, E.; Martín, M.P.; Schröder, M.; Behnke, H.-D.; Seemüller, E. 'Candidatus Phytoplasma pini', a novel taxon from Pinus silvestris and Pinus halepensis. Int. J. Syst. Evol. Microbiol. 2005, 55, 303–307. [**Google Scholar**] [**CrossRef**]

85. Śliwa, H.; Kaminska, M.; Korszun, S.; Adler, P. Detection of 'Candidatus Phytoplasma pini' in Pinus sylvestris trees in Poland. J. Phytopathol. 2008, 156, 88–92. [**Google Scholar**] [**CrossRef**]

86. Kamińska, M.; Berniak, H. Detection and identification of three 'Candidatus Phytoplasma' species in Picea spp. trees in Poland. J. Phytopathol. 2011, 159, 796–798. [**Google Scholar**] [**CrossRef**]

87. Kamińska, M.; Berniak, H.; Obdrzalek, J. New natural host plants of 'Candidatus Phytoplasma pini' in Poland and the Czech Republic. Plant Pathol. 2011, 60, 1023–1029. [**Google Scholar**] [**CrossRef**]

88. Ježić, M.; Poljak, I.; Šafarić, B.; Idžojtić, M.; Ćurković-Perica, M. 'Candidatus Phytoplasma pini' in pine species in Croatia. J. Plant Dis. Prot. 2013, 120, 160–163. [**Google Scholar**] [**CrossRef**]

89. Valiunas, D.; Jomantiene, R.; Ivanauskas, A.; Urbonaite, I.; Sneideris, D.; Davis, R.E. Molecular identification of phytoplasmas infecting diseased pine trees in the UNESCO-protected Curonian Spit of Lithuania. Forests 2015, 6, 2469–2483. [**Google Scholar**] [**CrossRef**]

90. Costanzo, S.; Rascoe, J.; Zhao, Y.; Davis, R.E.; Nakhla, M.K. First report of a new 'Candidatus Phytoplasma pini'-related strain associated with witches'-broom of Pinus spp. in Maryland. Plant Dis. 2016, 100, 1776. [**Google Scholar**] [**CrossRef**]

91. Huang, S.; Tiwari, A.K.; Rao, G.P. 'Candidatus Phytoplasma pini' affecting Taxodium distichum var. imbricarium in China. Phytopathogenic Mollicutes 2011, 1, 91–94. [**Google Scholar**] [**CrossRef**]

92. Paltrinieri, S.; Martini, M.; Pondrelli, M.; Bertaccini, A. X-disease-related phytoplasmas in ornamental trees and shrubs with witches'-broom and malformation symptoms. J. Plant Pathol. 1998, 80, 261. [**Google Scholar**]

93. Jomantiene, R.; Valiunas, D.; Ivanauskas, A.; Urbanaviciene, L.; Staniulis, J.; Davis, R.E. Larch is a new host for a group 16SrI, subgroup B, phytoplasma in Ukraine. Bull. Insectol. 2011, 64, S101–S102. [**Google Scholar**]

94. Davis, R.E.; Dally, E.L.; Zhao, Y.; Lee, I.-M.; Jomantiene, R.; Detweiler, A.J.; Putnam, M.L. First report of a new subgroup 16SrIX-E ('Candidatus Phytoplasma phoenicium'-related) phytoplasma associated with juniper witches' broom disease in Oregon, USA. Plant Pathol. 2010, 59, 1161. [**Google Scholar**] [**CrossRef**]

95. Gupta, M.K.; Samad, A.; Shasany, A.K.; Ajayakumar, P.V.; Alam, M. First report of a 16SrVI 'Candidatus Phytoplasma trifolii' isolate infecting Norfolk Island pine (Auraucaria heterophylla) in India. Plant Pathol. 2010, 59, 399. [**Google Scholar**] [**CrossRef**]

96. Valiunas, D.; Jomantiene, R.; Ivanauskas, A.; Sneideris, D.; Zizyte-Eidetiene, M.; Shao, J.; Zhao, Y.; Costanzo, S.; Davis, R.E. Rapid detection and identification of 'Candidatus Phytoplasma pini'-related strains based on genomic markers present in 16S rRNA and tuf genes. For. Pathol. 2019, 49, e12553. [**Google Scholar**] [**CrossRef**]

97. Cai, W.; Shao, J.; Zhao, Y.; Davis, R.E.; Costanzo, S. Draft genome sequence of 'Candidatus Phytoplasma pini'-related strain MDPP: A resource for comparative genomics of gymnosperm-infecting phytoplasmas. Plant Dis. 2020, 104, 1009–1010. [**Google Scholar**] [**CrossRef**] [**Green Version**]

98. Khan, J.A.; Sing, S.K.; Ahmad, J. Characterisation and phylogeny of a phytoplasma inducing sandal spike disease in sandal (Santalum album). Ann. Appl. Biol. 2008, 153, 365–372. [**Google Scholar**] [**CrossRef**]

99. Teixeira da Silva, J.A.; Kher, M.M.; Soner, D.; Nataraj, M. Sandalwood spike disease: A brief synthesis. Environ. Exp. Bot. 2016, 14, 199–204. [**Google Scholar**] [**CrossRef**]

100. Murali, R.; Rangaswamy, K.T.; Nagaraja, H. Occurrence of spike disease in sandal plantations in southern Karnataka and cross transmission studies between sandal spike and Stachytarpheta phyllody through Dodder (Cuscuta subinclusa). Int. J. Curr. Microbiol. App. Sci. 2019, 8, 1041–1046. [**Google Scholar**] [**CrossRef**]

101. Mondal, S.; Sundararaj, R.; Rao, H.C.Y. A critical appraisal on the recurrence of sandalwood spike disease and its management practices. For. Pathol. 2020, 50, e12648. [**Google Scholar**] [**CrossRef**]

102. Rao, M.S.; Muniyappa, V. Epidemiology of sandal spike-disease. In Tree Mycoplasmas and Mycoplasma Diseases; Hiruki, C., Ed.; University of Alberta Press: Edmonton, AB, Canada, 1988; pp. 57–58. [**Google Scholar**]

103. Lee, I.-M.; Gundersen-Rindal, D.E.; Davis, R.E.; Bottner, K.D.; Marcone, C.; Seemüller, E. 'Candidatus Phytoplasma asteris', a novel phytoplasma taxon associated with aster yellows and related diseases. Int. J. Syst. Evol. Microbiol. 2004, 54, 1037–1048. [**Google Scholar**] [**CrossRef**] [**Green Version**]

104. Kirdat, K.; Sundararaj, R.; Mondal, S.; Reddy, M.K.; Thorat, V.; Yadav, A. Novel aster yellows phytoplasma subgroup associated with sandalwood spike disease in Kerala, India. Phytopathogenic Mollicutes 2019, 9, 33–34. [**Google Scholar**] [**CrossRef**]

105. Ramachandran, S.; Kirdat, K.; Mondal, S.; Reddy, M.K.; Thorat, V.; Rishi, R.; Yadav, A. A threat to sandalwood cultivation in the naturalised Marayoor sandalwood reserve (Kerala, India) through single and mixed phytoplasma infections. Phytopathogenic Mollicutes 2020, 10, 89–95. [**Google Scholar**] [**CrossRef**]

106. Dijkstra, J.; Lee, P.E. Transmission by dodder of sandal spike disease and the accompanying mycoplasma-like organisms via Vinca rosea. Neth. J. Plant Pathol. 1972, 78, 218–228. [**Google Scholar**] [**CrossRef**]

107. Sastry, K.S.M.; Thakur, R.N.; Gupta, J.H.; Pandotra, V.R. Three virus disease of Eucalyptus citriodora. Ind. Phytopathol. 1971, 24, 123–126. [**Google Scholar**]

108. Marcone, C.; Ragozzino, A.; Seemüller, E. Detection of an elm yellows-related phytoplasma in eucalyptus trees affected by little leaf disease in Italy. Plant Dis. 1996, 80, 669–673. [**Google Scholar**] [**CrossRef**]

109. De Souza, A.N.; de Carvalho, S.L.; da Silva, F.N.; Alfenas, A.C.; Valverde Zauza, E.A.; Carvalho, C.M. First report of phytoplasma associated with Eucalyptus urophylla showing witches' broom in Brazil. Phytopathogenic Mollicutes 2015, 5, S83–S84. [**Google Scholar**] [**CrossRef**]

110. Salehi, M.; Esmailzadeh Hosseini, S.A.; Salehi, E. First report of a 'Candidatus Phytoplasma asteris' related phytoplasma associated with eucalyptus little leaf disease in Iran. J. Plant Pathol. 2016, 98, 175. [**Google Scholar**]

111. Azimi, M.; Farokhi-Nejad, R.; Mehrabi-Koushki, M. First report of a 'Candidatus Phytoplasma aurantifolia'-related strain associated with leaf roll symptoms on eucalyptus in Iran. New Dis. Rep. 2017, 35, 4. [**Google Scholar**] [**CrossRef**] [**Green Version**]

112. Baghaee-Ravari, S.; Jamshidi, E.; Falahati-Rastegar, M. "Candidatus Phytoplasma solani" associated with Eucalyptus witches' broom in Iran. For. Pathol. 2018, 48, e12394. [**Google Scholar**] [**CrossRef**]

INDEX

www.ingramcontent.com/pod-product-compliance
Lightning Source LLC
LaVergne TN
LVHW062214200926
848484LV00006B/1388